# Operational Safety of Dams and Reservoirs

## Understanding the reliability of flow-control systems

Institution of Civil Engineers publishing

# Operational Safety of Dams and Reservoirs

## Understanding the reliability of flow-control systems

Desmond ND Hartford
Gregory B Baecher
P Andy Zielinski
Robert C Patev
Romanas Ascila
and
Karl Rytters

Published by ICE Publishing, One Great George Street, Westminster, London SW1P 3AA

Full details of ICE Publishing sales representatives and distributors can be found at: www.icebookshop.com/bookshop_contact.asp

**Other titles by ICE Publishing:**

*Risk and Uncertainty in Dam Safety.*
Desmond ND Hartford, Gregory Baecher. ISBN 978-0-7277-3639-0
*Hydraulic Gates and Valves in Free Surface Flow and Submerged Outlets.*
Jack Lewin. ISBN 978-0-7277-2990-3
*Floods and Reservoir Safety, Fourth edition.*
Institution of Civil Engineers. ISBN 978-0 7277-6006-8
*A Guide to the Reservoirs Act 1975, Second edition.*
Institution of Civil Engineers. ISBN 978-0-7277-5769-2

www.icebookshop.com

A catalogue record for this book is available from the British Library

ISBN 978-0-7277-6121-7



ICE Publishing is a division of Thomas Telford Ltd, a wholly-owned subsidiary of the Institution of Civil Engineers (ICE).



Cover image: Lady Evelyn Dam owned by Ontario Power Generation, Toronto, Canada.

Commissioning Editor: Laura Balchin
Production Editor: Rebecca Norris
Market Development Executive: Elizabeth Hobson

FSC
www.fsc.org
MIX
Paper from responsible sources
FSC® C013056

Typeset by Academic + Technical, Bristol
Index created by James Helling
Printed and bound in Great Britain by TJ International Ltd, Padstow

This volume is dedicated to the memory of Dr Michael J Casey PE – friend and colleague.

# Contents

# Acknowledgements

*Sponsors and sponsors' representatives*
BC Hydro, Stephen Rigbey
Elforsk – Swedish Electrical Utilities R&D Company, Cristian Andersson
Ontario Power Generation, Tony Bennett
Vattenfall, Lars Hammar, Urban Norstedt (ret.)
US Army Corps of Engineers, Robert C. Patev, Andy Harkness

*Core working group and project manager*
Romanas Ascila, Vattenfall
Gregory Baecher, University of Maryland
Mona Bechai, MOBEC Engineering
Desmond Hartford, BC Hydro
Robert Patev, US Army Corps of Engineers
Karl Rytters, Sekond AB
P. Andy Zielinski, Ontario Power Generation

*Analysis team*
Tyler Atkinson, Ontario Power Generation
Michael Casey, George Mason University
Qianli (Shally) Deng, University of Maryland
Mikael Mathiesen, SWECO
Niklas Mattsson, Uppsala University
Rebecca McAfoos, University of Maryland
Andrew Sander, US Army Corps of Engineers
David Strajt, BC Hydro
Yibing Zhang, Ontario Power Generation

*Reviewers*
John T Christian
Peter Mason
Slobodan Simonovic

# About the authors

**Desmond ND Hartford** is Principal Engineering Scientist at BC Hydro reporting to the Director of Dam Safety. His responsibilities at BC Hydro span the development of risk management policy, risk-informed decision processes, and risk-informed management systems for BC Hydro's Hydropower and Dams risk profile. Dr Hartford is also responsible for introducing systems engineering techniques into the analysis of functionality of river systems and associated hydropower sub-systems; guiding the application of engineering principles-based assessment of spillway gate reliability comparable to those of the nuclear safety case regime; and leading BC Hydro's research and development activities for dam safety. He presently holds the position of Adjunct Professor in Civil Engineering at the University of Western Ontario providing industry guidance on a new research project to develop a prototype systems engineering model suitable for application in BC Hydro's natural and operating environment. Dr Hartford has received several national and international awards and honours, is an overseas member of the Royal Swedish Academy of Engineering Sciences, and in 2011 was recognised as one of the 20 most influential engineers in the world of Hydropower and Dam Engineering.

**Gregory B Baecher** is Glenn L Martin Institute Professor of Engineering at the University of Maryland. He is a geotechnical engineer with 40 years' experience specialising in asset and risk management for water resource development projects, dam and levee safety, hydropower and navigation. He is the author of five books on risk, safety and management of civil infrastructure, and over 200 technical publications. He is a member of the US National Academy of Engineering, recipient of the USACE Commander's Award for Public Service for his contributions to the levee system risk analysis of New Orleans following Hurricane Katrina, and recipient of the Panamanian National Award for Science and Technology Innovation for his contributions to enterprise risk management at the Panama Canal. Dr Baecher has worked extensively with US federal agencies on dam and levee safety and coastal protection works.

**Przemyslaw A Zielinski**, is Manager, Technology and Dam Safety at Ontario Power Generation (OPG) in Canada. Since the late 1980s Dr. Zielinski has been involved in numerous studies and assessments of safety of OPG dams. At OPG Dr Zielinski is responsible for developing risk management programmes for OPG portfolio of 240 dams. He was instrumental in drafting the regulatory framework for risk-informed dam safety assessment and decision-making in the province of Ontario. He is currently serving on the Dam Owners Advisory Committee for the Ministry of Natural Resources and Forests – regulator of dam safety in Ontario and on the Tennessee Valley Authority (TVA) Dam Safety Independent Review Board (IRB) which acts in an advisory role to the TVA's Dam Safety Officer. In 2010 he was elected as the Fellow of the Engineering Institute of Canada. Since the early 2000s, Dr Zielinski has been actively involved in the works of International Commission on Large Dams (ICOLD) and he is one of the principal authors of ICOLD Bulletin on Dam Safety Management. In 2013 he was elected as Vice Present of ICOLD and he is the current chair of the ICOLD Committee on Dam Safety. Dr Zielinski is also an active member of Canadian Dam Association where he co-chairs the Dam Safety Committee. In addition to his research in field of risk assessment and risk-informed decision-making in dam safety, he currently holds the position of Adjunct Professor at the System Design Engineering Department of the University of Waterloo.

**Robert C Patev** is National Risk Advisor at the Risk Management Center, Institute for Water Resources, U.S. Army Corps of Engineers. His duties have included serving as the National Risk Advisor for the US Army Corps of Engineers and he currently provides technical oversight and direction to the USACE Dam and Levee Safety Program, Asset Management Program and Major Rehabilitation Program. During his career with the USACE, Mr Patev has worked primarily on the development of state-of-the-art methodologies for risk and reliability assessments of USACE Civil Works Infrastructure. He has also developed USACE guidance documents and software to support state-of-the practice in Event Tree Analysis, Fault Tree Analysis, Bayesian Weibull Analysis, Expert-Opinion Elicitations, Uncertainty Analysis and Human

Reliability Analysis. Mr Patev also served as the USACE technical lead for the Operational Risk and Reliability Assessments for the Hurricane Katrina Interagency Performance Evaluation Team (IPET). He has also has worked on many international infrastructure projects, including risk and reliability assessments for Autoridad del Canal de Panama.

**Romanas Ascila** has worked in hydropower development and dam engineering in Sweden and several overseas locations since 2000. After his graduation, he worked for 10 years as a consultant and the head of department where he has been involved in many aspects of hydropower planning and dam construction projects. His major activities consisted of hydraulic design of spillways, dam safety evaluations, risk assessments for embankment and concrete dams, design of dams and spillways, hydrological and hydraulic modelling, power calculations, optimisation and GIS-analysis. Mr Ascila was deeply engaged in the development of the Swedish National Program for coordinated emergency preparedness planning where he was responsible for the hydraulic modelling, dam failure simulations, and inundation mapping. He was a member of several domestic and international research projects with responsibility of dam-break studies, development of the flood mapping methods, and computational aspects. For the last six years he has been working as a dam engineering specialist and systems engineering manager at Vattenfall, the biggest Swedish hydropower utility. Mr Ascila is responsible for safety management at several high hazard dams and the development of the reliability program of the spillway function for the whole Vattenfall's dam portfolio. He is also responsible for the development of the Master Plan for the Lake Vänern and Göta River valley where the dam safety and river basin sustainability goals must be addressed in the context of the changing climate and rapid social growth in the area. Mr Ascila has a particular interest in systems engineering, systems dynamics, CFD, numerical modelling, and, especially, simulation.

**Karl Rytters** is a consultant for hydropower development, dam safety, water resources, hydrology and hydraulic engineering. Since the early 1970s he has performed feasibility studies, design or implementation in more than 25 countries of schemes up to 3600 MW capacity and 800 m in head, with dam heights up to 170 m, spillways up to 50 000 $m^3$/s capacity with a special interest in hydraulic structures and sediment engineering.
Mr Rytters has performed research and development for dam safety and numerous dam safety assessments of existing dams nationally and internationally including peer reviews and panel work. He has been responsible for layout design of waterways, including main spillways of 5800 $m^3$/s at 90 m head, fuse plug spillway 3500 $m^3$/s, bottom outlet 275 $m^3$/s at 60 m head, emergency spillways for 4500 $m^3$/s, desilting facilities and surge chambers for 480 MW. He has been responsible for hydrological studies in more than 10 countries, including flood, sediment and river hydraulics studies.

# Foreword

This book arose out of a concern to better understand the reliability of discharge facilities at dams, particularly spillway systems which are so critical to dam safety. The authors brought a range of perspectives, training and experience to their task. The resulting book represents a landmark in the evolution of risk assessment for dams by providing a platform for wider adoption of a systems-based approach to the analysis of risks. The approaches described in this book could represent a step-change improvement for dam safety risk analysis, allowing owners and operators to better recognise and understand the risks posed by the complex interaction of natural and man-made systems.

The authors' thesis is that the unwanted consequences of dam operations are generally the result of unusual combinations of usual events rather than of extreme events, which are the focus of traditional design methods. Examples of actual incidents at dams and other complex facilities are cited in support of that view.

Through the two decades up to the mid-1970s a series of alarming dam failures and incidents, such as with Vega de Tera Dam (Spain), Malpasset Dam (France), Vajont Dam (Italy), Baldwin Hills, Lower Van Norman, and Buffalo Creek Dams, along with the newly constructed Teton Dam (all in the United States) caused growing concern. The result was a renewed focus on dam safety, beginning around the late 1970s with systematic inspection programs in a number of countries to assess the safety of existing dams.

Through the early 1980s it became evident that many dams were in need of improvement, especially in their capacity to safely pass floods. Among other influences, the high cost of retrofitting dams, along with increasing demands by lay managers for justification of large capital expenditures with no new productive use of water, resulted in growing dissatisfaction with the initial three class categorisation of dam failure consequences as the main determinant of a need for safety improvement. The search for a more compelling justification motivated a number of countries to explore the benefits of risk analysis, long practiced in other hazardous industries.

Initially dam engineers struggled to adapt the techniques of risk analysis to dam safety. Criteria for the tolerability of public safety risks, a main driver for improvement, were a particular challenge. Over time and with experience the needed elements came together and the risk-based approach gained acceptance. In June 1997 the first international meeting on risk assessment for dams was held in Trondheim, Norway in conjunction with the HYDROPOWER 97 conference. The International Commission on Large Dams (ICOLD) published *Risk Assessment in Dam Safety Management A Reconnaissance of Benefits, Methods and Current Applications* in 2005. In New South Wales, Australia, the regulator of dam safety developed a risk-based *Risk Management Policy Framework for Dam Safety*, which was endorsed by the state Government in August 2006, establishing a tolerability framework for public safety risks as a basis for determination of safety levels for dams.

In retrospect we see that a risk-based approach has considerably deepened the understanding of dam safety. For example, the need for comprehensive hazard analysis extended the focus beyond flood capacity, prompting research into failure modes less amenable to analysis, such as internal erosion and piping. Human factors and automated controls became better recognised as hazards. Also the need to account for the total sample space in comparing risks with tolerability criteria revealed that the major risks posed by dams do not arise from the extreme events traditionally used for design but from more moderate events. Many other insights became evident.

Despite much progress, difficulties remain. Though human factors have been recognised as a significant hazard for well over a decade, they are still not usually integrated into dam safety risk analyses. The same observation can be made for monitoring and data transmission systems. The complex interactions between the various influencing factors have generally not been modelled. Whilst uncertainty is the very essence of risk-based approaches, its analysis has largely been confined to sensitivity testing of key input parameters.

In this book on the reliability of flow-control systems, the authors have provided a means for overcoming these problems.

The flood discharge reliability of gated spillways in current risk analyses has mainly been assessed as the probability of gates operating as intended on demand.
The authors demonstrate that the problem is more complex. They have concluded that the risk analysis techniques currently available are not up to the task of properly identifying and estimating the risks of discharge facilities. This realisation is not a repudiation of the risk-based approach but points to the need to extend its techniques.

The recognition by the authors that combinations of events are critically important sources of incidents and failures means that interactions must be captured in analyses. Moreover the need to model time effects is well demonstrated in the book. The authors propose that the way to incorporate these aspects is by taking a systems-based approach as developed in other fields such as nuclear power, aviation and space exploration. Thus the authors propose extending the current approach to analysis by estimating the need for discharge, and then analysing the reliability of the release function, of the conveyance function and of the energy dissipation function, with operation of other dams on the watershed, with external disturbances, with human factors and with data acquisition and control systems all possibly affecting reliability. Once the system has been defined and the influences on performance are identified, simulation using techniques such as Monte Carlo analysis can reveal the risks and the uncertainties.

The authors acknowledge that the book points the way forward and is a platform for further development, rather than a treatise on established practice. Systems-based analysis is not yet widely applied for dams and no doubt much remains to be learned. Arguably dam safety analysts have reached the threshold of systems-based analysis through the practice of risk analysis over recent decades. When we look back it is evident analysts were initially ill-equipped to embark on risk analysis for dams but those involved have progressively learned by experience. It is reasonable to expect a similar outcome with systems-based analysis.

Owners and operators need to properly understand the risks posed by dams and related facilities. As the authors have observed, that task is becoming progressively more difficult with more complex organisation structures, greater reliance on automated data capture and control, aging dams and ever-changing understanding of natural events. Much is at stake, including people's lives and property, the owner's business viability and its corporate reputation. In the contemporary context, a well-educated populace is increasingly likely to call to account those who cause harm through an insufficient level of care. Owners need to do what can reasonably be done to minimise risks and can ill afford to miss risks. Simulations of the Lower Mattagami Project and of the Göta River System are presented in the book to show how systems-based analysis can identify risks of both mal-operation and failure and is able to reveal risks which would not otherwise have been recognized. A systems-based approach seems to us to be a defensible way of meeting the public and corporate duties of owners.

The book is well written and clear. In making the case for a systems-based approach real examples are drawn on wherever possible. A large number of relevant and useful sources have been cited, including many which set out the understanding of systems-based analysis as developed in other industries. Whilst the focus of the work remained the reliability of discharge facilities, it is our belief that the concepts set out in the book could have application in other facets of dam safety in greater or lesser degree.

We strongly recommend the book to dam owners, their professional advisers, safety regulators and other interested parties.

*Leonard McDonald*
*Past Chairman*
*NSW Dams Safety Committee*
*Australia*

*Kaare Høeg*
*Past President of ICOLD*
*Norway*

*September 2016*

# Preface

The emphasis on spillway gate reliability improvements that followed the Folsom Dam gate failures led to increasing use of fault tree analysis (FTA) in the safety assessment and remediation of spillway gate systems. A quantitative statement of the probability of gate failure was one of the useful features of these fault tree analyses. Discussions between Mr Urban Norstedt, the former Dam Safety Controller at Vattenfall AB and Dr Desmond Hartford of BC Hydro which centred on how best to maximise the benefit and value of these fault tree analyses, raised two additional questions as follows:

What is the probability of demand for the spillway gates to be brought into service?

What is the probability of operational failure of the spillway?

Elforsk – Swedish Electrical Utilities R&D Company – commissioned a feasibility study (pre-study) aimed at identifying a means of addressing the above questions. Two issues became immediately obvious during the feasibility study:

The probability of demand for the spillway to be brought into service is higher than the probability of flood events requiring discharge, and,

The probability of operational failure of the spillway is dependent on a great deal more than flood discharge.

The feasibility study was carried out in the context of the gearing up by the dam owning industries to address the technical and financial challenges of their aging portfolios of dams and changing operational conditions. These changes are occurring both in upstream and downstream development, increasing expectations for safety and cost efficiencies and ever more stringent global environmental performance.

This feasibility study was timely as the dams industry was beginning to focus on the fact that a major fraction of large modern dams – especially hydropower dams – were constructed in the middle decades of the 20th Century. These are now aging with many of the key functional components reaching the end of their operational life. Many have suffered deferred maintenance, and many require remedial attention even if their maintenance has been adequate. This deferred maintenance is especially true of spillways, conduits, and other flow-control systems. At the same time, there have been gradual – and on occasions, not so gradual – changes in hydrology, land use, and other environmental conditions. These changes are felt most strongly on the same spillways and flow-control systems. Many dam owners face dramatically higher costs of maintaining discharge functionality with attendant impacts on the safety of their dams.

Dam owners must prioritise spillway system repair and upgrading in view of the large costs involved. Some analytical means is needed through which the most critical dams and system components in a portfolio can be identified, and cost-effective plans can be developed for their remediation. For the most part, however, such analytical tools are not available. This situation is unsurprising even given the maturity of water resource engineering. Other safety-driven industries such as chemical processing, aerospace engineering, and even other sectors of electrical power generation, face a similar situation: traditional engineering analysis tools have evolved in response to design needs, not operational needs. Traditional engineering analysis tools focus on the capacity necessary to meet specified demands and on compliance with codes or regulations. They mostly ignore operational and systems effects, including the interplay of human operators, modern sensor and control systems, and randomness.

The approach to the analysis and understanding of operational safety of dams and reservoirs developed in this volume starts from the observation that the dominant risks to be managed in dam safety derive not from unique events but from adverse combinations of more usual events. One might think of these as *unusual combinations of usual conditions*. The timing of such events and their magnitudes can create conditions to which the dam had not been designed and which were not accounted for in preparing operational plans and strategies. These include the systems interactions among management policy, procedural factors, instrumentation and SCADA systems, operational and maintenance practices, design flaws, construction compromises, deterioration, outside disturbances, and many other things that are often overlooked at the time of design. The result is that many systems failures do not fit within a traditional engineering-analytic framework and a new systems framework is needed.

How does the present approach differ from and complement earlier approaches to reliability and dam safety? Fundamentally this approach goes beyond the traditional approach which considers ultimate structural and ultimate hydraulic capacities treated separately, to one that considers all ways that the dam can fail in its containment and conveyance functions, including those conditions where structural and/or hydraulic capacity can be exceeded. Traditionally, for example, analysing spillway capacity and flow control has meant sizing spillways for design conditions: What is the reservoir inflow that a spillway must accommodate; can gates and chutes withstand the pressures and flow velocities to which they will be exposed; can the kinetic energy of flows be dissipated before re-entering the downstream channel? In conventional practice, answering these questions has focused on independent and unique loads: What is the largest flow to be accommodated, what is the worst ground shaking an earthquake might impose, or what is the largest debris or ice load that might accumulate on a gate structure? Even contemporary risk analysis approaches, such as failure modes and effects analysis and probabilistic risk assessment, are of the same character. They concentrate on pre-determined chains of events leading to an accident or failure, and are usually just probabilistic interpretations of existing practice.

This approach is reflected in ICOLD Bulletin 61, *Dam Design Criteria* (ICOLD 1988), which says that,

> [The] engineer attempts to create a structural form which together with the foundation and environment will, (i) perform satisfactorily its function without appreciable deterioration during the conditions expected normally to occur in the life of the structure, and (ii) will not fail catastrophically during the most unlikely but possible conditions which may be imposed.

The first condition pertains to operation and maintenance, not safety. Only the second condition considers risk and uncertainty of failure. However, failures or significantly adverse performances do occur in other than extreme conditions. The traditional approach is not necessarily wrong; it just focuses on things other than the systems behaviour and how most incidents and failures actually occur. The evolving notion of risk analysis, and the thinking about how risks arise, motivates a new perspective.

In this volume, we make a clear distinction between *practicable* in the sense of physically feasible and *practical* in the sense of achievable under given circumstances. The term practicable is an adjective meaning, feasible, 'can be done'. The term practical is also an adjective but it pertains more to implementation (doing) than to the feasibility of an idea.

An approach may be practicable, but under certain circumstances may not be practical. For example, an approach that is practicable for, and has been implemented by, a large dam owner may not be a practical option for a small dam owner. In this regard, the term practicable does not have limitations associated with capacity and capability whereas the term practical does. This limitation is important as it applies to both organisations and to personnel, with the result that what is eminently practical to one organisation or individual may not be practical to another.

The 'new approach' we outline in this volume may not strike the journeyman engineer as immediately practical, just as computational fluid dynamics or critical path scheduling did not strike him or her as immediately practical in the last generation. But both of those methods are today valued and routinely applied in practice, and we anticipate that the 'new approach' laid out here, and the extensions by others that grow out of it, will enjoy a similar fate.

The book is organised in four sections. The first section lays out the problem and provides an overview of the proposed analytical approach using stochastic simulation of dam and reservoir operations. It explains why a need exists, and outlines a solution for *next-generation risk analysis*. The second section addresses the inputs to and constraints on reservoir operations. It considers hydrologic inflows, downstream outflow requirements, and external disturbances that may affect operations. The third section examines the engineered components of flow control. It considers spillways, SCADA and control, outflow conveyance and energy dissipation, and the importance of humans in the loop. The last section lays out a future analytical approach based on stochastic simulation of operations. It introduces the concept of a next-generation of safety analysis tools for dam safety based on understanding of operational factors.

This next-generation of safety analysis is not a replacement for our existing probabilistic tools predicated on failure-modes analysis, event and fault-tree models, and the characterisation of unique events. Those tools have proven themselves to be important and valuable, and have profoundly changed the practice of management of the safety of dams over the past two decades. These advancements are reviewed in many guidance documents, for example in ICOLD *Bulletin 130*, *Risk Assessment in Dam Safety Management*; in the more recent *(US) Federal Guidelines for Dam Safety Risk Management*, or in the predecessor volume to this one, *Risk and Uncertainty in Dam Safety* (ICOLD 2005; FEMA 2015; Hartford & Baecher 2004). The causes of many dam and reservoir failures in recent history can be traced to floods, earthquakes, or internal erosion in earth embankments. These are important failure modes.

The next-generation risk analysis tools for dam safety proposed in the present volume address operational risk. While unique events are important considerations in dam safety, it is also an empirical fact that a large number of accidents and failures – possibly even most – do not arise from such events but from routine operations. Even those accidents and failures that might be attributed to unique loads often might have been avoided were it not for operational factors: a sensor system reported a misleading measurement, an operator overlooked an important precursor, or a spillway gate was unavailable due to routine maintenance. Our existing Probabilistic Risk Analysis tools are valuable but they provide little or no help to the dam safety engineer trying to understand operational failures.

The original motivation for the five years of research leading to the current volume arose from a simple but revealing observation. The simple expression for the probability of failure in dam safety risk analysis is usually expressed as a product of the marginal probability of a demand on the system and the conditional probability of system response given the demand: $P(failure) = P(demand)P(failure \mid demand)$. The *demand*, for example, might be a reservoir elevation (pool level) caused by flooding.

What is this *P(demand)*? Is it a static probability that can be associated with an external natural process like floods or earthquakes? Yes, in some cases it is. But more often, the probability of demand is generated dynamically by the functioning of the system itself. The pool level (and thus, load) on a spillway gate depends dynamically on how the dam and reservoir as a system has been operated up to that time, and on how the dam and reservoir as a system responds. Time is essential, as is the possibly complex interaction of many operational factors. The critical chains of events leading to an accident or failure may have an emergent character, meaning that they are not obvious or easily identified *a priori* but arise from the interactions of the system itself. An analyst creating an event tree would likely never conceive of them, and if he or she did, the important element of time might be impossible to account for.

*P(demand)* is dynamic not static. In this regard, safety is not a function of a static comparison of demand and response, but a function of system controls. This led to the realisation that new analytical tools were needed for us to understand operating risk in hydropower and dam safety, and thus the present volume.

*The authors.*

## References

FEMA (2015) *Federal Guidelines for Dam Safety Risk Management*. Washington DC: Federal Emergency Management Agency.

Hartford DND and Baecher GB (2004) *Risk and uncertainty in dam safety*. London: Thomas Telford.

ICOLD (1988) *Dam Design Criteria: The Philosophy of their Selection*. Paris: International Commission on Large Dams.

ICOLD (2005) *Risk Assessment in Dam Safety Management*. Paris: International Commission on Large Dams.

# Part I

# Dam systems and modelling

**Operational Safety of Dams and Reservoirs**
ISBN 978-0-7277-6121-7


http://dx.doi.org/10.1680/osdr.61217.003

# Chapter 1
# Flow control failures in dam safety

The motivation for the current book arose from a simple observation: The probability of failure in dam safety risk analysis is usually expressed as a product of the marginal probability of a demand on the system, *P*(demand), and the conditional probability of system response given that demand, $P(failure \mid demand)$.

But what is this *P(demand)*? Is it a static probability that can be associated with an external natural process like floods or earthquakes? In some cases it is. But more often the probability of demand is a dynamic probability responding to the functioning of the system itself. The pool level, and thus the load on a spillway gate, depends dynamically on how the dam and reservoir as a system has been operated up to that time. Time is essential, as is the possibly complex interaction of many operational factors and seemingly independent events. An analyst creating an event tree would likely never conceive of them all, and if he or she did, the important element of time might be impossible to account for.

Algebraically, this relation is expressed as

$$P(failure) = P(demand) \times P(failure \mid demand) \tag{1.1}$$

The *demand*, for example, might be a reservoir elevation (pool level) caused by high inflows. To the extent that *P*(*demand*) is dynamic, it is not a function of a static comparison of demand and response, but a function of system controls. This observation led to the realisation that new analytical tools were needed to understand operating risk in hydropower and dam safety.

## 1.1. Operational safety

Dams, along with their associated spillways and other waterways, are built to retain and control the flow of water for purposes of power production, water supply, navigation, recreation, flood mitigation and possibly other things. *Waterways*, in this context, means all facilities for water passage from a reservoir, whether outlets, channels, conduits, such as penstocks and tunnels, or spillways. Waterways are themselves complex structures or systems comprising structural, mechanical and electric subsystems, such as gates and control equipment.

*A dam system* – in contrast to just the dam itself – comprises the body of the dam along with the various waterways past the dam, and usually with accompanying mechanical and electrical equipment for on-site operational control. There could be a powerhouse included. Today, the mechanical and electrical equipment increasingly includes real-time sensors and supervisory control and data acquisition (SCADA) control systems. In a broader context, the *dam system* may also be considered to include the whole of the reservoir and its surrounding drainage, communication links and the human organisation responsible for operating the system, including on-site operators, the dispatch centre and company policy makers. The state and nature of these many components of a dam system do not remain static during the lifetime of the system because of wear, ageing and maintenance, as well as changes to the surrounding infrastructure and to the society that is being served.

Current engineering approaches to dam safety are mostly based on probabilistic risk analysis (PRA). These methods were surveyed in the earlier, accompanying volume (Hartford and Baecher, 2004). PRA primarily address the capability of a dam to withstand extreme loads, such as the demand caused by the design flood and the spillway's capacity to pass that flood, or the demand caused by the design earthquake and the dam's capacity to withstand resulting ground shaking. These are important considerations, and PRA has served the dam community well by rationalising many uncertainties involved with the loads on and performance of dams. PRA has proven especially useful in appraising design and rehabilitation decisions and which design loads and corresponding factors of safety must be chosen. It has also proven useful in screening for dam safety across a collection of dams, by identifying those structures whose relationship of prospective loads to probable capacity is most questionable.

In contrast to the consideration of extreme loads versus structural or geotechnical capacities, experience has shown that many dam failures, and perhaps the majority of dam incidents, do not result from extreme geophysical loads but rather from operational events. These incidents and failures occur because an unusual combination of reasonably common events occurs, and that unusual combination of events has a malicious outcome. For example, a moderately high, but nowhere near extreme, reservoir inflow occurs; the sensor and SCADA system fail to provide early warning for some unanticipated reason; one or more spillway gates are unavailable due to maintenance, or an operator makes an error, or there is no operator on site and it takes a great while for one to arrive; and the pool was uncommonly high at the time. This chain of reasonable events, none of which by itself particularly dangerous, can in combination lead to an incident or even a failure.

While, in principle, one could accommodate an unusual sequence of operational events in a PRA analysis, it is unlikely that anyone would. PRA suffers from the limitations that only specifically identified and enumerated chains of events enter the analysis. An unforeseen or unusual combination of usual conditions that is not specifically identified and enumerated does not influence the outcome of a PRA. As a result, the availability on-demand, reliability, durability, resilience and maintainability of most flow-control systems remain uncertain. Maintenance policies change over the life of a dam. This makes it difficult to integrate flow control into static, risk-based approaches for assessing the safety of dams, to assess their ability to meet performance goals and to find good corrective measures where inadequacies exist.

The consideration of incident or failure scenarios resulting from operational chains of events not specifically identified requires a new approach. This approach sees dams as systems and includes the effects of successive or sudden changes of state due to operational and maintenance activities, human and organisational factors, and laws, policies and procedures, all of which occur in varying environmental conditions. It must attempt to capture 'emergent behaviours' of the dam system. Emergent behaviours are those patterns that arise through the interactions of system components and which are not necessarily exhibited by the components themselves. They are, to a large extent, unidentifiable *ex anti* but are observed in the system's operating interactions. Thus, decompositional engineering tools such as event-tree and fault-tree analyses are poor at uncovering them.

## 1.2. Operational incidents and failures

Schultz (2015) notes in an article about the possibility of a magnitude 9+ earthquake on the Cascadia subduction zone in the Pacific Northwest of the USA that 'certain disasters stem from many small problems conspiring to cause one very large problem. For want of a nail the war was lost; [and] but for 15 independently insignificant errors the jetliner was lost.' This subsection describes dam incidents or failures that follow this systems-behaviour rule.

Modern, well-designed and constructed dams fail at an historical rate of between $10^{-4}$ and $10^{-5}$ per dam-year (ICOLD, 1995). The historical rate of dam incidents – in which complete containment of the reservoir is not lost – is considerably higher, but high-quality statistics are not tracked for incidents. The International Commission on Large Dams (ICOLD) comments that there have been about 300 reported incidents at the 36 000 large dams listed in the *World Register of Dams* (ICOLD, 1984), or a rate perhaps 10 times higher than the rate of failures.

The terms *failure, accident* and *incident* are used across the industry in varying ways to describe the performance of dams. In the present book we adopt a variant of Vattenfall's usage:

- *Failure* – an event that results in damage to the dam facility or its surroundings. An uncontrolled release of the contents of the reservoir through collapse of the dam or some part of it.
- *Incident* – an event which puts the dam system in a hazardous state that by control actions taken by the personnel or the system itself does not develop to a failure.
- *Deviation* – errors, weaknesses or deficiencies in the physical systems, control and monitoring systems or organisational routines.

The distinction between *accident* and *incident* is not always well explained, therefore the need for clear definition. Some usages more or less equate 'accident' with 'failure', while others more or less equate 'accident' with 'incident'. In the present book we attempt to avoid the term 'accident' in favour of 'incident' or 'failure', as the case demands. This is consistent with the UK's Department for Environment, Food and Rural Affairs' (Defra) use of the terms (Defra, 2011).

About a third of dam failures are listed as being due to 'inadequate spillway capacity', which is a catch-all designation including things as distinct as extreme floods, sensor or SCADA malfunction, maloperation and human error. The other two-thirds of the failures are listed in roughly equal proportion as being due to either internal erosion, or slides and other miscellaneous causes. Among the approximately one-third of dam failures that involve inadequate spillways, incidents and failures are seldom easily attributable to a single event or cause. The following recent failures suggest the complexity of the causes and the systems nature of incidents and failures. In fact, most operational incidents and a large number of dam failures are caused not by extreme conditions but by the unforeseen or unusual combination of fairly usual conditions.

In each of many recent cases, as in so many others, the incident or failure did not result from an extreme design load alone, but from an unfortunate and unforeseen combination of more or less routine things. As Skog (2009) notes in relation to the failure of the Noppikoski Dam, Sweden, such problems could occur at most dams under similar circumstances. This is particularly the case where gated spillways are involved. This characterisation is common in failures of technological systems, as documented by Dörner (1997), Perrow (1999), Leveson (2012) and others; it is as true of dam safety as much as it is of nuclear power plants, spacecraft, industrial facilities and even financial systems (NRC, 2007). Yet, current engineering analysis methods do not fully capture these modes of adverse performance, and thus the latter are left out of dam safety risk assessment and of spillway-upgrade studies.

### 1.2.1 Elko Dam, British Columbia, Canada, 21 June 2013

The Elko Dam is located in the Canadian Rockies of eastern British Columbia. This 15.5 m high concrete dam was built in 1924 by the East Kootenay Power Company on the Elk River, and is located approximately 16 km upstream of Lake Koocanusa. The powerhouse has two Francis turbines capable of producing 12 MW of electricity. The power was first used by the Sullivan Mine in Kimberley and supplied the towns of Fernie, Cranbrook and Kimberley. Today, it is owned and operated by BC Hydro. The powerhouse is now out of service pending possible long-term site redevelopment.

The dam has a gated spillway structure in mid-river with two radial gates. A free overflow section extends to the left abutment on one side, and a six-bay stoplog structure extends to the right abutment. Concrete cut-off walls extend into both soil abutments. An upstream rack house on the right abutment houses the intake gate and trash rack hoists for a single penstock. Immediately downstream from the rack house there is a concrete penstock air vent.

The dam can only pass less than the 100-year flood before being overtopped, even with all discharge facilities operated as designed. Upon overtopping, the soil abutments would be expected to eventually fail when erosion extended beyond the concrete cut-off walls. Although costly, abutment failure would not be catastrophic, as there are no people or public property at risk downstream from the dam.

There are two anomalous operational aspects to Elko Dam: With stoplogs removed, debris tends to be drawn towards the right abutment and rack house, causing damage to the structures. Standing operating orders dictated that stoplogs are to be left in at all times in order to reduce flows on this side of the river. However, once in flood conditions, stoplog removal becomes unsafe due to lack of proper access.

The penstock is not expected to remain full of water once the intake gates are closed, due to leakage and the fact that a downstream drain valve to allow for sediment flushing is stuck permanently open. Due to the high water levels in the right abutment, an empty penstock is in danger of possible flotation. Thus, the operating order dictates that the intake gates remain open at all times. However, this allows for surging of floodwaters from the penstock air vent, with a resultant risk of bank erosion.

Both of these contribute to the flood risk at Elko. The dam was slightly overtopped in 1995 during a flood estimated to have of the order of a 50-year return period. Following the 1995 flood, some remedial works were attempted. An extension collar was put on the air vent; however, for reasons unknown this was removed at some later date. A new monorail beam was installed to help in removal of the west-side flashboards and stoplogs, but this system was found to be generally problematic.

The early summer runoff in June 2013 produced moderately high discharge in the river (Treharne, 2013). Elko Dam was overtopped on 22 June 2013 during a flood that was later estimated to have of the order of a 100- to 200-year return period (Figure 1.1). The right abutment was overtopped, resulting in significant bank erosion along and downstream from the rack house. This was exacerbated by overflow from the penstock vent. Crews worked to install lock-blocks to divert the flows from the areas of main damage and protect the intake structure as much as possible. Fortuitously, floodwaters receded before the erosion progressed beyond the concrete cut-off wall, so the abutment did not fail.

### 1.2.2 Taum Sauk, Missouri, USA, 14 December 2005

In the case of the 2005 overtopping of the embankment at the Taum Sauk pump storage project in the US state of Missouri (Figure 1.2), the forensic investigation reported the following contributors to the failure (Hendron *et al.*, 2005):

> The pressure transducers that monitored the reservoir water levels became unattached from their supports, causing erroneous water-level readings. The emergency backup-level probes were set at an elevation above the lowest points along the parapet wall, and thus they failed in their protection role because this enabled overtopping to occur before the probes could trigger shutdown. The normal operating high water level of 1 ft below the top of the parapet wall was too near the top of the wall to allow for any mistakes of maloperation. Visual monitoring of the upper reservoir water levels was almost non-existent, and there was no systematic 'ground-proofing' recorded of the relationship between the top of

Figure 1.1 Elko Dam incident, British Columbia, 2013. (Courtesy of BC Hydro)

the wall and associated water levels actually being achieved. There was no overflow spillway to safely carry accidentally over-pumped water downstream and below the dam.

### 1.2.3 Folsom Dam, California, USA, 17 July 1995

The chronology of the gate failure at the Folsom Dam in 1995 was described in the forensic report as (Todd 2009):

> About 8:00 a.m. on July 17, 1995, Gate No. 3 was being opened to maintain flow in the river during a power plant shutdown. The operator attempted to open the gates to effect a routine release. This involved starting the electrical hoist motors, which in turn engaged the lift chains and cables. Normally, the gate is allowed to stop at six inches and one foot automatically. The auto-stop function was overridden at the 2-foot level, as per normal procedure.

Figure 1.2 Taum Sauk Dam failure, Missouri, 2005. (Courtesy of US Geological Survey, Mid-Continent Geographic Science Center)

Figure 1.3 Spillway gate failure at Folsom Dam, California, 1995. (Courtesy of US Bureau of Reclamation)

The trunnions (at least at gate No. 3) were inadequately lubricated and inadequately protected from water and the weather, and thus were corroded (inadequate maintenance). When the attempt to open the radial gate was made, friction at the trunnion pin was transferred as bending moment into the gate arms (Figure 1.3).

> As the gate opening approached 2.4 feet, the gate operator felt an 'unusual vibration' and he stopped the gate hoist motor. As the operator turned to check the gate, he saw the right side of the gate swing open slowly, like a door hinged on the left side and saw water pouring around both sides of the gate leaf. The time from the operator's initial awareness of the vibration to observing gate displacement and uncontrolled flow of water was estimated to be no more than 5 seconds.

Design flaws made the gate increasingly susceptible to failure (large-diameter pin, size of radial gates, mechanics of pin friction). The gates had not been inspected and exercised on a regular schedule. The gate sensors did not indicate a problem (there was no sensor for strut bending). The operator was not able to stop the gate once trouble was suspected. Stoplogs could not be installed because provisions had not been made for them in the original design and construction, which was completed in the early 1950s.

### 1.2.4 Noppikoski Dam, Sweden, 7 September 1985

At Noppikoski Dam, Sweden, a total breach occurred on 7 September 1985 (Figure 1.4). The chain of events leading to the failure arose because it became impossible to open the spillway stoplogs, and as a result the earthen embankment overtopped and eroded (Skog, 2009).

The failure occurred after a period of higher than normal precipitation and a high groundwater level. When the operators attempted to lift the stoplogs to allow discharge from the reservoir, a mechanical fault was encountered with the hoisting equipment. The incident occurred during the night and, given the weather conditions, it was difficult to use the access roads due to flooding and erosion.

The telephone at the dam went dead, possibly because of a break in the lines. Electrical transmission poles fell, resulting in a blackout at the dam. A mobile crane was located with which to manually raise the stoplogs, but the situation arose on a Saturday and no able local crane operators could be found.

Helicopters were unable to fly due to the weather; and the helicopter radio could only communicate with the regional air force base – but not on Saturdays, when the communication line was unmanned. An unforeseen high downstream water level prevented normal opening of a gate in the diversion tunnel. Complications when trying to drain the tunnels arose because the raised upper water level surpassed the intake gate, filling the tunnels. Finally, the staff at the site were exhausted, and there were difficulties finding additional qualified personnel to work 24-hour shifts during a weekend.

Figure 1.4 Noppikoski Dam failure, Sweden, 1985. (Courtesy of Patrick Regan, personal communication)

### 1.2.5 Lower San Fernando Dam, California, 9 February 1971

The Lower San Fernando Dam – now replaced by the Los Angeles Dam – was the terminus of the main Los Angeles aqueduct, supplying 80% of the city's water. Early in the morning of the 9 February 1971, a magnitude 6.5–6.7 earthquake occurred in the foothills of the San Gabriel Mountains, causing a large slide in the upstream slope of the dam. After the failure, only a narrow wedge of soil remained to hold back the water in the reservoir and to prevent flooding of the densely populated downstream neighbourhoods of 80 000 people (Figure 1.5).

Had the reservoir been at full pool, a significant aftershock occurred, or the slide been only somewhat larger, a complete failure of the dam would have ensued. However, fortunately, none of these things did occur, and significant loss of life and property

Figure 1.5 Lower San Fernando Dam near-failure incident, California, 1971. (Courtesy of US Geological Survey)

damage downstream were avoided. Residents in an 11 square mile area downstream were evacuated for 3 days while the reservoir drained. The dam could not be repaired and was later replaced by the Los Angeles Dam. The 1971 earthquake and dam incident led to the creation by Governor Edmund G. Brown of a Governor's Earthquake Council as part of the (California) Seismic Safety Act of 1971.

### 1.2.6 Carsington Reservoir, Derbyshire, UK, 1984

Sometimes, a malicious chain of events leading to failure does not occur during operations but during design. Perhaps the final cause of failure is categorised in databases as 'design mistakes', yet the chain of smaller events culminating in the fatal design mistakes has the same character of systems interactions as do operational incidents.

Carsington Reservoir is located in the River Derwent basin in England, and is part of a water-storage project operated by a regional water authority. The dam was a zoned earth fill 35 m high and 1225 m long. Water was to be diverted from the river during high flows in winter, and then released back into the river during summer low flows. Construction of the dam took place from 1982 to 1984. The dam was nearly complete in the early summer of 1984.

During construction, concerns were raised by the contractor about the stability of the embankment. An inquiry was made and a report released reviewing the soil engineering parameters used, the slope stability analyses conducted by the engineer, and the resulting factors of safety used for design. The report concluded that: 'a revised design was essential if the dam was to be completed with confidence and safety'. A revised design was to be prepared in the winter of 1983–1984 but, due to animosities between the parties, no meeting to discuss the report was held, and no revisions to the design were made (Kennard and Bromhead, 2000).

Tension cracks were first observed along a 65 m length of the crest of the embankment on 4 June, 1984. A day and a half later, a 400 m section of the upstream slope slid to a maximum horizontal displacement of 15 m (Figure 1.6).

The causes of failure were traced in the forensic investigation to a series of design and engineering factors. Different factors of safety were used for the upstream and downstream slopes, and neither factor was suitably high. None of the assumed slip surfaces that the engineers analysed passed through the upper part of the core or the upstream fill–foundation interface. During construction, the contractor found the requirements for testing and removing clay from the foundation to be inconsistent, resulting in weak material being left in place, and the third-party report recommending revisions was left without response. The forensic report (Coxton, 1986) concluded,

> The slide of a large part of the core and upstream shoulder of Carsington Dam in July, 1984, took place against an unusual background. One of the largest earth-fill dams in Britain, it was designed by a team that appeared to show limited geotechnical expertise. Far too few shear strength tests were undertaken, unrealistic slip surfaces were analysed,

Figure 1.6 Carsington Dam failure, Derbyshire, UK, 1984. (Courtesy of Tony Waltham Geophotos)

instrumentation data was not utilised; and the design factor of safety was not adequately specified. The dam incorporated an oddly-shaped core which maximised the length of potential slip surface through this comparatively weak element of the dam cross-section.

Towards the end of the second earth-moving construction season, as the dam embankment rose towards 5 m of its final height, movements in the fill and the engineers' concern about a high pore pressure and no consequential change of design, caused the contractor to seek an assessment of the dam's stability from its specialist consultant, an experienced earth-fill dam designer. In a report reviewing stability, undertaken by both authors with assistance from a number of colleagues, the eventual failure mode was identified, and likelihood of failure, due to low factors of safety, was correctly predicted. Appropriate corrective measures were identified. This could have been a 'near miss' case – if the original designers had discussed technical issues. In the event, strained relationships arising out of other issues prevented action being taken to rectify matters.

Further earth-fill was instructed to be placed in the third construction season, and the avoidable failure took place just as the embankment dam was being completed to full height. Post-failure investigations were carried out and published by others, including a governmental report and several technical papers. The dam has been rebuilt satisfactorily on the same site. There are several lessons to be learnt from Carsington, including that clear channels for the debate of technical issues must be provided, so that they can be debated free from commercial or personality related issues, without fear of litigation.

Other analyses of the Carsington Dam failure were presented by Skempton (1985) and Rowe (1991).

### 1.2.7 Ka Loko Reservoir, Hawaii, 2006

To say that a dam failed because it was 'overtopped' is like saying that the patient died because he stopped breathing. Yes, at the end that is exactly what happened, but why did it happen? The explanation is usually more complicated: the patient was in a car accident, the airbag failed to function properly, the ambulance was delayed by traffic, the arriving emergency personnel were tired from a long working shift, and by the time the patient finally got to the emergency room ... he had stopped breathing.

Ka Loko Dam on the Waialau River in Hawaii overtopped on 14 March 2006 following 40 days of rain, releasing 150 000 $m^3$ of water in a 12 m wave into the valley below, killing seven and causing US $50 million in damages (Figure 1.7). It was the most damaging dam failure in Hawaiian history. The 'root cause' of failure was concluded to be the lack of a functioning spillway. The chain of events is more complex, however, and outside the normal realm of engineering practice.

Figure 1.7 Ka Loko Dam failure, Hawaii, 2006. (Photo by Christopher P. Becker, Wikipedia Commons)

During the post-failure investigation it was learned that the dam had never been inspected by the State Department of Land and Natural Resources (DLNR), as required every 5 years under Hawaiian law. The private-citizen owner of the dam, a wealthy and influential car dealer, had not permitted inspectors on the property, despite requests from the DLNR.

An allegation arose from a neighbour to the dam site that in late-1997 the owner had illegally graded the land at the reservoir with the intent of developing lakefront houses. A 15 m high hill next to the dam had been graded and the soil used to fill in the dam's concrete spillway. The neighbour allegedly wrote to the owner warning of the 'dangerous situation' but received no response to two letters. The owner claimed not to have received the neighbour's letters.

The neighbour subsequently notified the DLNR by letter, enclosing photographs. The neighbour maintained that the state agency never responded to his letter and photos at the time (1998). The letters and photos were later provided to the post-failure investigation team.

An anonymous complaint to the Kauai County Engineering Department led to the DLNR issuing an order to the owner to 'stop work immediately'. But the owner did not stop. As county documents revealed in the post-failure investigation, the inspector was subsequently instructed by, then mayor, Maryanne Kusaka to 'stay away from [the owner]'. The owner was quoted by the *Hawaii Reporter* as saying that he had given a contribution of US $9000 to the mayor hoping he would get the harassing inspectors off his back. The mayor did not report the contribution, as required by Hawaiian law, and denies receiving it (Zimmerman, 2012).

At trial, the owner maintained that over the years he had tried to bring the condition of the dam to the attention of the state's Public Utilities Commission, but that he had no authority to repair the dam because he was not the water system manager. 'Tragically, however, my constant calls over the years were ignored,' the owner testified (Zimmerman, 2014). The owner ultimately spent six weeks in jail for recklessly endangering the lives of the seven victims, the sentence having been reduced for medical reasons, on testimony of the prison physician, from seven months.

### 1.2.8 Tous Dam, Valencia, Spain, 20 October 1982

During a 24-hour period starting on 19 October 1982 and continuing into the 20th, more than a half a metre of rain fell along the Mediterranean coast of Spain south of Valencia, centred on the Júcar River basin. The towns of Alcira, Algemesí and Antella, and parts of Valencia itself were flooded.

Tous Dam, a 70 m tall and 400 m wide rockfill structure with an impervious core, abutted concrete blocks on either bank from an earlier concrete dam design that was never completed. A second stage of the project was envisioned, which would have raised the crest to 142 m.

Three vertical-lift gates were located along the right abutment (Figure 1.8). The dam was (and still is after reconstruction) located several kilometres upstream of the affected towns. It is the last flood-control dam along the river. The drainage into Tous Dam is about 18 000 km$^2$. Since the 14th century the basin has been known for catastrophic flooding, due to climatological conditions and warm seawaters along the coast (Alcrudo and Mulet, 2007). Total rainfall over the drainage during the storm of 21–22 October was about 600 million m$^3$, while the capacity of the reservoir was 120 million m$^3$. The planned maximum capacity with the second stage complete would have been above 500 million m$^3$ (Segura, 2009).

The elevation of the pool rose rapidly, exceeding the normal operating level of 84 m. The closed-gate elevation at 87.5 m (sill at 77 m) was overtopped, and then the crest elevation at 98.5 m was exceeded, with the reservoir pool ultimately attaining an elevation of 99.5 m. Downstream, 200 000 people were affected, with 100 000 evacuated from their homes. Eight people were killed by direct consequence of the flood and the direct financial cost was some €300 million.

The rain started on the evening of 19 October. By 1:00 a.m. on 20 October the Júcar Water Authority announced to the public that the river flow had increased to 260 m$^3$/s and was expected to reach 1000 m$^3$/s later in the afternoon. At the time, only mechanical rain gauges were available in the basin, and these were few in number, so there was no early warning of the impending flood. The normal average flow was 40 m$^3$/s. According to Serra-Llobet *et al.* (2012):

> Meanwhile, the engineer in charge of the dam was alerting regional authorities about the extreme situation and the imminent risk of collapse. However, nobody trusted him until it was too late. Around 2:00 a.m., the electric system of the Tous Dam broke down. When the engineer arrived to the site at 9:00 a.m., the water had already reached the only remaining power generator. He could not manually open the gates to empty the reservoir, so water began to overflow the spillway. He alerted the authorities of the delicate situation of the dam at 9:30 a.m. and called for a new generator. However, flooding blocked the vehicle that was carrying this piece of equipment. When it finally arrived at 2:00 p.m., it was too late. Despite the warnings of the engineer, who insisted on taking emergency measures, the governor of

Figure 1.8 Tous Dam failure, Valencia, 1982. (Estrela, 1999)

> Valencia did not ask the local populations to evacuate. At 4:30 p.m., the Júcar Water Authority reported that the water level was only one metre below the rim of the dam. At 5:00 p.m., the [river] had already reached and flooded the village of Sumárcer, the first village located downstream of the dam, and was on its way through other towns, although the flood had not yet reached its maximum height.
>
> Local mayors and authorities found it difficult to communicate with the civil government due to saturation of telephone lines. However, some local authorities managed to confirm the seriousness of the situation through the police brigade (Guardia Civil) and went into the streets with loudspeakers to warn residents of the impending flood. This action prevented a far greater death toll.

The result of the Tous failure was significant for Spanish dam safety. Spain became among the first countries in Europe to legislate on a national scale, including for periodic safety reviews. A new Tous Dam was constructed around the concrete monoliths of the old dam. No other new embankment dam in Spain has been constructed with gated spillways, and at least one fixed autonomous energy group plus one portable energy group is required to be permanently placed at every dam. In Spain this has been called the 'Tous syndrome' (I. Escuder-Bueno, personal communication, 2015).

### 1.2.9 Other incidents involving flow control and gate failures

While incidents involving flow control are common, failures involving flow control such as those just described are less so (Lewin, 2001). In the past few decades, a number of outlet incidents have been reported (Table 1.1). The statistics on such incidents are anecdotal, however, so empirical frequencies by type are not available.

Table 1.1 Recent outlet failures

| Dam | Year | Location | Description | Reference |
|---|---|---|---|---|
| Folsom | 1995 | USA | Trunnion pin corrosion | USBR (1996) |
| Washi | 1967 | Japan | Dynamic structural instability | Lewin and Ballard (2001) |
| –* | | Sweden | Controls | Lagerholm (1996) |
| –* | | Sweden | Debris accumulation | Lagerholm (1996) |
| Tarbela | 1992 | Pakistan | Stuck during lowering | Kahn and Siddique (1994) |
| Seton | 1988 | Canada | Wire lifting ropes broke on radial gate | Lewin and Ballard (2001) |
| Terzaghi | 1995 | Canada | Vertical-lift gate jammed open during testing | Lewin and Ballard (2001) |
| Bray Weir | 1994 | England | Vertical-lift gate jammed open during testing | Lewin and Ballard (2001) |

–*Names omitted for reasons of commercial confidentiality.

A variety of other dams have experienced malfunctions of gates due to debris or ice accumulation, hoist and brake failures, or failures of the gate-control system. Lewin *et al.* (2003) provide a list of many of these, including six dam failures caused by gates failing to open upon demand.

Lagerholm (1996) presented the results of a survey of major dam owners in Sweden on the rates of faults or incidents with spillway gates. The statistics indicate annual gate fault rates of the order of 0.1 for common gate types such as radial or vertical roller gates.

Reliability assessments using fault-tree analysis have been conducted for the Thames Barrier and Barking Creek barrier, which are parts of the London River Thames defences (UKAEA, 1987). This study led to an estimate of unavailability per gate-demand of $1.55 \times 10^{-4}$, or a return frequency unavailability of 1 : 560 demands. A similar study has been conducted of the Venice mobile barrier system (Lewin and Scotti, 1990). In the Netherlands, the Rijkswaterstaat has conducted a reliability analysis of the New Waterway Storm Surge Barrier (van Iperean, 1984), which returned an unavailability of $1.0 \times 10^{-3}$, or a return frequency of 1 : 1000 demands.

Lewin (2001a), in his seminal textbook on hydraulic gates, devotes only 16 pages to the topic of risk and reliability analysis, despite the fact that he has been among the pioneers of the analytical approach. Lewin and his colleagues, however, have provided a summary of recent gate failures (Table 1.2).

## 1.3. Contemporary dam safety risk analysis

Dam safety decision-making generally falls into one or more of the following categories:

- standards-based decision-making
- risk-informed decision-making
- probabilistic risk analysis.

Decisions based on these respective premises require an ascending order of sophistication from standards- to values-dominated decisions regarding information, analyses, corporate-level decisions and regulatory environment. Many complex and important dam safety decisions involve portions of all three processes at different stages. A decision-making framework, which explicitly incorporates and elaborates on these processes, allows engineers, owners, regulators and stakeholders to determine the appropriate actions and tools to implement and sustain dam safety decisions given a wide variety of situations. These approaches to dam safety decisions are discussed at greater length in Hartford and Baecher (2004).

### 1.3.1 Standards-based decision-making

Standards-based decisions have been the most commonly used in the analysis of flow-control systems at dams, although spillway configuration and dimensioning was one of the earliest dam safety problems to be addressed by risk analysis.

Standards-based decisions are essentially decisions based on engineering principles and norms that employ a form of design checking against stated criteria. While historically such criteria were mostly associated with structural analysis, today similar engineering criteria have evolved for the consideration of hydrology and hydraulics, foundations, abutments and other components of the dam system (ICOLD, 1988). These engineering criteria have also evolved in parallel with advances in dam engineering and the development of advanced analytical modelling.

The common design check against deterministic engineering standards is the factor of safety (*FS*). This is the ratio of the capacity (strength) of the dam or its components to the demands (loads) placed upon it:

$$FS = \frac{\text{Company}}{\text{Demand}} \tag{1.2}$$

A factor of safety greater than 1.0 suggests that the dam is sufficiently strong to withstand the specified load. Typically, a factor considerably larger than 1.0 is specified to account for the influence of uncertainty in both the calculated capacity and the specified demand. For example, the (US) Federal Energy Management Commission (FERC, 1991) specifies required factors of safety for the downstream and upstream slope stability of embankment dams under long-term, steady-state seepage and normal pool. The factor of safety, although a calculated construct, is related to the physical properties of a dam, in that the larger the factor of safety the greater the capacity of the dam to withstand the applied loads.

Quantitative engineering standards are usually promulgated by regulatory authorities, even though they are typically taken from industry practices, by standards-setting professional organisations approved by government, by the industry or, more indirectly, in terms of guidance provided by non-governmental organisations such as the national member bodies of

**Table 1.2** Summary of published information on failures of spillway gates

| Failure category | Type of failure | Reference | Country |
|---|---|---|---|
| Dam failure due to gates failing to open (six failures) | Malfunction of gates | Reynolds and Hindley (1994) | Russia |
| | Power supply | Utillas *et al.* (1992) | Romania |
| | Power supply | Narayana Murty (1979) | India |
| | Power supply (fatalities) | Water World (1982) | Spain |
| | Vibration of outlet gate | Sagar and Tullis (1970) | India |
| | Power failure | Reynolds and Hindley (1994)<br>Diacon *et al.* (1992) | Romania |
| Failure of gate installation (not causing dam failure) | Power supply, gates overtopped, swept out, two instances | Lewin (2001b) | Central American Republic |
| | Power supply, reservoir level rises to within 0.5 m of dam crest | Lewin (2001b) | Central American Republic |
| Structural failures of spillway gates (five failures) | Trunnion friction | USBR (1996) | USA |
| | Ice loading/fracture | Freishist *et al.* (1976) | Russia |
| | Vibration | Yano (1968) | Japan |
| | Vibration | Noble and Lewin (2000) | UK |
| | Debris accumulation | Lagerholm (1996) | Sweden |
| Spillway blockage due to debris | Jordan River | Lewin and Ballard (2001) | Canada |
| | Revelstoke | Lewin and Ballard (2001) | Canada |
| | La Joie | Lewin and Ballard (2001) | Canada |
| | Alouette | Lewin and Ballard (2001) | Canada |
| | Lachite Dam (1987) | Lewin and Ballard (2001) | Canada |
| | Moreona (1916) | Lewin and Ballard (2001) | USA |
| | Lake Lynn Dam (1985) | Lewin and Ballard (2001) | USA |
| | Heiwake (1951) | Lewin and Ballard (2001) | USA |
| | Palagnedra Dam (1978) | Lewin and Ballard (2001) | Switzerland |
| | Coamo River Dam (1985) | Lewin and Ballard (2001) | Puerto Rico, USA |
| | Vinkelfallet (1992) | Lewin and Ballard (2001) | Sweden |
| Hoist failures | Damaged hoist | Ellis and Dulin (1992) | Hungary |
| | Hoist motor failure | Ellis and Dulin (1992) | USA |
| | Hoist rope failure/jammed gate | Ellis and Dulin (1992) | South Africa |
| | Hoist chain failure | Lemos and Lemos (2000) | Portugal |
| | Hoist chain failure | Lemos and Lemos (2000) | Canada |
| | Limit switch failure | Lagerholm (1996) | Sweden |
| | Hoist rope seizure | Sagar and Tullis (1970) | Canada |
| | Hoist rope failure | Watson (1997) | Canada |
| | Hoist failure | Hobbs and Azaredo (2000) | Australia |
| | Hoist rope failure/gate jammed | Narayana Murty (1979) | Pakistan |
| | Hoist failure/gate jammed | Sagar and Tullis (1970) | Canada |
| | Brake failure/runaway gate | Noble and Lewin (2000) | UK |
| | Brake failure/runaway gate | Lagerholm (1996) | Sweden |
| Controls | Overtopping of dam | | Spain |
| | Condensation of electrical contacts | | Slovenia |
| | Inadvertent opening | | France |
| Seals leakage and gate operation problems | Ice formation | Watson (1997) | Canada |
| | Ice formation | Lagerholm (1996) | Sweden |
| | Ice formation | Bockerman and Wagner (1999) | USA Q1 |

Source: Lewin (2001a) and Lewin *et al.* (2003).

ICOLD. Current dam safety practice is usually predicated on the rare occurrences of extreme loads, such as unlikely but possible reservoir inflows or powerful seismic events. ICOLD Bulletin 61, *Dam Design Criteria – Philosophy of Choice* (ICOLD, 1988), prescribes that:

> [The] engineer attempts to create a structural form which together with the foundation and environment will (i) perform satisfactorily its function without appreciable deterioration during the conditions expected normally to occur in the life of the structure, and (ii) will not fail catastrophically during the most unlikely but possible conditions which may be imposed.

The second condition speaks directly to the concern for the performance of the dam under extreme loads; the first condition requires consideration of interactions and combinations of more usual events, and may actually be more difficult for which to design. In practice, the first condition is most often approached in an unstructured way or based on the experience of the designer and accepted rules of thumb in dam design. In many cases, dam safety guidelines are silent on the matter of engineering performance standards for dams and their components intermediate between normal loading and extreme flood and earthquake conditions. This is one reason why dam engineers are unwilling to talk about requirements for the management of dam systems, because they are often unsure of the desired criteria, especially regarding challenging issues such as internal erosion or electronics, and even less so regarding human reliability. Engineering standards for dam safety that are now embodied in laws and regulations evolved from engineering norms of dam design, which themselves evolved over many years based on experience, and more lately engineering analysis.

For some extreme load events, such as hydrological inflows, and for some dam sites, historical data exist from which to judge the frequencies of extreme demands. These may then be used to assign load criteria. However, for other events and for other sites, such data may not exist. In such cases, ICOLD Bulletin 61 suggests that 'an intelligent assessment of evidence has to be made'. Presumably this means that expert judgment must be relied upon.

As with other engineering methods, engineering standards-based decision-making has itself evolved in recent years. Increasingly, there is a movement to replace global factors of safety as per Equation 1.1 with partial factors for each major uncertainty in a calculation. In the simplest case this might mean one factor on capacity in the numerator and a different factor on demand in the denominator. These separate factors are chosen in some proportion to the degree of uncertainty in each respective parameter. This approach is something of a pseudo-probabilistic method of safety checking.

The deterministic standards-based approach to dam safety has changed little since the US National Research Council studies following the Teton Dam failure (NRC, 1983, 1985). The focus remains on the physical structure, not on operations, data collection, communications or operations. Loading scenarios – floods, earthquakes and normal operations (which in the context of the mid-1980s meant internal erosion, but absent quantitative criteria) – are assessed separately. Analytical criteria for the internal erosion of embankment dams have improved somewhat since that era, but still today operational factors, SCADA system errors and human factors have little place in engineering-standards based assessment of dam safety.

### 1.3.2 Risk-informed decision-making

Risk-informed decision-making is different from engineering standards-based decision-making in that the focus is on the level of protection to the public from the hazardous dam and reservoir. In contrast, in standards-based decision-making the hazards are the natural and other conditions that threaten the dam. In fact, however, this risk-informed approach is rarely applied in current practice. What often passes for a risk-informed approach might more accurately be described as traditional standards-based thinking with a probabilistic veneer.

Risk-informed decision-making implies taking into consideration a probabilistic description of the natural and other hazards imposed on, and the fragility of, the dam system, as well as the quantitative consequences of incidents or failures, in making decisions about dam safety, in a way that is focused on the totality of the level of protection to the public. Within this context, 'risk' is taken to be the expected consequences of incidents or failures; that is, the product of the probability of an incident or failure, and the consequent consequences of that incident or failure. Risk-informed decision-making involves balancing the expected economic, social and environmental costs of a dam safety risk against the costs of risk reduction, at least in a qualitative way.

The risk assessment practices for dams that emerged in the latter part of the 20th century went some way towards risk-informed decision-making. However, as with partial safety factors, the practice is a pseudo-risk-informed process, as it is essentially the traditional standards-based approach for 'floods', 'earthquakes' and 'normal' operating conditions with a quasi-probabilistic characterisation of the hazards and the dam response. One purpose of this book is to provide a means of advancing the practice to a risk-informed form.

Figure 1.9 Bow-tie risk management model

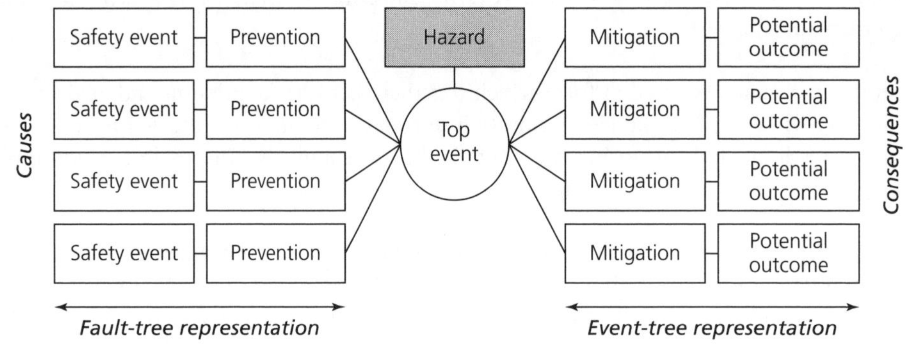


### 1.3.3 Bow-tie model

Risk-informed decision-making can be exemplified by the *bow-tie* concept of reliability and risk modelling, which originated at Royal Dutch Shell in the 1970s. This approach grew out of the chemical processing and oil and gas industry, and is now widespread in technological sectors, including dam safety. The bow-tie, so called for its resemblance to that item of apparel, represents the undesired outcome of an incident or failure at its centre, with prevention measures represented by a fault tree to the left, and mitigation and consequence outcomes represented by an event tree to the right (Figure 1.9).

The bow-tie representation attempts to show the link between preventive actions, hazards, and mitigation responses in a qualitative way that allows for interactions to be identified and rational decisions to be made. To construct the bow-tie model, it is necessary to create a model of how the system works. This typically requires submodels, such as schematic diagrams, influence diagrams, engineering drawings, plans and so on.

Upstream of the bow-tie, to the left-hand side, is the traditional risk matrix of potentially adverse events. To the downstream side, to the right-hand side, is the organisational risk management scheme. So, one reason why this representation of risk-informed decision-making has been successful is that it integrates threat assessment with risk management. It also meshes with the UK Health and Safety Executive (HSE) risk management paradigm (HSE 2001). Yet, the representation is qualitative.

### 1.3.4 Failure modes and effects analysis

Failure modes and effects analysis (FMEA) is a form of risk-informed analysis that is used to map out the consequences of specific events that can occur during the operation of an engineered system, and to use this information to identify and prioritise necessary actions. FMEA can be applied in several forms and for a number of purposes. In its simplest application it can be a free-standing technique to give a structured understanding of the failure modes applicable to the components of an engineered system, or it can be an integral part of a more comprehensive probabilistic analysis of the risks associated with multiple integrated systems.

FMEA was developed originally for design purposes, but now finds application in the analysis of potential for failure of existing systems. It also finds application as part of a wider asset management process that deals with the ongoing satisfactory output of the system under consideration. The use of FMEA is no longer restricted to engineered systems and is now used in a diverse range of societal activities, healthcare management being an example.

### 1.3.5 Probabilistic risk analysis

Risk-based decision-making differs from risk-informed decision-making in that it relies on the quantitative evaluation of the probabilities of incidents and failures, and of their corresponding consequences, in order to calculate quantitative risk. In the literature of technological risk management, risk-based decision-making is often referred to as *probabilistic risk analysis* (PRA). The principal methodologies of PRA are fault-tree and event-tree analysis. The former is more common in nuclear and chemical plant safety. The latter is more common in dam safety and civil infrastructure risk analysis.

Fault-tree analysis (FTA) is a technique the mathematical foundation of which is well developed and which has been applied extensively in reliability and safety assessments for a wide range of engineered systems such as missile launch systems, chemical process facilities, nuclear power plants, dams, control systems and computers. In addition, the software and the databases available for conducting FTA are sophisticated and add significantly to the efficiency of performing a risk analysis.

The fault tree is a graphical construct that shows the logical interaction between the elements of a system the failure of which, either individually or in combination, could contribute to the occurrence of a defined undesired event such as a system failure.

Fault trees offer the analyst the capability to construct a logic model of a system that is visual, and therefore is easy to view and read, and that provides a qualitative and quantitative insight to the system's operations and reliability.

It is important to note at the outset that FTA is one of many tools available to the risk analysis team. In a risk analysis for a dam system, various methods will generally be used to build a logic structure to analyse the expected future performance. As such, FTA will simply be one of the methods used. In the course of the risk assessment it is important to coordinate how an FTA for a system fits into the overall risk analysis model. This theme is critical to the risk analysis in general and to the FTA in particular.

Event-tree analysis (ETA) is one of the techniques available to the engineer conducting a reliability or safety analysis for a dam. It is an apparently straightforward endeavour that finds widespread application in many industries and businesses. It is an inductive type of analysis that, unlike FTA, is not supported by an extensive theoretical basis. ETA is the most widely used form of analysis in risk analysis for dam safety, although the lack of a theoretical basis means that the correctness of these constructs may be difficult to determine.

An ETA is an analysis process the essential component of which is the event tree. The event tree is a graphical construct that shows the logical sequence of the occurrence of events that is visual, and therefore is also easy to view and read, and also provides a qualitative and quantitative insight to the system's operations and reliability.

Current dam safety practice, both in the traditional deterministic form and in the more modern PRA form using fault trees and event trees, is still usually based on the rare occurrences of extreme loads, such as unlikely but possible reservoir inflows or powerful seismic events. Adding PRA to the evaluation does not change this situation at all. As an example, the Canadian Dam Association (CDA) guidelines for dam safety risk analysis presume extreme floods and earthquakes to be probabilistically independent events (CDA, 2007). Each has some probability of occurring in any given year, and each has some probability of leading to an incident or failure. This is the same whether in standards-based evaluation or in PRA. Indeed, from a geophysical view, these natural phenomena are likely to be probabilistically independent. The occurrence of one does nothing to change the probability of occurrence of the other.

From an operational and safety view, however, earthquakes and floods are not independent. If an earthquake occurs and causes serious damage to a dam system, it may take a year or more for repairs to be completed. An annual flood will occur during that time, and even a 10-year or 100-year flood may occur. So, when looked at from an operational or systems vantage point, these hazards are not independent in their effect on dam safety. Typically, removing major weaknesses takes 5–15 years from identification to full remediation. Tongue River Dam in Montana, USA, is an example, and not a particularly extreme one – it took some 15 years to complete the repair after an erosion cavity was found in the embankment after construction (Miller *et al.*, 1999). During this 15-year period the dam was abnormally susceptible to flood or earthquake damage, even though the failure modes might naively be considered independent.

The Tongue River Dam case provides an excellent illustration of the way the traditional focus on 'extreme floods, earthquakes and normal operating conditions', as re-framed in the contemporary risk assessment methodology, is somewhat lacking, as the risk analysis concluded that the dam was vulnerable to extreme floods but that it was essentially no more likely to suffer from the internal erosion failure mode than any other well-designed and constructed dam. In reality, the dam was undergoing internal erosion into the abutment, a situation that was only revealed by the construction works to augment the spillway capacity for extreme floods.

## 1.4. Flow control

Dam systems for flow control comprise a broad set of components such as structures, equipment, sensors, communication facilities, personnel, management arrangements and policies that enable the handling of water flows through the reservoir and past the relevant dam to the downstream reach of river (Leveson, 2011). In this sense the totality created by man between the river upstream and the river downstream may be considered a flow-control system. The dam body is an integral part of this system and, depending on the objectives of the modeller, it may or may not be included in the purpose-built system model for flow control being the object for certain types of system failures caused by sliding, internal or external erosion, or excessive settlements.

The term 'system' is used here as it is commonly used in risk analysis, and its definition and limits in any particular instance is a matter of modelling convenience. A system is understood to denote a set of interacting, interrelated and interdependent elements that form a complex whole for the purposes of analysis. Where relevant, it can include not only one dam but a cascade of dams within a river basin. Modelling of such a system intends to simplify the reality of the complex whole in such a way that a realistic resemblance of the salient features of the system are retained and can be scientifically analysed. Determining what these salient features are is a crucial step in the application of the method, which will vary according to the specific interest of the modeller.

Every dam system for flow control is unique, but as a generalisation the principal components include (Figure 1.10):

Figure 1.10 Conceptual logic or influence diagram of flow-control functions in dam systems

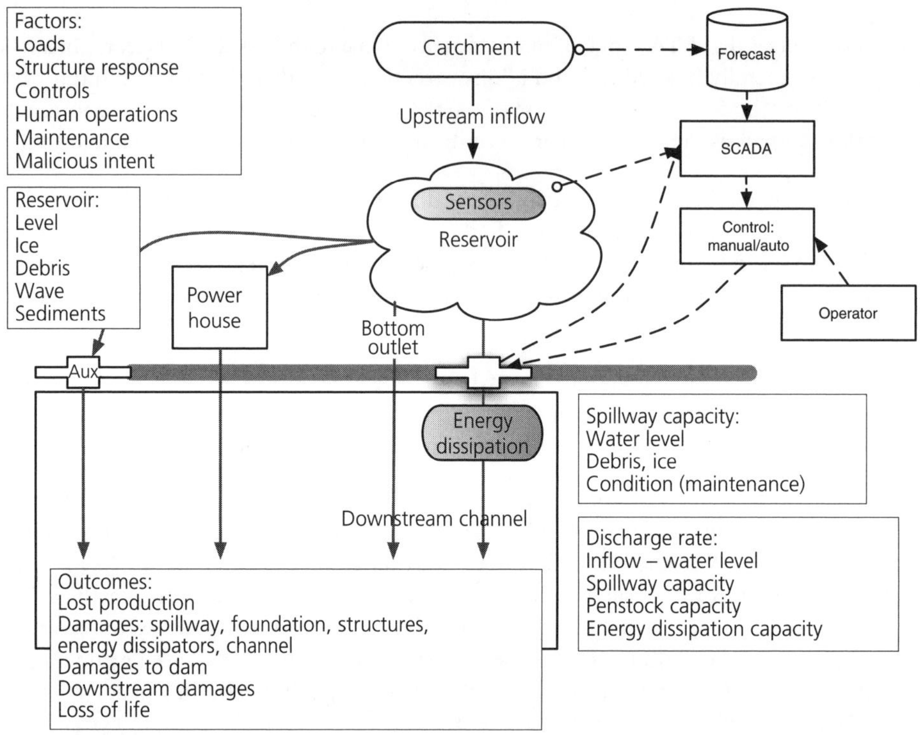


### 1.4.1 The catchment, reservoir and dam

Discharges from upstream reservoirs and natural precipitation in the (local) catchment result in inflows to the reservoir. The reservoir is ponded behind a dam to serve as a buffer for time-varying upstream inflows in order to harmonise the availability of water or electricity with the demand (Figure 1.11).

The normal retention level in the reservoir is variously called the *upper storage level* (USL) and *full supply level* (FSL) or the 'total water volume retained gross storage'. The minimum retention level is the *minimum operating level* (MOL) and the water volume retained at MOL is the *dead storage*. These elevations are normally fixed legally by issued permits.

At MOL the dam cannot (or is not permitted to) release further water downstream than the current inflow. The volume of water that can be stored between FSL and MOL is typically referred to as the 'live storage'. There may also be a flood surcharge of temporarily stored water during larger floods, if the reservoir level exceeds the USL. The freeboard is the vertical distance between the highest retention level and the top of the dam or the top of the core.

Of the multitude of dam types, most would fail on (significant) overtopping. Embankment dams may fail already when the dam core is overtopped. For a variety of reasons the top of the core is almost never where shown on construction drawings.

### 1.4.2 Spillways and other waterways

Water may be drawn from a reservoir through structures designed for free-surface flow or closed-conduit flow. Most (but not all) waterways built for discharge of large flows, such as flood spillways, are of the former type. Low-level (bottom) outlets and power intakes are, on the other hand, (usually) designed for closed-conduit flow in some part. Discharge facilities are adapted to different operating requirements; some can be regulated, others not; some are temporary, others permanent; there are service and auxiliary spillways, sediment outlets, fish passages, navigation canals and locks, etc.

In many discharge facilities, gates and valves acting as movable water barriers, may actively control and regulate the amount of water drawn from the reservoir. Gated spillways generally permit the use of a larger live storage than do ungated spillways, which are often economically favourable. On the other hand, gates are critically sensitive components of dam systems, and gated waterways cannot be expected always to be available on demand.

Due to their passive nature, ungated (free overflow) waterways generally provide higher reliability of operation than do gated facilities; discharge in accordance with predetermined, physically based rules is initiated without any human or other interference or action being required.

Figure 1.11 Catchment of the W.A.C. Bennett Dam and reservoir on the Peace River in northern British Columbia, Canada. At 183 m high, it is one of the world's highest earth-fill dams. Construction of the dam began in 1961 and finished in 1968

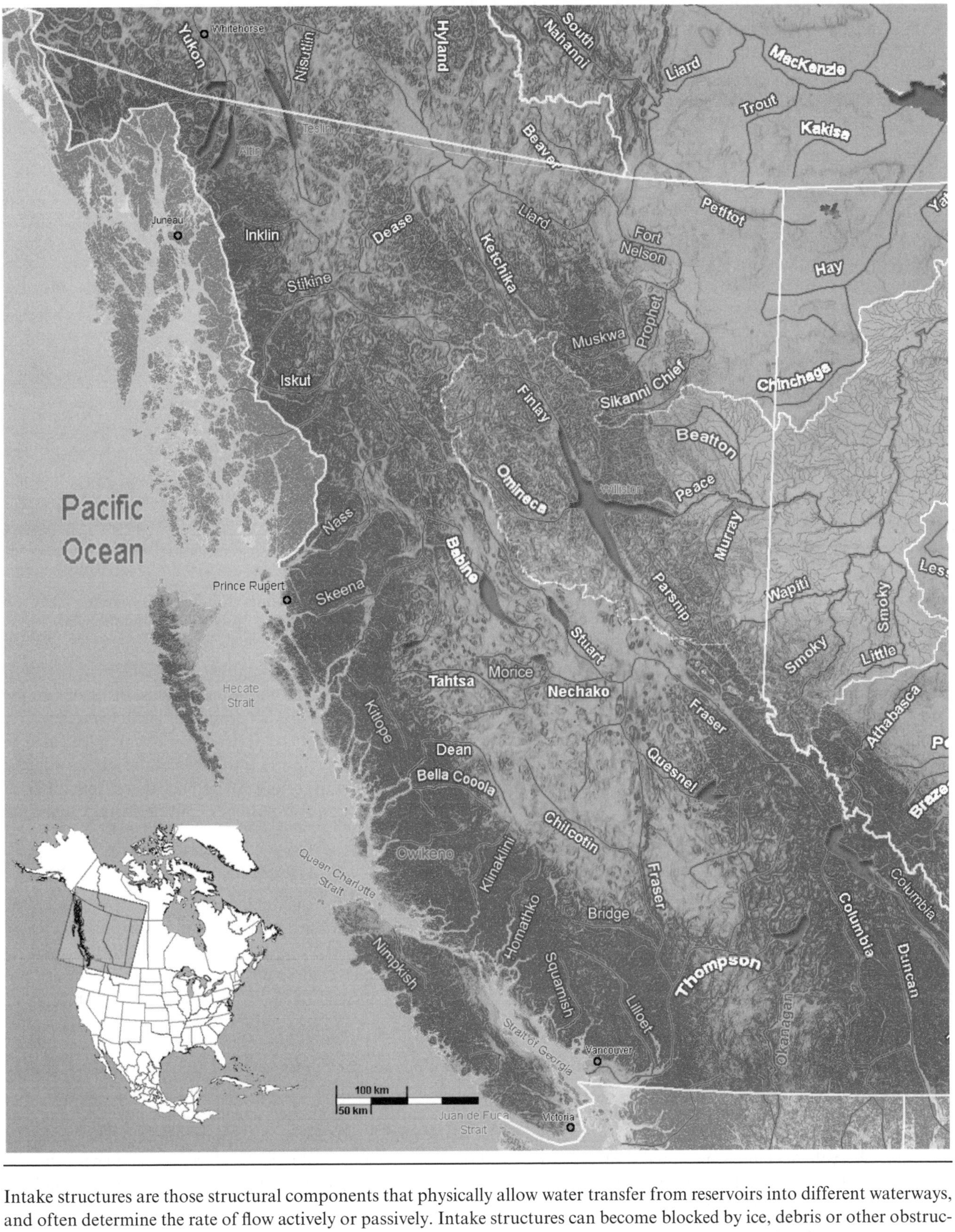


Intake structures are those structural components that physically allow water transfer from reservoirs into different waterways, and often determine the rate of flow actively or passively. Intake structures can become blocked by ice, debris or other obstructions. Intakes to power lines normally incorporate trash racks to protect turbines and cooling-water intakes.

Once water flows past the intakes it needs to be conveyed past the dam and to the river downstream, which may entail both horizontal transportation and substantial vertical drops between pool elevation and tail water. A conveyance system is

Figure 1.12 Conveyance and energy dissipation systems: Brilliant Dam on the Kootenay River near Castleger, British Columbia. Commissioned in 1944. (Originally posted to Flickr by BriYYZ, Creative Commons license)

required to serve this function. The conveyance system may be a chute, canal, culvert or tunnel (Figure 1.12). The vertical drop turns potential energy into kinetic energy in spillways, and the conveyance typically contains a portion with high-velocity flow, posing the prospect of high hydraulic stresses (erosion) and the potential for cavitation.

Excess kinetic energy generated by the vertical drops must be kept under control in the conveyance and be dissipated to the extent that the flow can be returned to the river in a safe manner. This is necessary to prevent erosion of the downstream river channel, and to prevent erosion of the toe and foundation of the dam itself. A variety of devices have been implemented to dissipate hydraulic energy through turbulence in hydraulic jumps or impact or diffusion into larger water bodies (plunge pools). Energy dissipators are regularly subject to high forces. Water is released to rivers and downstream dams from different discharge facilities beside spillways, such as hydropower stations, bottom outlets, fish passages, and diversion canals and tunnels.

### 1.4.3 Operational control

Dam systems are operated in accordance with strategies specifically developed for each individual dam system to keep the system safe, while respecting environmental and other legal regulations and providing the particular services it was created for, whether power production, water supply, flood control or other. Reservoir operation strategies always need to balance production benefits with dam safety. Typically, operational strategies will change over the life of dam systems in response to additional statistical data becoming available, developing societal demands as well as market and policy changes.

Modern dams usually are equipped with automated SCADA equipment, sometimes referred to as 'distributed control systems'. These combine sensors with industrial controllers, computers and data storage capabilities, and together with communication links typically facilitate remote or even automatic control of components of the flow-control system. The consequences of failure of SCADA systems can be dramatic. As a result, hardware and software for SCADA systems in dam and reservoir operations are usually ruggedised to withstand temperature, vibration and voltage extremes, and are enhanced by having redundant hardware and communications capabilities. Programming errors and component failures may still incapacitate SCADA systems.

SCADA systems provide for human-operator control, both remote control from dispatch centres and local on-site control. Operators are always important in dam operations, including flow control. As a result, human operators may also cause mistakes or introduce errors of commission or omission into the operation of dam systems, either without intention or, less commonly, out of malice. Human operators may also take actions that may be, or that they believe to be, in concert with operating policy, yet which may lead to maloperation.

## 1.5. Dams as engineered systems

> Dams as systems: A common view of a dam is to think of the major water retention structure(s) (the dam) that was built to create the reservoir. However, it is more realistic to view the dam as a system; a system of structures (natural and man-made), the reservoir and its boundaries, mechanical and electrical components, and operators. More importantly, the successful performance of a dam system depends on the aggregate satisfactory performance of the system that prevents a failure and uncontrolled release of the reservoir.
>
> (NPDP, 2015)

The physical flows of water in a dam system are – to use the terminology of Perrow (1999) – more or less 'linear' and more or less 'loosely coupled'. So, from the normal accidents point of view, the complexity of flow control in dam systems is simpler and easier to manage than are many technological systems. This is illustrated in Figures 3.10 and 3.11 in Chapter 3, which lay out the interactions between the physics-based parts of a dam system and the corresponding communications (i.e. sensors and SCADA) and controls (i.e. operations) parts of the system. In contrast, the communications and controls parts of the dam system are non-linear and some are more tightly coupled, and the interaction introduces significant feedback.

Yet the goal of the new approach of treating dams as systems is ambitious, and the PRA risk- modelling approaches that have evolved over the past four decades – since the Reactor Safety Study of the early 1970s (NRC, 1975) – are inadequate to the task. These include engineering methodologies such as FMEA, fault trees, event trees and other engineering reliability methods (Kumamoto and Henley, 1996). The difficulty in these methods is that they divide the risk analysis task into a number of specific chains of events that might occur. For a complex system that operates dynamically over time, the combinatorial number of end leaves in these analyses becomes almost infinitely large. For analytical tractability, interactions or dependencies between various paths through the chains of events are ignored.

The 'new approach' is based on developing a purpose-built model of the relevant dam system, and random simulation of its operation in time, in the way pioneered by von Neumann in the mid-20th century, and today commonly known as the Monte Carlo method. This approach can take into account interconnections and dependences between different components of the system, changes over time in their state, as well as the influence on the system of organisational limitations, human errors and external disturbances.

The Monte Carlo method has proven powerful in shedding light on a broad spectrum of systems engineering problems, ranging from nuclear physics to transportation logistics to mathematical and financial systems. It is a category of computational algorithm that uses (pseudo-) random calculations to generate a statistical sample of a computational result. From this sample, frequentist statistical theory can be invoked to make estimates of system outputs along with quantitative assessments of the imprecision errors in those estimates. Because the statistical error in the sample estimates depends only on sample size (i.e. the number of iterated trials) and not on the complexity of the system being sampled, highly complex systems can be dealt with while achieving reasonably precise outputs.

The advent of inexpensive modern computers has made the method accessible even on the desktops of practicing engineers. The method is particularly useful for simulating complicated systems with many coupled degrees of freedom (e.g. fluid dynamics, complex materials, weather patterns, transportation and actuarial science). It is also well adapted to systems involving significant uncertainty either in inputs or in structural relationships. Monte Carlo methods are especially suited to modelling systems having: (i) many components, parameters, and system states; (ii) many interactions between the components; (iii) uncertainty in parameter values and interaction formulae; and (iv) stochastic outputs. All these things apply when modelling dam systems.

The following section discusses the foundations of systems thinking about flow control for dams, the operation of such systems and the disturbances that influence such systems.

## 1.6. The need for a new approach

A new approach is required, as current engineering practices do not and cannot address the character of some of the most probable causes of incidents and failures, which are the unforeseen combination of rather usual conditions. That is, most incidents and failures occur not because an extreme event occurs (e.g. a flood or earthquake), but rather because a series of more common events occurs, which in their unfortunate and unexpected combination leads to an adverse outcome – an incident or failure.

This unfortunate combination may not be of exactly 'usual events'. Incidents and failures most commonly occur due to combination of events, which individually are within the range of the design events, but for this dangerous combination to

**Table 1.3** Dam gate system events and the resulting consequences (McCann, 2012)

| | None or unknown impact | Uncontrolled release | Dam failure |
|---|---|---|---|
| Structural | 19 | 4 | 3 |
| Gates inoperable | 10 | 2 | 2 |
| Control system error | 0 | 4 | 0 |
| Operator error | 0 | 2 | 4 |
| Cable or chain failure | 3 | 0 | 1 |
| Hoist mechanism | 4 | 0 | 0 |
| Power supply | 0 | 0 | 1 |

Source: McCann (2012).

develop there is most likely a combination of not really usual events, perhaps 'rather usual' is a better description. It may not be possible for an incident or failure to occur if all components, and therefore events, are in a perfectly normal state. Some conditions must be in the range of 'not quite usual' – for example, a 50-year flood, lack of required maintenance, slightly incompetent personnel or organisation, bad instructions or policies, a power blackout, or the like – and yet not be extreme or malicious individually.

In a review of the performance of spillway gates and associated operating equipment reported to the National Performance of Dams Program (NPDP), McCann (2013) reported that apart from the failure of spillway structures themselves, the performance of hydraulic systems is important, even during normal operations. These systems are affected by maloperation, control-system errors and a host of interactions between factors. Table 1.3 summarises 59 flow-control incidents identified in the NPDP database, which occurred before 1995 at US dams, and grouped by apparent cause. In each case, a causal chain of contributing factors leads to each apparent final cause, and thus to the subsequent incident.

In English common law, there are usually two causes for a harm to occur: the cause-in-fact, and the proximate cause. The originating cause, or cause-in-fact, is sometimes called, the 'but-for cause'; that is, 'but for' this cause the incident would not have occurred. For example, 'but for the intense rainfall the dam would not have overtopped'. This does not assign culpability: The rain is not morally culpable. Of course, many other events may have been required for this but-for cause itself to cause the incident or failure. The legal description of blame or responsibility is whether an event is close enough to the adverse consequence that it is culpable. This is the proximate cause, and is potentially different from the but-for cause. The main thrust of the proximate cause is direct causation; that is, there are no intervening causes between an event and the resulting incident or failure. For example, 'the stoplogs were not lifted from the spillway, causing the dam to overtop'. The unforeseen combination of pernicious events may lie in this second category of cause.

### 1.6.1 Evolution of risk analysis in other technical fields

The observation that most incidents and failures occur not from extreme loads but through more common yet unforeseen sequences of events is now widely recognised. Perrow (1999) argues that ever increasing complexity in technological systems makes failures inevitable. He states that complex rather than linear interactions among components lead to systems performance that is difficult to understand. Flow control in dam systems displays both sorts of interaction. Perrow also argues that tight versus loose coupling between components increases the chance of incidents. Tightly coupled systems are rigid. They provide little slack between what happens at one component and at another. Loose systems provide more buffer and opportunity for intervention. Tight coupling increases the likelihood that operator intervention will make things worse rather than better, as the nature of an emerging situation may well not be understood immediately.

Many dams may be considered to be tightly coupled systems because they have little flexibility and have time-dependent operational needs. That is, a dam cannot usually wait around while disturbances are attended to in an orderly fashion. Delays in operations may not be tolerable. Water flows into the reservoir and must be dealt with. Characteristics of tightly coupled systems are: delays in processing are not possible; production sequences are invariant; little slack exists in methods, equipment or personnel; and substitutions of supplies, equipment or personnel are limited. All of these things increase the likelihood of incidents. In tightly coupled systems, expediencies cannot be used to save a situation: Redundancy, segregation, diversity and defence in depth must be designed into the system from the start (Rigbey, 2013).

Rasmussen *et al.* (1990) and Leveson (2012) argue that the chain of causality leading to an incident is often opaque. The causal path to an incident will be 'prepared by resident conditions, which are latent effects of earlier events or acts' (Rasmussen *et al.*, 1990). Thus, the length of the causal path depends to a large extent on the stopping rule for seeking contributing causes:

Do the causes stop with technical faults or with operator 'errors', or do they extend back to management and design decisions? Regan (2010) makes this same argument but with specific reference to dams:

> [...] chain-of-event models, such as a typically defined failure mode or risk assessment event tree, oversimplify the causes of incidents and exclude many systemic factors and non-linear interactions. The failures of Teton, Silver Lake and Taum Sauk all had contributions from systemic factors and non-linear interactions that were unrecognised prior to the failure. Our current dam safety programs are, in essence, trying to determine the safety of the dam by examining a few components of the dam, one component at a time. [...]
>
> Analysing individual components against a prescribed standard is insufficient to assure the safety of dams. Reducing the risk associated with dams to a level that is as low as reasonably practicable requires evaluating the dam as a complex system with interactions of sub-systems that may be difficult to recognise. We must include subsystems that are not normally included in typical standards-based or risk-informed decision-making processes. A systems approach to dam safety provides the opportunity to identify places where risk can be reduced in a cost effective and timely manner.

To achieve this goal of a new approach, one needs to take a systems view on the analysis of function and failure of flow control in dam systems. This is contrary to recent decades' separation of analysis according to different fields, that is, analysis being divided between different specialists. However, it is more in accordance with the approach used when most dams were planned in earlier times. The specialities and methods of analysis used today are still absolutely required but need to be supplemented with an improved overview of how things come together and influence each other. Equally important is an improved overview and understanding of the hierarchy of the requirements (i.e. the system purpose) affecting the whole system and its components.

In this book we have called this view of how things come together – including requirements (purpose), functions and components – a 'systems view'. This does not mean a reversion to old times when this overview was largely achieved by 'engineering judgement', and it demands more, not less, scientific study. Also, the evolving notion of how risks arise motivates a new perspective and a new mindset. Risk analysis is predicated on a systematic decomposition of all things that might go awry, and as a result it forces us to think about systems performance. The 'new approach' laid out in the present book attempts to develop this systems-engineering approach into a practical methodology for understanding and analysing dam systems and dam safety.

The new approach needs to be in agreement with systems thinking. It needs to capture real, evidence-based causes of incidents and failures. Among other things, this means that the new approach must address instrument and control systems, system state and maintenance practices and schedules, the dynamics of the system, time histories of system states, times to repair damages, times for consequences to manifest, human reliability, and other heretofore neglected but important aspects of performance, such as the dam owner's maintenance regime. To achieve this, the entire flow-control system at dams needs to be considered, including not only spillways but also all other waterways, including low-level (bottom) outlets and power intakes. In a dam system there are special ways and arrangements for how things are connected – forms, hierarchies, delay and feedback mechanics, systems dynamics, mathematically challenging nonlinearities and so on – which lead to simulation approaches. Of course, it is not only the technical parts of the system that are important: organisation, social context and human reliability are also major parts of flow-control systems.

### 1.6.2 Next-generation risk analysis

This next-generation of risk analysis is not a replacement for the existing PRA tools predicated on failure-modes analysis, event- and fault-tree models, and the characterisation of extreme events. These tools have proven themselves to be important and valuable, and have profoundly changed the practice of dam safety risk management over the past two decades. These advances have been reviewed in many guidance documents, for example in ICOLD Bulletin 130, *Risk Assessment in Dam Safety Management* (ICOLD, 2005), the more recent (US) *Federal Guidelines for Dam Safety Risk Management* (FEMA, 2015) or the predecessor volume to this book, *Risk and Uncertainty in Dam Safety* (Hartford and Baecher, 2004). The causes of many dam and reservoir failures in recent history can be traced to extreme floods, earthquakes or internal erosion in earth embankments. These are important failure modes.

The next-generation risk analysis tools for dam safety proposed in the present book address operational risk. While extreme events are important considerations in dam safety, it is also an empirical fact that a large number of incidents and failures – possibly even most – do not arise from extreme events but from routine operations. Even those incidents and failures that might be attributed to extreme loads often might have been avoided were it not for operational factors: a sensor system reported a misleading measurement, an operator overlooked an important precursor, or a spillway gate was unavailable due to routine maintenance. Our existing PRA tools are valuable but they provide little or no help to the dam safety engineer trying to understand operational failures.

## 1.7. The organisation of this book

This book has the following objectives:

- to consolidate available knowledge on the function and failure of dam systems in managing the safe discharge of inflows and the control of water retention for productive uses
- to establish a new approach for analysing and assessing the properties of dam systems in a more holistic way
- to facilitate knowledge transfer to the engineering community.

The book is organised in four parts.

Part I provides background and the motivation for a systems approach to dams and flow control. Chapter 1 discusses the need for a new holistic approach for analysis of dam systems, what requirements should they fulfil, how they are configured, how they function and fail, and the use and limitations of contemporary methods of risk analysis for dams. Chapter 2 introduces system control and operation from an organisational management perspective. It introduces the concept of an 'organisational accident' and the way that organisational influences can lead to the development of latent hazards, which remain hidden until they combine with real-time events that together culminate in incidents. Corporate organisation, human operators and their reliable – or unreliable – performance in changing organisational environments are also addressed. Chapter 3 discusses the foundations of systems thinking about dam systems, the operation of such systems and the natural disturbances that influence the systems. Chapter 4 discusses Monte Carlo simulation approaches to modelling and analysing dam systems, with emphasis on flow control, and describes how these approaches allow the analyst better to understand the interactions between components, operating procedures and disturbances. Taken together, the chapters in Part I set the foundations for the more detailed discussions in Parts II and III.

Parts II and III narrow in on the disciplinary topics within dam safety and flow control. Chapter 5 addresses hydrology and reservoir management. Chapter 6 addresses reservoir operations at the implementation level. Chapter 7 addresses a broad spectrum of disturbances that may interfere with the flow control in dam systems (e.g. floating debris, landslides, earthquakes or grid outages). Chapter 8 addresses mechanical and electrical systems used for gates and valves, gate reliability, maintenance and the role of human factors in gate performance. Chapter 9 deals with the water-conveyance function of spillways and waterways.

Part IV presents a simulation modelling approach to flow control. Chapter 10 presents details of the simulation approach illustrated by means of two examples that integrate the disciplinary considerations laid out in Parts II and III.

### REFERENCES


Alcrudo F and Mulet J (2007) Description of the Tous Dam break case study (Spain). *Journal of Hydraulic Research* **45(suppl 1)**: 45–57.

CDA (Canadian Dam Association) (2007) *Dam Safety Guidelines*. CDA, Montreal, Canada.

Coxton RE (1986) *Failure of Carsington Dam*. Report to the Secretary of State for the Environment. HMSO, London, UK, 176pp.

Defra (Department for Environment, Food and Rural Affairs) (2011) *Lessons from Historical Dam Incidents: Technical Summary*. Environment Agency, Bristol, UK.

Diacon A, Stematiu D and Mircca N (1992) An analysis of the Belci Dam failure. *Water Power and Dam Construction* **44(9)**: 67–72.

Dörner D (1997) *The Logic of Failure: Recognizing and Avoiding Error in Complex Situations*, 1st edn. Basic Books, New York, NY, USA.

Ellis RO and Dulin RV (1992) *Wanapum Development, Spillway Gate Hoist Failure*. ASCE Waterpower, Nashville, TN, USA.

Estrela T (1999) Hydraulic modelling of the Tous Dam break. In *Fourth CADAM (Concerted Action in Dam Break Modelling) Workshop*, Zaragoza, Spain.

FEMA (Federal Emergency Management Agency) (2015) *Federal Guidelines for Dam Safety Risk Management*. FEMA, Washington, DC, USA.

FERC (Federal Energy Management Commission) (1991) Embankment dams. In *Engineering Guidelines for the Evaluation of Hydropower Projects*, Ch. 4. FERC, Washington, DC, USA.

Freishist AR, Rozina ID and Rakhmanov AL (1976) *Operating Experience Gained with Flat Hydraulic Gates under Winter Conditions*. Gidrotekhnischeskoe Stroitelo'stvo, USSR, No. 4. English translation. Plenum, New York, NY, USA, pp. 348–353.

Hartford DN and Baecher GB (2004) *Risk and Uncertainty in Dam Safety*. Thomas Telford, London, UK.

Hendron AJ, Ehasz JL and Paul K (2005) *Taum Sauk Upper Dam Breach*. Federal Energy Regulatory Commission, Washington, DC, USA.

Hobbs G and Azaredo D (2000) Spillway gate reliability. *A Conference on Dams: the benefits and concerns of dams*. Cairns.

HSE (Health and Safety Executive) (2001) *Reducing Risks, Protecting People – HSE's Decision Making Process*. HMSO, London, UK.
ICOLD (International Commission on Large Dams) (1984) *World Register of Dams*. ICOLD, Paris, France.
ICOLD (1988) *Dam Design Criteria – Philosophy of Choice*. ICOLD Bulletin 61. ICOLD, Paris, France.
ICOLD (1995) *Dam Failures – Statistical Analysis*. ICOLD Bulletin 99. ICOLD, Paris, France.
ICOLD (2005) *Risk Assessment in Dam Safety Management. A Reconnaissance of Benefits. Methods and Current Applications*. ICOLD Bulletin 130. ICOLD, Paris, France.
Kennard MF and Bromhead EN (2000) Carsington Dam – the near-miss which became a bulls-eye. In *Forensic Engineering*. American Society of Civil Engineers, San Juan, Puerto Rico, pp. 102–111.
Khan KA and Siddique NA (1994) Malfunction of a spillway gate at Tarela after 17 years of normal operation. *Proceedings of the 16th ICOLD Conference, Durban* **Q71**, **R27**: 411–428.
Kumamoto H and Henley EJ (1996) *Probabilistic Risk Assessment and Management for Engineers and Scientists*. IEEE Press, New York, NY, USA.
Lagerholm S (1996) Safety and reliability of spillway gates. *Repair and Upgrading of Dams*, Stockholm, Sweden ICOLD Proceedings Supplement. ICOLD, Paris, France.
Lemos FO and Lemos CH (2000) Hydrodynamic forces on falling spillway gates. *Proceedings of the 20th ICOLD Congress, Beijing, China* **Q79**, **R11**: Vol. 4, p. 157.
Leveson NG (2011) Applying systems thinking to analyze and learn from events. *Safety Science* **49(1)**: 55–64.
Leveson NG (2012) *Engineering a Safer World: Systems Thinking Applied to Safety*. MIT Press, Cambridge, MA, USA.
Lewin J (2001a) *Hydraulic Gates and Valves in Free Surface Flow and Submerged Outlets*, 2nd edn. Thomas Telford, London, UK.
Lewin J (2001b) Failures and vulnerabilities of reservoir control structures. In *Learning from Construction Failures: Applied Forensic Engineering* (Campbell P (ed.)). John Wiley, New York, NY, USA, pp. 168–193.
Lewin J and Ballard G (2001) *Preliminary Reliability Assessment of Spillway Gate Installations: Portland District – Hills Creek Dam Demonstration Risk Assessment, Huntington District Demonstration Portfolio Risk Assessment, and Fort Worth District Portfolio Risk Assessment*. Project No. 388. US Army Corps of Engineers, Washington, DC, USA.
Lewin J and Scotti A (1990) The flood prevention scheme of Venice: experimental module. *Water and Environmental Journal* **4(1)**: 70–77.
Lewin J, Ballard G and Bowles DS (2003) Spillway gate reliability in the context of overall dam failure risk. *US Society on Dams Annual Lecture*, Charleston, SC, USA.
McCann M (2012) Performance of hydraulic systems. In *National Performance of Dams Program*. Canadian Geotechnical Society – Western Quebec Division Meeting. Montreal: National Performance of Dams Program. Available at: https://npdp.stanford.edu/sites/default/files/reports/performance_of_hydraulic_systems.pdf (accessed 01/06/2016).
McCann MW (2013) System Engineering and Risk-Informed Management of Civil Infrastructure. Presentation at the US Nuclear Regulatory Commission Flood Forum, Rockville, MD, USA.
Miller DJ, Smith K and Volpe RL (1999) Piping and internal erosion control measures at Tongue River Dam, Montana. *Proceedings of the 19th USCOLD Annual Meeting and Lecture*, Atlanta, GA, USA.
Narayana Murty TVS (1979) Failure of the Machhu-II Dam. *Indian Journal of Power and River Valley Development* **March**: 54–67.
NRC (National Research Council, USA) (1975) *Reactor Safety Study*. US Nuclear Regulatory Commission Washington, DC, USA.
NRC (1983) *Safety of Existing Dams: Evaluation and Improvement*. National Academies Press, Washington, DC, USA.
NRC (1985) *Safety of Dams: Flood and Earthquake Criteria*. National Academies Press, Washington, DC, USA.
NRC (2007) *New Directions for Understanding Systemic Risk: A Report on a Conference Cosponsored by the Federal Reserve Bank of New York and the National Academy of Sciences*. National Academies Press, Washington, DC, USA.
Noble M and Lewin J (2000) Three cases of gate vibration. *British Dam Society, Proceedings of the 11th Conference* (Tedd P (ed.)). Thomas Telford, London, UK.
NPDP (National Performance of Dams Program) (2015) Dams as systems. See http://npdp.stanford.edu/ (accessed 01/06/2016).
Perrow C (1999) *Normal Accidents: Living with High-Risk Technologies*, updated. Princeton University Press, Princeton, NJ, USA.
Rasmussen J, Nixon P and Warner F (1990) Human error and the problem of causality in analysis of accidents. *Philosophical Transactions of the Royal Society of London, Series B – Biological Sciences*, **327(1241)**: 449–462.
Regan P (2010) Dams as systems – a holistic approach to dam safety. In *30th Annual USSD Conference on Dams: Collaborative Management of Integrated Watersheds*, Sacramento, CA, USA.
Reynolds P and Hindley M (1994) Double dam flood failures. *Water Power and Dam Construction* **46(9)**: 2.
Rigbey S (2013) BC Hydro spillway gate reliability program. In *HydroVision International 2013*, Denver, CO, USA.
Rowe PW (1991) A reassessment of the causes of the Carsington embankment failure. *Géotechnique* **41(3)**: 395–421.

Sagar BTA and Tullis JP (1970) Problems with recent high-head gate installations. *Proceedings of the International Hydraulic Research Symposium*, Stockholm, Sweden, Paper F1.

Schultz J (2015) The really big one. *The New Yorker*, 20 July. See http://www.newyorker.com/magazine/2015/07/20/the-really-big-one (accessed 01/06/2016).

Segura F (2009) La Pantanada de tous, 25 anys després. *VIII Jornades d'Estudis de Cullera*, 23–25 November. Ajuntament de Cullera, 10–40.

Serra-Llobet A, Tàbara JD and Sauri D (2012) The Tous Dam disaster of 1982 and the origins of integrated flood risk management in Spain. *Natural Hazards* **65(3)**: 1981–1998.

Skempton AW (1985) Geotechnical aspects of the Carsington Dam failure. *Proceedings of 11th International Conference on Soil Mechanics and Foundation Engineering*, San Francisco, CA, USA, pp. 2581–2591.

Skog G (2009) *The Dam Failure at Noppikoski.* Svenska Kraftnät, Stockholm, Sweden.

Todd RV (2009) *Dam Safety Best Practices: Trunnion Friction Radial Gate Failure*, US Bureau of Reclamation, Denver, CO, USA.

Treharne A (2013) Elko dam update – situation has improved. *The Free Press (BC)*, 22 June. See http://www.thefreepress.ca/news/212625101.html (accessed 01/06/2016).

UKAEA (1987) *A Reliability Assessment of the Thames Tidal Defences*. Safety and Reliability Directorate SRS/ASG/31447.

USBR (US Bureau of Reclamation) (1996) *Forensic Report on Spillway Gate 3 Failure, Folsom Dam*. USBR, Denver, CO, USA.

Utillas JL, Gamo A and Soriano A (1992) Reconstruction of the Tous Dam. *Water Power and Dam Construction* **44(9)**: 55–65.

van Iperean A (1984) Design of the new waterway storm surge barrier in the Netherlands. *Hydropower and Dams* **May**: 66–72.

Water World (1982) Overtopped Spanish dam collapses as spillway gates stay shut. *World Water* **5(11)**: 8.

Watson MA (1997) Spillway gates: will they open safely? *ICOLD, 19th Congress*, Florence, Italy.

Yano K (1968) On the event of gate destruction of the Wachi Dam. *Disaster Prevention Research Institute*, *Annals of Kyoto University* **11-B**: 1–17.

Zimmerman M (2012) Wealth, power, influence keep Pflueger from facing trial for Kauai Dam breach deaths; plea deal reportedly in the works. *Hawaii Reporter*, 7 September. See http://www.hawaiireporter.com/wealth-power-influence-keep-pflueger-from-facing-trial-for-kauai-dam-breach-deaths-plea-deal-reportedly-in-the-works (accessed 01/06/2016).

Zimmerman M (2014) Six weeks behind bars for seven lives? Pflueger released from prison months ahead of schedule. *Hawaii Reporter*, 29 November. See http://www.hawaiireporter.com/six-weeks-behind-bars-for-seven-lives-pflueger-released-from-prison-months-ahead-of-schedule/123 (accessed 01/06/2016).

Operational Safety of Dams and Reservoirs
ISBN 978-0-7277-6121-7


http://dx.doi.org/10.1680/osdr.61217.027

# Chapter 2
# Management, control and operation

The focus of this chapter is on reservoir and river operations, which typically pertain to how decisions are made with respect to water passing through the system to deliver intended products and services downstream. System operation is the *modus operandi* of the organisation that owns and operates the system. It is recognised that the 'owner–operator model' is not the only feasible organisational arrangement, but the term is used as a convenient means of conveying the notion that both ownership and operator decisions together constitute the operation of the system.

This view highlights the fact that the system state and system response are determined by a large number of factors in addition to the hydraulic operations regime. The net result from an operational reliability and dam safety perspective is the concept of the *organisational accident*, or systemic causes, of accidents: These are the latent conditions in a dam owning or operating organisation that result in an unforeseen combination of fairly usual conditions, which in turn result in incidents. The capacity of physical and operational processes to respond to an adverse condition at a dam is influenced by past conditions and past decisions that have a bearing on dam safety and operational control in the present.

## 2.1. System operation

The issue of what we mean by system operation can be addressed in the following two questions:

1 Given the overall objectives of the system, how much water must be released through each waterway at any time?
2 Given the decision on how much water must be released, what needs to be done and how, and when should the decision be implemented?

The term *waterway* is used here to include each piece of flow-control or discharge equipment, or path by which water can flow.

The first question presents a problem of decision-making under uncertainty. The uncertainty arises from: (i) natural process uncertainty influencing both the supply of available water, and production objectives (i.e. demand); and (ii) uncertainty about the prospects of successful implementation of the decision.

In practice, treatment of part (i) of question 1 involves the development of detailed inflow–outflow management rules. More sophisticated approaches, such as decision support systems, are available and, while not widely used, may help inform the more traditional rules-based approach to inflow–outflow management.

Question 2 presents the problem of conditioning the system such that it can successfully respond to various and conflicting demands. The introduction of internal conflicting demands, and the need to make trade-offs between them, increases operational uncertainty and may also increase technological uncertainty.

Part (ii) of question 1 is typically not addressed in operational practice, although it could be addressed in a sophisticated decision support system. In this regard, question 2 provides the basis for the inputs to part (ii) of question 1. How, and the extent to which, questions 1 and 2 are addressed in the operation of a dam–reservoir system are organisational matters.

In practice, the answers to the two questions are set within the context of the operational management regime of the owning–operating organisation. The following addresses aspects of the operational context, whereas how the above questions are dealt with is considered in Chapter 6.

### 2.1.1 Organisational practices

The matters of 'how' and 'extent to which' are complex matters that reflect a wide range of organisational views over considerable periods of time. For example, financial decisions that influence the design stage of an asset may eventually emerge as reliability issues in operational practice years in the future. Furthermore, political considerations concerning public perceptions and expectations when they influence operational decisions may provide the short-term objectives while concealing longer term latent safety hazards (e.g. holding water behind a dam to alleviate downstream flooding may put the system in an operational state that renders it incapable of withstanding the effects of the as yet not appreciated but imminent larger inflow). Similarly, maintenance and operational decisions long in the past, and often forgotten, influence present-day assumptions and actions concerning the operation of equipment in ways that can have unexpected outcomes now and in the future.

Set in this organisational context, system operation goes far beyond the boundaries of dam engineering and dam safety assessment, and it also goes beyond the domain of what are conventionally termed 'decision support systems'. This is because determination of the system state involves immediate, short, intermediate and long-term management direction from the top of the dam-owning organisation, with various aspects of implementation occurring down through all organisational levels and across multiple organisational functions and operations.

The complexity of the 'whole-system' operational process is determined by operational arrangements within the dam-owning organisation, and by the people who are responsible for implementing the operational arrangements within the various parts of the organisation. Within these arrangements, the personality traits of individuals can have important influence on the stability of the system, as subjective personal-values-based decisions about 'what is good' and 'what is good enough' often enter the decision-making process. The individuals involved apply their own risk assessment process in relation to their involvement with the functioning of the system.

This feature is difficult to capture and virtually impossible to model. Furthermore, there may be a long delay, perhaps years, between 'action' and resulting 'influence'. These features render the process of dam, reservoir and river system operation vulnerable to what Reason (1997) calls the 'organisational accident'.

### 2.1.2 Management practices

The dam safety management practices that have emerged in recent years (ICOLD, 2011) focus on physical and operational barriers intended to prevent hazardous conditions from progressing to failed states (Figure 2.1). The upper part of Figure 2.1 – according to the paradigm of Reason – represents the barriers required to prevent hazardous conditions from emerging within the physical system. These are the components of the active failure. The lower part of the figure represents activities put in place by the organisation to ensure the adequacy of the barriers. These are the components of the latent conditions.

Figure 2.1 Focus of dam safety management controls (bow-tie model). (ICOLD, 2011)

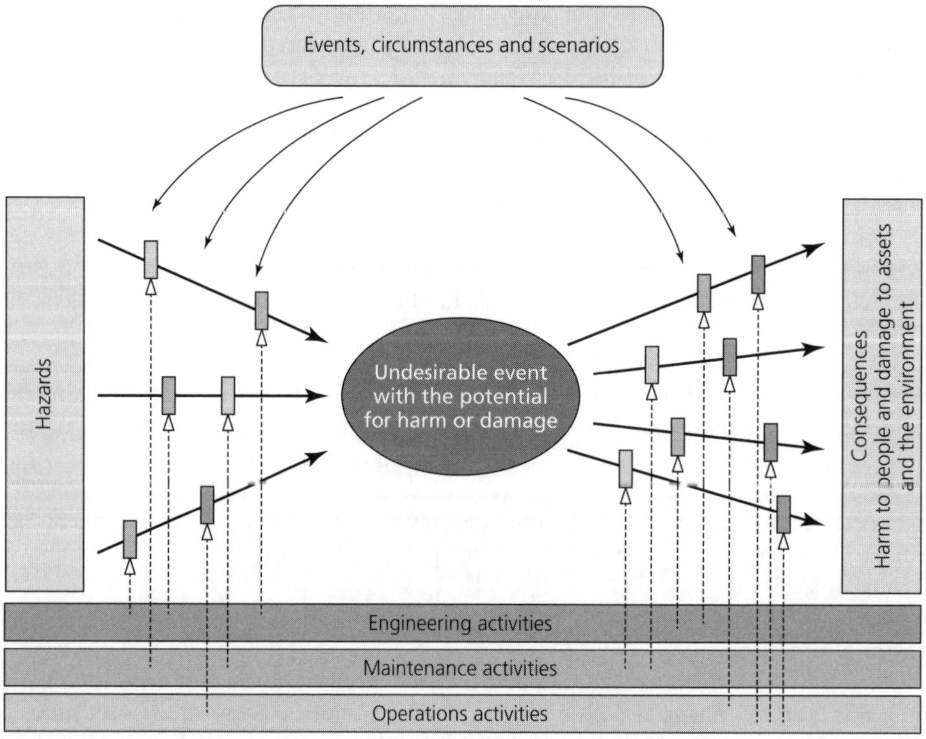

Figure 2.2 Stages in the development and investigation of an organisational failure. (Reason, 1997)

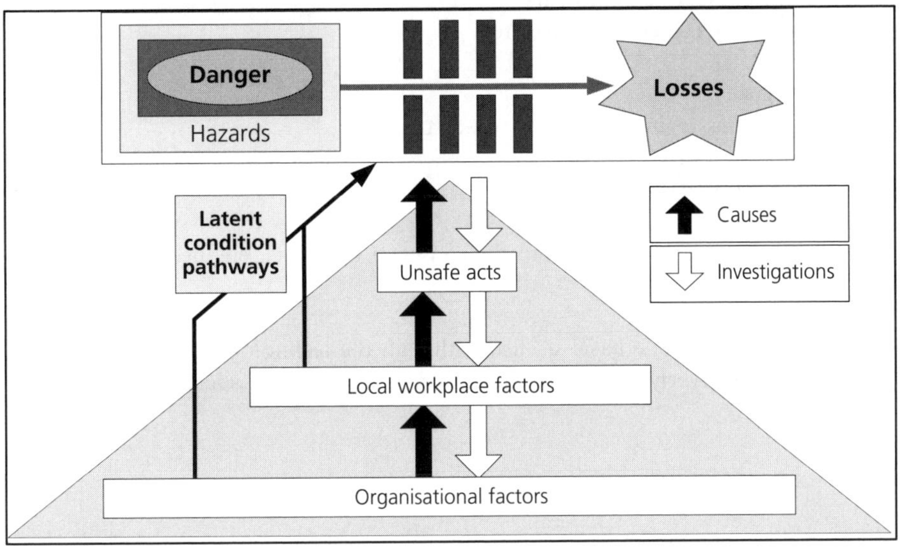


Reason extended this thinking, as illustrated in Figure 2.2, to follow the causal path that an organisational failure typically follows, and the corresponding stages of the subsequent investigation into that failure. The rectangular box at the top represents the main elements of an event (active failures), while the triangular shape below represents the systems producing it (latent conditions). This lower part of the model has three levels: the person (unsafe acts), the workplace factors (error-provoking conditions) and organisational factors. The black upward arrows indicate the direction of causality, and the white downward arrows indicate the investigative steps. These latent conditions include those where decisions made in the past by the owner, and sometimes a long time in the past, can control how real-time operations are carried out in the present.

Reason makes two organisational factor distinctions between active failures and latent conditions:

- Active failures usually have immediate and short-lived effects at the point that they occur, while latent conditions may remain dormant for a long time without discernible effects until they interact with local circumstances in a way that results in a failure. Active failures tend to be unique to a specific event.
- Active failures usually occur at the human–system interface, whereas latent conditions develop at the upper levels of the organisation and within the related design, production, contracting, regulatory and governmental organisations. Latent conditions can contribute to a number of different accidents if they remain undiscovered and uncorrected.

Modern incident investigation focuses not just on the active failures but also on the latent conditions in order to reveal systemic causes of the failures. Reason subsequently updated his view, recognising that active failures are not necessary preconditions for an accident and that long-standing latent conditions are sometimes all that is required for an accident to occur (Reason *et al.*, 2006).

Others have built on the work of Reason. Most notably, these extensions include the functional resonance analysis method (FRAM) of Hollnagel (2012), and the system-theoretic accident model and processes (STAMP) of Leveson (2012).

The contributions to latent conditions are many in the organisation, but can be broadly outlined as including the following:

- licensing arrangements
- societal expectations (including political expectations in the past and present)
- the organisation's social responsibility (including corporate values and principles)
- risk appetite (strategic and operational risk)
- the organisation's strategies and policies
- organisational culture
- organisational arrangements
- management and procedural arrangements (including asset management arrangements, and the maintenance and replacement regime)

- human resourcing and competence (including compensation and rewards)
- budgeting, financing and investment arrangements
- system reliability and availability targets and measures
- human factors
- design of the operations regime
- implementation of the operations regime (including forecasting)
- operator error in real-time operations
- failures in the safety assurance process.

Within this modern view of the nature of systems and the causes of accidents, engineered systems are recognised as being embedded in social systems (Miles, 1973).

Underlying every technology is at least one basic science, although the technology may be well developed long before the science emerges (e.g. glassmaking). Overlying every technical or civil system is a social system that provides purpose, goals and decision criteria.

As Leveson points out:

> Effectively preventing accidents in complex systems requires using accident models that include that social system as well as the technology and its underlying science. Without understanding the purpose, goals and decision criteria used to construct and operate systems, it is not possible to completely understand and prevent accidents.

In recognition of the social dimensions of systems safety, the matters of

- licensing arrangements
- societal expectations (including political expectations in the past and present)
- the organisation's social responsibility (including corporate values and principles)

are presented in Sections 2.1.3 and 2.1.4, which introduce the concepts of an organisation's approach to managing the dam system, which can in turn be related to ICOLD Bulletin 154 (ICOLD, 2010: 154). A means of modelling the more engineering–operational–technological dimensions of the subject is introduced in Section 2.4, based on Hollnagel's FRAM.

Against this background it is important to recognise that the public and political environment may also contribute to the latent conditions. For example, while the notion of the hazard potential of dams and reservoirs is well established and central to dam engineering and dam safety, dams are frequently not considered as hazardous entities in the public domain, especially if the benefits of dams, such as water supply, environmental flows, power generation and recreational water sports, are valued attributes. Furthermore, reservoir–river system operation, which involves the control of vast amounts of hydraulic energy on a continual basis, is a hazardous process in the sense that process failure can lead to release of the hazardous agent (the water), often with a catastrophic outcome. Under conditions at which energy production is at its greatest, loss of control can be most threatening, and recovery from loss of control most difficult if the design and implementation of the totality of the system controls does not accommodate these circumstances.

Some elements of the system controls are more critical than others under normal operational conditions. However, less critical elements can become critical to hydraulic control if the system transitions to a state that renders them so. Thus, a decision to defer maintenance of a redundant feature may be the final causal factor in loss of control if it is called into service. Situations such as this typify the way that 'unusual combinations of not uncommon conditions' can arise during operation within the normal operational parameters, and can lead to loss of control and system failure.

Dams and reservoirs are not 'fully engineered' systems in the sense that they depend in part on natural systems; thus, dam engineering and dam safety assessment are difficult to capture in simple rules and procedures. Dams and reservoirs, by their very nature, are not deterministic systems, and their operation cannot be defined using deterministic procedures.

### 2.1.3 Legal regime and licensing

The legal regime and licensing of the jurisdiction in which the dam is located govern the *modus operandi* of the owner-operator. Legal regimes for the storage and release of water have existed since ancient times (the *Code of Hammurabi*, *c.* 1780 BCE), although the punishments for violations have changed. However, although the legal regime and licensing ultimately govern the methods of operations, they do not necessarily control how an organisation operates in an absolute sense.

The actual operations are controlled by the 'directing mind' of the dam owning–operating organisation. The term 'directing mind' is used to describe the thought processes that guide the methods of operations of the organisation.

The legal regime varies from country to country, with two general concepts being common: a prescriptive concept and an objectives-based concept. The prescriptive concept sets out what is to be done, and may include precise instructions of how implementation is to occur, whereas the objectives-based concepts sets out what outcome should at least be achieved, without specifying if this will be sufficient and without defining how the objective should be achieved. These two concepts are formalised in the two generally known legal systems: the Roman system, with its origins on the European mainland, and the common law system, with its origins in England. Arguably, however, outside of the core Anglo-Saxon world, most legal systems are some kind of mixture of these two.

There are a number of important differences between these regimes that lead to different ways of determining what is 'safe enough'. Of particular relevance is the difference between the Roman system, where everything that is not explicitly forbidden is allowed, and the common law system, where what is not explicitly allowed is forbidden, unless it can be justified, where necessary in court (Ale, 2005). This leads to totally different interpretations between the two systems as to the meaning of the terms 'as low as reasonably achievable' (ALARA) and 'as low as reasonably practicable' (ALARP).

Within these regimes, licensing arrangements generally define responsibilities and boundaries for the operation of dams and reservoirs. These licensing arrangements generally guide the focus of those responsible for directing the dam-owning organisation. For instance, if the matter of flood control were specifically excluded from a particular license then the 'directing mind' would be expected to operate the system in a manner that does not involve reduction of downstream inundation. However, a factor such as social responsibility could determine that, although perhaps not required, outflow modification of floods during a flood event is appropriate.

### 2.1.4 Societal expectations and owner's social responsibility

The societal expectations with respect to the development, ownership and operation of dams have changed dramatically over the last 30 years. The most significant changes have occurred with respect to the design of new dams. In the modern context, the arrangements for the design or modification of a dam or the design of the operational arrangements involves some type of public consultation process in the context of adaptive management (NRC, 2005).

The outputs of this process provide the essence of the reservoir–river objectives for the system and the requirements of the operational regime (Figure 2.3). Essentially, this is that stage at which the different, and often competing, engineering, economic, environmental and social parameters are brought together to create the overall system objectives and constraints. Ideally, the objectives for the system are best cast in terms of a single statement of user need if possible, and the system operational requirements characterised in some type of hierarchical format with what might be termed 'key capability requirements' (must-have attributes) at the highest level, with capability requirements (should-have attributes), elective attributes (should have if reasonably achievable) and discretionary attributes (nice to have if reasonably achievable) beneath.

The operational challenge at this point is to transform these broadly stated objectives, requirements and attributes, which can be considered in light of three water management actions ('store', 'pass' and 'divert'), into dynamic hydraulic operations control functions. These dynamic hydraulic operations control functions are achieved through management actions that themselves draw on various other capabilities, such as engineering, environmental sciences, manufacturing and production processes, etc., all of which are integrated with a management system of some type.

The influence diagram shown in Figure 2.4 suggests the complexity of store–pass–divert interactions for the case of the three-reservoir cascade in British Columbia making up the Bridge River Power Project. This simplified flow of water along the many waterways leading from Downton reservoir at the top, through Carpenter Reservoir in the middle, and into Seton Lake at the bottom, is shown in Figure 2.5. The influence diagram itself is limited to the store–pass–divert relationships only in the uppermost reservoir behind La Joie Dam.

The main horizontal flow line represents the inflow into Downton Reservoir that is stored behind La Joie Dam. Releases from the dam flow downstream to Carpenter Reservoir. Local inflows at Carpenter add to these waters, and subsequent flows are directed downstream in parallel to the Lower Bridge River and to Seton Reservoir.

Below the main horizontal flow line are represented the various waterways leaving La Joie Dam and constituting its total discharge. Above the main horizontal flow line are represented potential consequences attending those releases. Among these consequences are spillway damage and undermining, spillway blockages, accumulating maintenance requirements and seepage through the dam. This influence diagram provides the means for graphically indicating positive and negative feedbacks among the store, pass or divert decisions.

Figure 2.3 Key decision points in planning and project development. (WCD, 2000)

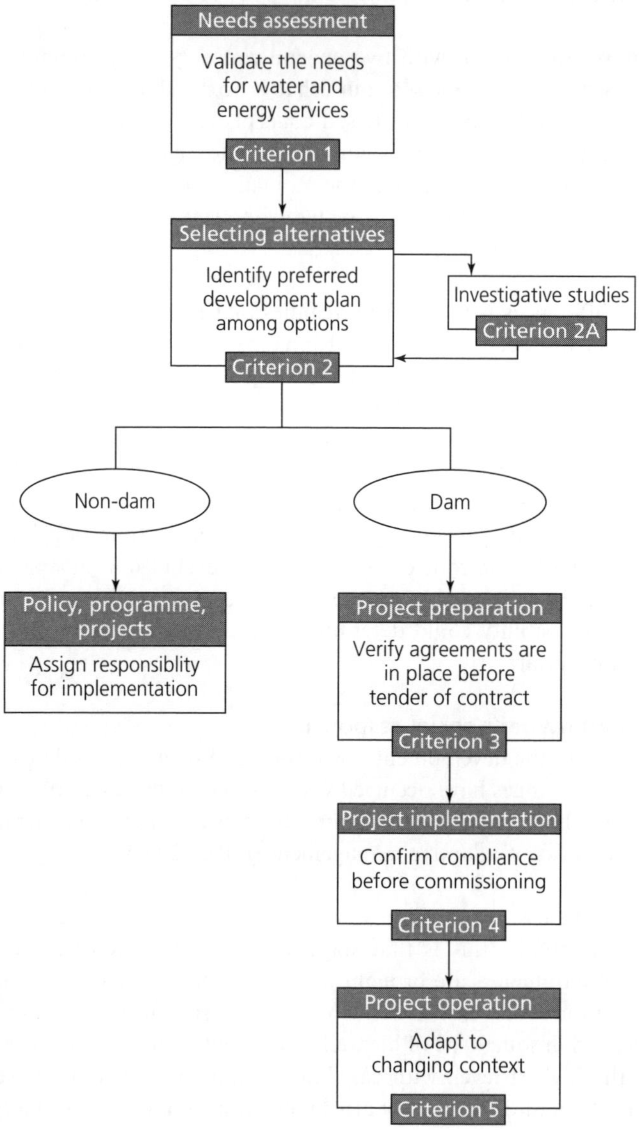


## 2.2. Operational objectives

The essence of operating objectives as embodied in a water-use plan is:

- How much water needs to be released?
- How is the water released?
- When is the water released?

Dams and reservoirs are very significant assets built for a purpose. They are best managed in terms of an asset management system with associated asset management processes (Figure 2.6). Rarely is there a single purpose of a reservoir, although there are numerous examples of dams built with the only objective being to control and mitigate the floods. Such flood-control reservoirs often remain empty for prolonged periods of time and fill only during flood periods, attenuating flood waters. There are also reservoirs built strictly for irrigation, and in many countries there are thousands of small dams built and operated by famers to provide water for crops. Some dams are built for recreation, with the only objective being to capture water during freshet and maintain a water level during the rest of the year for boating, swimming or fishing. Some dams are built strictly for hydroelectric generation, and their only purpose is to maximise power output.

The dashed ellipse at the performance point in Figure 2.6 defines the capability of the asset system to function normally and respond to the effects of disturbances.

What happens most often, however, is that a single-purpose dam changes in the way it is operated over time because other goals are added due to growing demands. In many cases the construction of the dam invites further development downstream,

Figure 2.4 Simplified model of the operational controls of La Joie Dam (LAJ) and Carpenter Lake Reservoir two-reservoir store–pass–divert system part of the Bridge River Power Project. HCV1, HCV2, hollow cone valves; PRV, pressure relief valve; TRZ, downstream dam TRZ. (Courtesy of BC Hydro)

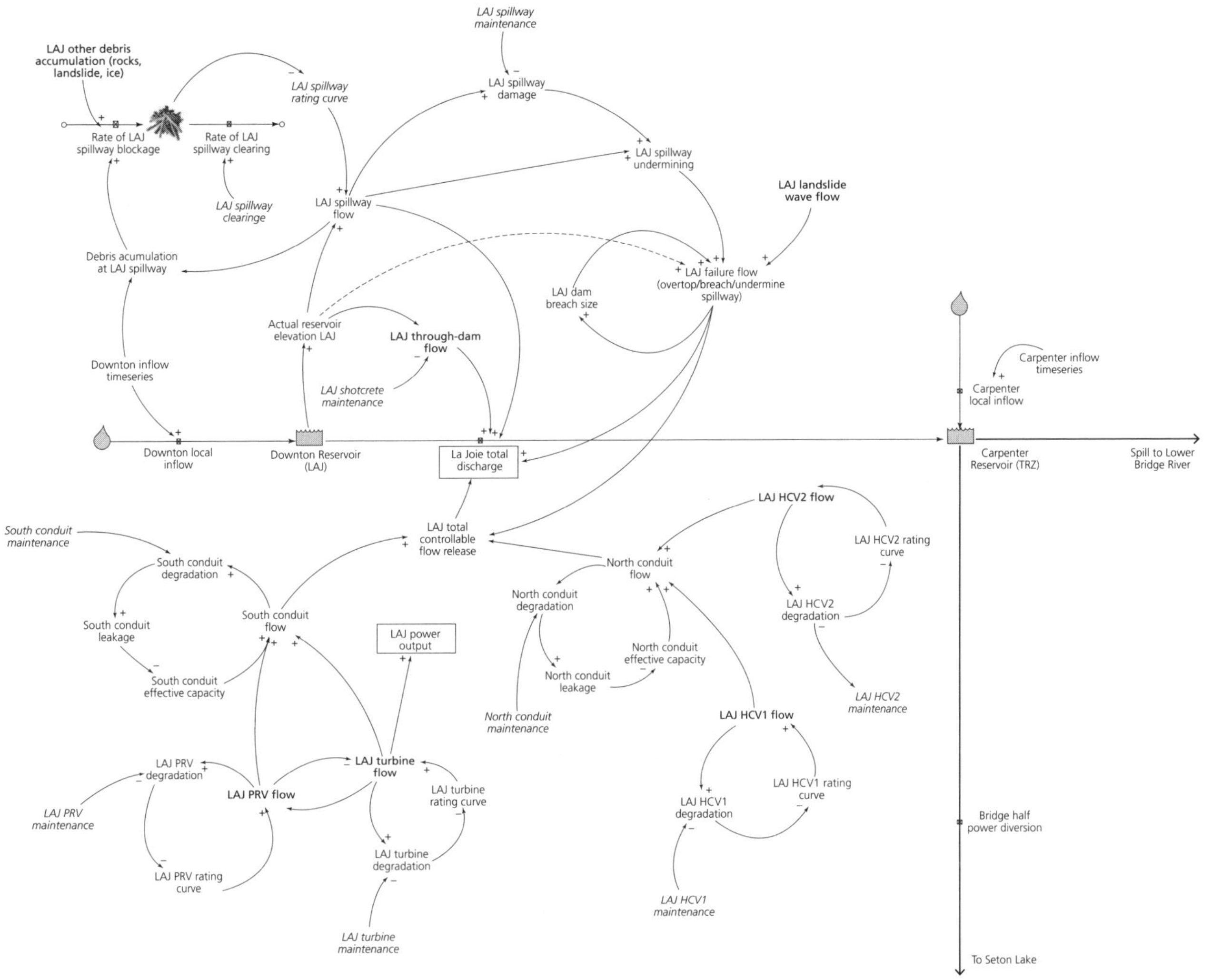


and the expectations of riparian communities change. Where once there was no demand for flood control, now communities have been built in the downstream floodplain which require protection. The dam could have been built with the only goal to generate power but now the communities demand that the dam should also provide recreation. The list of possible purposes of a reservoir is long (Figure 2.7).

### 2.2.1 Reservoir storage

The storage behind a dam is commonly divided into zones, as illustrated in Figure 2.8. In a typical reservoir the entire available storage might be divided in three zones as illustrated in the figure. The exclusive-capacity zone is established for a single purpose. Most often this space is dedicated to flood control, although one can find many examples of reservoirs that have the exclusive-capacity zone dedicated to navigation or hydroelectric generation.

The multi-purpose capacity zone may serve a wide variety of other purposes, as listed in Figure 2.7. For a single-purpose reservoir the two upper zones can be collapsed into one to serve the single dedicated purpose. The purpose of the inactive zone (sometimes also called the 'dead zone') is to maintain a minimum pool level and provide storage for accumulating sediment. While this three-fold division is common, sometimes an additional 'surcharge zone' is created to account for additional storage between the top of the exclusive capacity and the maximum design water level (USACE, 1991).

### 2.2.2 Water use plan

The water use plan is a statement of the role of the dam and reservoir in the regional water resource system. This plan states the objectives of the facility and the constraints under which it operates.

**Figure 2.5** The Bridge River Power Project, including La Joie Dam, Carpenter Reservoir behind Terzaghi Dam, and Seaton Lake behind Seaton Dam. BR1, BR2, generating stations; HCV, hollow cone valve; LLO, low-level outlet; LLOG, low-level outlet gate; NI, natural inflow; PRV, pressure relief valve. (Courtesy of BC Hydro)

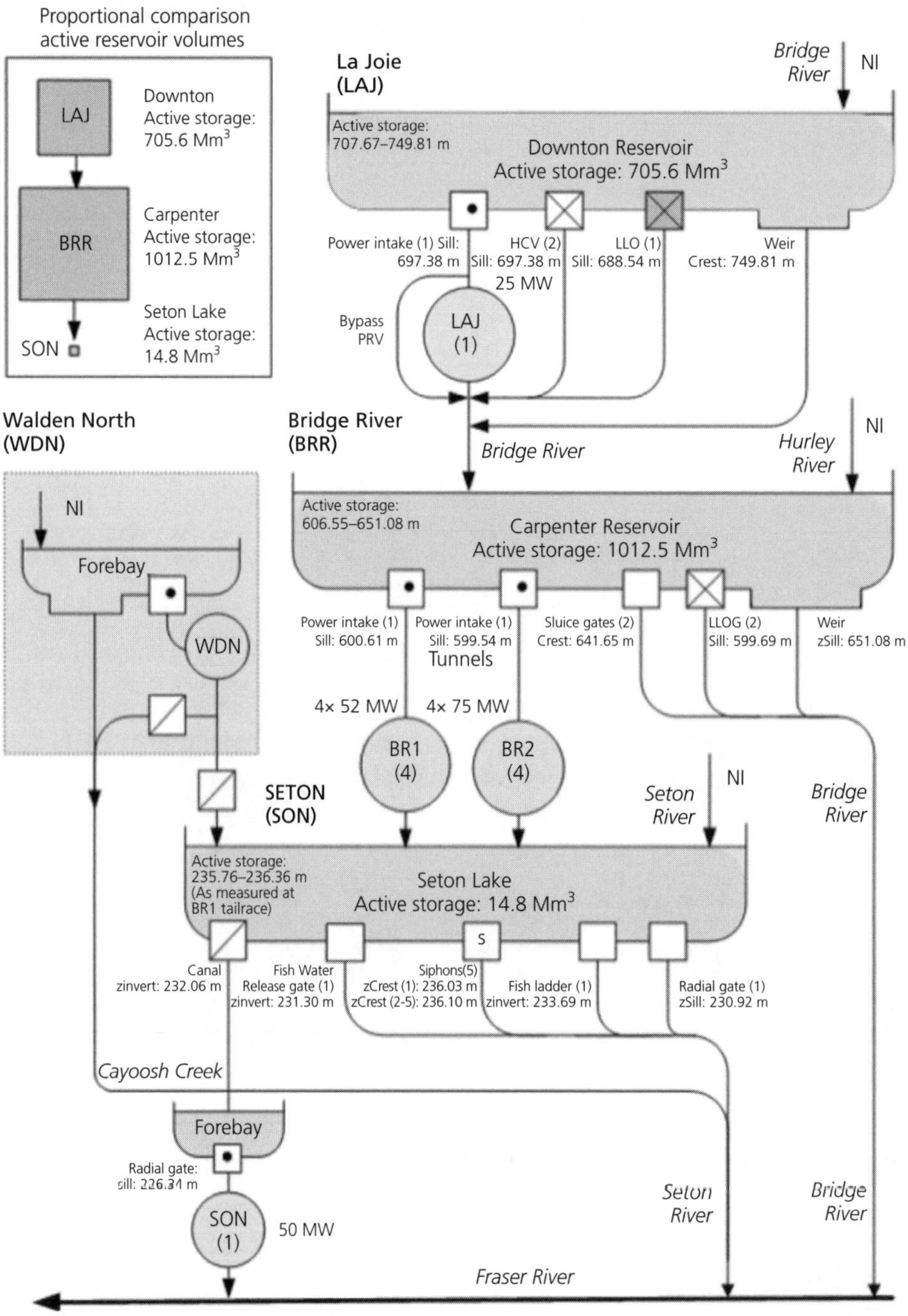


These constraints will include, for example, the necessary power it is scheduled to generate, the qualities and quantities of downstream water releases for other purposes (e.g. ecological, water supply or recreational), and flood routing requirements.

The water use plan typically summarises the hydrology and discharges of the river system and reservoir, and categorises the schematic flow configuration and waterways (Box 2.1 and Figures 2.9 and 2.10).

### 2.2.3 Operating concepts

The operation of a single-purpose reservoir is much simpler than that of a multi-purpose one. Consider a reservoir dedicated strictly to flood control. The most effective operation of the reservoir is to keep it empty at all times except for times of heavy inflows, which if released would cause adverse impacts downstream. If inflow exceeds a threshold, excess inflow can be stored, and only the amount up to the threshold would be released.

Figure 2.6 Overview of the asset management process (PAS 55). (BSI, 2004)

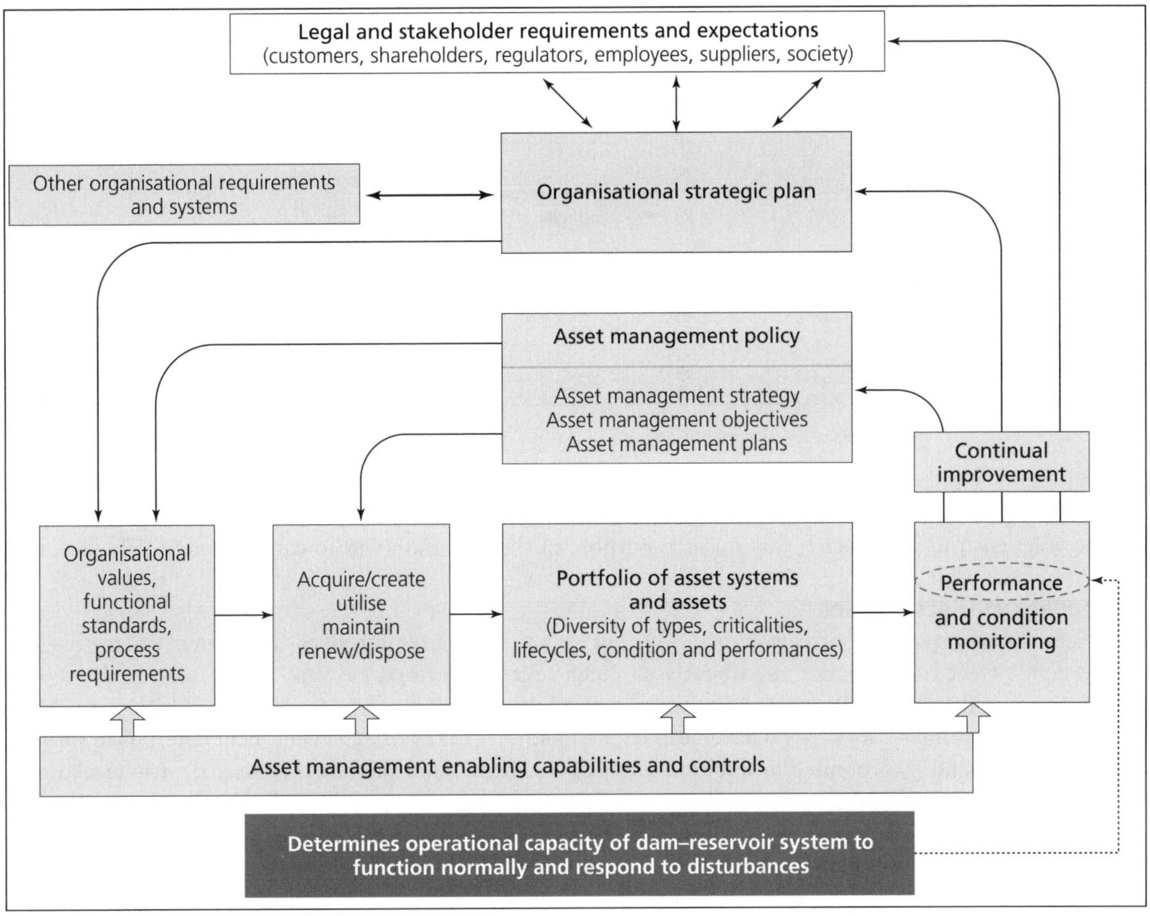


Figure 2.7 Purposes of reservoir projects authorised by the US Congress. (Adapted from USACE, 1997)

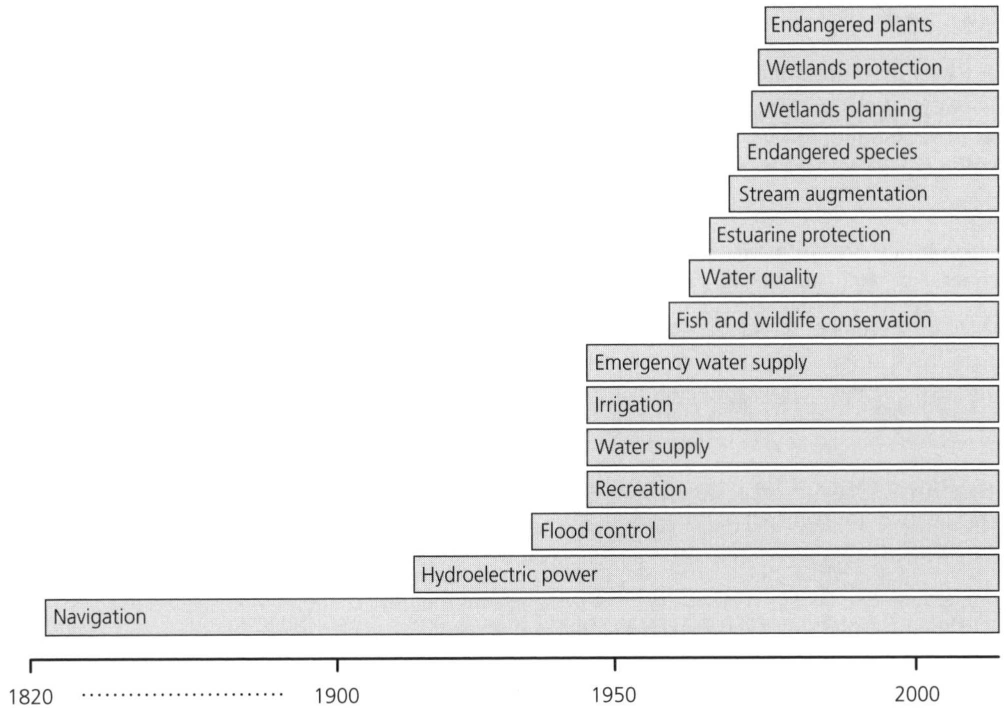

Figure 2.8 Typical storage allocation in a reservoir. (Adapted from USACE, 1997)

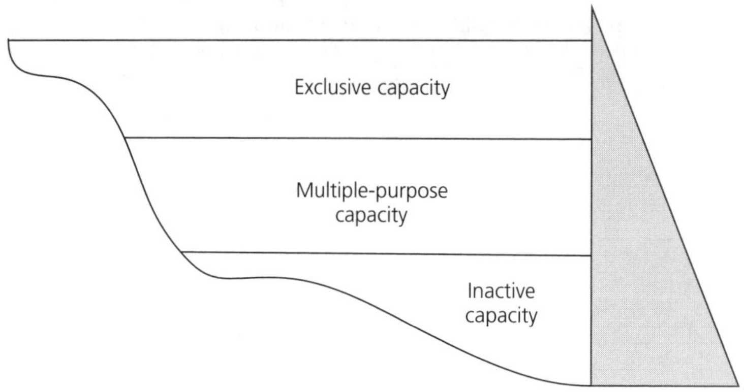


However, even such a simple case gets more complicated in practice. Every reservoir has a finite storage and the inflows are uncertain. Therefore, following the simple rule explained above, inflows can lead to damages that would be avoidable if the operating rule was constructed in such a way that the purpose of the operation was to capture only the highest inflows.

## 2.3. Operational strategies

Dams and reservoirs are typically constructed to achieve one or more primary objectives. The primary objective(s) may have secondary, tertiary or even lower order sub-objectives, which together form the overall objectives for the reservoir–river system. It may be that the full spectrum of objectives is defined at the outset, perhaps in a hierarchical way. Alternatively, it may be that once the primary objective(s) have been defined and the means of achieving them determined, other potential functions are identified and incorporated as additional objectives. Once defined, the objectives, their interrelationships and interdependencies together define the operational regime of the system.

### 2.3.1 Reason for the system and its operational regime

Water management systems differ from many other productive systems in that they constitute human-altered natural systems that utilise a naturally occurring resource without fundamentally altering the physical properties of that resource (i.e. the water), although characteristics of it, such as its potential energy, temperature, soluble and suspended contents, and the like may be changed.

Box 2.1 Excerpts from BC Hydro's Cheakamus Project Water Use Plan (BCHydro, 2008)

Under an order issued on 6 April 1999, BC Hydro is required to release into the Cheakamus River the greater of 5 m/s or 45% of the previous 7 days' average inflows to the reservoir. The remaining portion of the water may be diverted to the powerhouse, to a maximum of 65 m/s, to produce electricity, and is released into the Squamish River. Inflow that cannot be used by generating facilities or stored in the Daisy Lake Reservoir is to be released down the Cheakamus River.

Emergencies and dam safety requirements take precedence over the constraints outlined in this water use plan. Emergencies include actual or potential loss of power to customers. Dam safety requirements for operations, including procedures for surcharging the reservoir and undertaking special drawdowns of the reservoirs, are outlined elsewhere.

From 1 October to 31 December the target maximum reservoir level will be limited to an elevation of 373.5 m to provide additional storage space in the reservoir, to assist in managing high inflow events. During the routing of flood events, BC Hydro will coordinate operations with the district of Squamish.

*Maximum flow releases*: BC Hydro shall release a minimum flow from Daisy Lake dam of

- 3 m$^3$/s from 1 November to 31 December
- 5 m$^3$/s from 1 January to 31 March
- 7 m$^3$/s from 1 April to 31 October.

BC Hydro shall release additional flow from Daisy Lake Dam beyond that noted above when required to maintain a minimum flow measured at the Brackendale gate according to a defined schedule.

Ramping rates for increasing and decreasing releases shall be constrained by a predetermined table of values.

Figure 2.9 Schematic flow configuration for the Cheakamus system. CMS, Cheakamus Dam; NI, natural inflow; NPRF, non-power-release facility. (BC Hydro, 2008)

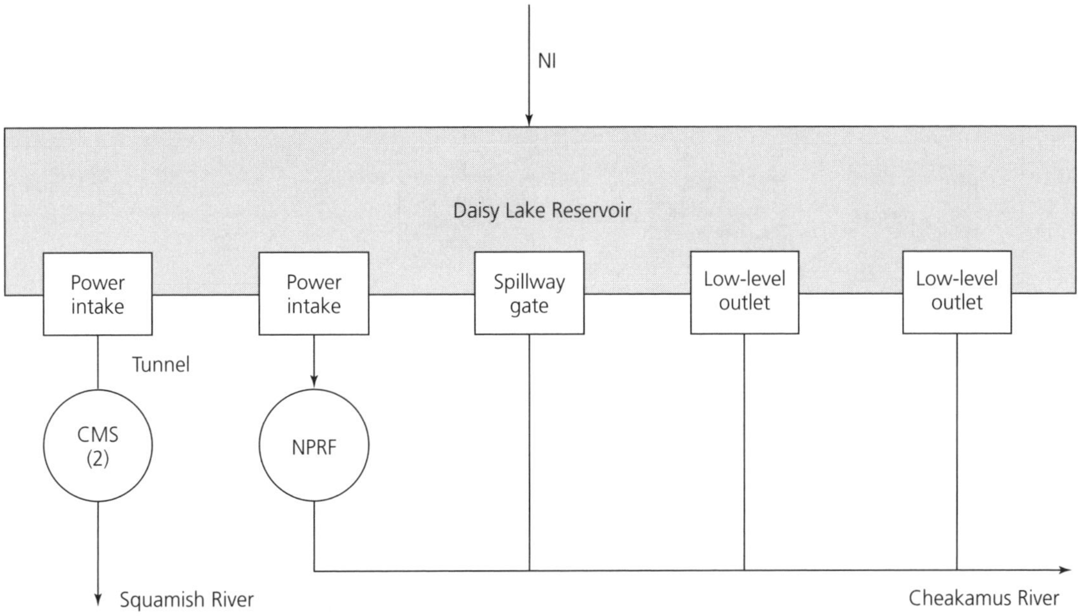


In the modern context, the reservoir–river and the operational regime for a water management system are determined in part by the owning organisation and in part with the consent of society. The boundary between societal controls and the owner's responsibilities is broadly defined in licensing and regulatory arrangements. However, there is not a distinct separation between the influence of external societal factors and the owner's internal operational system. Rather, there are interdependencies and feedback paths between the external influences and the internal controls. Some of these are dynamic, while others may have delayed impacts.

The conceptual flows and feedbacks in the reservoir–river system are suggested in Figure 2.11. The main horizontal paths are the flows from upstream to downstream. Some of these are through production waterways, some through spillways, some through the dam, and so forth. Influencing these flows are a large number of considerations, shown as bubbles and boxes with arrows suggesting lines of influence. This is similar in concept to the influence diagram shown in Figure 2.4 but focuses on requirements and constraints.

Figure 2.10 Cheakamus Dam impounding Daisy Lake, British Columbia. (BC Hydro, 2008)

Figure 2.11 Flows and feedbacks in operational regime (concept)

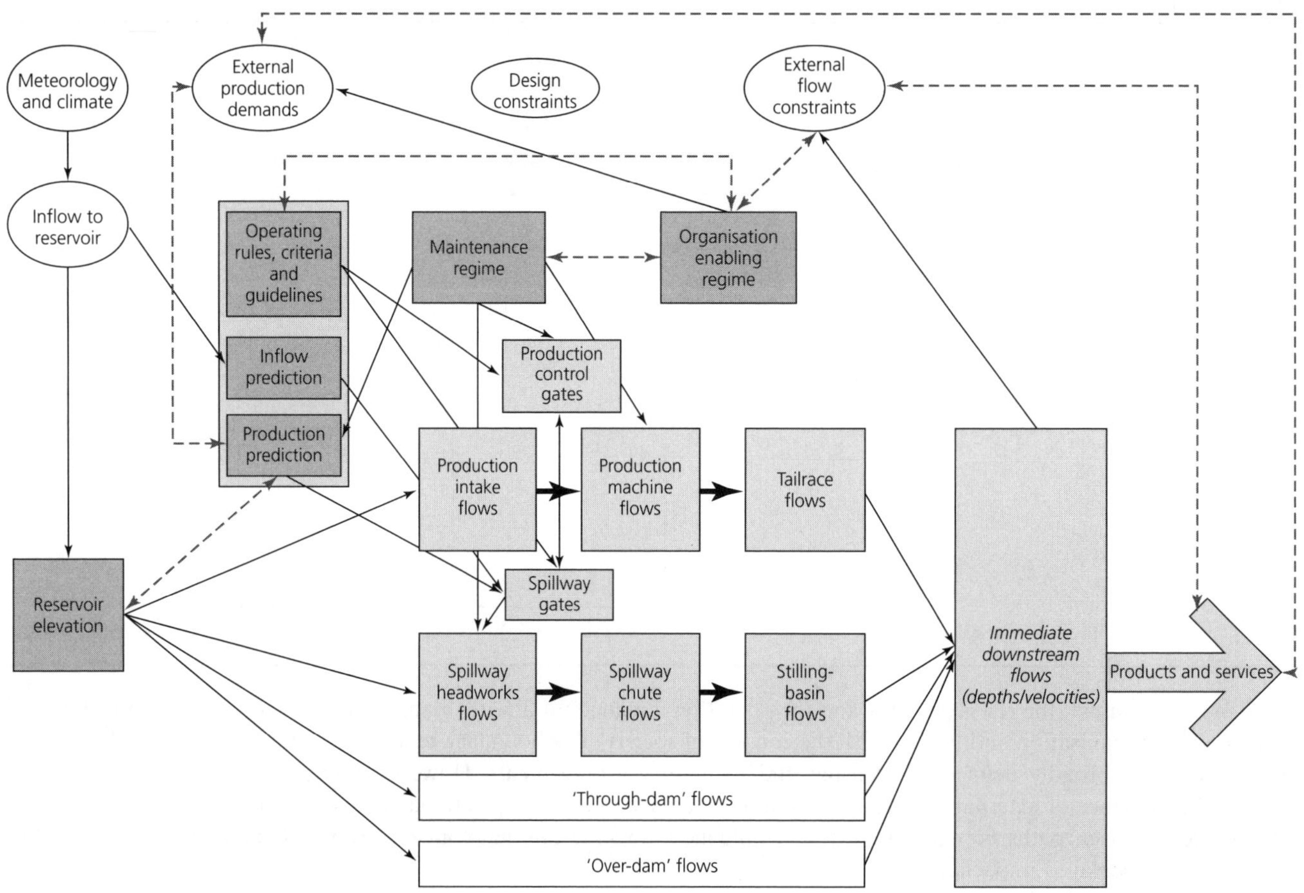


### 2.3.2 Whole life-cycle water asset management

The management of physical assets such as dams, hydraulic production systems and supporting infrastructure has a long history, arguably dating back to Egyptian and Mesopotamian times when water system assets were managed by means of robust design. More formal methods of asset management emerged in the 1970s, and systematic approaches to physical asset management have emerged over the past 20 years or so. In the modern context, whole life-cycle asset management can be depicted as illustrated in Figure 2.12. The activities within the dotted boundary represent in-service asset operations, which from the perspective of a management system hierarchy, is commonly considered.

The concept shown in Figure 2.13 from PAS 55 can be transformed into the well-established management system framework described in ICOLD Bulletin 154 on dam safety in the operational phase pass (ICOLD, 2010). The PAS 55 illustration of an asset management system that takes account of the external and internal factors and constraints that govern the *modus operandi* of a dam-owning organisation is shown in Figure 2.6. (Note: PAS 55 – In 2004 the UK Institute of Asset Management (IAM), in conjunction with the British Standards Institution (BSI), developed PAS 55, the first publicly available specification for optimised management of physical assets. This proved very successful, with widespread adoption in utilities, transport, mining, process and manufacturing industries worldwide. The 2008 update (PAS 55:2008) was developed by 50 organisations from 15 industry sectors in 10 countries. The International Standards Organisation (ISO) accepted PAS 55 as the basis for the development of the ISO 55000 series of international standards (BSI, 2014).

The culture and traditions of the organisation, together with the prevailing engineering practices, operational arrangements and the societal expectations of a dam–reservoir system, provide the overarching framework for the functional performance and reliability of the system.

The 'directing mind' of the owning/operating organisation will typically have some flexibility concerning operational choices within legally binding parameters. Excursions outside the legally binding parameters are always possible due to some breakdown of the physical or organisational control process, which may have legal consequences depending on the outcome and the legal enforcement regime. These breakdowns can be expected to involve both active failures (e.g. operation during the passage of floods, or dam incident management issues) and latent causes, such as deferral of maintenance for reasons of resource constraints or organisational culture (including matters such as staffing competence, staff complement and bonus structures).

Figure 2.12 Whole life-cycle asset management cycle

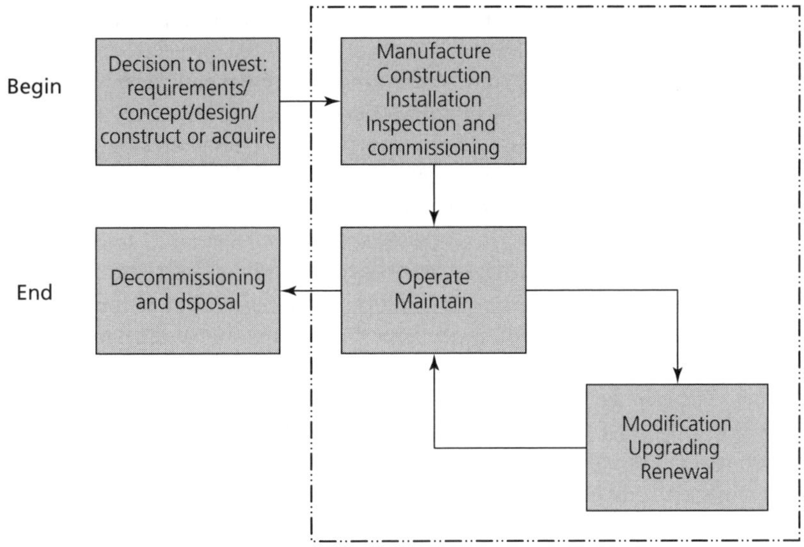


These latent conditions are an inevitable part of any organisation and they are not necessarily the result of bad decisions. Resources are rarely equitably distributed across organisational activities, and the distribution of resources may be based on sound commercial arguments. However, these inequities create quality, reliability or safety problems for someone somewhere in the organisation at some later point in time.

The 'directing mind' of an organisation can, and usually does, influence the design, construction, operation and maintenance of the system over the whole life cycle, as this is where the control of financial resources and expectations of the organisation are determined. Embodied in this is the risk appetite of the organisation, which itself is fashioned by the style of the upper echelons of the organisation, and how these upper echelons respond to outside influences.

As judgements are made at all levels of the organisation, how individuals at various levels interpret the organisational risk appetite may also be an influencing factor. Individuals may introduce personal values into decision-making. This can

Figure 2.13 Hierarchy of controls in an asset management system (PAS 55). (BSI, 2004)

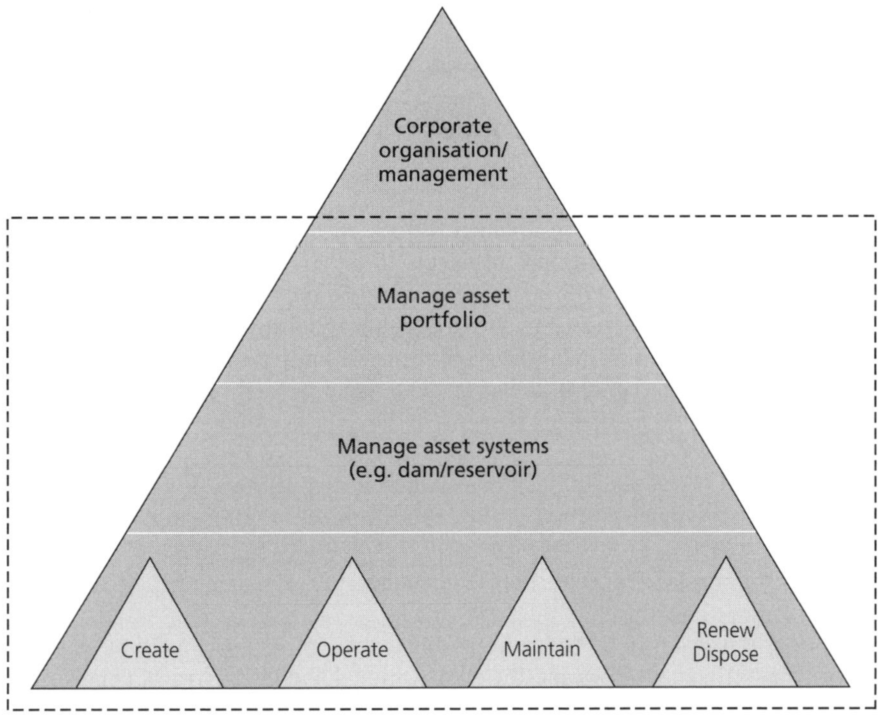

happen at all levels of the organisation. For example, the chief financial officer might set a budget for operation and maintenance costs, with the expectation that the budget is not to be exceeded. Such a budgeting process is often set in the absence of a full analysis of all the implications of the decision, and will set in motion a chain of decisions at lower levels of management. This implementation, typically done in terms of individual projects or initiatives that are justifiable in their own right, may lead to an unstable system state when all the interdependencies between elements, functions and controls come into play. However, it is not just budgetary restrictions that can lead to unfavourable system states. Other factors are inappropriate reward and compensation structures, a culture with characteristics that lead to unfavourable management and work practices (e.g. 'blame and train' safety management, poor appreciation by non-technical executives of their role in ensuring the integrity of the technological and built systems), etc.

## 2.4. Qualitative modelling of dam safety management activities

In the modern context, safe management of operational activities built on the concept of control processes (feedback loops) is built into the human, technology, organisational and, more recently, information systems to ensure continued safe operation of the system as a whole. Accident analysis has provided many insights into, and is arguably a basis for, qualitative management of safety related activities, with barrier-based methods such as bow-tie analysis (see Figure 2.1) being related to the failed barrier method of the 'Swiss cheese' type of accident investigation model (upper part of Figure 2.2). This relationship follows naturally from the purpose of accident analysis, which is to learn from the accident in order to prevent its recurrence. Barrier-based methods of safety management provide a useful means of addressing the problem of loss of flow control in dam and reservoir systems.

In a general sense, barriers can be characterised in different ways (Svenson, 1991), although the idea of barriers in the management of risk has earlier origins. Barriers can be defined simply in terms of equipment, built entities or rules that can stop the development of an accident. Alternatively, a distinction between three types of barriers – passive, active and procedural barriers – may be made. One way is with regard to their temporal relation to an actual or hypothetical accident. Typically, barriers may be considered to be preventive or protective. The distinction between preventive and protective barriers is relative to the occurrence of the accident-initiating event. Barriers that are designed to work before a specific accident event takes place serve as preventive measures. Such barriers are supposed to ensure that the accident does not happen, or at least to slow down the development of conditions that may result in an accident. Barriers that are intended to work after a specific initiating event has taken place serve as means of protection. These barriers are intended to shield the environment and the people in it, as well as the system itself, from the adverse effects of the accident.

Barriers may be either active or passive and are not necessarily physical in nature. If a barrier is active, it involves one or more functions, the results of which achieve the purpose of the barrier. If a barrier is passive or inactive, it means that it serves its purpose by its presence rather than by actively doing something. With respect to prevention, an active barrier, such as a warning alarm, may hinder an action being carried out, while a passive barrier, such as lock on a door, may block access to a dangerous area. In relation to protection, an active barrier, such as a fire-suppression system, has the function of reducing the consequences, whereas a passive barrier, such as a fire wall, contains or holds the damage-causing agent.

Overall, for the purposes of this book, the process of engineering the system to safely retain water and pass flows through and around the dam in a controlled way benefits from the use of qualitative barrier analysis and the various other related methods, of which two of several are briefly outlined below. In general, the full suite of barrier types and uses can be applied in various ways and in various places to any flow-control system.

The human–technology–organisation (MTO) process, which focuses on the interaction between humans, technologies and organisations (Lundberg *et al.*, 2009), is one development related to the barrier concept. It was developed for accident and incident analysis, and further developed for improving accident investigation, safety, quality and efficiency within companies and organisations. The process has been widely used by Swedish utilities and regulators to describe knowledge and analytical techniques that focus on human and organisational factors and their relationship with nuclear safety. MTO was introduced in Sweden after the Three Mile Accident as a concept similar to the 'human factors' concept developed in the USA. It was the intent that the explicit mention of the three interrelated elements in the concept – human, technology and organisation – would stimulate a comprehensive 'system view' of nuclear safety. This view should go beyond a strict technological perspective to recognise and highlight human and organisational factors as important moderators of risk. In recent years, MTO has been extended to include the information systems involved in the control process, resulting in the acronym MTOI.

MTO is associated with at least three different (but related) domains:

- *MTO as a set of analytical techniques.* In this domain the MTO concept focuses on the methods that analyse the relationships between humans, their activities and the organisational and technological context in which these activities take place.

- *MTO as a human factors specialist domain.* In this domain the MTO concept is foremostly perceived as a specialist domain, supported by knowledge of human factors, psychology and other human-related sciences.
- *MTO as a metaphor for system thinking about safety.* In this perspective the MTO concept is viewed neither as a set of specialist domains nor as a set of specific methods, but as a general attempt to develop a safety-culture thinking that focuses on the entire socio-technical system (including technology, human factors and organisational issues).

Hollnagel's FRAM (Hollnagel, 2012), which can be related to MTO, is used here to illustrate the way in which the systemic approach can be applied in a qualitative way to both physical assets and operational activities, as would be set out in a management system. The purpose is to illustrate not just the FRAM process but also the way of thinking and qualitative modelling of all the factors relevant to discharge function reliability for dams. The thinking involved in the following utilisation of the FRAM model provides an approach that can be developed and applied across the creation of the analytical layer associated with the construction of dynamic system models, as illustrated in Chapter 10. How this can be achieved is developed by means of one small but important aspect of operational control, specifically spillway gate testing.

### 2.4.1 Qualitative modelling of spillway gate maintenance and testing activity

Inspection, testing and maintenance of spillway gates is an essential element of the operational management of dams and reservoirs, as it is the means of assurance of the relevance and accuracy of spillway gate reliability parameters used in a spillway system reliability analysis. The broader application is in the reliability of the totality of the discharge function, and the example outlined below is just as applicable to the maintenance, inspection and testing of the hydraulic production systems. It is also applicable to any other function that relies on maintenance, inspection and testing, including human competence.

The four principal steps in a FRAM analysis (Figure 2.14) are:

1. Identify essential system functions, and characterise each function by means of six basic parameters (based on the structured analysis and design technique).
2. Characterise the (context dependent) potential variability (using a checklist).
3. Define functional resonance based on possible dependencies (couplings) between functions.
4. Identify barriers for variability (damping factors) and specify required performance monitoring.

*Step 1.* Identifying essential system functions, and characterising each function by six basic parameters. The functions are described through six aspects, in terms of their input (I, that which the function uses or transforms), output (O, that which the function produces), preconditions (P, conditions that must be fulfilled to perform a function), resources (R, that which the function needs or consumes), time (T, that which affects time availability) and control (C, that which supervises

Figure 2.14 Unit of a FRAM model. (Hollnagel *et al.*, 2006)

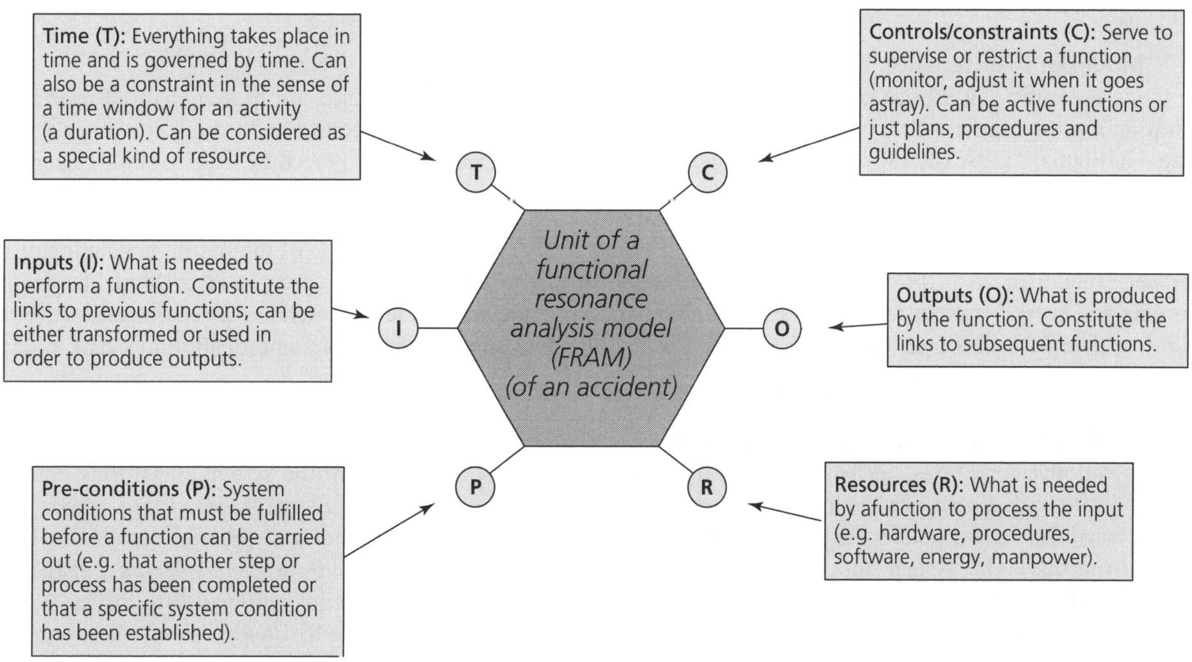

or adjusts the function), and may be described in a table and subsequently visualised in a hexagonal representation. The main result of this step is a FRAM 'model' with all basic functions identified.

*Step 2.* Characterisation of the (context dependent) potential variability through common performance conditions. Eleven common performance conditions (CPCs) are identified in the FRAM method and these are used to elicit the potential variability: (1) availability of personnel and equipment; (2) training, preparation and competence; (3) communication quality; (4) human–machine interaction and operational support; (5) availability of procedures; (6) work conditions; (7) goals – number and conflicts; (8) available time; (9) circadian rhythm and stress; (10) team collaboration; and (11) organisational quality. These CPCs address the combined human, technological and organisational aspects of each function. After identifying the CPCs, the variability needs to be determined in a qualitative way in terms of stability, predictability, sufficiency and boundaries of performance.

*Step 3.* Defining the functional resonance based on possible dependencies/couplings between functions and the potential for functional variability. The output of the functional description of step 1 is a list of functions, each with its six aspects. Step 3 identifies instantiations, which are sets of couplings between functions for specified time intervals. The instantiations illustrate how different functions are active in a defined context. The description of the aspects defines the potential links between the functions. For example, the output of one function may be an input to another function, or produce a resource, fulfil a precondition, or enforce a control or time constraint. Depending on the conditions at a given point in time, potential links may become actual links, hence producing an instantiation of the model for those conditions. The potential links between functions may be combined with the results of step 2, the characterisation of variability. That is, the links specify where the variability of one function may have an impact, or may propagate. This analysis thus determines how resonance can develop between functions in the system. For example, if the output of a function is unpredictably variable, another function that requires this output as a resource may be performed unpredictably as a consequence. Many such occurrences and propagations of variability may have the effect of resonance; the added variability under the normal detection threshold becomes a 'signal', a high risk or vulnerability.

*Step 4.* Identifying barriers for variability (damping factors) and specifying required performance monitoring. Barriers are hindrances that may either prevent an unwanted event taking place or protect against the consequences of an unwanted event. Barriers can be described in terms of barrier systems (the organisational and/or physical structure of the barrier) and barrier functions (the manner by which the barrier achieves its purpose). In FRAM, four categories of barrier system are identified: (1) physical barrier systems block the movement or transportation of mass, energy or information; (2) functional barrier systems set up preconditions that need to be met before an action (by human and/or machine) can be undertaken; (3) symbolic barrier systems are indications of constraints on action that are physically present; and (4) incorporeal barrier systems are indications of constraints on action that are not physically present. Besides recommendations for barriers, FRAM is aimed at specifying recommendations for the monitoring of performance and variability, to be able to detect undesired variability. The following is derived from Hollnagel (2005).

A FRAM type representation of the maintenance and testing procedure could be represented in terms of a series of functional units (modules).

### 2.4.2 Model of maintenance, inspection and testing

In the modern context, all aspects of dam safety management in the operational phase can be systematised in terms of an organisation's management system, which at the detailed level of maintenance, inspection and testing could be of the form illustrated in Figure 2.15. A FRAM type model of the testing stage of the management procedure (step 4) could be represented as shown in Figures 2.16 and 2.17.

There are obviously links between all of the functions in Figure 2.17 through one or more of the six fundamental parameters and between one or more of the six fundamental parameters in Figure 2.18. There may even be several links between functional units and several reasons for links between fundamental parameters. These links define the relationships between the functions. Figure 2.18 illustrates the functional tasks and functional subtasks between some of the links that might arise between steps 4 and 6 of the management procedure.

## 2.5. Stable and unstable system states

Normal operational conditions can be considered to be the stable system state where the system transforms inflows into productive outflows in a controlled manner in accordance with the design intent. Ideally, the stable system state is the state that the physical system, the public, the environment, the organisation and its operational staff become attuned to. Hydraulic deviations from this stable system state should result in some form of adjustment within the *modus operandi* of the various entities involved with the system. Sometimes the hydraulic deviations will be minor, resulting in little or no discernible change in the state (physical, organisational or thought) of the affected entities. Other hydraulic deviations may be major, as occur during a large flood that demands or results in very significant changes in the entire state of the system and

Figure 2.15 Maintenance management procedure within a management system

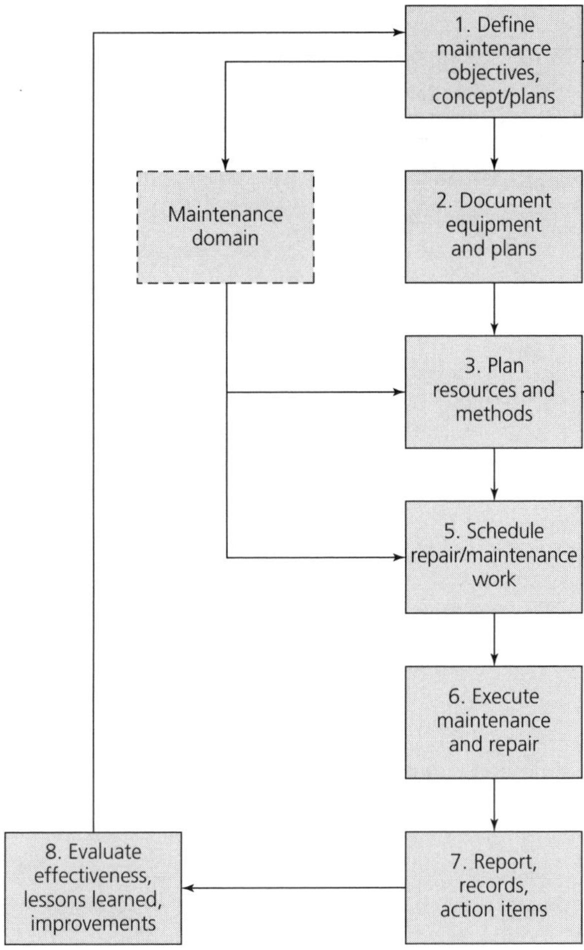


affected entities. Control over the reservoir volume and outflow must also be maintained for all normal stable, deviant stable, abnormal and unstable system states.

A (trial) distinction is made here in an effort to unravel some of the considerations that a system may exist in and/or pass through during various operating conditions that might occur during the life cycle of the system, which can be described broadly as follows:

- *Normal stable.* The system and its subsystems, functions, processes, products and services are functioning entirely as envisaged by the design, the owner–operator and all stakeholders (including the environmental elements that have achieved a new state of equilibrium through adaptation).
- *Deviant stable.* The system overall, in terms of its functions, processes, products and services, is functioning in a stable manner, but one or more of the subsystems, subfunctions and/or processes are not in the 'normal state'. Such conditions have been broadly divided into 'internal deviant stable' and 'external deviant stable';
  - *Internal deviant stable* system states include planned outages due to maintenance activities or forced outages of the type for which the system is fault tolerant. They occur within the organisational and operational regime without any external manifestation of a deviation from the norm. These system states can become accepted as being normal over time within the organisation ('normalisation of the deviant', 'behavioural drift'), which may result in a subtle and unnoticed transition of the system state to a less than desirable condition.
  - *External deviant stable* system states include high-flow situations where the inflows and outflows are above the annual average and even larger than recent memory, but within the operational parameters as defined in terms of the licensing arrangements. Some stakeholders may view these system states as abnormal insofar as their perspective is concerned. These perspectives may influence the operation of the system. As is the case of the subtle changes in the internal state described above, organisational and operational responses to these perceptions can also lead to drift in the understanding of 'normal', usually with unknown longer term implications for system operation.

Figure 2.16 FRAM-type model of the maintenance and testing procedure

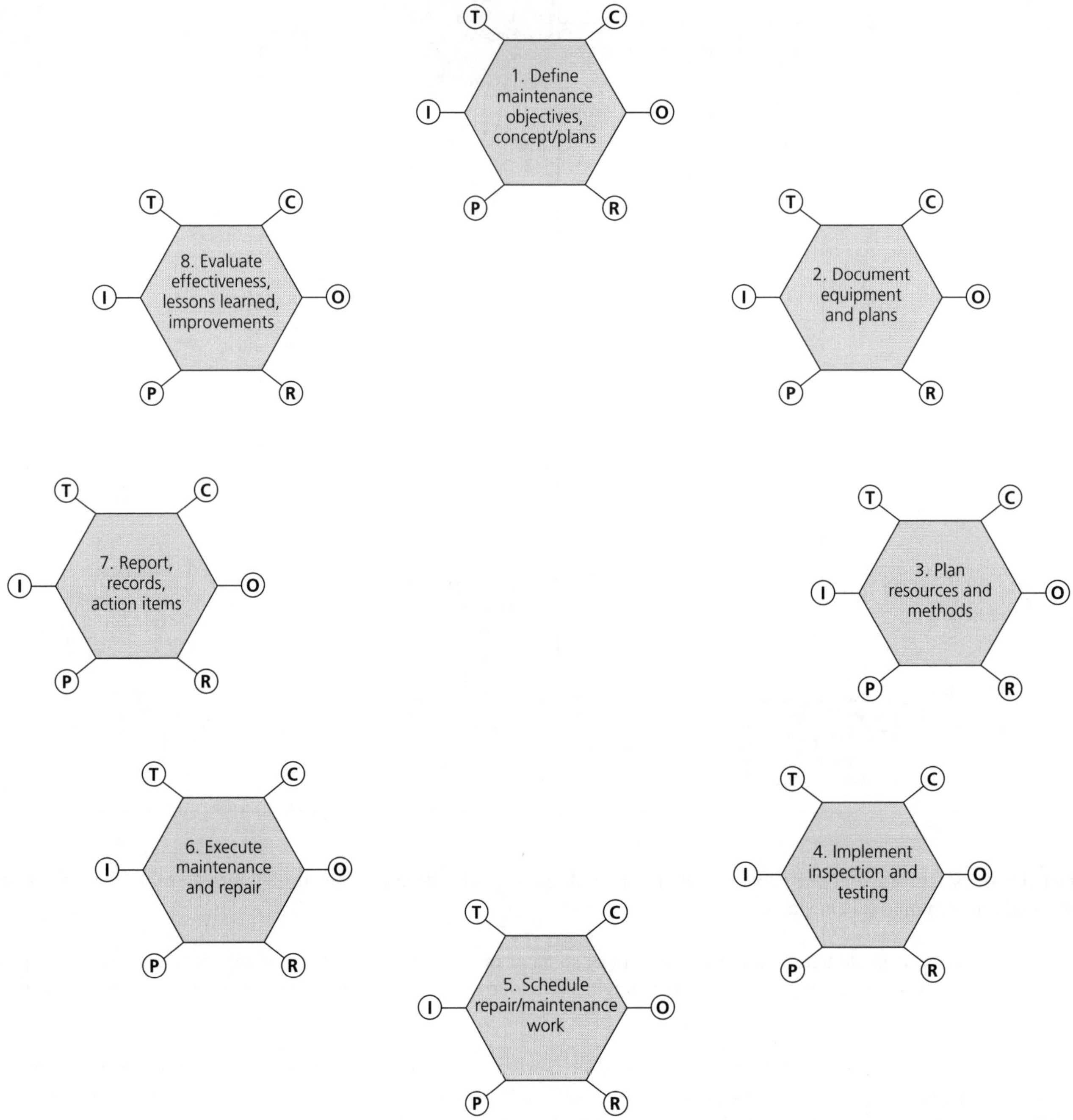


- *Abnormal.* The system exhibits behaviour that requires a change in the operational mode of the system. It may or may not involve an immediate change in the outputs of the system, but could result in a change in the outputs over time. Importantly, the system state remains under the control of the owner–operator, although this control may include external inputs and considerations. Internally such a condition typically involves a need to deviate from the normal operating plan (normal stable and deviant stable). Externally such a condition may manifest itself as an advisory communication from the owner–operator or a change in the outputs of the system. In many cases, some abnormal system states as viewed internally may not have an outward manifestation until they escalate to the stage where an external change in the outputs occurs or when the owner–operator issues an advisory communication (or both, and not necessarily simultaneously).
- *Unstable.* The unstable system state is that where the owner–operator has either partial control or has lost control of the performance and functions of the system. The transformation to an unstable state may be dramatic, as would be associated with a large earthquake or a major production fault, or it might go almost unnoticed for some time until a very serious and even catastrophic condition develops. The unstable state may or may not be controllable by the operator, and it may be a previously anticipated state for which an emergency intervention plan exists.

Figure 2.17 Model of the testing procedure

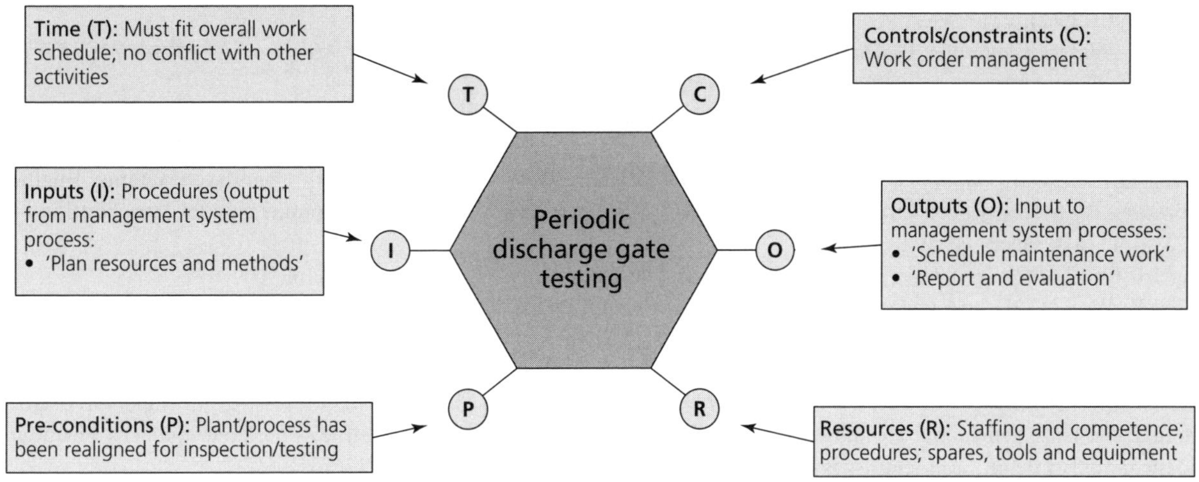


On the other hand, the unstable state may escalate to a catastrophe, where the rate of escalation to the failed state can vary between years (e.g. internal erosion failure of an earth dam) and almost instantaneous (e.g. seismically induced liquefaction failure).

Different parts of the system and the system as a whole can exist in different states at the same time. For example, part of the system, such as the production facility, may transition from stable to an abnormal state, as could occur under production fault conditions, while the overall system state transitions to an internal deviant state.

In some cases, an operator of a reservoir might be faced with having to deal with the simultaneous occurrence of a production facility fault, a high inflow, an external disturbance and a new performance expectation. Such conditions can arise during large floods, when the production facility might be forced to shut down, the inflows bring an associated large quantity of debris, and there is a need to condition the outflows to avoid exacerbating an emergency condition downstream (e.g. a landslide that blocks the river channel downstream causing flooding in the community between the dam and the landslide).

Figure 2.18 Model (limited) of the functions and dependencies involved in the testing process

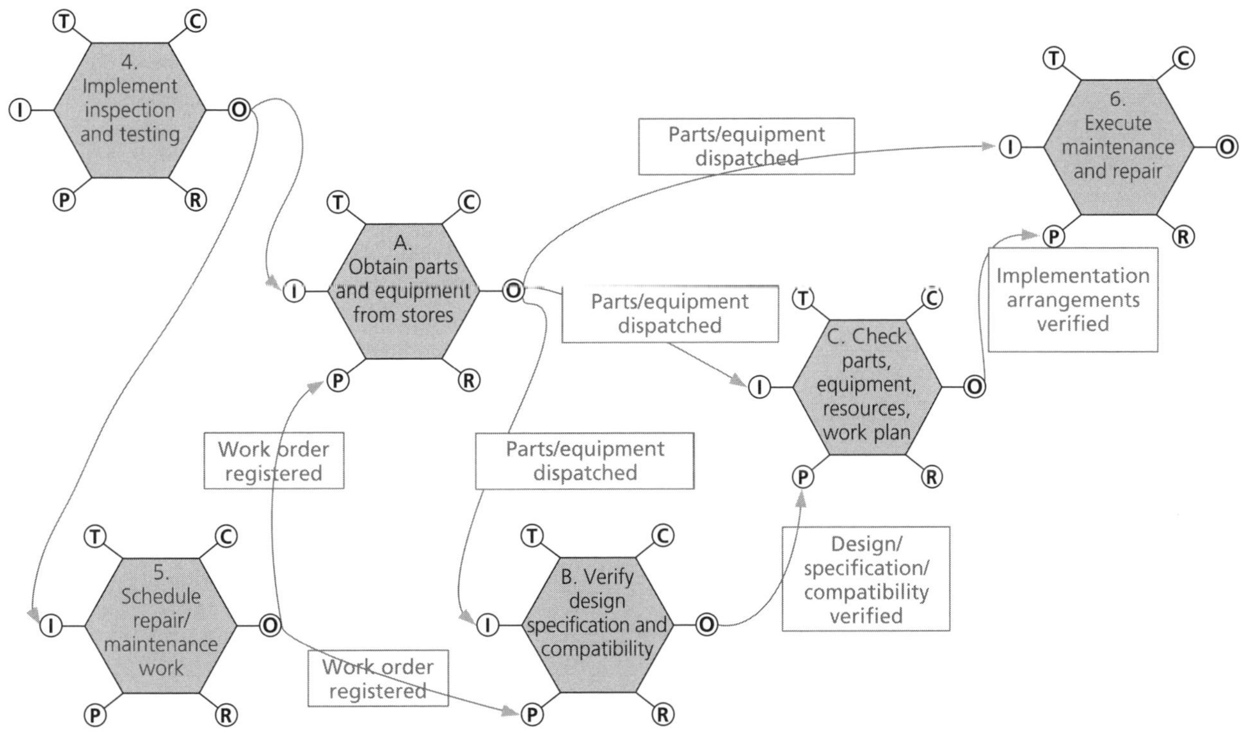

## REFERENCES

Ale B (2005) Tolerable or acceptable: a comparison of risk regulation in the United Kingdom and in the Netherlands. *Risk Analysis* **25(2)**: 231–241.

BC Hydro (2008) *Cheakamus Project Water Use Plan, Cheakamus River Channel Morphology (Year 1)*. BC Hydro, Vancouver, Canada.

BSI (British Standards Institution) (2004) PAS 55: Asset management. BSI, London, UK.

BSI (2014) BS ISO 55000 Series: Asset management. Overview, principles and terminology. BSI, London, UK.

Hollnagel E (2005) *A Functional Resonance Accident Model: Method and Examples*. Cognitive Systems Engineering Laboratory, University of Linköping, Linköping, Sweden. See open-psa.org/joomla1.5/index2.php?option=com_sobi2&sobi2Task (accessed 29/06/2016).

Hollnagel E (2012) *FRAM – The Functional Resonance Analysis Method*. Ashgate, Farnham, UK.

Hollnagel E, Woods DD and Leveson N (2006) *Resilience Engineering: Concepts and Precepts*. Ashgate, Farnham, UK.

ICOLD (International Commission on Large Dams) (2011) *Dam Safety Management: Operational Phase of the Dam Life Cycle*. ICOLD Bulletin 154. ICOLD, Paris, France.

Leveson NG (2012) *Engineering a Safer World: Systems Thinking Applied to Safety*. MIT Press, Cambridge, MA, USA.

Lundberg J, Rollenhagen C and Hollnagel E (2009) What-you-look-for-is-what-you-find – the consequences of underlying accident models in eight accident investigation manuals. *Safety Science* **47(10)**: 1297–1311.

Miles RF (1973) *Systems Concepts: Lectures on Contemporary Approaches to Systems*. John Wiley, Chichester, UK.

NRC (National Research Council/Royal Society of Canada) (2005) *Review of the Lake Ontario–St. Lawrence River Studies*. National Academies Press, Washington, DC, USA.

Reason JT (1997) *Managing the Risks of Organizational Accidents*, 1st edn. Ashgate, Farnham, UK.

Reason J, Hollnagel E and Paries J (2006) *Revisiting the 'Swiss Cheese' Model of Accidents*. Eurocontrol Experimental Centre, Brétigny-sur-Orge, France.

Svenson O (1991) The accident evolution and barrier function (AEB) model applied to incident analysis in the processing industries. *Risk Analysis* **11(3)**: 499–507.

USACE (US Army Corps of Engineers) (1991) *Optimization of Multi-purpose Reservoir System Operations: A Review of Modeling and Analysis Approaches*. USACE, Washington, DC, USA.

USACE (1997) *Hydrologic Engineering Requirements for Reservoirs*. USACE, Washington DC, USA.

WCD (World Commission on Dams) (2000) *Dams and Development: A New Framework for Decision-Making*. WCD, London, UK. See: http://www.internationalrivers.org/resources/dams-and-development-a-new-framework-for-decision-making-3939 (accessed 01/06/2016).

**Operational Safety of Dams and Reservoirs**
ISBN 978-0-7277-6121-7


http://dx.doi.org/10.1680/osdr.61217.047

# Chapter 3
# Systems and systems engineering

Dams are engineered systems that are set in a natural environment, and as such the dam system, comprising the dam and appurtenant structures, reservoir, foundations, abutments, etc., is an engineering altered natural system. These are not merely a collection of components but complexes of interacting parts, subject to a variety of disturbances, and operated by human agency. In the past, engineers have concerned themselves with understanding how the component parts of dam systems operate individually and how the components interact with one another. The goal has been to understand the risks facing those systems, and how to reduce the risks or to mitigate their effects. The emerging paradigm today, however, includes not just how to treat risk and mitigate consequences, but also how to accommodate risk.

Dams and their associated flow controls are highly complex systems of engineered structures, natural processes and human operation. They behave in complex ways that are not amenable to simple decompositional analysis, and thus need to be understood in a systems engineering context.

The key considerations in a systems engineering approach are:

- understanding and management of system requirements
- the functional arrangements and capabilities of the system
- how these capabilities are achieved
- the environment in which the system functions.

The capabilities of the system, specifically, are the products and services that the system produces (e.g. water for irrigation, hydropower or navigation). How the capabilities are achieved means, in operational terms, 'how the process works'. The environment in which the system functions is that domain, physically, socially and organisationally, within which the system's products and services are produced.

Traditional engineering education is usually deterministic, and little attention is paid at variability residing within the system components and loading setups and conditions. Variability is a central concept within reliability engineering, which requires a sufficient degree of attention for the evaluation of the risks posed by the dam systems. Due to complicated arrangements of the components within a dam system, evaluation of the variability of individual components is not sufficient for reliability or safety analysis. Many different components interact to achieve the required capabilities, and their combined variability is of interest for evaluation. Analysis of the variability of interacting components cannot be carried out using an analytic approach, and requires systemic methods (holistic approach) involving system models and probabilistic simulation techniques.

## 3.1. Objectives of the systems approach

The systems approach to assessing the performance and safety of dams is familiar from the perspective of dam design, and yet unfamiliar from the perspective of dam safety assessment. There are perhaps few examples in historical dam design where a classical systems approach has been used as in other modern branches of engineering. On the other hand, dam designers often implicitly use essential parts of systems engineering without being aware that they might be doing so. In contrast, however, dam safety assessment has seldom benefited from a systems approach. The hallmark of the superior designer has been the ability to balance the many aspects of dam operation and performance into an integrated whole. That integrated whole comprises the natural siting of the dam, its hydrology and geology, the physics of water containment and the control of discharges and production, and the monitoring and control of operations. The interdependencies of these many facets of the dam are never far from the mind of the chief designer.

In contrast, post-construction dam safety assessments more often rely on checklists. Will the design spillway capacity pass the assumed peak reservoir inflow? Will the maximum earthquake of concern cause no significant harm to the dam or its ancillary structures? Will icing on the spillway gates in the most severe winter cause no harm to the hoisting system? The assessments focus on checks of the isolated design parameters, checks of extreme loadings or environmental conditions, and checks of construction quality (ICOLD, 1988).

Regan (2010) summarises criteria checking and contemporary dam safety risk analysis as follows:

> Determining whether a plant is acceptably safe is not possible by examining a single valve in the plant … (Marais *et al.*, 2004). This may seem intuitively obvious but much of current dam safety decision-making is not appreciably different. Criteria-based decision-making processes analyse a few specific components such as the dam body as a whole to determine if it meets applicable criteria under various loading conditions and the spillway to determine if it will safely pass the inflow design flood. Risk-informed processes do essentially the same thing, except they estimate a probability of occurrence for the failure and, rather than just compare the results of an analysis to a specific criterion, include an evaluation of consequences in assessing if the risk is tolerable. In either case we are essentially trying to determine the safety of a dam by examining a few components of the dam, one at a time, in isolation from other components.

The endeavour of dam design involves more than the 'checks of design parameters' that are central to the safety assessment process. ICOLD Bulletin 61, *Dam Design Criteria – Philosophy of Choice* (ICOLD, 1988), captures the essence of the design endeavour (Box 3.1). On the other hand, dam safety assessment endeavour traditionally focuses on hydrological and structural design parameters, as described in ICOLD Bulletin 59, *Dam Safety – Guidelines* (ICOLD, 1987: Chs 3 and 4). Typically this involves checking that the dam will not fail catastrophically due to the most extreme conditions imposed by flood and earthquake hazards, and that the factors of safety under normal conditions will conform to established engineering norms. Thus, the essence of dam safety assessment as captured in various dam safety practices can be summarised by the description given in Box 3.2.

**Box 3.1** Design endeavour: essence of the distinction between dam design and safety assessment

The objective of dam design is to create a structural form which together with the foundation and environment will, most economically:

a. perform satisfactorily its function without appreciable deterioration during the conditions expected normally to occur in the life of the structure, and
b. not fail catastrophically during the most unlikely but possible conditions which may be imposed.

ICOLD Bulletin 61, *Dam Design Criteria – Philosophy of Choice*, p. 23 (ICOLD, 1988)

**Box 3.2** Safety assessment endeavour: essence of the distinction between dam design and safety assessment

The safety of a dam manifests itself in being free of any conditions or developments that could lead to its deterioration or destruction. The margin which separates the actual conditions of a dam, or the conditions it is designed for, from those leading to its damage or destruction is a measure of its safety. To be safe, therefore a dam has to be supplied with appropriate reserves, taking into account of all reasonably imaginable scenarios of normal utilisation and exceptional hazard which it may have to withstand during its life.

ICOLD Bulletin 59, *Dam Safety – Guidelines*, p. 13 (ICOLD, 1987)

In recent years, the dam safety community has embraced failure modes and effects analysis (FMEA) in an attempt to understand the risks posed at a particular dam. This approach has sometimes been called 'potential failure modes analysis' in the dam safety literature but, while this differs little in principle, it may differ in practice, from the FMEA widely used in other fields. (The key difference in practice is the categorisation of failure modes in terms of subjective assessment of the 'credibility' of the failure mode by the PFMA team under the guidance of the facilitator, whereas in terms of the established FMEA process all physically admissible failure modes are included regardless of 'perceived subjective likelihood'.) The FMEA approach is a useful tool for understanding dam safety, and especially in attempting to tease out from large amounts of data, observations and expert opinion how a specific dam might be threatened and how it might behave when subject to extraordinary demands. However, this technique too only generates linear causal chains that may lead to an accident or failure, and as a consequence differs in no important way from criteria checking. Assigning probabilities to the events in the chain in no way changes this limitation.

Thus, in current practice, dam design focuses primarily on functions and functionality, whereas dam safety assessment focuses primarily on the capacity of the dam to resist the forces that act on it up to and including the conditions associated with the most extreme loads due to flood, earthquake and other conditions. Typically, dam safety practice does not consider conditions that are intermediate between 'normal' and 'extreme', as it is assumed that capacity to withstand the norms and the extremes implies capacity to withstand all intermediate conditions. It is essential to take into account the variability between 'normal' and 'extreme' and the randomness of this when assessing the safety and reliability of dam systems.

Regan (2010) makes another point in his discussion of dams as systems. In earlier times, dam tenders attended to dams. The dam tender's job was to ensure the safe and efficient operation of the dam. The tender was physically present at the site, and understood through first-hand experience how the dam behaved (to the extent permitted by experience), and how aspects of the dam interacted as a system. In this light, the dam tender shared something of the intuitive sense of the designer about the dam as a system. Today, the exigencies of business have mostly removed tenders from dams, which are tele-operated by attendants from sometimes hundreds of kilometres away. Decisions are made on the basis of instrumental readings and supervisory control and data acquisition (SCADA) systems. The ever more frequent use of cameras as complementary surveillance equipment to some extent bridges this reliance on sensors, although a camera is nothing but a sensor.

These decisions may not have the safe performance of the dam as the overriding factor. Similarly, many other activities may not be focused solely on dam safety. The intuitive sense of systems behaviour from a dam-safety perspective may be lost. A centralised group of attendants may simultaneously be responsible for many dams. In addition, as Perrow (1999) in his theory of 'normal accidents' might argue, the complexity introduced by the operator having to rely on remote sensor systems rather than on direct and immediate observation may itself introduce additional ways for things to go wrong.

## 3.2. Dam safety from a systems perspective

### 3.2.1 The systems approach

The systems approach attempts to consider all the physical and functional interrelationships between the parts of the dam and reservoir, and to combine the analysis of the parts in their functional and spatial interrelationships in a unified structure. Rather than separating the analysis into physical parts, analysing the performance of each part separately and then recombining the results to provide an overall statement of expected performance; the systems approach attempts to address the interactions of these physical and operational parts. The systems approach attempts to capture the dynamics of dam functions and dam safety philosophies in a more thorough way than is allowed by decompositional analysis.

The intention to treat all conditions relevant to safety and performance analysis involves an expansion of the traditional design-check concept. The approach includes internal factors that pertain to the management, operation and maintenance of the dam, in addition to design loadings, which in themselves result in changes in the required functional performance of the dam. The operation of the dam, the reservoir and their productive functions are considered together as they are interrelated.

The systems approach also considers disturbances in the production function that change hydraulic forces and other conditions acting on the dam. Similarly, inspection and maintenance regimes – when coupled with design philosophy and design basis – determine the rate at which wear and tear due to operational activities and environmental factors and the effects of ageing degrade the performance capacity of the dam and its functions. Viewed in this way, systems-focused dam safety involves a much broader, structured and tractable view of all factors that contribute to the safe functioning of dams than does the traditional load-resistance philosophy of the traditional design-check approach.

Systems engineering today is a recognised branch of engineering, even with dedicated academic departments at major universities, but it is a relatively recent entry into the major divisions of civil engineering practice. It is generally thought to have originated at Bell Laboratories in the USA during World War II, and since that time has seen wide application in the defence, aerospace and electronics industries (De Weck *et al.*, 2011). Systems engineering is inherently interdisciplinary, focusing on the life-cycle performance of complex engineering projects, from design through operations and disposal. It comprises both technical- and human-oriented disciplines such as feedback control processes, industrial engineering and organisational studies. Dam safety and the interdependencies between dam safety and production functions is the latest domain to begin to incorporate the systems approach.

In the past two decades, systems engineering has broadened from its original setting in defence and aerospace to touch many other fields. Important among these new fields of application of systems thinking are civil and environmental engineering, especially as related to water resources and power generation.

Having originated in the defence and aerospace sectors, systems engineering has tended to emphasise requirements definition, the documenting and management of requirements, design synthesis, system validation, verification against the stakeholder requirements and production of physical systems. However, as noted by Rhodes and Hastings (2004), as a field, this emphasis is sometimes criticised for being too focused on processes and not enough on the systems themselves. That is, systems engineering is sometimes criticised for concerning itself too much with requirements and management, and not enough with system properties and behaviours. The emphasis in this book with respect to dam systems is the reverse: The emphasis is on systems properties and behaviour, with little consideration to design. That is, in the present book systems engineering is treated as an engineering science, not as a management science. With that objective in mind, model-based systems engineering (MBSE) is emphasised in the present volume. The MBSE approach has the best ability to facilitate complex engineered systems. This method is the most suitable one for understanding the behaviour and potential malfunctioning of the engineered dam system.

Systems engineering, especially when focusing on systems properties and behaviour, promotes the use of modelling and simulation to validate assumptions and to understand the interactions between system inputs, outputs, components and disturbances. The use of models and simulation enhances early identification of potential failure modes and functional shortcomings, and sometimes of unanticipated systems behaviours, which may not have been obvious *a priori*.

### 3.2.2 Models and simulations

Models and simulations are important in systems engineering. These logical or mathematical constructs allow us to obtain information about how an engineered physical system, such as a dam and power plant, behaves in the field, without having to physically test a real-life prototype. The use of modelling and simulation in engineering has been well recognised for many years, but these tools have advanced dramatically over the past two decades with the ever-advancing capabilities of computer technology.

Models and simulation are generally cheaper than experimenting with full-scale prototypes, and also safer if near-failure conditions are to be investigated. They allow experiments to be done with extreme environmental conditions, such as earthquakes and hydrological events that might only be hypothetical in the field. In this sense, models and simulations might be more realistic than field experiments, as they allow complete flexibility in setting environmental parameters that might be difficult to reproduce in prototype tests. Finally, models and simulations are almost always cheaper and quicker to conduct than field tests. This allows more comprehensive 'what-if' testing and the analysis of many alternatives.

A mathematical model is any set of meaningful quantitative relationships between system inputs, outputs and (environmental) disturbances. The transition between the inputs and outputs is the functioning of the system itself plus any sensing, operation and control that transpire while the system is operating. These relationships can be as simple as accounting for mass input and output quantities, slightly more complex using statistical correlations, or even more complex using differential equation approximations. They may or may not incorporate operating rules and human operator performance.

Simulation models, at least in the sense used here, are a special sort of system model. In the sense that we used the term 'simulation model' in this book, we mean a stochastic process model that mimics the probabilistic occurrence of inputs, disturbances, systems physics and operating actions over time, and that generates a corresponding stochastic output of response variables such as downstream flows and physical system damage states. A further discussion of modelling approaches is given in Chapter 4 and a specific discussion of stochastic simulation is provided in Chapter 10.

### 3.2.3 Monte Carlo method

The broad class of stochastic (or random) simulation methods that use randomly generated number inputs to calculate random outputs is commonly known as the Monte Carlo (MC) method. This is the method used in this book, and detailed in Chapter 10. The MC method is generally credited to the mid-20th century mathematician Stanislaw Ulam, who worked at the Los Alamos National Laboratory during World War II developing computational methodologies for the design of weapons systems, specifically the Manhattan Project. The modelling methodology was initially developed to handle complex integrations but soon found application in stochastic process modelling.

The method proved valuable in allowing approximate solutions to be obtained to complex problems that could not realistically be solved analytically. John von Neumann at Princeton's Advanced Study Institute was an early champion of these methods, and used the early ENIAC computer at the University of Pennsylvania to obtain practical solutions to MC analyses. The MC method was classified as a national security secret during the War ('Monte Carlo' was its code name) and was only declassified in 1949 (Metropolis and Ulam, 1949).

The essence of MC simulation is that it uses statistical sampling to estimate systems performance. By sampling a large number of times, statistical estimation theory can be used to approximate the output behaviour of the system to any desired precision. Thus, a specific set of input parameters, a deterministic vector or matrix of values, is generated. This deterministic set of parameter values is entered into the model and an output calculated. The output may be either deterministic or probabilistic. This calculation generates one specific realisation (i.e. sample observation) of the system performance. Then, another specific set of input parameters is generated according to a selected probability distribution, and again a specific realisation of the system performance is calculated. This is repeated a large number, say $n$, of times. From this procedure one obtains a sample of size $n$ of the system performance given the selected probability distribution of the inputs. From this sample, statistical theory is used to estimate means, variances and other properties of the output, and with calculable precision.

The distinction between MC simulations and 'what-if' or sensitivity calculations is important. In 'what-if' calculations relatively few deterministic scenarios of input parameters are used and correspondingly few sets of outputs are generated. Scenarios might be chosen, for example, as the best estimate, a pessimistic estimate and an optimistic estimate, and these lead to corresponding outputs. In contrast, an MC analysis might use tens of thousands of inputs and calculate tens of thousands of outputs. This generates a probability distribution of model behaviour and calculable levels of probable error. It is common for MC analyses to generate a much narrower range in the outputs than do sensitivity calculations because in the former the scenarios are weighted by their relative probability of occurring whereas in sensitivity calculations they are not. Low-probability outcomes in MC analysis end up as rare events and are treated accordingly.

### 3.2.4 Practicalities of a systems approach for dams

The question that arises from systems thinking is how does one identify and define all operational and physical conditions? Central to the approach proposed here is the premise that dynamic modelling and simulation of systems provides an operational basis to resolve this question.

The philosophy of systems engineering as a whole has two essential attributes:

- structural performance and resilience
- functional performance and resilience.

Structural performance and resilience pertain to the ability of the dam to withstand the forces that are applied to it and to maintain the structural support and integrity required for the functions of the dam and reservoir. Functional performance and resilience pertain to processes, products and services that the dam is intended to provide. Specifically, the dam is intended to retain the stored volume and to pass all flows through and around the dam in a controlled manner.

In simple terms, the dam has containment and conveyance functions in support of one or more hydraulic processes. The conveyance functions can be further subdivided into diversion flows for productive purposes, such as power production or irrigation, and release flows whereby the water is passed directly from upstream to downstream for safety, environmental or social purposes.

The range of the volumes of water to be stored is divided into 'live storage' and 'dead storage'. The rate of change of storage, the rate at which water is diverted and the rate at which water is passed directly downstream are all interrelated and must be considered throughout the design process, including in the determination of the structural design basis.

The systems approach also gives consideration to the influence of disturbances to one or more functions for reasons that can be external or internal to the system. The possibility of one or more combinations of both external and internal disturbances, ranging from those that occur essentially simultaneously to those that occur at different times but in ways such that the effects of the disturbances combine, are also considered.

## 3.3. Systems models

Systems are ubiquitous in nature and in all aspects of human endeavour, so they are usually taken for granted. Seldom is a definition entertained of what systems are or how they function. A simple distillation from von Bertalanffy (1969) is that:

> A system is a set of elements standing in interrelation. Here the term 'element', which can have many meanings and interpretations, pertains to being a 'fundamental constituent'.

For the present purpose, and in line with the definition above, Forrester (1968) proposed an operational interpretation of a system as 'a grouping of parts that operate together for a common purpose'. These definitions, where the interpretation provides a less abstract understanding, provide a distinctly different paradigm of what dam safety is about when compared with the familiar perspective of 'load-carrying capacity' of the individual parts. In Box 3.3 we expand a little on Forrester's interpretation to propose a seminal definition of a system as relevant to dam engineering.

The dual use of the term 'safe' is to highlight that the dam, reservoir, and waterway system should be safe to work on and around (i.e. internally safe), while not posing an undue threat to the safety of people, the environment or property due to the manner in which the containment and conveyance functions are performed (i.e. externally safe).

**Box 3.3** Definition of 'system' for the purpose of dam safety

A system is a set of interrelated components acting to achieve defined goals. For dam safety, those defined goals are to provide safe, efficient and effective containment and conveyance of water to achieve one or several practical objectives, such as the provision of power, protection against flood, support for navigation or the supply of potable water.

Dams and spillways as systems can be conceived on many levels:

- dams by themselves
- dams and production facilities (the entities that deliver the products and services)
- dams and production facilities in the context of reservoirs and watersheds
- dams, production systems, reservoirs and watersheds, in the context of entire river systems.

This need arises because, to avoid failure (defined at whatever level, from elements of the dam to the entire river system), all parts of a system must function together in a harmonious way (Figure 3.1).

As these systems exist and function within a natural environment, they must continue to work even when subjected to disturbances that originate externally to the system. An orchestra provides a useful analogy, whereby rather than having individual players playing the music in their own way, the composer (by means of the score) and the conductor (by means of providing the interpretation of the score and the control of the orchestra) together provide the features to unify the functions and outputs of the individual entities in the orchestra to provide the expected results. For dams, reservoirs and river systems, the designer of the dam and the operational regime provide the features and functions to operate the hydraulic system in a controlled way.

### 3.3.1 Natural systems

There are, however, important ways in which the orchestra analogy does not apply, specifically a dam and reservoir is a hazardous installation and, unlike the functioning of an orchestra, which is almost entirely controlled, a dam and reservoir is a human-altered natural system. This alteration is not restricted to the spatial arrangement of the natural environment because the storage of water alters the hazard profile of the region. For example, the natural flooding regime is altered by

**Figure 3.1** Model of a multiple river system, with individual lake subsystems

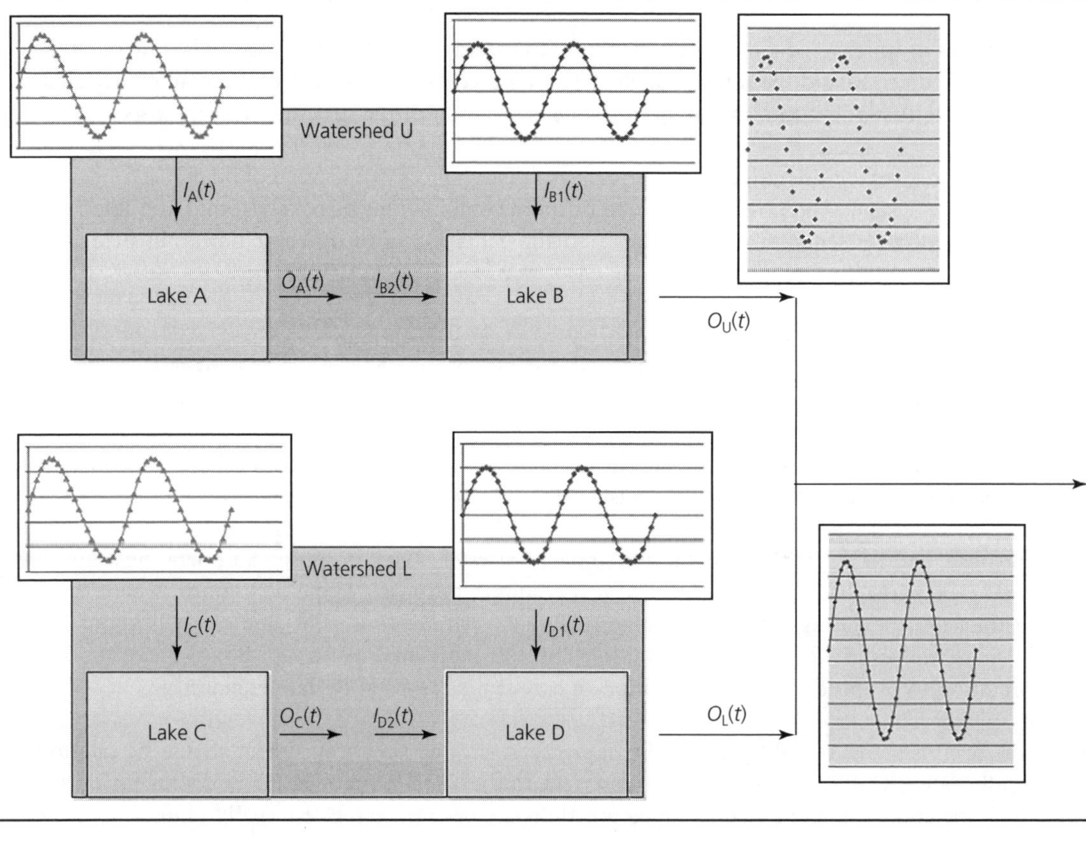

the size and operation of the reservoir(s), and landslide hazards that impact reservoirs create hazardous conditions that did not exist before the storage was created (e.g. the Vaiont Dam incident).

The dam and reservoir together present a hazard that did not exist prior to their construction. As such, the design and systems operation are more complex than similar activities in a controlled environment. The result is that the engineering of dams, reservoirs and river systems involves more uncertainty and less control than can be achieved in engineered systems such as nuclear power plant, petrochemical plant and aircraft.

The principles governing the behaviour of natural systems are not widely understood. This situation has changed somewhat in recent years in that unifying concepts and principles now exist, and there is a considerable literature on specific types of systems. Unfortunately, there is still little literature on systems theory and its application to civil engineering in general.

Contemporary literature on systems thinking focuses on ‘wholes’, thereby demanding a broad view of the system of interest, as can be exemplified by looking down at the Earth’s surface from an aircraft (Checkland, 1981; Richmond, 2004). However, this ‘whole’ or ‘holistic’ view lacks vertical resolution, as detail is omitted. This view of the whole is essential to determine the essential factors or elements relevant to the holistic view. The thinking associated with the ‘essential elements’ view is quite different from the commonly sought after high-resolution view that people typically adopt and as can be obtained by focusing on the detail (the ‘near the ground view’ in an aircraft analogy).

### 3.3.2 Complexity and simplicity

To think in systems terms requires the ability to trade the high resolution of depth and the greater certainty that comes with it against the low resolution of breadth and the comprehensive understanding of the ‘whole’ that it brings.

Holistic conceptions provide horizontally extended but vertically restricted views, whereas the ‘divide and conquer’ conceptions, which form the majority of conventional engineering, provide the alternative horizontally restricted but vertically extended views (Figure 3.2).

There are various views on how to proceed from this point in the thinking involved in drawing out the essential details. However, the description of systems should be sparse or parsimonious (Checkland, 1981; Richmond, 2004). Richmond refers to the second level of thinking as ‘systems as cause’ thinking. This fundamental functional model reflects the operational objective of a dam; specifically, to retain the ‘captured’ volume (containment, i.e. store), and to pass flows through and around the dam in a controlled way (conveyance, i.e. divert and pass) (Figure 3.3).

Systems can be referred to as ‘complexes’ of ‘elements’, and systems theory and systems analysis have emerged to provide a means of understanding these ‘complexes’, which are described by mathematically non-trivial relationships. This understanding of the characteristics of the ‘complexes’ does not eliminate the need to understand the characteristics of the individual elements, which is normally achieved by means of classical scientific or reductionist analysis. Yet, the elements are not the subject of systems theory. Systemic processes such as feedback mechanics, delay, decay, arrangements and the like treat elements in the same manner in any kind of system. The analysis of systemic processes is the primary concern of systems thinking. Elements can be anything. The idea is that elements are affected by systemic processes in similar ways, regardless of what kind of system they reside in.

Figure 3.2 Perspective on the essential features of holistic thinking and focused thinking

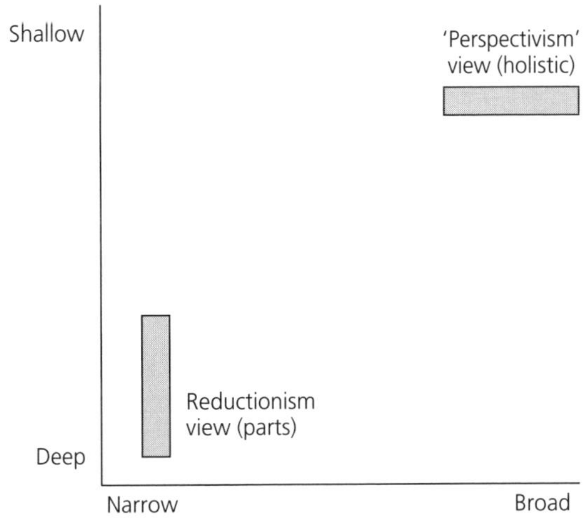

Figure 3.3 Fundamental form of a functional model of the 'systems of interest'

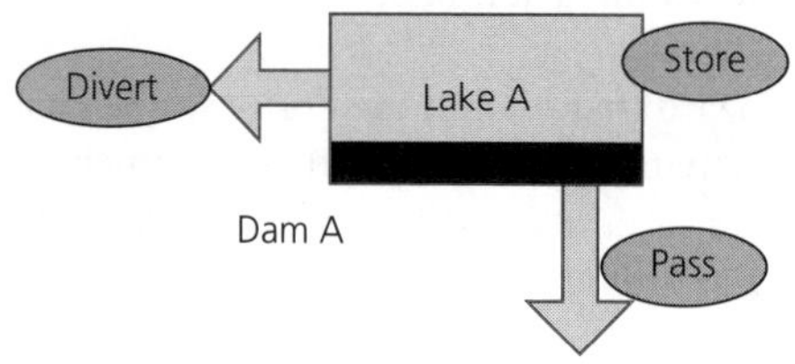


Figure 3.4 Distinctions and relationships between 'complexes'

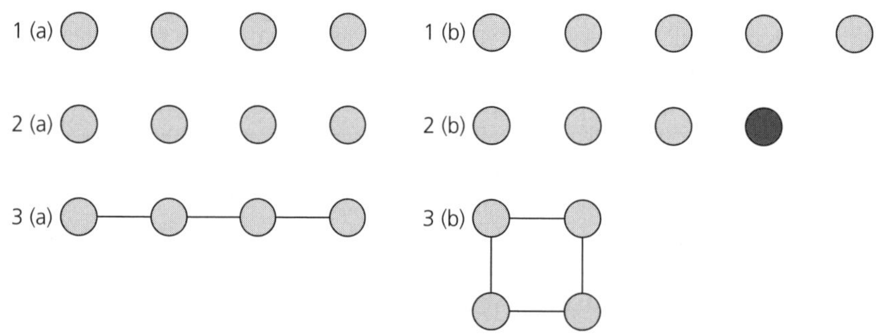


When dealing with 'complexes' of 'elements', a complex can be considered in the ideas represented by the arrangements of elements, as illustrated in Figure 3.4. The complexes are described by three attributes:

- the number of elements
- the type of each element
- the relationship between elements.

More specifically: the number of elements of the same type may be different but there are no relationships between them; the number of elements is the same but the elements are different and there are no relationships between them; or the elements are the same but the relationships between them are different.

The three examples can be grouped into two forms: summative and constitutive. In Figure 3.4, type 1 and type 2 complexes are both of the summative form, while type 3 complexes are of the constitutive form. Physical characteristics such as weight are of the summative form, whereas chemical characteristics are of the constitutive form. Figure 3.5 represents the relationship

Figure 3.5 A simple linear system

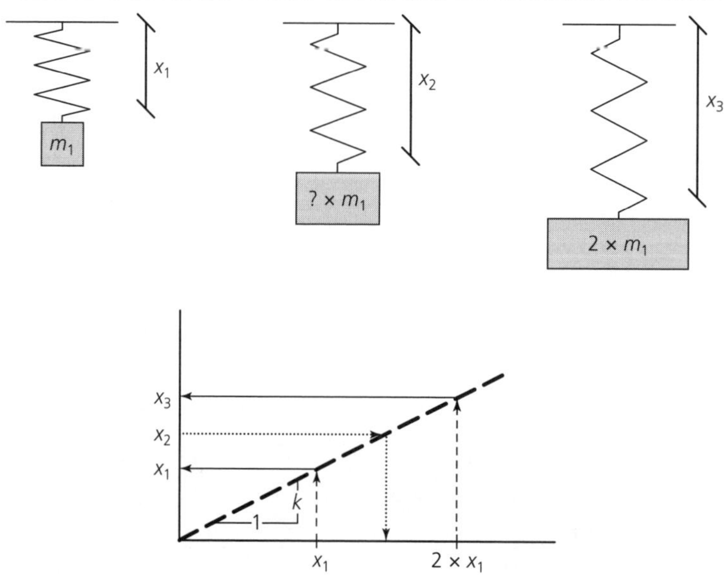

between two physical properties, the mass $m$ and the elastic constant $k$ of the spring; this relationship is a linear summative constitutive relationship.

### 3.3.3 Linear systems and superposition

The properties of the spring–mass system cannot simply be determined by summing the physical properties of the mass and the physical properties of the spring, each determined in isolation. The *principle of superposition* holds for linear systems, but for non-linear systems this old reliable principle, which is the basis of so much classical engineering analysis, does not apply.

The spring–mass system concept is particularly important in civil engineering, and the theory of elasticity plays a dominant role in the analysis of the performance of dams. Essentially, much of the common practice of safety analysis involves the assumption that the 'entity subject to analysis' behaves as a linear system. One advantage of restricting the modelling assumptions to 'linear elastic' is that, usually, the behaviour remains well within the failure envelope. In recent years, non-linear analysis of the behaviour of dams has been developed to a considerable degree. However, these non-linear analyses tend to be focused on the behaviour of individual entities such as a concrete dam.

In dam engineering a sand–gravel filter is a 'system', although it is not usually considered as such. The properties of this system depend not only on the constituent parts and their formative elements (i.e. the soil grains) but also on the physical relationships between the grains. All three of the factors listed above – number, type and relationship – matter. Some of the properties of the filter, such as weight, are summative; that is, the weight of the filter is equal to the weight of the sand fraction and the weight of the gravel fraction. From the perspective of seepage flows, the arrangement of the sand element relative to the gravel element matters, as do the relationships between the sand and gravel particles (Figure 3.6). Obviously, an experienced dam engineer knows that the sizes of materials and the arrangement of the elements are important, and the appropriate sizing and arrangement is defined in terms of the 'filter design criteria'; this analogue provides a useful illustration of the importance of the three features of 'complexes'.

The properties of the combination of sand and gravel that comprise the three different arrangements of 'filter materials' shown in Figure 3.6 are all similar in some respects and different in others. These differences are significant, as the properties that provide the filtration capability of the different combinations of sand and gravel are quite different. Similarly, the combination of the core–filter–drain seepage-control feature of an earth dam is also a system where the arrangement of the elements is of critical importance.

However, it is not just the arrangement of the elements that is important from a systems perspective, and while it is clear that the arrangement of the core–filter–drain system is linear in space and that passage of water through the system is also linear in space, whether or not the system behaviour overall is linear or non-linear remains to be determined.

Another example of the complexity of functional arrangements of many components is illustrated in Figure 3.7. A single part often has several functions assigned to it, making the system more than the sum of its parts. Mapping and understanding these functional interrelationships is one of the key parts of systems analysis and engineering. There are special tools available for this kind of work, such as Enterprise Architect, No Magic Draw and IBM Rational Rhapsody; the common standard implemented within these applications is SysML (systems modelling language).

In general, systems problems are more complex than the linear summative types of classical engineering analyses, which are frequently mathematically trivial (i.e. the mathematical relationships have a simple structure). The common but vague statement 'the whole is greater than the sum of the parts', which is commonly used with respect to explaining the difference between the systems approach and the classical approach, simply means that the constitutive characteristics of the whole are not

**Figure 3.6** Sand–gravel (fine–coarse) 'filter systems' having different flow-control properties

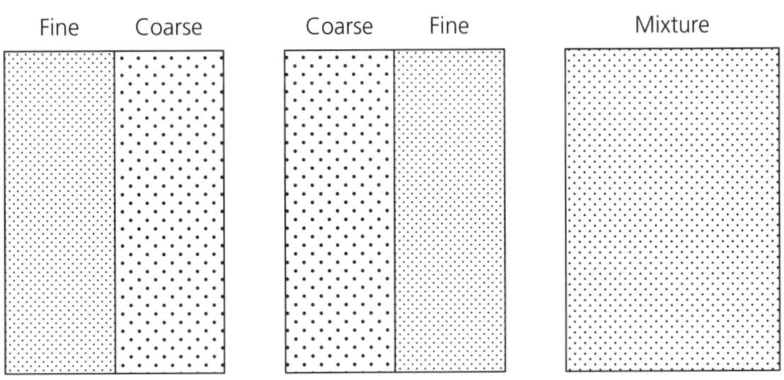

Figure 3.7 The parts and functional arrangement of a spillway gate (modified courtesy of AWMA Water Control Solutions)

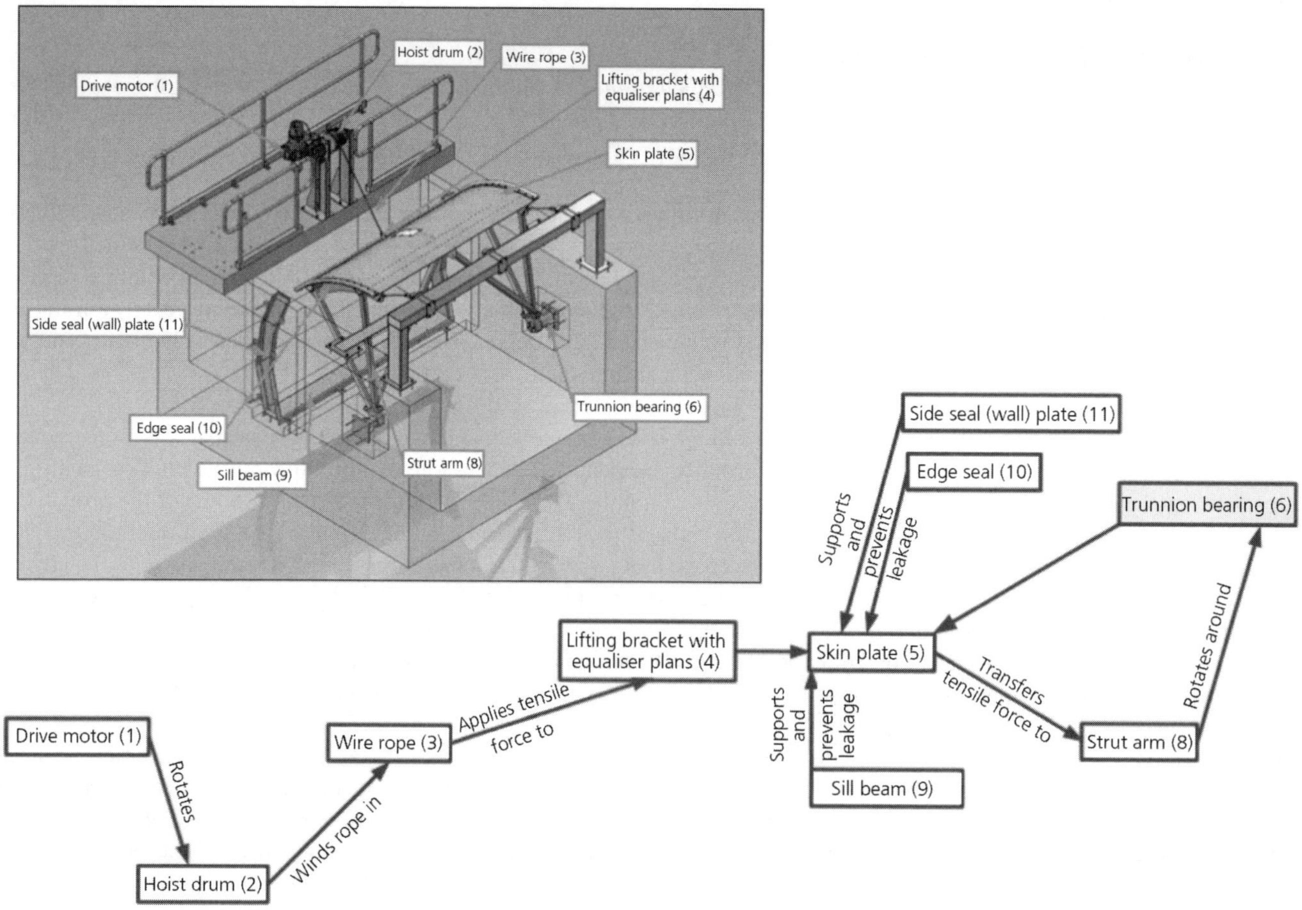


explainable from the characteristics of the individual parts alone.

### 3.3.4 Describing systems

Many complexities arise when describing systems, as different systems can meet the same objectives in different ways. A complication arises because some systems are outwardly the same from the perspective of meeting objectives, and comprise elements that are outwardly the same with respect to outputs but which function internally in different ways. One result is that the elements are not necessarily interchangeable, or may not accept the same inputs; cars with two-stroke, four-stroke and diesel engines are outwardly similar (propulsion and physical form) but internally different. Dams are even better examples.

Notwithstanding the difficulties associated with defining systems, it is necessary to create a practical representation that can be explained, reviewed and checked for its effectiveness and consistency. To do this, it is necessary to create a representation of the system by developing a model of its salient features. The identification of these salient features demands a broad and comprehensive view of the system and the problem at hand.

The term 'model' means different things to different people. Therefore, a clear statement is needed of what it means to refer to something as a model. Frequently, and not unlike the term 'uncertainty', the term model is used in a primitive sense, that is, its meaning is accepted by its usage but left formally undefined.

According to Forrester (1968): 'A model is a substitute for an object or system. A model can be of many forms and can serve many purposes.' More generally, a model is an abstract and simplified conceptual representation of the workings of a system of objects in the real world, and often includes mathematical or logical objects and relationships representing the objects and relationships in the real-world system. It is constructed for the purpose of explaining the workings of the system or predicting its behaviour under hypothetical conditions. Therefore, any set of rules and relationships that describe something is a model of that 'thing'. All models, be they a mindset, mathematical, physical or descriptive, can represent reality with varying degrees of accuracy and precision.

Figure 3.8 Spatial arrangement and flow model for parallel flow-control system (see also Figure 3.1)

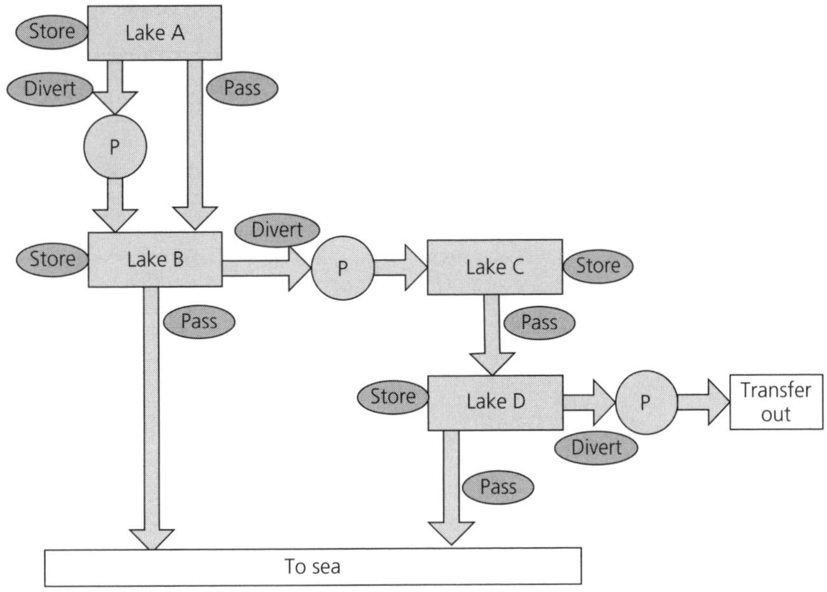


While the nature of systems may not be readily assembled into a unified structure, the spatial arrangement of the elements of a system is readily developed. The spatial arrangement of a system of dams, reservoirs and rivers as illustrated in Figure 3.1, and the arrangement including production facilities (P) can be illustrated as in Figure 3.8. A spatial model provides a generalised platform for modelling the flow-control systems considered in this book. The system can be considered as comprising four elements, each connected by waterways. The elements can just as readily be referred to as subsystems comprising smaller elements. The 'state of the system', where one way in which 'state' can be defined is as the 'volume of water in the system', can be described by means of the stored volume, the volume being passed and the volume being diverted. The functioning of the system means that the volume in the system changes over time, depending on the 'state' of each subsystem, as defined by how much water is passed into the lakes and through the waterways.

Dam and reservoir systems interface with the surrounding natural system, and therefore the modelling of the outflows from the spillway or waterway becomes an input to the natural system. These, in turn, interface with social and economic systems. Similarly, the inflow into the reservoir arises in the surrounding natural system, creating different types of relationships between the dam, the reservoir and the natural system from those that exist downstream of the dam.

### 3.3.5 Systems interactions and non-linearity

Methods of modelling the interdependencies between elements in systems provide a means of representing the functionality of the system in visual and mathematical ways. For the purposes of discussion, Figure 3.9 illustrates a model of a garden fishpond and the interdependencies and interactions between its elements.

The influences involved in the fishpond are represented as follows: (1) the arrows marked '(+)' indicate positive effects – 'the more of this, the more of that' and 'the less of this, the less of that'; (2) the arrows marked '(–)' represent negative effects – 'the more of this, the less of that' and 'the less of this, the more of that'; (3) the lines with circles on the ends indicate limitations – the variable in question can increase only to the extent that the variables influencing it are present; (4) the arrows on the peripheries of variables indicate influences within an individual variable; and (5) the arrows crossing the boundary of the system indicate influences outside the system. These relationships between the elements of a system extend the concepts of relationships between complexes illustrated in Figure 3.4 to include characterisation of the influence of the relationships.

To a first approximation, water flows in dam systems with little or no physical feedback. Hydrological inflows fill the reservoir, the level of which rises. Water is released downstream through the powerhouse to produce electricity, which in turn is output to the regional grid. When the pool rises more than is sustainable, the spillway gates are opened and water is discharged downstream. Depending on the discharge amount, issues may or may not arise with the physical stability of the downstream hydraulic conveyance, the energy dissipation structures and the downstream channel.

The addition of monitoring, control, operational policy and environmental requirements creates non-linearities where none previously existed. Most of these non-linearities are associated with the control actions. The more arrows that are attached to a physical flow-control element of the dam in Figure 3.10, the greater the opportunity for 'unusual combinations' of

**Figure 3.9** Model of a garden fishpond. (Dörner, 1997)

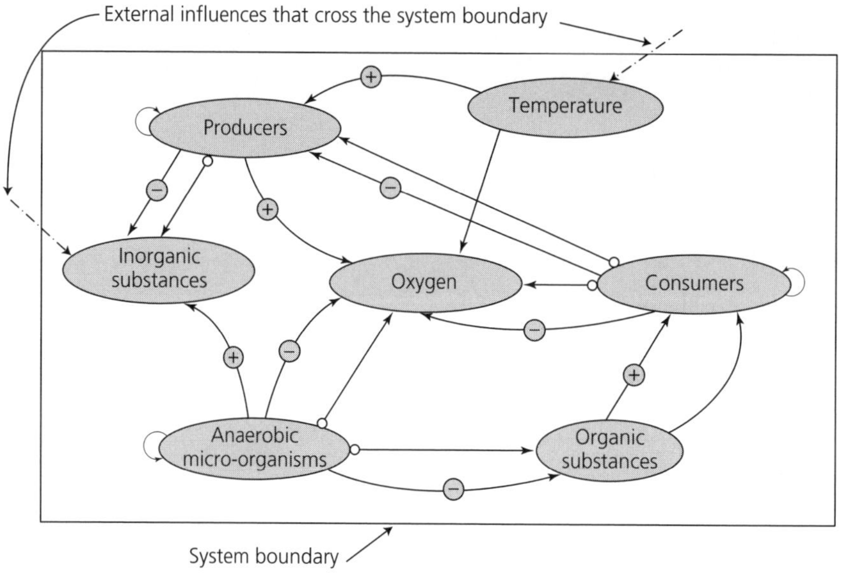


adverse events to occur. None of these control actions has anything to do with civil engineering per se, and while some are associated with mechanical and electrical elements of the dam they are generally not considered in dam safety assessment. These 'operational and hydraulic control' considerations are introduced in conceptual form in Figure 3.10.

In Figure 3.10, while the physical linear water flows model can be expected to be generally applicable (see Figure 3.8) from one dam to the next, the operational and management functions and organisational controls that permit the implementation of the reservoir operating rules can be expected to be different from dam system to dam system, and from owner to owner. This means that most of the 'management and procedural' functions and sources of non-linearity and internal hazard are unique to each dam system and each owner. The associated implication is that a significant portion of any dam system

**Figure 3.10** The flow of water in a dam system is a reasonably linear thing; the inherent non-linearity of dam systems is caused by monitoring, control and operating rules

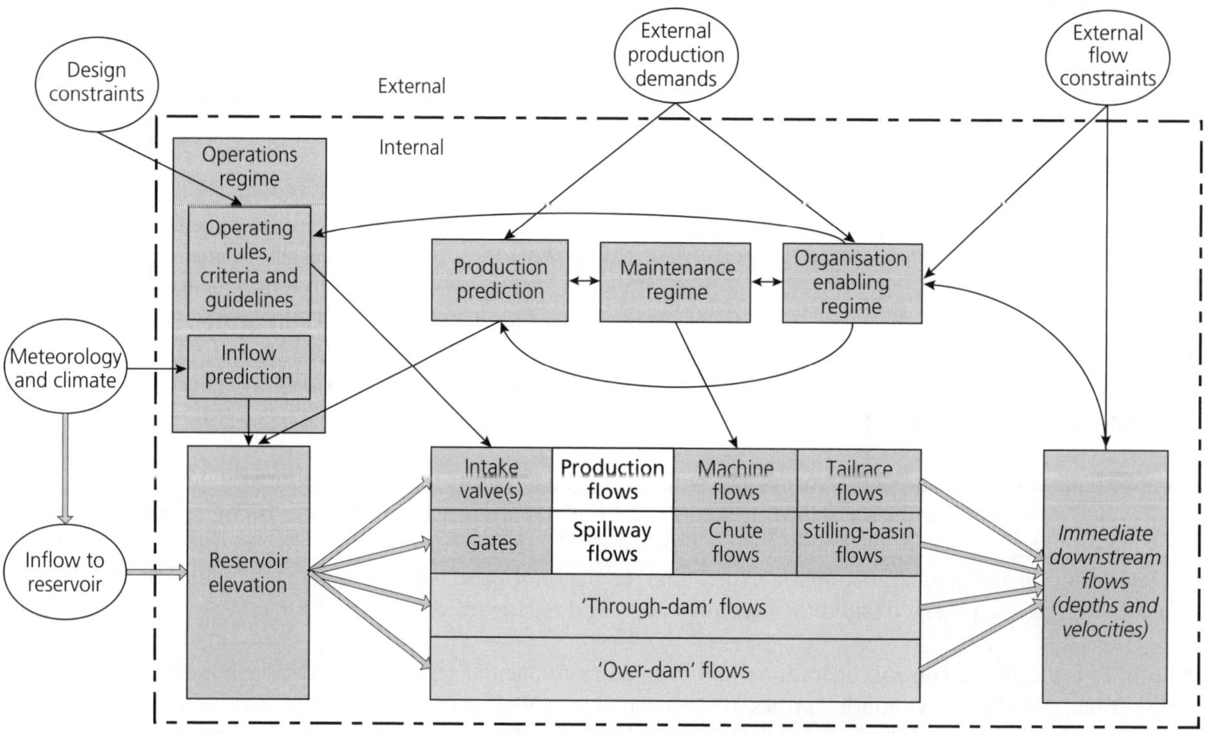

model and analysis will not, in general, be transferrable from one situation to the next. Therefore, one can expect that the model and analysis of any 'dam system' will be as unique to each dam system and owner as the dam system itself is unique. In other words, there is no such thing as a 'one-size fits all' (generic) model or analysis tool that can be transported from one dam system to the next and applied directly.

In terms of the established 'parts' view of dam systems and dam safety, the systems model in Figure 3.10 could be considered as a collection of 'parts', each of which must conform to certain norms or requirements. An alternative view is to consider the model in Figure 3.10 as comprising 'layers' of controls underlain by the physical linear layout and function of the civil structures that contain and convey the water in a linear way. The various elements and functions illustrated in Figure 3.10 can be considered as being layered over the spatial arrangement of the physical parts. These 'layers' of controls that overlay the physical containment and conveyance elements represent the constraints on and the demands that are expected of the dam system in terms of products and services.

Figure 3.11, which illustrates a 'systems model' of the Göta River system in Sweden, serves to demonstrate a model of a real system of a lake, river, sea, dams and production facilities. Clearly, Figure 3.11 is a simplification of a real river system and controls. The river flows begin at Lake Vänern, the largest lake in Sweden and the third largest in Europe (to the far left of the Figure 3.11), pass three dams/hydropower projects and then downstream to the city of Gothenburg, where the river flows into the Kattegat (Northern Sea) on the west coast. In addition to the three dams' hydropower projects, the river is an important route for commercial shipping and recreational watercraft. The river banks are prone to landslides, and the hydraulic control structures also play an important role in controlling the water supply in Gothenburg. The river system provides multiple products and services, with a range of social and environmental constraints and expectations.

There are concerns about whether the maximum discharge of 1030 $m^3/s$ is sufficient in a scenario where heavy rain floods Lake Vänern, causing considerable damage. Previously this was thought not to be possible in reality, but in 2001 the lake was flooded almost a metre above maximum level. In this situation, the Göta River was allowed a discharge almost 1200 $m^3/s$ for months, causing a significant risk of landslides. The subsystem model of the Lilla Edet hydropower project, which is the second of three dams in the cascade, extracted from the system model of the river is shown in Figure 3.11.

Clearly, while the water flow as modelled in Figure 3.11 passes in a linear way from Lake Värnen (on the left of the figure) to the Kattegat (on the right of the figure), there are various functional demands and interdependencies between the principal elements of the river system of the type illustrated in Figure 3.11. Thus, in reality, the 'linear system' attributes of the traditional river system/dam system view shown in Figures 3.1 and 3.8 may be transformed into non-linear systems by virtue of the various 'control functions' that are superimposed on the physical system.

In order to model the system in a realistic way, the analyst must understand all the control functions, where they are applied and the interdependencies and interactions that exist, or may exist, as a result of the dynamics of the system. These system dynamics can be further complicated by external static and dynamic influences over the system, which may be variously perpetual or occur randomly in space and/or time. The modelling endeavour, which because of uniqueness of situation must be carried out on a case-by-case basis in a manner that recognises that any general modelling framework that might be available (such as the simulation framework presented here), must be adapted and reconfigured to suit the individual case. Thus the modelling endeavour is largely intellectual in nature.

### 3.3.6 Mental (thought) models of systems

Much of the practice of dam engineering analysis involves models, and in particular mental models, where judgement has a significant influence on the analytical process. According to Forrester (1968), the human mind is adapted to building and using models that relate objects in space, and is excellent at manipulating models that associate words and ideas. However, the unaided human mind, when confronted with modern social and technological systems, is inadequate for constructing and interpreting dynamic models that represent changes over time in complex systems.

Dam and reservoir systems are even more complex than many civil engineered systems, as they embody dynamic (in space and time) natural physical, social, ecological and technological subsystems, with the result that the mind's mental modelling capacity is overwhelmed by the complexity of the functioning of the system as a whole. When faced with such complexity, uncertainty and lack of sureness, the natural human condition is to separate out what is the problem at hand from everything else and to focus only on 'what is important'. Unfortunately, this exercise involves a mental model of the trivial type, where it is concluded *a priori* that such separation is an appropriate means of solving the problem at hand.

Weaknesses of mental models include:

- weak definition
- poor identification of assumptions

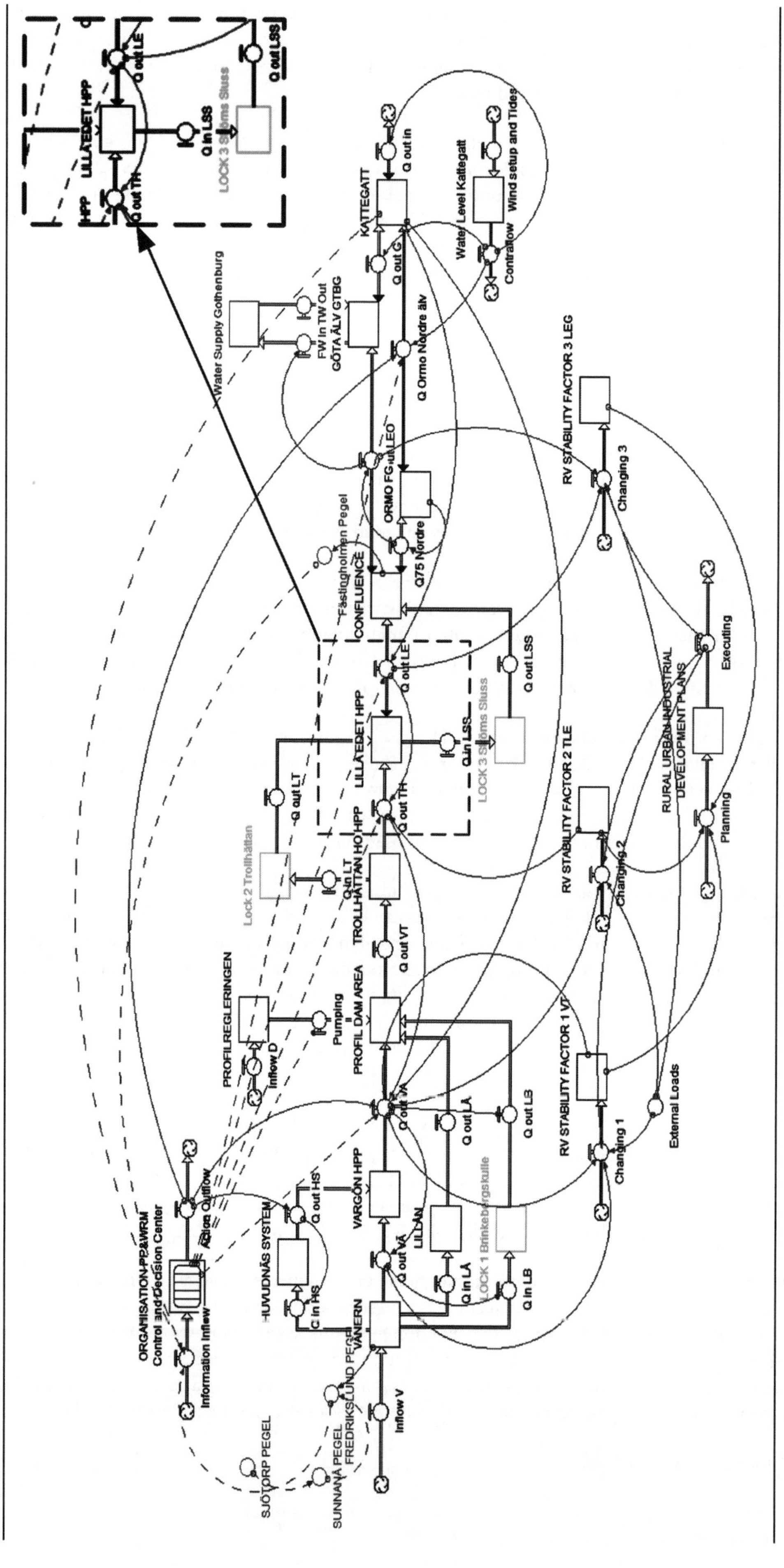


Figure 3.11 Systems model of the Göta River system

- inaccessible to others
- poor capacity to deal with dynamics.

The weaknesses in the mental modelling of dynamic systems can never be eliminated but can be alleviated by converting knowledge and understanding from mental models or ideas into models represented by explicit statements in the form of flow diagrams and equations. In performing the transformation the models become explicitly defined abstractions of reality. Judgements, for example, can be transformed into explicit representations through logic models relating the conclusion (judgement) to the evidence through the line of reasoning of the logic model. The above weaknesses provide further evidence of the need for modern engineering practices to externalise 'engineering judgement'. This can be achieved through the use of logical and mathematical methods of proving the validity of deductive and inductive arguments (Howson and Urbach, 2005). The process of constructing a formal model of the system reveals the mental image of the system, with general statements of size, magnitude, influence, etc., being given numerical values.

This process often begs the question of whether or not the model is 'right', even though this question cannot be answered because the validity of a model can never be proven in absolute terms and is necessarily relative. Rather than being inhibited by the issue of whether or not the model is right, the real value of modelling is the usefulness of the model, which can be assessed in comparison with the mental image or other abstract model that would be used instead. Uncertainty determines that there will never be perfect information. In reality, it is always necessary to work with information that is intermediate between these extremes. Thus models are assessed on a relative scale based on how they clarify our knowledge of and insight into systems.

Rather than being expected to produce 'right answers', the purpose of modelling is to clarify thought, to capture and record what is known, and to assess the consequences of the assumptions. The modelling endeavour renders the thoughts and knowledge accessible to others, and provides the opportunity to understand the behaviour of the system and its change in behaviour as implied by the model with the objective of an improved understanding of reality. A successful model improves the accuracy with which reality can be represented and understood by everyone involved. Thus the goal of the modelling endeavour is better understanding and communication, and improved management of the dam and reservoir system.

## 3.4. General systems engineering process

There are quite a lot of systems engineering methods and processes available, which are being used in different branches and organisations when it comes to engineering effort. To be able to establish a systems approach for dam system reliability management, there should be a systems engineering process established within the organisation. The processes being used range from very complex to extremely agile, and are tailored only to the key considerations within systems engineering – requirements management, functional system arrangements and capabilities, and implementation of these capabilities within the given environment. Systems engineering methods are often expressed as process models containing a set of inter-related activities. The most well-known process model is probably the Vee model and its numerous adaptations. This process model is a good candidate for starting off the development of the systems engineering process.

### 3.4.1 Process application on existing systems

The hydropower industry in countries such as Sweden, where the new construction of hydropower facilities is very limited, is facing different kinds of challenges in terms of managing of ageing dams, spillways and power stations. These challenges range from the need to adopt existing hydropower facilities to new technical and environmental requirements. Sometimes, the task of adapting and existing facility is more challenging than that of a completely new construction.

The construction of a systems engineering model, as illustrated in Figure 3.12 for an existing dam system, is begun by defining the components of the existing system. The next step is to allocate functions to the existing components and verify these against the design and other requirements. If the requirements are satisfied, the result is documented and no further steps are required. However, the reality is seldom so simple, and modifications are often required. In this case, the process would be accomplished by following the standard steps of systems engineering (Figure 3.12).

Figure 3.12 The systems engineering process for existing and new systems

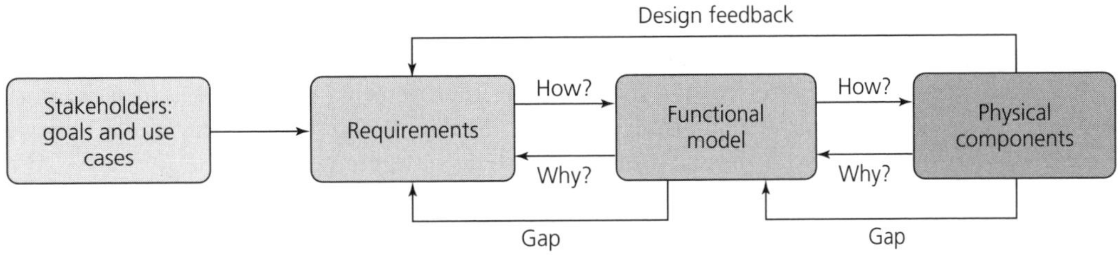

### 3.4.2 Application to new systems or modification of existing systems

When planning a new system or modification of the existing system, the systems engineering process can be accomplished using one of the available standard systems engineering process models. The essence of this approach lies in the sequence of the steps taken. Engineers and experts of different disciplines tend to begin their thinking at the component level. This is natural, because single components are easiest to understand. However, by doing so some important parts in the process are left out. The key principle in systems engineering is to use the 'top-down' approach, which means that stakeholders are involved in the initial phase of the process.

The first step is the careful analysis of stakeholders' requirements and needs, followed by integrated management of these requirements.

Requirements management is an important step. It includes the analysis and maintenance of stakeholder needs. At this stage the conflicting requirements are resolved, expressions are verified and corrected, and the validity of the requirements throughout the life cycle of the system is checked. For traceability purposes, links to the stakeholders' needs are established.

The next step is to allocate the requirements to the system and subsystem functions. This step involves several activities: the definition of the functional boundaries of the system in terms of properties and behaviour; the definition of every function to be performed by the system; the definition of the implementation constraints; and other measures related to ensuring the quality and maintenance of the critical performance parameters showing the fulfilment of the stakeholders' requirements, costs and risks associated with the project. At this stage, a specification of the system requirements and function is developed, taking into account critical qualities of the system, such as, among other things, safety, reliability, usability and maintainability.

Only at the end of the systems engineering process is the systems architecture designed and verified against the stakeholders' needs and requirements.

This systems engineering approach is a new way of engineering that enables an integrated approach, and sustains important functions of the system through the top-down perspective. By working according this method, a better integrated solution, in terms of usability, safety and reliability, social aspects and, certainly, costs, is usually achieved.

## 3.5. Example of system operational arrangements

We now introduce the concept of 'operating arrangements', which encompass a great deal more than the traditional 'operating rules' for reservoirs. The operating arrangements include all the organisational features, infrastructure, maintenance, demand forecasting and reservoir management, dam safety and personnel, as well as the reservoir operating rules. The integrity of these arrangements and all interrelationships are under management control through a 'management system' for dam safety in the operational phase. Details of such a management system for dams are contained in ICOLD Bulletin 154, *Dam Safety Management: Operational Phase of the Dam Life Cycle* (ICOLD, 2011).

The purpose is to provide sufficient somewhat real yet somewhat hypothetical information to exemplify and illuminate the complex range of physical, functional and operational management arrangements that are overlain on the physical arrangements of the dam system to effect the control of the process of transformation of natural inflows into controlled and productive outflows. For dam systems, it is usually necessary that all these physical, functional and operational arrangements perform as required in order to maintain reservoir and outflow control, and to ensure dam safety

To explain more specifically the development of operating rules under normal conditions, it is instructive to consider an example structure. For this purpose, we turn to BC Hydro's Cheakamus Dam on the Cheakamus and Squamish River system north of Vancouver (Figure 3.13). The project details and systems representations used in this section are drawn from BMT Isis Ltd, the engineering and risk management consultancy that provided support to BC Hydro, documented principally in BMT Isis (2008a,b) and modified to suit the purposes of this chapter.

The operating arrangements and generation schedule at the time of the study (2008) as modified for the purposes for this chapter are given in Boxes 3.5 to 3.8 in the following subsections. While the spatial arrangements of the system as a whole have not changed since the study, the spillway gate subsystem, including the associated gate operation infrastructure, has been variously replaced or reconfigured in accordance with BC Hydro's recently established 'spillway gate reliability principles', which form the basis of BC Hydro's major spillway gate improvements programme across its portfolio of dams with gated spillways. Therefore, while the spatial arrangements of the dam and reservoir today are as they were in 2008, all the spillway gate, reservoir operation and hydraulic control subsystems have been changed.

The following lays out illustrative considerations, expectations, objectives and operational arrangements of a dam system that need to be considered, both individually and collectively, in the analysis of the system. Safety and production must be

Figure 3.13 (a) Cheakamus Dam; (b) Daisy Lake and associated Cheakamus River and Squamish River systems. (Courtesy BC Hydro)

(a)

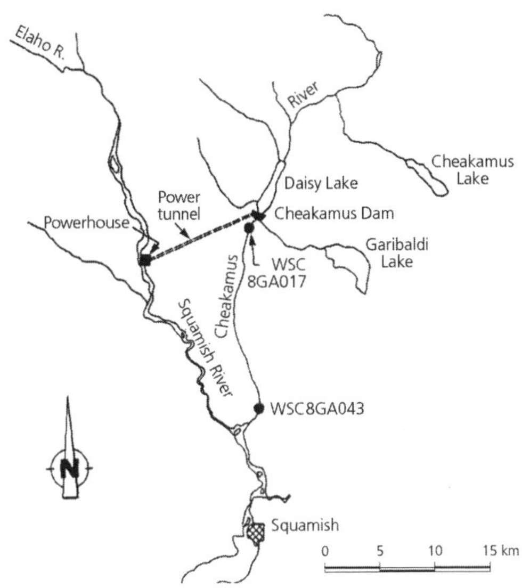


(b)

considered together in the context of prevailing laws, regulations, constraints and expectations, with production being the key to meeting all objectives and expectations. Terms have been modified and generalised to render them conceptually applicable to any dam, and therefore are not particularly informative regarding the specifics of the Cheakamus Dam and hydropower system.

### 3.5.1 General system considerations

Cheakamus Dam (sometimes referred to as Daisy Lake Dam) was commissioned in 1957. It impounds Daisy Lake and diverts water from the Cheakamus River basin through an 11 km long rock tunnel to a powerhouse on the Squamish River 340 m below the pool elevation of Daisy Lake (see Figure 3.13).

Cheakamus Dam is principally a hydroelectric project, although it impounds a reasonably large reservoir that is prized for the recreational opportunities it provides. It also provides limited flood protection, although this was not among the original design objectives.

The unique characteristics of Cheakamus Dam and its operation prior to the 2008 development of a systems engineering plan and the creation of operating arrangements are summarised in Box 3.4.

### 3.5.2 Safety considerations

The operating arrangements for Cheakamus Dam were informed by various safety considerations, the principal ones being:

- flooding downstream, with the associated risk of multiple deaths of members of the public
- sudden river surges threatening the safety of recreational users downstream
- worker safety hazards caused by call-outs to the dam under emergency conditions
- personal injury or death of personnel or the public at the dam site.

The main flooding risk downstream of the dam was due to:

- rapid uncontrollable floods due to spurious gate movement, high reservoir inflow, seismic events or malicious attack
- developing floods due to progressive structural failure of the dam or high reservoir in-flows
- river surge due to spurious operation of the spillway gates, minor structural failure of the dam, reservoir slope slippage or minor landslides into the reservoir.

These risks were, in turn, dependent on a number of factors, including reservoir, dam and river status sensor performance, high inflows, gate movement decision management, communications link performance from the area control centre (ACC) to the dam, gate system performance, and the adequacy of emergency and contingency arrangements.

**Box 3.4** The physical and functional characteristics of Cheakamus Dam prior to the development of the operating arrangements and generating schedule (pre-2008)

- Cheakamus Dam was not continuously attended. Local control required attendance.
- Spillway gates were operated remotely from the area control centre via a SCADA system.
- Autospill was not in place.
- An ultra-high frequency link provided communications from the dam to the microwave network.
- The controls for both spillway gates were housed in a control building on the concrete deck above the gates and relied on a master programmable logic controller (PLC).
- The lower level outlet gate (LLOG) and hollow cone valve (HCV) could only be operated locally.
- Two mobile diesel generators were on site, one auto-start and one manual start.
- The power supply from the Rainbow substation at Whistler relied on a pole-mounted transformer that was assessed, by BC Hydro staff, as being vulnerable to seismic activity and was undergoing improvements.
- All discharge operating gates combined were insufficient to spill the flows associated with the 1983 probable maximum flood (PMF) of 4173 $m^3/s$ without preconditioning the reservoir. The barriers could, however, spill 10-year to 1000-year floods.
- The spillway gates had a 2 m restriction when being opened under remote control, partly reflecting the limitations of the gate-position sensors and partly to prevent the river from overflowing its banks.
- The gates could not be opened if overtopped.
- There was a crested free overflow section that could serve as an emergency discharge if required.
- River flows had to be maintained and the reservoir could not be emptied to allow access for maintenance of the underwater structures and the mechanical interfaces of the gates.

### 3.5.3 2008 operating overview

The pre-2008 operating arrangements for Cheakamus Dam specified a tiered structure of control according to the set of rules given in Box 3.5. The rules specified that the ACC had principal responsibility for control of the spillway gates but that ACC exercised this responsibility under guidance from a manager within BC Hydro's reservoir operations group.

BC Hydro had prepared a statement of requirements for the operation of the dam, under which the pre-2008 normal operating rule was developed:

> BC Hydro has responsibility for the safe operation, maintenance, inspection and testing of the Daisy Lake dam spillway gates. The implications of a major accident at Cheakamus are such that the dam is assessed as 'Very high consequence' and there is a need to ensure that the spillway gates meet the appropriate standards for safety and reliability. The safety of the public and of BC Hydro personnel is of paramount importance but the water is a valuable resource and must also be managed efficiently to maximise the revenue earning potential of the Cheakamus Project. The spillway gates, together with their associated controls and infrastructure, must prevent unacceptable flooding for the defined range of seismic or weather driven events. They must also be able to retain safe levels of water in the reservoir whilst ensuring acceptable river conditions downstream under normal operating conditions. They must be subject to the management, operational procedures, standards, inspection and support that ensure the required levels of safety are sustained throughout their operational life.

**Box 3.5** Pre-2008 Cheakamus Dam operating arrangements: normal operation overview

*1. Overview*

Under normal inflow conditions, defined by the Water Use Plan (WUP), fish, and recreation releases, gate operations are undertaken by the ACC. BC Hydro operations are organised by region; the ACC with responsibility for Daisy Lake Dam which is the Lower Mainland Control (LMC). LMC operates the dam in accordance with instructions from Generation Operations: Manager, Planning, Scheduling and Operations (PSOSE).

Operating Orders for the activities on the dam are covered in Local Operating Orders (LOOs), Generation Operating Orders (GOOs) and System Operating Orders (SOOs).

(Courtesy of BC Hydro)

These requirements implied that the operator should have had at least the following capabilities, which were reflected in later parts of the reservoir operation instructions:

- a means of monitoring the reservoir level
- a means of monitoring the spillway gate positions
- a means of monitoring the river flow immediately downstream of the dam
- a means of controlling the reservoir level
- a means of controlling the downstream river flow
- infrastructure to implement planning, scheduling and operating shift engineer (PSOSE) decisions about the movement of barriers
- a spillway gate system that is sufficiently reliable to deliver the demanded spill rates under all defined conditions.

The pre-2008 normal operating arrangement and generating schedule had to balance revenue generation against safety, where safety included plant and equipment, downstream and upstream riparians, and the public. These reservoir operation instructions were for guidance; they were not mandatory and could be overridden in an emergency. Furthermore, the operation instructions did not anticipate mechanical or electrical reliability, human errors, maloperation or extreme hydrological events. Operating decisions that affected operator or public safety were made under the ACC's jurisdiction. In cases where increased spill rates could threaten safety downstream, for example, conflicts were resolved by the ACC.

### 3.5.4 Reservoir levels

The outflow discharges from Daisy Lake Reservoir are shown schematically in Figure 3.14. Water for power generation at the Squamish River hydropower plant is withdrawn into a tunnel upstream of the right abutment of the dam. This outflow is transported to the Squamish River and does not re-enter the Cheakamus River downstream of the dam.

Three waterways flow from the dam into the downstream Cheakamus River channel: (1) a gated surface spillway with two radial (Tainter) gates; (2) a lower-level sluice-gated conduit (lower level outlet gate (LLOG)); and (3) a hollow cone valve (HCV) outlet, which is always open to maintain low-oxygen-content freshwater environmental flows. The spillway operation gates (SPOGs) can be operated remotely by the ACC, but the LLOG sluice gate and the HCV outlet can only be operated manually on site, requiring local personnel. The spillway gates, which can be operated independently of one another, have a 2 m restriction on their openings under remote control, principally to reduce the probability of downstream damage or flooding due to rapid release of water from the reservoir. This limit can be overcome on site. The demand placed by inflows to Daisy Lake is summarised in the flood-frequency chart shown in Figure 3.15, which gives seasonally averaged inflows in normal years.

The objectives for the reservoir level (Box 3.6) accounted for the available spill capacity in the four effluent waterways, and also for the seasonally varying inflow. The aim of the objectives for the reservoir level was to minimise the amount of water spilled into the downstream river channel, and thus to maximise the water passing through the powerhouse compared with that passing into the downstream Cheakamus River channel.

Figure 3.14 Water flow control arrangement for Daisy Lake Reservoir. CMS, control management system; HCV, hollow cone valve (for environmental flows); LLOG, lower level outlet gate; SPOG, spillway operation gate with two gates. (BC Hydro, 2011)

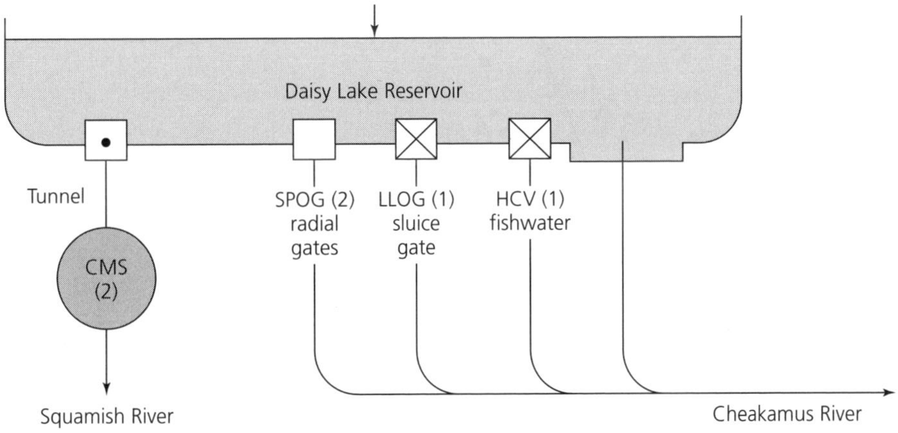

Figure 3.15 Daisy Lake Reservoir inflow frequency. (Courtesy of BC Hydro)

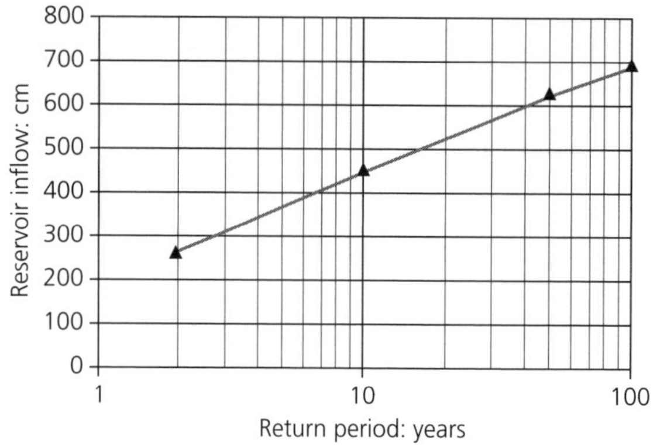


Box 3.6 Cheakamus Dam operating regime: normal operation reservoir levels

*2. Reservoir Levels Objectives*

**2.1** The Daisy Lake Reservoir is normally operated between a maximum of around 378 m elevation and a minimum of around 366 m elevation.

**2.2** During the months of October to December, when occurrences of heavy and prolonged rainfall are most severe, the preferred operating range for the lake is 370 m to 373.5 m to minimise spill potential.

**2.3** During the spring freshet (May–August), inflow into Daisy Lake Reservoir normally exceeds the hydraulic capacity of the turbines, therefore, Daisy Lake Reservoir is usually drafted as low as possible before the onset of the freshet to accommodate the high expected inflow with minimum spill.

(Courtesy of BC Hydro)

### 3.5.5 Downstream flows

Similarly, the aims of the objectives for the reservoir level were to control downstream flows to ensure public safety, to ameliorate damages due to flooding and erosion, and to maintain environmental flows (Figure 3.16). The pre-2008 averaged seasonal outflows of the dam are shown in Box 3.7. In addition to safety, the monitoring of downstream flows was motivated by the last two considerations above: controlling erosion and maintaining environmental flows. Full-bank discharge was set at 450 cms, but left to the determination of BC Hydro's Lower Mainland Control Centre. The operating rules also cite a Water User Plan under the Provincial Water Act, which specified minimum downstream flows that the operator had to attain.

Figure 3.16 Daisy Lake Reservoir outflow to Cheakamus River, by month. (Courtesy of BC Hydro)

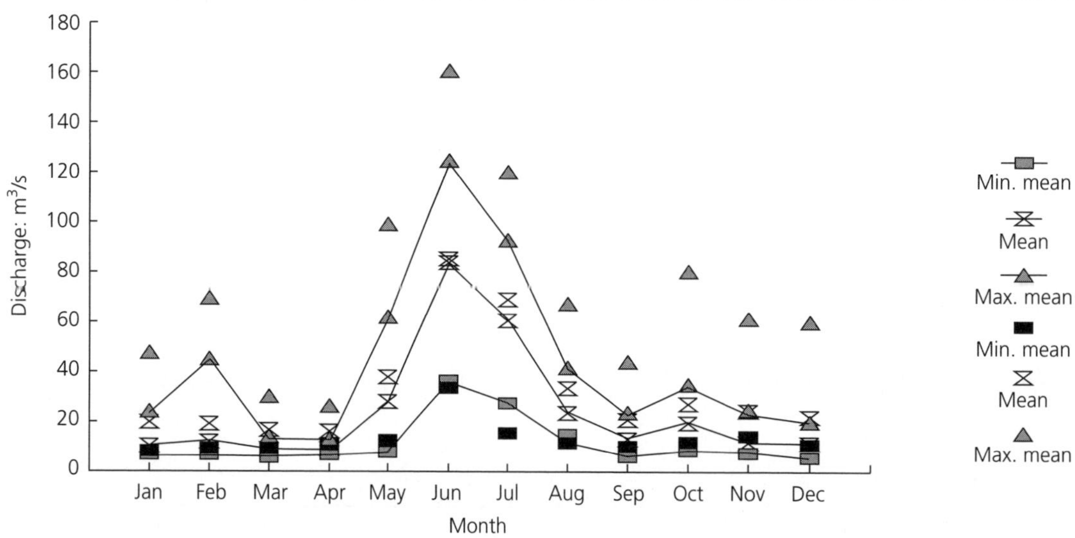

**Box 3.7** Pre-2008 Cheakamus Dam outflows: normal operation monitoring downstream flows

*3. Monitoring Downstream Flows*

3.1 The WSC [Water Survey of Canada] gauge, on the Cheakamus River, was used to monitor the river discharge from tributary inflows. The amount released at the dam to load up the river to bank-full discharge could be determined by monitoring the gauge.

3.2 The Cheakamus River channel is very mobile. As a result, the bank-full discharge was constantly changing. The October 1997 estimate of bank-full discharge was 450 cm. Actual bank-full conditions of the river were determined by the Coastal Generation Manager.

For environmental reasons, primarily involving fish protection, the operational regime specified the minimum flow requirements from Daisy Lake Reservoir at the WSC gauge.

*4. Operation of Discharge Facilities*

4.1 Pre-2008 operation of these facilities was as follows:

- The Cheakamus generating station (on the Squamish River), the dam's hydro-generator and the free-crest weirs were operated as conditions permit.
- The HCV was always open, as needed, to meet the minimum flows for fish health.
- The LLOG was operated either fully open or closed. Discharge via this gate was employed only after the discharge capacity of the spillway gates was fully utilised.
- The two radial spillway gates could be operated in the following modes: (1) Remote Control – gate openings were set by supervisory control by the LMC ACC; and (2) Local Manual – gate openings could be set manually on-site at the dam.
- Under Remote Control mode, the Spillway Radial Gates could be opened to a maximum of 2.0 m each which was controlled by the software at the ACC. Local (at-dam) manual operation was required to release flows in excess of the 2.0 m/gate opening.

(Courtesy of BC Hydro)

The objectives for operating the discharge facilities specified that the HCV was always open, to provide environmental flows, and that the LLOG was operated either fully open or fully closed. The low level operation gate was only to be opened when the full capacity of the two SPOGs had been fully achieved. The SPOGs were to be operated either remotely by the ACC or locally by on-site operators, as required (Box 3.8).

**Box 3.8** Pre-2008 Cheakamus Dam during floods: High reservoir in-flow operation of discharge facilities

*5. High Reservoir In-flow conditions*

5.1 Operating Responsibility for routing flood in-flows via the Spillway Radial Gates, which could be opened to a maximum of 2.0m each, remained with the LMC until this maximum opening was reached. Once this condition was reached Operating Responsibility was transferred to the Coastal Generation Manager as dictated by System and LOOs under flood routing conditions, by the LMC.

5.2 The changeover procedure for the operation of the spillway gates was detailed in the Local Operating Order.

5.3 To meet the prescribed flood releases for spillway gate openings in excess of 2.0m the Coastal Generation Manager or his/her delegate had responsibility for implementing gate movements in accordance the LOOs. The Coastal Generation Manager took restrictions contained in the LOOs into consideration. The gate openings were determined by local observations and inflow predictions provided by Generation Operations.

5.4 The Coastal Generation Manager or his/her delegate reported to LMC for entry into the Cheakamus Project Log all opening and closing operations of gates, valves, low level sluice, etc., other than for turbine control, along with the time for each operation.

5.6 The Manager of Coastal Generation or his/her delegate is responsible for liaising with and providing flood inflow routing information, where required, to the local public, local municipal governments, local fishery agencies and local residential, industrial, and commercial entities.

5.7 Under emergency conditions where flooding upstream or downstream of a project is imminent, the Manager of Coastal Generation or his/her delegate shall take necessary action to address the emergency.

(Courtesy of BC Hydro)

Figure 3.17 Control and operational tiers for Cheakamus Dam

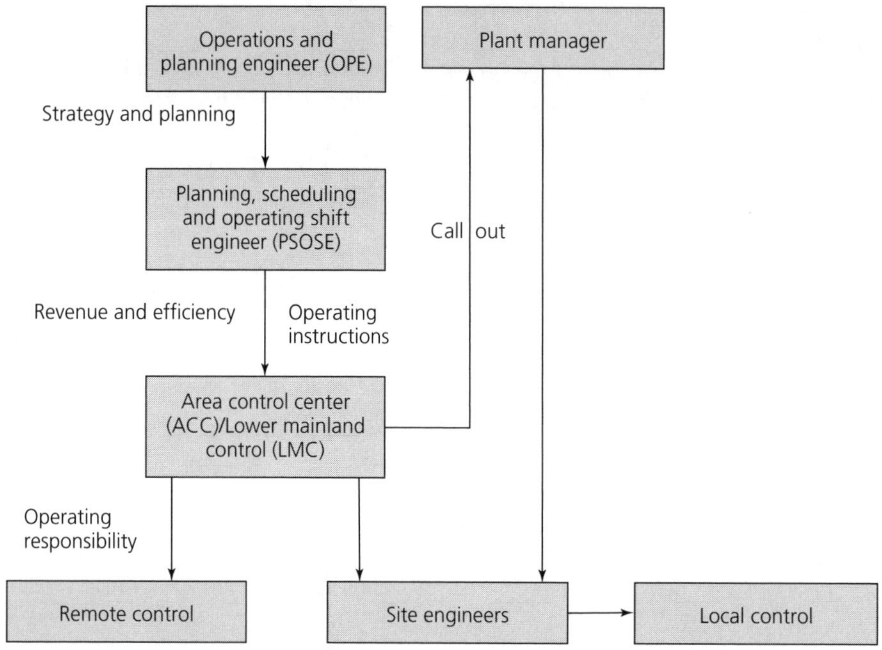


### 3.5.6 Responsibilities

Cheakamus Dam was not usually staffed with on-site operators, so the outlet works and diversion to the Squamish power station were operated remotely from an ACC by means of a SCADA system. Local control required that an operator be present on site. There was no autospill capability at the dam.

Three tiers of responsibility and control relative to the principal objective of hydropower generation (Figure 3.17) affected operations at Cheakamus Dam:

- *Planning tier*. This tier decided what the power generation targets for the reservoir should be, based on market demands and hydrological inflows, and based on the readiness of the plant and environmental circumstances. Decisions with respect to the planning tier were the responsibility of the operations and planning engineer (OPE) and staff, who generated strategic planning guidance for the operation of the dam.
- *Grid operations tier*. This tier specified the moment-by-moment power demand, combining planning targets with spot market demand for electricity. This tier accepted strategy guidance from the OPE as input and generated detailed plans to maximise revenue and efficiency. The grid operations tier was the responsibility of the PSOSE.
- *Plant operations tier*. This tier responded to the grid operator's demands for power production. It accepted planning and scheduling input from the PSOSE and generated detailed operational rules and direction at the area or plant-group level. It was the responsibility of the ACC.

Operating control of the dam was undertaken either by a remote control centre, as in the case of an unstaffed dam like Cheakamus, or by on-site operators and engineers. The ACC communicated operating information to the plant manager (who typically was not on site), who in turn provided direction to site engineers and other staff. The ACC had operating responsibility for gate operation under normal conditions but was not responsible for making decisions to spill water, rather simply following instructions from the PSOSE. The ACC was not involved in planning or in determining trade-offs with regard to gate openings and closings.

The OPE was responsible for long-term planning, including weather-related forecasting and seasonal data, operating constraints and orders, and policy guidance.

The PSOSE was responsible for the generation schedule as planned by the OPE, and had the goal of maximising revenue from power generation. The latter included, balancing the costs and requirements for spilling water, electricity prices in the utility market, and operational factors including down time.

### 3.5.7 Data communications and evaluation

The management chain of responsibility and control at Cheakamus Dam was supported by two communications systems: the operator and plant information system (OI/PI), for data capture and dissemination; and the SCADA system, which provided monitoring and alarm functions for the ACC. The SCADA system also provided remote manual control (Figure 3.18).

The OI/PI provided historical and real-time data by which the PSOSE decided when to open and close gates. The data captured by the OI/PI system included those from the data collection platform (DCP), which monitored factors related to reservoir inflows (e.g. depth of snow).

The SCADA included a set of display dashboards by which the ACC could operate the spill gates at all sites in its area. The SCADA could not be accessed by the PSOSE, who had to rely on the lower mainland control (LMC) to take control action.

Limitations inherent in the data communication system were:

- An ultra-high frequency link provided communications from the dam to the microwave network via a single station.
- The controls for both spillway gates were housed in a single control building on the concrete deck above the gates and relied on a single master programmable logic controller (PLC).
- The power supply from the local substation relied on a pole-mounted transformer that was assessed, by BC Hydro staff, as vulnerable to seismic activity. Plans were being implemented to bring the transformers to ground level.
- Commands from the ACC were received, via the ultra-high frequency link, via a modem and a single remote terminal unit (RTU). The RTU passed the data to a single PLC, which interpreted the data and instructed the PLCs associated with each gate. The gate drive PLC controlled the variable-frequency drive, which regulated the speed of the winch motor.
- The 480 V AC distribution system received power from the hydrogenerator, or through a transformer and an automatic changeover switch from the 25 kV local substation. The hydrogenerator was defective, so power for the distribution system was primarily from the local substation via overhead cables. In the event of a loss of power supply from the local substation, a main mobile diesel generator was provided to automatically start and supply the distribution system through an automatic changeover switch. A further, temporary, mobile diesel generator was also provided to supply power should the main diesel generator fail. This second diesel generator required local start-up.
- The distribution system distributed power through circuit-breakers to the dam's user systems.

## 3.6. Operational arrangements and instructions under fault conditions

The distinction between 'failure' and a 'fault' can be important. When a fault is defined, the analyst should document what the undesired state of the component is and when its occurrence is unfavourable. The lesson here is that all failures are faults, but all faults are not failures. In a fault-tree analysis a distinction is made between the existence of a fault and its occurrence. The term 'existence' is used to denote that a fault has occurred and will continue to exist if the component is not repaired or replaced. The 'occurrence' of a fault denotes that a fault has occurred and can be repaired. The difference between the occurrence and existence of a fault is not critical in the construction of a fault tree but is important in the quantification of event probabilities.

A disturbance to a dam system due to, say, a landslide, an earthquake or a weather event may or may not cause a fault. If a disturbance does cause a fault, then in most cases the fault will persist until it is rectified (occasionally, a fault might correct itself, but most will remain). A fault can be an inherent characteristic of a component or element of the system, as many components have inherent weaknesses that often cannot be eliminated (a light bulb in a switching room burning out is a simple case of a fault that, in itself, will not lead to a failure). The existence of a fault does not imply or necessarily lead to a failed condition for the system, as there may be sufficient redundancy and segregation in the system to accommodate the fault. However, something as simple as a lighting fault due to a bulb burning out could be a contributing factor in a system failure if the lighting fault inhibited or delayed an essential and time-sensitive switching action in an emergency. If 'fault tolerance' does exist, this system robustness can only be determined through systems analysis.

### 3.6.1 Reservoir operation under conditions of discharge function fault

In the above discussion of the operational arrangements for Cheakamus Dam, as modified for this chapter, there was no mention of reservoir operation and control under conditions of 'normal inflows when discharge function is disabled'. Furthermore, there was no account of disturbances to the system due to earthquake or other natural hazard, or loss of discharge function due to an external disturbance. This is not unusual. For most operational regimes, operating procedures and specific 'rules' are determined for conditions of the 'as-designed' system in a fully functional state. The system function issues that arise are not simply confined to the 'functional state' of the physical subsystem or component as is typically the case in engineering analysis, they also apply to the transfer functions between the physical and procedural functions.

One question that might come to mind is: 'What is the operational arrangement for reservoir operation with the discharge facilities under fault conditions?' Another is: 'Who is responsible for the design and for the maintenance of the operational arrangement for fault conditions?' Often, these arrangements will be 'reactive' in nature, and they are rarely provided in detail. In addition, those responsible for implementing these arrangements will not be familiar to the responsible staff as they are normally busy with their regular duties.

Figure 3.18 Control and operational tiers for Cheakamus Dam, including data acquisition and support. (BMT-Isis, 2008b)

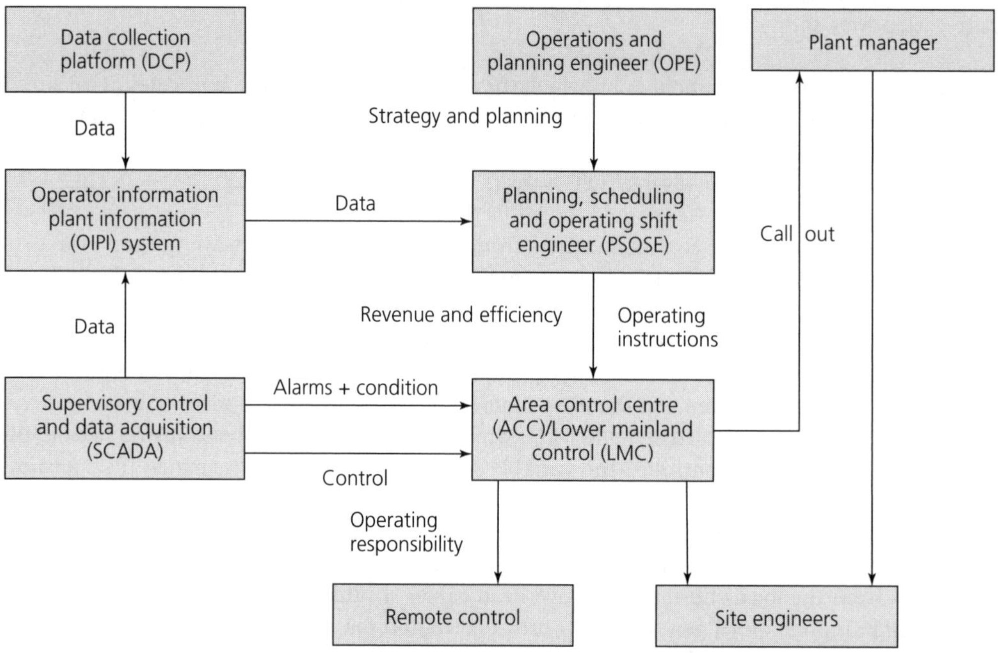


In principle, the lines of control may be clear (Figures 3.18 and 3.19) but the functions to be carried out at each point in the control chain and the transfer from one control point to the next may not be implemented appropriately.

In addition to having appropriate procedures to follow for circumstances where the discharge function is under fault conditions, it is important for the design of the procedures to properly accommodate the operational circumstances during which the procedures are to be implemented. In this regard it is important to consider common-cause failures from the perspective of both equipment and procedures. For example, an earthquake may disable the discharge functions of a dam but, even if there is a response plan available to be implemented, implementation may not be successful because the earthquake has rendered it impossible for the manual operation to occur. In such cases, the disturbance (earthquake) affects all physical parts and all functions, including those functions that are carried out by individuals at various points in the decision chain.

Figure 3.19 Management, procedural and human functions for Cheakamus Dam. (BMT-Isis, 2008b)

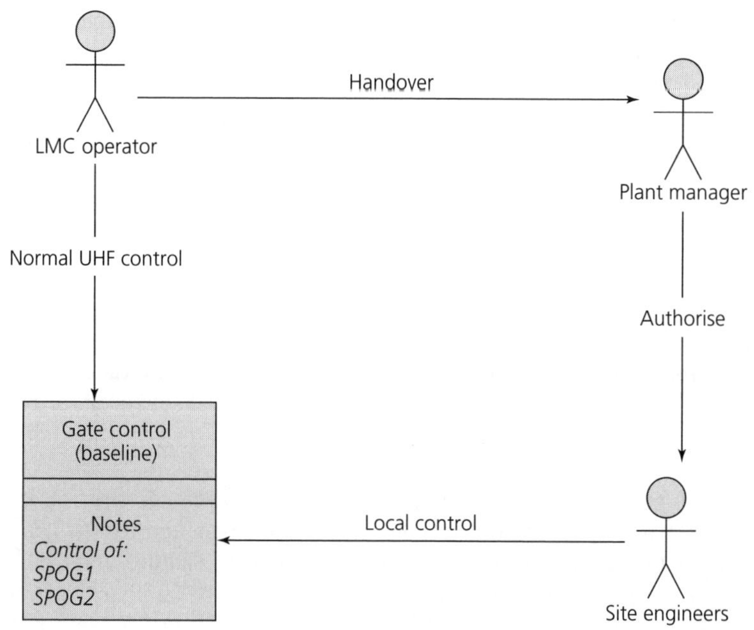

Throughout all this, and unless the entire system is designed to fail to a safe condition, management preferences, operational trade-offs and human factors must also be accommodated in the analysis of the system reliability.

## 3.7. Dam safety in relation to systems

Dam engineering and dam safety assessment, in common with most engineering activities, have evolved out of the classical scientific method of analysis. 'Scientific' analysis requires that an entity to be examined (i.e. the dam and reservoir) can be resolved into its constituent parts to the extent that the characteristics and properties of each part can be explained, and that the parts can be reassembled such that the analysis of the parts when combined reveals the properties of the entity as a whole.

The dam–reservoir system is 'reduced' into its constituent parts in a systematically structured way, and the entities are grouped to identify interactions and interdependencies between levels. This may be taken to a more detailed level in the treatment of FMEA and the progression of failure processes (see Hartford and Baecher 2004).

Figure 3.20 shows the separation of the identifiable components that are physically and functionally separable. This separation is only possible if there are no interactions or interdependencies between parts that result in the parts being functionally inseparable. Typically, once independence is demonstrated, or when the interdependencies with other elements are recharacterised as 'influences' that can be modelled as part of the separated entity, each component is analysed in isolation. For example, the performance of the impervious barrier can be determined by determining the individual performances of the dam core and the interface between the dam body and the foundation (e.g. grout curtain) and then analysing them in a manner in which the elements together control the seepage flow regime overall.

When viewed in this way, the important question that arises is whether or not the methods of classical engineering can still be applied in the usual way if the principle of superposition does not apply? To answer this question, it is necessary to resort to systems analysis, which does not begin with the classical engineering assumptions of superposition and linearity.

Figure 3.20 Analytical separation of a dam owner's portfolio into subsystems (parts) for further analysis

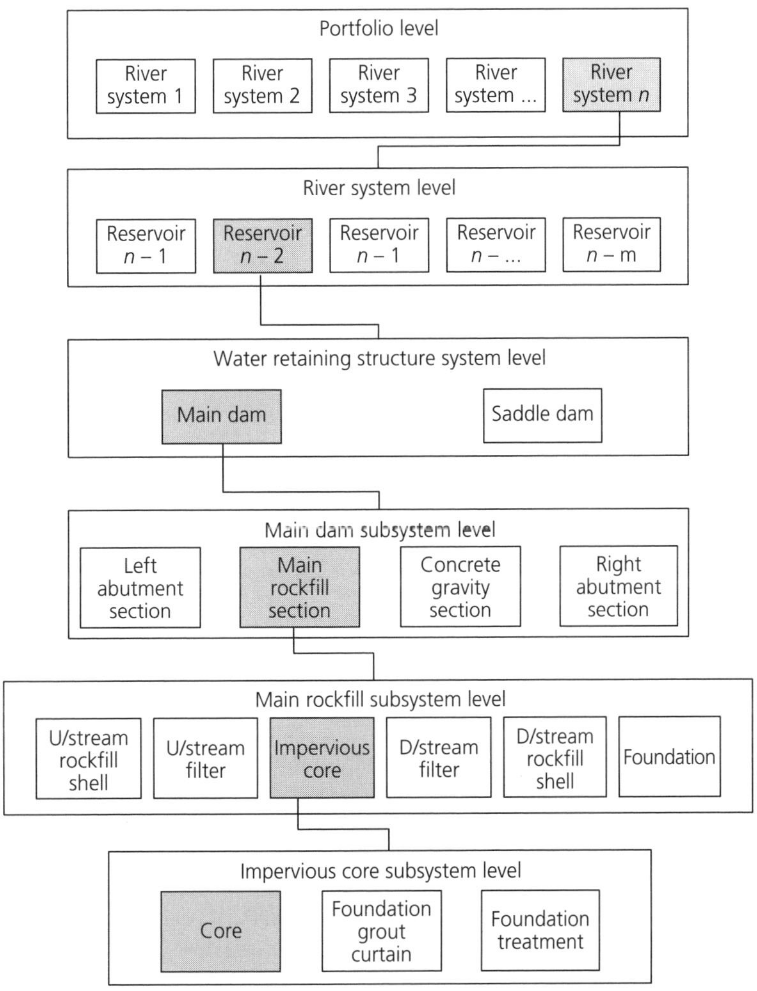

In traditional dam engineering analysis, which follows the reductionist philosophy of science, if components are assumed to be physically and functionally separable, once each component has been demonstrated to have adequate performance, then the safety of the dam as a whole is considered acceptable. If any component is found to be potentially deficient, then steps are taken to correct the deficiency or to modify the dam to compensate for the effects of the deficiency. In this view, the dam system comprises the sum of the parts, and the system characteristics are arrived at through summing the analytical results obtained for each of the parts. The principle of superposition obtains.

The questions that are answered by means of classical engineering, that are then combined through superposition to give the performance of the system as a whole, are of the form:

- What is the demand imposed on this 'part'?
- What is the performance capacity of a 'part'?
- What is the relationship between capacity and demand for this 'part'?

### 3.7.1 Classical science and systems

Classical science involves resolving entities into isolatable parts that can be handled in terms of the various fields of the sciences. The approach has proved to be highly successful in explaining a wide range of phenomena. Classical engineering reflects this philosophy through resolution of the engineering problem into engineering domains that are underpinned by the respective sciences (e.g. physics). Classical science in operation requires that two general conditions apply:

- The interactions between the parts are either non-existent or so weak that they can be neglected. This condition is necessary in order for the analysis of the individual parts to actually be performed logically and mathematically independently of each other.
- The relationships describing the behaviour of parts are linear (see Figure 3.5). This condition is necessary to ensure that the equations that describe the behaviour of the parts are of the same form as the equation describing the whole system, thereby permitting the summative process required to 'put the parts back together'.

A linear relationship $f(x)$ is one that satisfies both the property of additivity

$$f(x+y) = f(x) + f(y) \tag{3.1}$$

and homogeneity

$$f(\alpha x) = \alpha f(x) \tag{3.2}$$

These together provide the conditions for the principle of superposition to apply.

The principle of superposition is essential to much of the scientific endeavour and engineering practice, and a great deal of conventional engineering analysis and design depends on the assumptions about its appropriateness. The matter of the validity of the principle of superposition is not only relevant to engineering analysis and design, it is the essence of the classical scientific method, which is used widely in all scientific and most societal endeavours, including accounting and finance (consolidated accounts), healthcare and medicine (surgical specialisations), management systems and project-management practices (work breakdown structures). However, the entities that reductionist methods are applied to are typically complex systems, with the result that there may be gaps between the assumptions made in the analysis and the actual functioning of the entity.

The conditions of superposition and linearity do not apply to systems (specifically defined as 'parts in interaction') where:

- the interactions between the parts are 'strong', or
- the interactions between the parts are 'non-linear'.

This renders the 'systems problem' as a problem of the limitations of scientific analytical procedures. Therefore, the methodological problem of systems theory is to provide for problems that, compared with the analytical summative ones of classical science and engineering, are of a more general nature. It is from this concept that the semi-metaphysical statement 'the whole is greater than the sum of the parts' emerged. This statement simply means that the constituent characteristics of the system are not explainable from the characteristics of the parts analysed in isolation.

An absence of the ability to formulate an integrating structure for engineered systems not only adversely affects the ability to solve the engineering problem at hand, it also inhibits learning from experience, as without structure it is difficult to use past experience to educate for the future. This observation suggests a different approach, than has been the case up to now, is required to the use of case histories of dam failures and incidents, namely one that leads to systematic recording and analysis of case histories within a broad framework that is integrated across individual cases. This will make the information accessible to others without imposing the requirement to read and internally reason through each case history individually. The systematic process will also reveal gaps or internal inconsistencies in the accounts of the case histories.

As systems are integrated wholes, comprising elements and their functions, which are necessary for the objectives of the 'whole' to be realised, they are 'reducible' in a scientific sense. This means that systems are amenable to analysis, which generally means 'breaking something down into its components'. Precisely how this process of analysis is carried out is of importance in systems because the functions and outputs of the system are different from the functions and outputs of the constituent elements.

## 3.8. Summary

Systems and systems thinking in dam performance analysis requires the analyst to understand how the dam system works and how it might fail to work in its entirety. This implies that the analyst needs to understand the breadth and depth of the functions that together result in appropriate performance under all conditions within the 'design envelope'. This does not imply that the systems analyst must understand the fine details of the methods of performance analysis of each part, as this can be achieved by the domain specialists working within the analysis team. In many respects, with the passage of time, while the physical arrangement of the dam system may be much as it was when it was designed, the ways in which operational control of the all the functions is achieved may have changed substantially over time. Furthermore, the demands on and the expectations of the performance of dam systems can also be expected to change over time in response to environmental, social and political factors. Thus, not only can the dam system be expected to function differently, and often in a more complex way, as changes are made over time, but also the products and services provided by the dam system often change over time, changes that are manifested in terms of additional constraints and expectations.

Systems engineering has developed to address the increasing complexity of engineered and built systems through using a growing body of principles and techniques for predicting system behaviour. The systems engineer takes an 'overall view' focused on specifying the performance of the system, creating a combination of components or elements to meet specifications, describing the characteristics of components and their interconnections, and evaluating and assessing system performance. The systems engineer leaves the design of the components and connecting links to the domain specialists in order to avoid diverting attention from the external characteristics of the system, including performance, reliability and cost.

### REFERENCES

BC Hydro (2011) *Generation Operating Order Daisy Lake Dam*, BC Hydro, Vancouver, Canada.

Bertalanffy LV (1969) *General System Theory: Foundations, Development, Applications*. George Braziller, New York, NY, USA.

BMT Isis (2008a) *BC Hydro Safety Assessment Report: Daisy Lake Dam*. Report to BC Hydro, Vancouver, Canada.

BMT Isis (2008b) *Systems Engineering Assessments for Daisy Lake Spillway*. Report to BC Hydro, Vancouver, Canada.

Checkland P (1981) *Systems Thinking, Systems Practice*. Wiley, Chichester, UK.

De Weck OL, Roos D and Magee CL (2011) *Engineering Systems Meeting Human Needs in a Complex Technological World*. MIT Press, Cambridge, MA, USA.

Dörner D (1997) *The Logic of Failure: Recognizing and Avoiding Error in Complex Situations*, 1st edn. Basic Books, New York, NY, USA.

Forrester JW (1968) *Principles of Systems*. Pegasus Communications, Waltham, MA, USA.

Hartford DND and Baecher GB (2004) *Risk and Uncertainty in Dam Safety*. Thomas Telford, London, UK.

Howson C and Urbach P (2005) *Scientific Reasoning: The Bayesian Approach*. Open Court, Chicago, IL, USA.

ICOLD (International Commission on Large Dams) (1987) *Dam Safety – Guidelines*. ICOLD Bulletin 59. ICOLD, Paris, France.

ICOLD (1988) *Dam Design Criteria – Philosophy of Choice*. ICOLD Bulletin 61. ICOLD, Paris, France.

ICOLD (2011) *Dam Safety Management: Operational Phase of the Dam Life Cycle*. ICOLD Bulletin 154. ICOLD, Paris, France.

Marais K, Dulac N and Leveson N (2004) *Beyond Normal Accidents and High Reliability Organizations: The Need for an Alternative Approach to Safety in Complex Systems*. MIT Press, Cambridge, MA, USA.

Metropolis N and Ulam S (1949) The Monte Carlo method. *Journal of the American Statistical Association* **44**: 335–341.

Perrow C (1999) *Normal Accidents: Living with High-Risk Technologies*. Princeton University Press, Princeton, NJ, USA.

Regan P (2010) Dams as systems – a holistic approach to dam safety. In *30th Annual USSD Conference on Dams: Collaborative Management of Integrated Watersheds*, Sacramento, CA, USA.

Rhodes D and Hastings D (2004) *The Case for Evolving Systems Engineering as a Field within Engineering Systems*. Massachusetts Institute of Technology (MIT), Cambridge, MA, USA.

Richmond B (2004) *An Introduction to Systems Thinking with STELLA*. ISEE Systems, Inc., Lebanon, NH, USA.

**Operational Safety of Dams and Reservoirs**
ISBN 978-0-7277-6121-7


http://dx.doi.org/10.1680/osdr.61217.075

# Chapter 4
# Modelling dam and reservoir systems

Consider the 2005 failure of the Taum Sauk Upper Reservoir near Lesterville, Missouri, which was introduced in Chapter 1 (Regan 2012):

> On December 14, 2005, at about 5:10AM the dam overtopped during a pumping operation. The water level alarms did not sound because both alarms had to trigger to sound an alarm (after being rewired in parallel) and the high-high water alarm was about two feet higher than the lowest point on the wall as the settlements of the constructed embankment wall had been ignored. Due to the lack of a spillway the parapet wall overtops. The water falls 10 feet onto the earthen embankment rapidly eroding the material and undermining the parapet. The wall overturns unleashing a 10 feet wall of water that rapidly erodes the remaining embankment. It took only about 12 minutes to drain the reservoir. Peak outflow was estimated at 8200 $m^3/s$ (more than the Mississippi River above its confluence with the Ohio River). Luckily a downstream park and campground was empty due the time of year and 5 people in a house survived even though the house was swept from its foundation.

The components of the reservoir system worked as designed; that is, the failure was not caused by the unreliability of the gauges, alarms, pumps or other components. The automated filling system for the reservoir continued to pump water into the full pool because the gauges – which were operating – were not physically where the software 'model' of the facility presumed them to be, and so the embankment was overtopped and the accident occurred.

The individual components worked as specified, but together they created a hazardous system state. Reliability analysis uses a bottom-up approach (e.g. failure modes and effects analysis (FMEA)) to evaluate the effect of component failures on system function, while a safety case requires a top-down approach that evaluates how hazardous states can occur from a combination of both incorrect and correct component behaviour. This might be the proper behaviour of a component at an improper time or under the wrong environmental conditions.

As described in Chapter 1, current engineering practises do not address many common causes of accidents and failures, which are *unforeseen combinations of usual conditions*. Indeed, in recent decades the most likely causes of fatalities associated with dams have more often been to do with sensor and control systems, human agency and inadequate maintenance than with extreme loads such as floods and earthquakes.

In Chapter 2 we went on to argue that there is a need to think about dams and flow control in a systems perspective if one is to grasp these more common accidents and failures. The present chapter introduces a modelling methodology for understanding flow-control functions at dams that is based on systems thinking, and lays out the basics of that modelling approach.

## 4.1. Performance and reliability of dam systems

The modelling approach to flow-control functions and dam systems attempts to bring together the systems aspects of engineering and operational concerns in a way that emphasises their interactions. Traditional engineering modelling and analysis of dam systems separates major disciplinary concerns into stand-alone vertical analyses. Hydrology is considered to the extent that the frequencies of reservoir inflows are determined. The capacity of spillways and other waterways is considered to the extent that they are large enough and stable enough to accommodate specified discharges. The mechanical and electrical performance of gates and valves are considered to the extent of their availability on demand. None of these independent considerations – as important as they are for design – gets beyond the notion of a 'root cause' of an accident or failure to the issue of unforeseen combinations of conditions. There usually is not a root cause, but rather a combination of causes, which might not be problematic if it were not for their interaction.

A successful modelling approach to dam systems must achieve the things discussed in the preceding chapters; specifically, it must grapple not just with the isolated performance of disciplinary components but also with how those components interact.

This is a tall order. It is not something that has traditionally been done in modelling dam systems. In addition to dealing with the physics of the dam system, the modelling approach should deal with how information is generated on the state of the dam and reservoir at any moment, with how that information is communicated to automated systems or human operators, and with how decisions are made either by automatic controllers fed by the supervisory control and data acquisition (SCADA) system or by human operators interpreting the data in order to decide whether to open or close waterways or to effect flow-control functions in other ways. Thus, the modelling approach should bring together the river basin hydrology, the routing of reservoir inflows through the reservoir system, the operating rules and the human factors associated with operating the spillway and other waterways, the outflow systems, the hydraulics of the outflow and the energy dissipation, and conditioning the discharge to the downstream river channel. In the past, modelling approaches have not done these things.

So, the present goal of modelling dam systems is to understand how the interactions between the system's components, communications and control combine to affect performance and the potential for accidents and failures. How might complicated chains of events combine to affect the ability to control flows? These chains of events might, for example, involve some natural disturbance, its measurement, the reaction of operators and the physical behaviour of hydraulic components. Each time an accident or failure occurs in the field, the chain of events preceding it is more or less unique. As a result, simple statistical history offers little help in informing our understanding of these causal chains. How can the more likely of these causal chains be identified? How can their likelihood of occurrence be assessed? These are significant challenges for modelling and analysis.

### 4.1.1 Dams as systems

The emerging philosophy across many industries for assessing the safety of man-made facilities is based on an underlying principle of system safety. The often used argument is that, even if a dam is decomposed into its primary subsystems (earth embankment, sluiceway, gates, powerhouse and headworks, etc.) and components (riprap, shell, filter, core, etc.), there is little or no interaction between the components and subsystems. Thus all components can be considered separately, and the safety of the entire dam system can be derived from the information about the reliability of individual components.

This interpretation misses two points: More often than not there are multiple dams within a single watershed, and the operation of any single dam may affect the safety of other structures. For this reason, hydrology and water management have always considered 'river systems', where interactions between individual reservoirs (system components) are of the utmost importance. The term 'dam system' extends beyond the physical components and technical aspects to include social, managerial, organisational human and political factors, which together create a complex dynamic system with complicated interrelations and feedbacks. To a large extent the concept of dam systems has been explicitly recognised in Defra (2013a,b) where the aspects listed in Table 4.1 are included in the definition of a reservoir system.

**Table 4.1** Dams as systems

| Aspect | Definition |
|---|---|
| Physical features | ■ The reservoir system, including the surrounding hillsides, the dam(s) and its abutments and foundations, and other reservoirs in cascade<br>■ All appurtenant structures<br>■ Electromechanical equipment<br>■ All instrumentation and communication systems<br>■ Any other natural or man-made physical features relevant to the safe operation of the reservoir |
| Operational features | ■ Maintenance, monitoring, surveillance and inspection procedures<br>■ Any manuals and software needed to operate automated or remote control systems for reservoir operations upon which safe reservoir operation depends, including:<br>– inflows and discharges<br>– information such as inflow flood forecasts<br>– management systems<br>– communications and decision protocols |
| Human factors | ■ Include maintenance, monitoring and surveillance, supervision and inspection<br>■ All management aspects of the owner/undertaker or operating organisation on which safe reservoir operation depends |

While the concept of a 'dam system' has been adopted as a general principle (Defra 2013a,b), it has not been extended to all elements of a broader system comprising all dams and reservoirs (interrelated through physical, operational and human-agency-based links) within the watershed under consideration. Thus many important dynamic interactions and feedbacks are either not accounted for or are only partially included in the analysis during the development of loading scenarios.

### 4.1.2 Reliable performance

The function of a dam system is to retain water volumes for productive uses and to pass flows in a controlled way. This definition is consistent with, and a subset of, the dam safety management objectives proposed by ICOLD (2010):

> (1) Control the release of damaging discharges downstream of the dam through controls embedded in the normal operating regime of the dam; (2) restrict the likelihood of events that might lead to a loss of control over the stored volume and the spillway and other discharges; and (3) mitigate through on-site accident management or emergency planning the consequences of such events if they were to occur.

Thus it is argued that the dam safety community recognises that safety depends on more than engineering, and that accidents and failures are complex processes that may include design errors and omissions, construction issues, operations limitations, and maintenance and surveillance (Bowles, 2007).

The performance of such a system is reflected in its ability to meet the time-varying demands placed upon it by the river, to provide the beneficial functions for which the water is retained (i.e. meeting power requirements, providing flood control, supporting navigation and other things), and to meet the desired downstream flow conditions (providing environmental flows, maintaining low-energy flows and the like). Any one or all of these things could be a design requirement. The factor of safety (the ratio of capacity to demand) or the probability of success of meeting any of these requirements constitutes a design check. The factor of safety was commonly used in an earlier age, while probabilities of success (or, conversely, of failure) are more commonly used in the current age of risk analysis. Philosophically, they are the same.

The *reliability* of a dam system is the dependability with which it meets its performance requirements. More formally, reliability is the probability that the dam system functions to the specifications, and continues to provide the services intended when demands are placed on the system. This is something more than meeting individual design requirements. The reliability of the dam system depends, of course, on the hydrology and on the physics of the dam components. However, it also depends on sensor and data systems, operating rules, and the vagaries of human and natural factors. The success of a dam system depends on keeping in check the probability that a chain of events occurs that leads to a loss of control over retained water volumes and flows. In other words, success depends on the reliability with which the dam system performs its function.

Thinking about dams in a systems context illuminates the issue raised in Chapter 1. Contemporary practices, based on specified load criteria, are necessary for bounding design decisions, but are insufficient for determining reliability as defined in the preceding paragraph.

For example, dams in seismically active regions are designed so that the main structure and its appurtenant works will not pose a threat to public health and safety if subject to a maximum design earthquake. Yet this criterion says nothing about the dam's ability to continue its flow-control functions after an earthquake, or about what might happen to the dam if a large, post-earthquake flood were to occur before the earthquake damages had been repaired.

The question is: 'How can systems engineering considerations be captured in an analytical modelling approach?' Prior to the fact, the probability of the simultaneous occurrence of multiple operating variances in some pernicious combination might be thought to so low as to be negligible, yet the history of large systems failures suggests that such combinations do occur.

This reality is akin to winning the National Lottery: The probability of any one ticket winning the grand prize may be vanishingly small, and yet one of them is the winner. The probability of any particular sequence of simultaneous variances can be small, in fact so small that it may be dismissed as not being a serious concern. Yet, for a complex system such as a dam, the number of possible combinations and interdependences of variances rises exponentially with the size of the system. The number of possible combinations of unfavourable events is correspondingly as large as the probability of any one of them occurring is small. The probability of at least one combination of pernicious events actually happening can be non-negligible.

### 4.1.3 Complexity and emergent behaviour

In any large engineered system, the salient features reflecting complexity emerge when individual parts interact. That is, patterns arise out of a multiplicity of relatively simple interactions, and those patterns may be absent in the behaviour of the individual components; as a result, they may not be predictable by considering the properties of the components by themselves. These unpredicted patterns are said to be *emergent behaviour*. It is the interactions between the parts of a dam system, and thus this emergent behaviour, perhaps even more than the reliability of the individual components, that controls systems behaviour. The components and subsystems of a dam that allow power to be produced and controlled are all interdependent, and their interactions increase as operations become more involved.

The realisation that infrastructure systems are complexly interconnected and interdependent is not new but this fact has been highlighted in large-scale infrastructure failures of recent years (Brown *et al.*, 2004). The catastrophic failures of the power grid in the north-east of the USA in 1965, and again in 2003, and of the blackout in the south-west of the USA in 2011 illustrate the interconnectedness of large-scale infrastructure systems and the ability of failures within one system, such as electrical power, to affect other systems, such as transportation or telecommunications. They also illustrate how a series of unrelated and relatively unimportant human errors and component weaknesses can combine to cause large-scale system failure in ways that were not easily predicted from component behaviour. These lessons apply to dam systems as well as to other technological systems.

The importance of systems interactions and emergent behaviour has grown as infrastructure systems have become larger and ever more sophisticated. Among the factors cited as characterising these changes in the nature of infrastructure reliability are (Nelson, 2009):

- increased sophistication of controls
- greater variety and expectations of reliability
- interconnection of ageing systems into larger networks
- high-efficiency operations leading to reduced redundancy
- poorly understood interdependencies
- deregulation, consolidations, downsizing, and reduced reserves and capacity
- too few trained professionals to service operational needs
- controls requiring recalibration of management judgement.

All of these things apply to dam systems.

### 4.1.4 Operational factors

Most work in dam safety risk analysis addresses the reliability of system components and their interactions under random design loadings. Yet, as noted by Regan (2013), 'Failures are seldom the result of a single "root cause"'. Most failures result from a combination of environmental, engineering and human events or factors. The probabilities associated with these more common events may be much larger than the probability of design loads, even allowing for the probability of their joint occurrence. These combinations of events involve not only system components and external loadings, but also SCADA, communications, operations and control functions.

For example, the annual probability of a design flood or maximum design earthquake might be on the order of $10^{-3}$ or $10^{-4}$. The fragility of a component (i.e. the conditional probability of failure under a design load) might be $10^{-2}$ or lower. So, the probability of the load happening and the component failing might be $10^{-5}$ or even smaller. In contrast, if three much more common events happening together would also cause an accident, and if each of these events had a probability of 0.1 on some day, then their joint probability could be as high as $10^{-3}$, even if they were all probabilistically independent. Of course, they are likely not to be probabilistically independent, and their joint probability is likely to be larger than the product of their three marginal probabilities (e.g. sensor errors and maloperation are each, arguably, more probable on a cold icy day than on a warm dry day).

In an international survey undertaken by Hobbs and Azavedo (2000), dam owners reported the 'causes' of gate accidents and failures as shown in Table 4.2. Most of these involve operational issues. Flow control in dam systems involves civil, mechanical, electrical, communications and control components, as well as organisational and human influences.

Folsom Dam, discussed in Chapter 1, is a concrete gravity structure designed and constructed by the US Army Corps of Engineers. It was begun in 1948 and was put into service under US Bureau of Reclamation management in 1956. On 17 July 1995, the gate 3 failed due to overstressing, releasing nearly 40% of the water from Folsom Lake before the gate could be repaired. Like many accidents or failures, this one involved what Reason (1997) calls 'latent conditions and active errors'. Latent conditions are those aspects of the system, such as design errors, inadequate maintenance or flawed operating rules, that develop over time, and may be present long before the accident or failure occurs. Active errors are those aspects of the accident or failure that involve human actions immediately preceding or during the failure.

Table 4.2 Survey of gate failure incidents. (Hobbs and Azavedo, 2000)

| Category | Recorded incidents (60 total) | Contribution: % |
|---|---|---|
| Human factors | Maintenance related (10)<br>Human error during gate operation (3) | 22 |
| Loss of motive power | Loss of mains (5)<br>Motor failure (2)<br>Failure of back-up power (2) | 14 |
| Control and communication | Failure of limit switches (7)<br>Position indicators (3)<br>Controls failure (3) | 22 |
| Gates | Rollers or bearing failure (4)<br>Rope failure (3)<br>Miscellaneous (3) | 17 |
| Hoist units | Screw thread nut failure/problems (2)<br>Mechanical failure of hoist equipment (3) | 8 |
| Miscellaneous | De-icing/heater failure (4)<br>Operation manuals (1)<br>Concrete movement/deterioration (2)<br>Trunnion anchor inspection problems (1)<br>Float-well design/control issue (1)<br>Truck lost load, damaging hoist units (1) | 17 |

Operational issues are important contributors to spillway accidents and failures, although they are seldom included in dam safety risk studies. Operational failures include, among other things (Regan, 2010a):

- debris blocks the spillway (perhaps with log boom failure) – dam overtops
- gates fail to open (hoists, chains, binding, electrical, remote communications) – dam overtops
- gates open inadvertently, creating life-threatening flows
- communication breakdown – no warning downstream
- loss of access to operate gates – dam overtops
- loss of release capacity (e.g. turbine load rejection) – dam overtops
- overfilling of off-stream reservoir due to loss of water level gauge function 0150 dam overtops
- reluctance to open gates and flood people out
- other external events.

The modelling of the reliability of dam systems needs to include operational issues.

## 4.2. Beyond current risk analysis

To effectively model dam systems, a way is needed that goes beyond traditional decompositional analysis and attempts to understand the interactions between components and systems engineering behaviours. Reliability analysis works well for purely physics-based systems and for understanding systems at the component level, but modern dams and their attendant flow-control functions are sufficiently complex to tax the limits of those methods. That is, traditional reliability methods of the sort found in current risk analysis, despite their strengths, are inadequate for tackling the complexity of modelling dams at a systems level.

### 4.2.1 Current risk models

Current reliability modelling, of the sort used as the basis for dam safety risk analysis, addresses a physical system that, typically, is abstracted as shown in Figure 4.1. The operating system comprises the physical components of the dam, powerhouse, spillways and other components. Each of these can be represented by physics-based models, for example of the mechanical operation of the gates and valves, the hydraulics of the outflow conveyances and energy dissipation structures, the power-generation waterways and other features. The inputs are the hydrological inflows to the reservoir, together with any pertinent characteristics and water volumes (as discussed in Chapter 5). The outputs are the hydraulic discharges to the downstream channel and their temperatures, sediment loads and other characteristics, and any other products such as power generation or flood control, all of which need to be conditioned to the given requirements (as discussed in Chapter 6). The disturbances (as discussed in Chapter 7) are those extraordinary events (such as earthquakes, earth and rock slides into the reservoir or

Figure 4.1 A physical system affected by disturbances, as used in current dam safety risk analysis studies

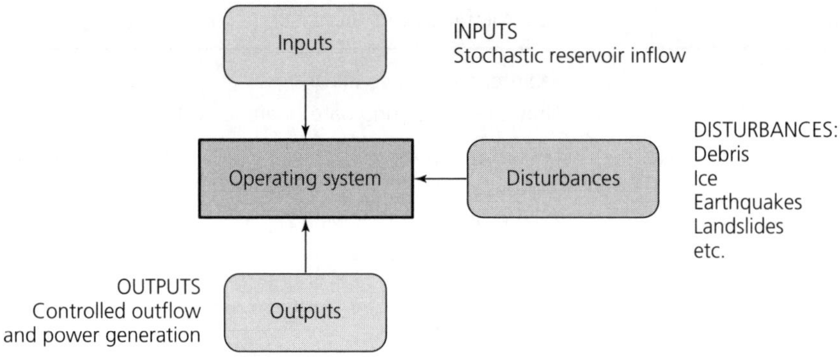


conveyance works, fire in the drainage area, debris inflow, or other unpredictable occurrences of not purely hydrological nature) that may affect dam operations or even dam safety.

This model – and the risk analysis approach based on it – has served the dam safety community well. It is the basis of current dam safety guidance such as that from the International Commission on Large Dams (ICOLD, 2005), the Australian National Committee on Large Dams (ANCOLD, 1994), the Canadian Dam Association (CDA, 2007) and the US Bureau of Reclamation (USBR, 2003), among others. This approach is capable of capturing the impacts of natural environmental loads and the physics-based performance of the engineered components constituting a dam. Within this paradigm, the dam is principally a linear system without feedback. Inputs, such as reservoir inflows, are dealt with by routing them through the powerhouse and other waterways, and they become outputs. With annual recurrence frequencies, a variety of disturbances may occur, and these may or may not affect the operating system's ability to transfer inputs to outputs. The annual probability that the operating system is unable to deal with the transfer of inputs to outputs, either because of the magnitude of the inputs or because of the impact of the disturbances, defines the risk.

The current approach to risk analysis is increasingly being used to rank order the importance of hazards, to compare risks posed at one facility with those posed at others, and to compare risks against absolute criteria of acceptability. The first two of these are reasonable because they compare like things –the probabilities that a facility adheres to certain design checks.

The third use, comparing risks against absolute criteria of acceptability, is more problematic. Acceptable risk criteria of the sort promoted by the UK Health and Safety Executive (HSE, 2001), and increasingly in the dam safety community (ANCOLD, 2003), are based on the historically observed frequencies and severities of industrial accidents. These presumably include accidents from all causes. The presumption is that risks currently accepted by society are *de facto* acceptable. Yet, current risk analysis estimates of the probability of accidents and failures overlooks the reality that many accidents and failures do not occur in ways that are analysed by these risk analysis methods. Thus, the probabilities so assessed are, in reality, lower bounds; they are useful but incomplete.

### 4.2.2 Communications and controls

The pace of technological change is increasing, and increasingly modern engineered systems incorporate digital technologies that supplant the traditional electromechanical systems of the past. Communications and controls are as important to the performance of modern flow-control functions as are gates and valves (Figure 4.2). Many recent accidents involving the unplanned openings or closings of spillway gates have been attributed to sensor and control systems (Regan, 2010b).

Figure 4.2 Role of communication and control in systems performance. (Adapted from Leveson, 2004)

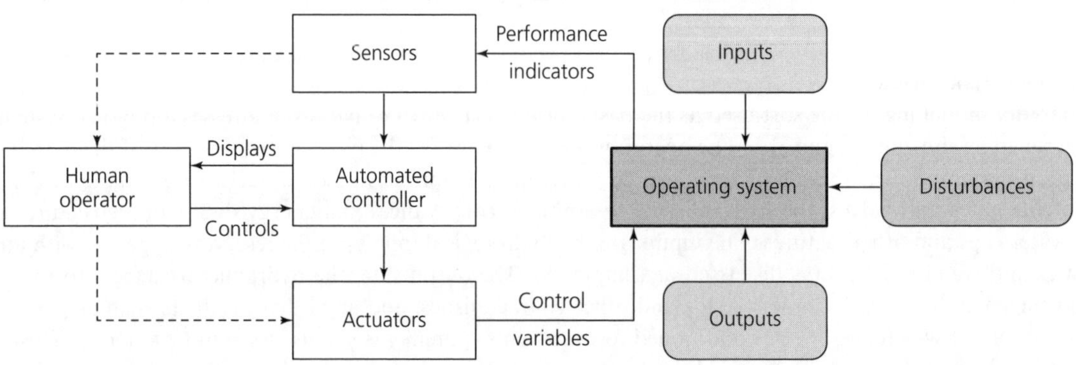

From a systems engineering point of view, the introduction of communications and controls begins to add non-linearity to dam system behaviour. Current risk analysis deals mostly with linear systems. That is, there are few feedback loops in the behaviour of a system such as the one represented schematically in Figure 4.1: inputs are translated directly to outputs, although they may be perturbed by random disturbances. Once sensing of the system is introduced, information on prevailing performance indicators is communicated to controllers and human operators, and control actions may be taken to change the state of the operating system on the basis of its current state. Non-linearities appear in the system.

Automated sensor and control systems, taken as a whole, are sometimes called 'supervisory control and data acquisition' (SCADA) systems. This is a term that comes from industrial control systems, which are computer-controlled devices that monitor and control a process. In the present case, that process is flow control. Typically, SCADA systems may control large sites or installations, and can be operated from a distance as part of a larger network of control systems. This allows for unmanned dam sites for which controls can be overseen from a central facility, possibly at a great distance – sometimes hundreds or thousands of kilometres away. These SCADA systems for controlling an individual dam may also be integrated with grid operations, which dictate production schedules, and with other facilities.

Just as the physical system in Figure 4.1 involves material flows of water from inputs through the operating system to outputs, the SCADA system involves information flows of data and instructions from sensors through automated controllers to actuator commands (see Figure 4.2). Most control commands are performed automatically by programmable logic controllers (PLC), which are specially programmed to the situation at hand. Each step in this flow of information is subject to reliability concerns, particularly as most sensors and PLCs are locally programmed.

SCADA systems usually host the emergency alarm capability for a facility, although this may be overridden by operator intervention. SCADA systems are also likely to be integrated, via internet communications, with corporate information systems, which means that they may be susceptible to malicious cyber-attack.

As the cost of unreliable performance of SCADA systems may be high, and may involve potential loss of life, as in the case of a spillway gate opening unexpectedly, a typical SCADA installation may be hardened to military-grade standards in order to protect it from extreme environmental conditions (temperature, moisture, storm, ground shaking or electrical interference). It is not unusual for there to be redundant SCADA components and communication channels in sensitive applications. The reliability of such systems is usually categorised according to traditional electronic reliability theory metrics (e.g. by the mean time between failures).

The remote acquisition of data and control of operations is a major benefit of SCADA systems. This telemetry of information flows may be performed over corporate IT networks, although increasingly the internet is used for such communications. The use of the internet has its own issues, both of reliability and of security (these issues are outside the present (specific) scope of discussion). The issues have led some organisations to move to satellite-based communications, which bypass the reliability issues of land-based networks as well as some security issues. The potential vulnerability of SCADA systems to malicious threat has been a major concern in the national security literature (EU, 2013; NCS, 2004; OAG Canada, 2012).

### 4.2.3 Human agency

Human behaviour and human intervention in control have always been a part of complex systems performance, but humans today increasingly share the control of systems with computer-aided devices. In the most extreme case, such as modern aircraft, the operator has, to a degree, ceased to operate the systems and has become more a 'systems monitor'. Difficulties of operator–computer interaction and communication arise with greater frequency, and human operators sometimes feel cut off from important information about systems functioning. This has led to new types of human operator issues, these often being issues of omission rather than commission.

The term 'human error' was for years the description used to categorise the negative impacts of human performance on complex technological system controls. Human error was abstracted as a branch of reliability, and studied to derive objective frequencies and models of human performance akin to the reliability models applied to mechanical or electrical components. However, this concept of operator error has evolved. The current view is that operator performance is a social and psychological process, not an objective technical one (Woods *et al.*, 2010). That is, human error is no longer seen as a human deficiency. It is seen as the result of the operational environment and social structure within which human operators take actions or make decisions. This revised understanding is reflected in the work of Norman (1981) and Reason (1990), among others, and applies equally to dam systems.

Human agency in spillway operations and flow-control functions is, as in most technological systems, a major consideration in the reliable performance of those systems. These issues are considered more fully in Chapter 8. In the past, however, human control actions have seldom been considered in dam safety risk. This is understandable. Dam safety risk analysis has not

Figure 4.3 Technological system within the context of operating staff and organisational management. (Adapted from Leveson *et al.*, 2009)

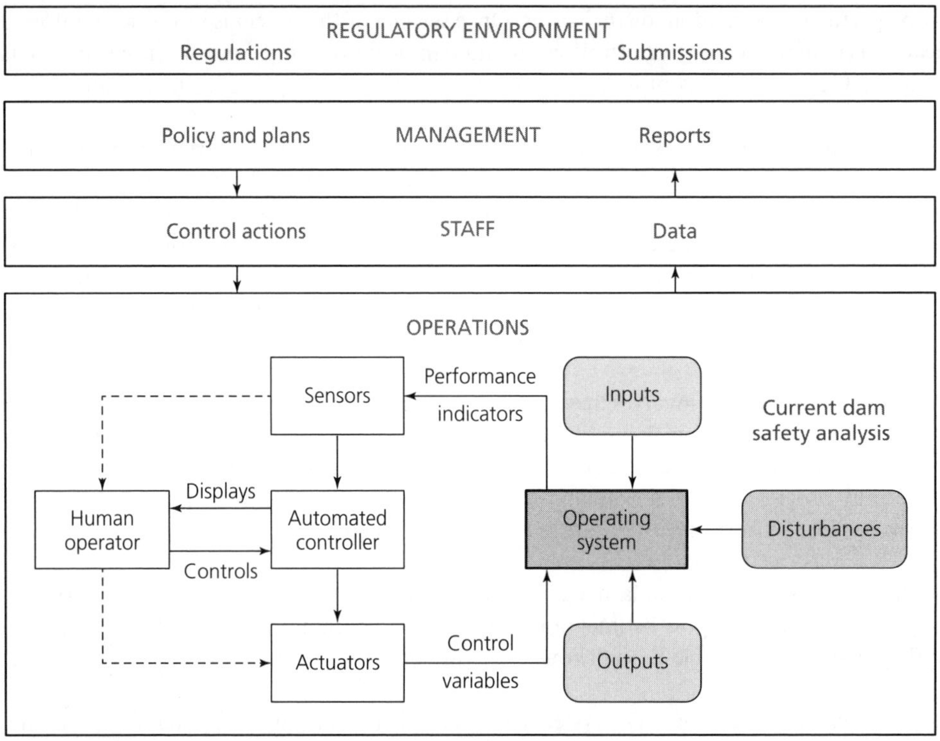


typically considered operations, and human agency is difficult to include in engineering models, except in a traditional reliability-based way, that is, represented as human error probabilities.

### 4.2.4 Organisational influences

Outside the layers of direct communications and control are organisational factors. These too operating in reciprocal ways: communications of information about the site flow up to the organisation, and control decisions flow back down. These are usually not operative in real time as an accident sequence is progressing, but are pre-existing elements of the environment in which a technological system operates (Figure 4.3). At each level in the organisation, information is fed upward, and requirements, policies or control actions are fed down.

At the highest level of the organisation, requirements are established or regulations are imposed from the outside. At the management level, policies or plans are developed that control the operational environment. At the off-site staff level, reference or indirect controls may be imposed. Each layer in the organisation imposes constraints on the layer beneath it. Moving in the other direction, operations data are sent upward to off-site offices. From there, summary reports are generated for management, either on a periodic basis or following incidents. From there, submissions aggregating operating performance over required periods of time are made to the regulatory layer.

Two channels of effective data and control communications are required in this structure: one channel carrying information upward, and the other carrying control instructions downward. Management generates safety policy, standards and resources to staff and on-site management. Flowing backwards, it receives status reports, risk assessment and accident summaries about the status of the safe operation of the facility.

### 4.2.5 Systems behaviour

Modern engineering modelling is based on the precept that mathematical equations, when combined with empirical observation, can be used to mimic or replicate salient features of natural or engineered phenomena, and thus to forecast the behaviour of physical situations. This has proven powerful for assessing the behaviour of a wide range of phenomena, from structural mechanics, fluid dynamics and the performance of electrical controls to the behaviour of natural processes such as hydrological conditions and geological forces. Great advances have been made by applying these deductive mathematical models to a host of engineering problems, and by using them to design and scale new constructed works.

Deductive mathematical models have proven less adept, however, at mimicking or replicating the behaviour of complex systems comprising a large number of interacting and phenomenologically distinct components. This, of course, is the situation with flow-control functions. The interactions between the many and differing components of these systems are complex and involve some degree of randomness; they typically do not lend themselves to simple mathematical representation. The feedback loops joining the behaviours of the many system components may be convoluted, are almost always non-linear and are arrayed in a combinatorially large set of interactions. They do not lend themselves to being represented by deductive mathematical formulae, and there is little reason to think that these interactions between the components of large systems necessarily follow traditional modelling schema. This phenomenon has been observed in attempts to simulate complex systems in a variety of fields (Wolfram, 2002).

Modelling flow-control functions requires complex systems approaches. Traditional engineering analysis has approached non-linear complex systems by assuming that for small disturbances such problems can be approximated by piecewise linear systems, thus simplifying the analysis. For flow-control systems, three things invalidate this approximation: First, the continuous calculus of small disturbances breaks down when dealing with discrete systems, necessitating the use of discrete (event-based) rather than continuous mathematics; second, feedback between sub-system behaviours may amplify or dampen performance in ways that become too expansive to represent explicitly; and, third, the interacting subsystems are of many and incommensurate types (mechanical, instrumental, data-epistemic, human performance, etc.).

The result of these many considerations is that traditional deductive mathematical models are inadequate for the task of capturing the salient behaviour of dam systems reliability. For this purpose, a new computational approach is required. This does not need to be invented wholly anew, but can be adapted from similar applications in other fields of systems modelling.

Leveson *et al.* (2009) have described a systems approach to understanding technological accidents as comprising the following characteristics:

- top-down systems thinking that recognises safety as an emergent system property rather than a bottom-up summation of reliable components and actions
- a focus on the integrated socio-technical system as a whole and the relationships between the technical, organisational and social aspects
- a focus on providing ways to model, analyse and design specific organisational safety structures rather than trying to specify general principles that apply to all organisations.

Leveson *et al.* argue that safety against accidents is not a property of the reliability of individual system components, such as the hoisting mechanism of a spillway gate, but of those components in the context of the whole and in the context of the operating organisation. They argue that reducing the probability of accidents occurring is a control problem rather than a problem of component reliability: 'Accidents occur when component failures, external disturbances ... or dysfunctional interactions between system components are not adequately controlled or handled'.

These interactions include not only the technological components but also the operational, organisational and regulatory systems that provide control. The interfaces at all of these organisational levels are characterised by communications of operating data and information moving upward, and control actions and instructions moving downward.

Regan (2010b) applied this systems thinking to dam safety, arguing, as Leveson *et al.* (2009) do, that the safety of a dam must be viewed in the context of the interaction of the dam and its operation within a larger organisational, communications and control environment. He argues that the failures of Teton, Silver Lake, Taum Sauk and other dams are not easily understood through the lens of traditional engineering risk analysis (e.g. event-tree or fault-tree models), which focuses on the reliability of the physical components of a dam in isolation from their integrations with communications and control systems. Indeed, it has been observed that most recent fatalities associated with dam operations have been caused by interactions of sensors, controls and maloperation decisions (Hartford and Zielinski, 2013).

## 4.3. Modelling approach

The first step in considering any system is to define its boundaries clearly. Components within the boundaries are part of the system, and those not within the boundaries are not. This may be a simple distinction, but deciding how exactly to set the boundaries may be anything but simple. There is no one correct answer; the choice depends on the objectives at hand.

### 4.3.1 Defining system boundaries

Consider the situation of an unstaffed dam in a remote location. Perhaps something misfunctions or otherwise goes wrong at the dam during a heavy storm. A work crew cannot reach the dam because the road network is impassable due to the storm. This was the situation at Noppikoski Dam when it failed in 1985 (described in Chapter 1). This unavailability of the road network affects the safety of the dam and its flow-control equipment. Should the road network be considered within the boundary of the 'system'? In a similar way, the regional power grid affects the demand on the dam and its power generation. If the grid goes down for some reason, power cannot be discharged into the grid, and power generation must stop. This has immediate impacts on the discharge capacity through the powerhouse waterways. Should the regional power grid be within the boundary of the 'system'? For practical analytical and modelling reasons, not everything can be inside the system.

Things that are not inside the boundary of the system, but still affect the system's functions are part of the environment or context within which the system operates. Those things outside the boundaries of the primary system may still be members of other systems and they may interact with the dam and its flow-control system, as both the regional road and power grids do. Interactions across the boundary reflect the connection of the system with its environmental context. Usually the catchment area is not considered part of the system (as it is not under direct control), but rather the system limit is set at the reservoir inflows.

The clear definition of the boundary is important because those elements within the boundary are presumably (or should be) under the direct control of the engineers and operators, and become elements of a systems model. Those elements outside the boundary are part of the environment or context surrounding the system, and are typically considered as exogenous forces or disturbances, dynamic boundary conditions or constraints, usually not under the immediate control of the engineers and operators. Their influence on the functioning of the system can be very important (as, for instance, the hydrology and meteorology of the catchment), but in the modelling and analysis the reverse influence of the primary flow-control system on the elements outside the boundaries is usually ignored.

### 4.3.2 Analysing the system

Modelling the system's reliability of flow-control functions in a modern dam involves: (1) characterising the performance of a spectrum of systems components, (2) following the dynamic interaction of these components over time, and (3) tracking the possible occurrence of external disturbances to the system that may perturb component performance. This creates a number of difficulties for the engineer:

- the components are many and complexly interrelated
- it may not be possible to analyse the components in isolation
- the demand functions on components are stochastic processes of time and possibly of space
- external disturbances (such as natural hazards) may affect component performance
- the pattern of interactions between the components over time may be important.

Alternative approaches to reliability analysis, such as stochastic process modelling, reliability block diagram representation, fault-tree modelling, or event-tree analysis, are powerful tools – and are useful in understanding the performance of individual components – but quickly mire under the combinatorial large number of outcomes when the reservoir, dam and waterways are taken as a system operating in a river environment.

### 4.3.3 Flow-control functions

Suitable functioning of the flow-control system is defined here to mean, 'retain water volumes for productive uses and to pass flows in a controlled way'. Operationally, this definition involves at least the following criteria of acceptable flow-control system performance:

- the provision of controlled discharges through the powerhouse train
- controlled release of discharges over the spillways and other waterways without damage to the structures
- controlled outflows in the downstream conveyance and energy dissipation system without damage to those structures
- acceptable discharge conditions to the downstream reach of river, including velocities, water levels, air content, sediment, temperature or other conditions that affect the suitability of the release water.

Thus, in the present context, spillway functioning means that (a) there is no loss of control of the waters or system, (b) damage is limited and controlled, (c) water is transported downstream in a safe manner, and (d), at worst, waters enter back into the downstream reach with the characteristics and damage potential no more adverse than those of a natural flood.

Damages to the outflow works and downstream structures are controlled by the actual discharge rates passing the flow-control system (Figure 4.4). These may depend on operating decisions or practical implementations. These actual discharge rates, in turn, depend on current capacities of the outflow waterways (spillways, power station waterways or other outlets) and the inflow circumstances, including disturbances such as debris and ice.

## 4.4. Modelling framework

The modelling framework for appraising the systems reliability of flow control involves four parts:

Figure 4.4 Flow-control system damage. The energy in the system flows from potential energy at the outlet, to kinetic energy in the outflow waterway, to being de-energised in the dissipation structures

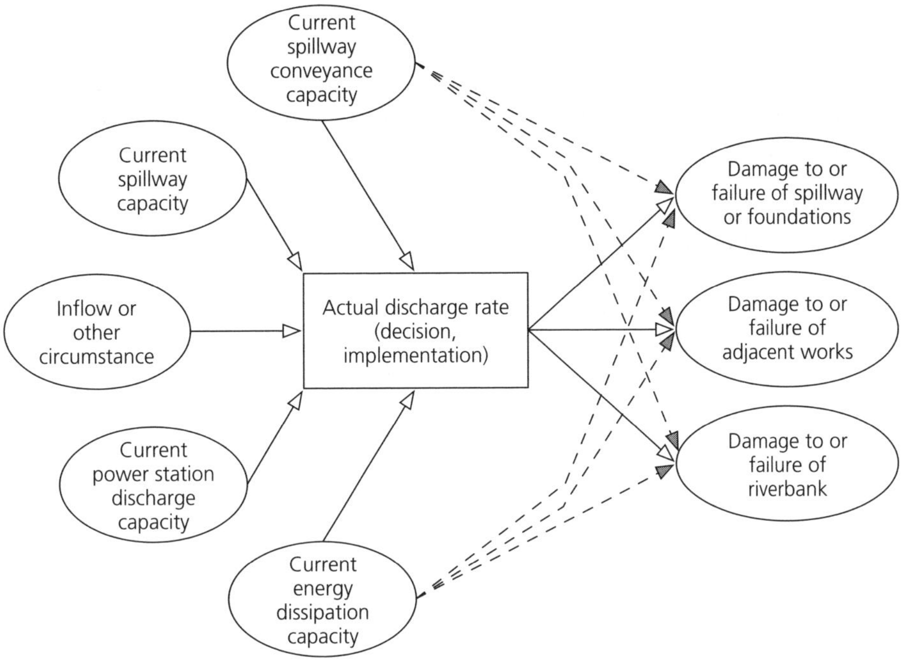


- simulation by which stochastic reservoir inflows are randomly generated through time using a Monte Carlo approach propagated through the reservoir–spillway–outflow system generating demand functions on all system components
- engineering modelling to infer the impact of spillway heads and discharges (demand functions) on the hydraulic structures accommodating the outflows
- component reliability analysis to ascertain the performance of individual components of the outflow works relative to defined failure modes and metrics
- Systems reliability assessment through which demands on and performance of flow-control systems components are convolved into annual exceedance probabilities of adverse performance and fault and failure sequences identified.

The hydrology of the river basin system may be simulated from a stochastic process of flows, or from physics-based models of runoff and routing, or in some other way as discussed in Chapter 5. The modelling framework is accommodating with respect to the simulation approach. The river basin may include other dams or facilities modelled as separate systems or just be the (natural) environment of the modelled dam. The catchment area provides the input driving the flow-control system.

The modelling framework conceives of the flow-control system as a number of components. The outputs of each component generally form the input to the next component in the system logic. The system is, however, usually not linear but contains feedback loops and interdependencies.

The reservoir inflow from upstream and the inflows from local drainage basins are routed through the reservoir to create a demand function on the spillway outflow structure. These simulated flows routed through the reservoir are sensed by SCADA and automatic control systems. The response of these SCADA and control systems is similarly simulated by Monte Carlo methods, and both the routed flow and the control system outputs are used to simulate the gate and spillway functioning. The outfall of the spillway then forms the demand function on the outflow and energy-dissipation components. Finally, the outflow to the downstream channel is simulated and compared against performance criteria.

At each component, the upstream demand function is applied to the operation of the component. Engineering reliability modelling is used to evaluate the performance of that component in relation to the demand function. The performance of the component is described in a fragility curve, expressing the component's probabilistic behaviour as a function of the demand placed upon it. The fragility curve describes performance in many ways and at different physical locations. For example, the demand on a series of spillway gates may involve hydraulic head, debris and ice loadings. The response of the gates may be characterised as structural performance and discharge capacity, and the responses may vary along the series of gates.

In parallel with the flow of water, the system water-control function is also dependent on the flow of information at a number of different levels to facilitate operative decision-making and execution of the desired operation by remote control or on-site action by operating personnel. At a higher level the allocation of resources and establishment of company policies for operation and maintenance is involved. These elements containing human response in different situations contribute significantly to the feedback loops and non-linear behaviour of the system as a whole.

The reliability models for each component will generally describe performances of more than one type, and may be of more than one analytical structure. For example, the performance of a lift gate may be analysed using fault-tree analysis of the structural loadings and response. The performance of the outflow conveyance may be analysed using hydraulic models of the fluid pressures, aeration and cavitation.

The analytical strengths of this simulation approach are that it: (1) supports a systems approach by accounting for the interactions between hydrological inflows, disturbances, hydraulic performance, and structural and geotechnical response; (2) provides a means of modelling and calculating the feedback performance of the various components of the system; and (3) supports a reliability scaffold for systems performance. In concept, this same simulation approach could be applied to a variety of other systems engineering and performance analyses, including such things as internal erosion in earth embankments, but that is beyond the scope of the present discussion.

Simulation allows for the many complex interactions of the flow-control system to be separated out and tracked over time in their dynamic interactions with one another (Ross, 2006). The simulation approach stands in contrast to other reliability analysis approaches such as reliability block diagram analysis or fault- and event-tree analysis. While parts of the flow-control system, such as gates and electrical equipment, have been successfully modelled using block diagram analysis (USACE, 2001), in general the complete flow-control system is too complex and dynamic for this approach to be widely successful. Event-tree methods have proved more successful in analysing flow-control reliability, but fail to capture the dynamic interactions of systems components and disturbances.

### 4.4.1 Stochastic simulation

The modelling framework is illustrated in Figure 4.5. Stochastic reservoir inflow is simulated as a random function of time. This random function may be based on historical data, may represent a forecast of future conditions based on climate-change scenarios, may rely on forecasts of upstream changes to the drainage basin, or may incorporate other predicted conditions.

These inflows are routed through the reservoir to create demand functions on the spillway and other water outlets, together with an operating policy defined on behalf of the owner to meet required objectives in terms of dam safety, power production, flood mitigation, etc. These demand functions define required discharges and gate positions, water heights and other factors over time. At any point in the simulation the effectuated parameters, the outflows and the reservoir levels define the loads, water heights, discharges and other conditions brought to bear on the dam and on different waterway structures, including intakes, chutes, penstocks, gates, turbines, tunnels, energy dissipation works and outlets.

*Systems analysis* can be defined as the use of scientific methods to determine preferred plans, design and operational strategies for complex systems. Two basic methods of systems analysis are commonly applied to water resources management and control, one using simulation techniques and one using optimisation techniques. Simulation models are computer models capable of describing how the system functions, and are usually used to predict what changes will result from a specific action. More precisely, simulation models provide the characterisation of the system response to a set of system inputs. Optimisation models, in contrast, are computer models that allow identification of the set of decision variables that correspond to the optimal value of an objective (or set of objectives) subject to the system constraints. Most often the established general objective relates to the best possible allocation of water resources within a given time period. In the present case, the reliability model is based on simulation, not optimisation.

### 4.4.2 Approaches to simulation

The generation of stochastic inputs to the system can be approached in several ways. The modelling framework accommodates any of these, and it is left to the user to decide which approach is most appropriate or preferred for a particular application.

The simpler and less computationally intensive way is to hypothesise a stochastic generation process of upstream discharges following some reasonable and 'known' probability distribution and having some reasonable and 'known' autocorrelation function in time and space (e.g. between tributaries) (Figure 4.6). The parameters of this stochastic process could, in principle, reflect historical conditions in the basin, and might or might not account for parameter uncertainty through an inverse probability weighting.

A perhaps more complex approach is to use stochastic meteorological modelling to generate precipitation events and then to route the rainfall–runoff relationships for these events into a time series of stream flows by use of a hydrological model or another form of routing. The advantage of the approach is that precipitation data are often available in much longer

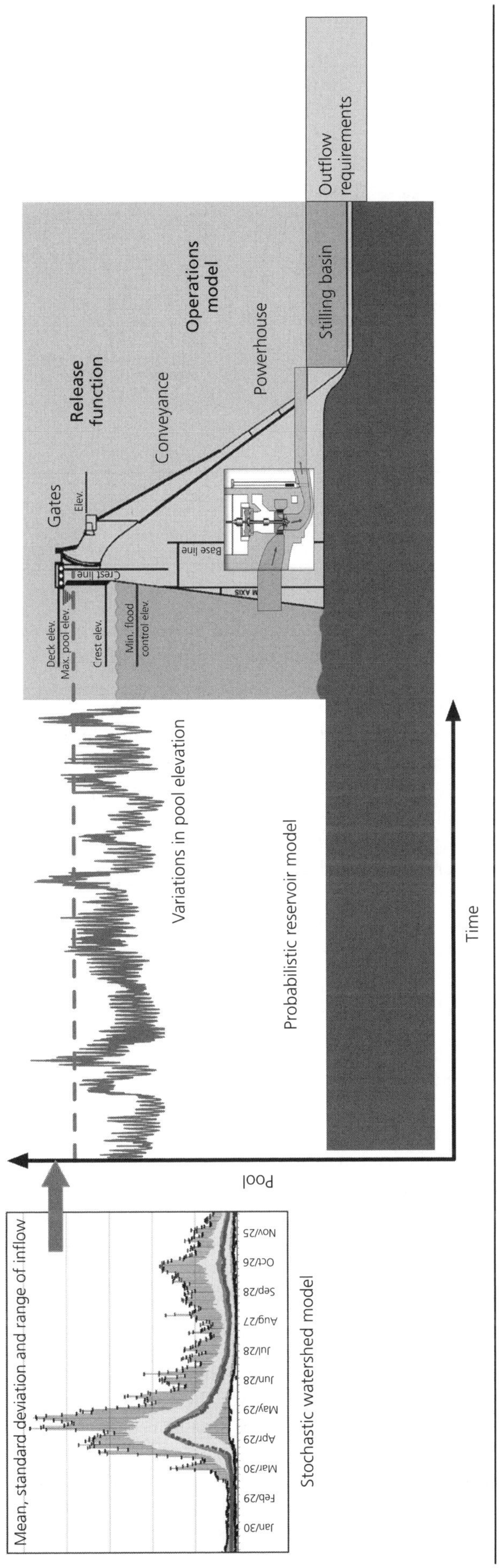


**Figure 4.5** Modelling framework for simulating flow control. (Dam section adapted from USACE, 2001)

Figure 4.6 Stochastic nature of reservoir inflows caused by climate, basin hydrology and routing of discharges through the river systems and upstream reservoirs

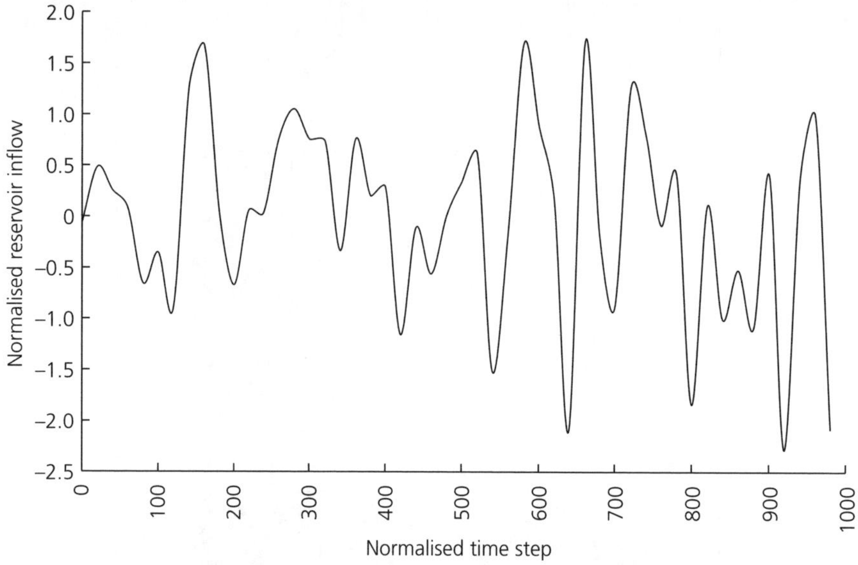


series than are stream flows. From the reliability modelling point of view, each of these approaches yields similar results, that is, a random time series of reservoir inflows from which the performance of the flow-control system can be modelled.

### 4.4.3 Hydrology

From a hydrological vantage point, a river basin may be seen as a collection of interrelated elements working towards some common objective. Typically, a natural river network is characterised as dynamic, because the elements interact and their states change over time. The elements may be subdivided into three groups: components, attributes and relationships. *Components* are the operating parts of a system, and consist of input, process and output. Each component may assume a variety of values defining the system state. *Attributes* are the properties or discernible manifestations of the components. *Relationships* are the links between components and attributes.

River networks – whether in a natural state or with man-made structures and human activities present – can be understood as systems. The components may comprise (Figure 4.7):

- local sub-basins (sub-watersheds)
- natural and artificial channels
- diversion and control structures (dams)
- reservoirs providing storage behind dams.

The components can be broken down into other subcomponents, and abstracted as flows in reaches and storages in reservoirs. A critical modelling decision regarding the level of analysis is, which are the components that together with their interactions constitute the system to be modelled? The objective is to have the complex network represented by an aggregate of simple components. The model of each component is a formal but simplified representation of its real counterpart. It can capture only some of its aspects of reality, and these should be those aspects that are most relevant for the purpose at hand.

River basins producing loads (inflows) on dam and flood-control systems may be represented by hypothetical stochastic models or by hydrometeorological models. Both of these are time-series models.

Time-series models belong to the family of stochastic models, that is, mathematical models with states changing over time according to probabilistic laws. Climatological processes over the watershed and the resulting hydrological process, like most natural phenomena occurring in natural systems, can be seen as behaving in a probabilistic manner.

Reservoir inflows, and consequently the elevation head on dams and outflow structures, are variable functions of time, which depend on the climate, basin hydrology, and routing through the river system and reservoir (Figure 4.8). From a modelling perspective, the reservoir inflows appear as a stochastic process, characterised by probabilistic moments, probability distributions and temporal autocorrelations.

Figure 4.7 Characterisation of a river network as flows and storages

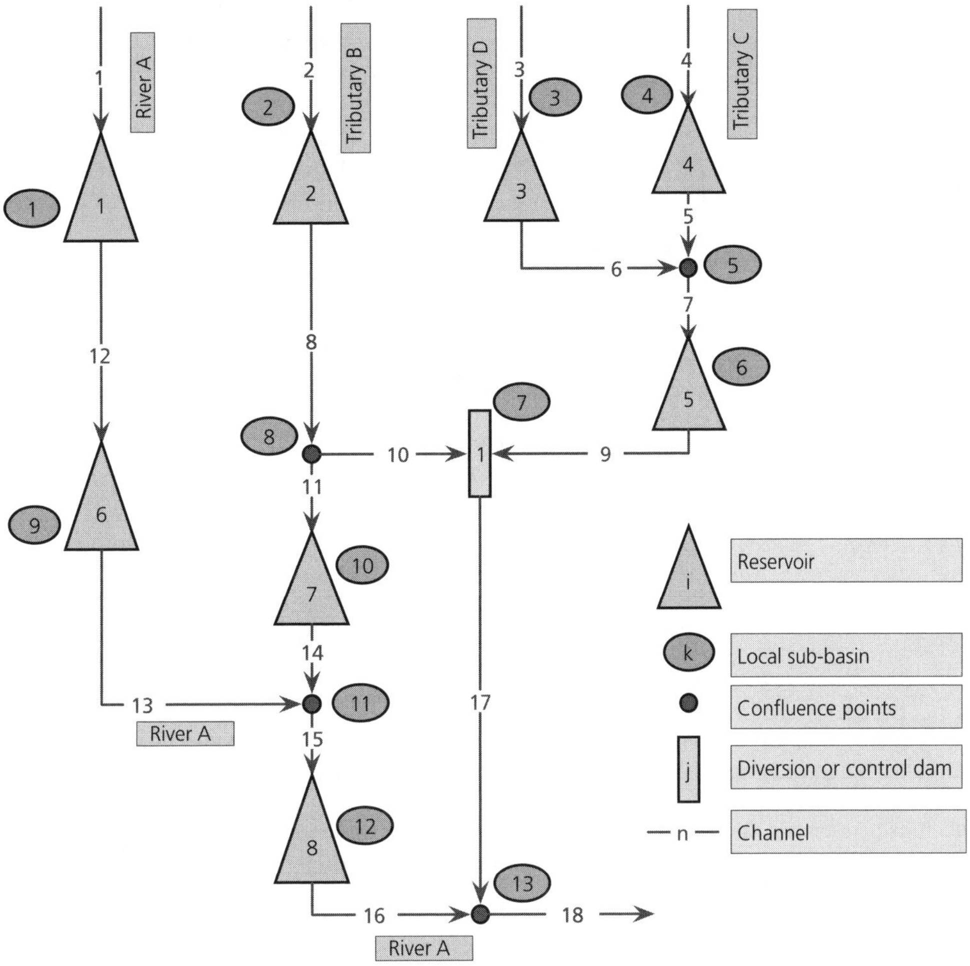


A dam system (i.e. a dam with all its structures and waterways) may also be seen as a flow-control system characterised by quantities that vary over time, including, for example, water levels, discharges and velocities at different locations within the waterways and reservoir. The response of such a dam or flow-control system to natural events and human interventions cannot be examined through experiments on the real system, but may be examined by performing experiments on a 'copy' of the system, where this 'copy' is constructed as a physical or mathematical model, in the present case a mathematical model (Figure 4.9). The model of the flood-control system comprises mathematical objects, equations, graphs and diagrams, and the behaviour of the system is examined through *simulation*.

Figure 4.8 Inflow and outflow relationship to the state variable of channel reach or reservoir

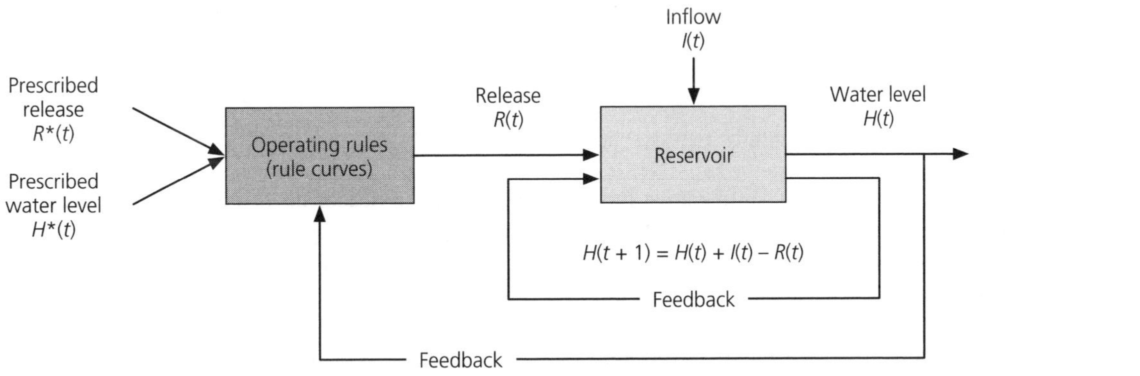

Figure 4.9 Stochastic time series of reservoir inflows in a simulation of Missinaibi River, Ontario

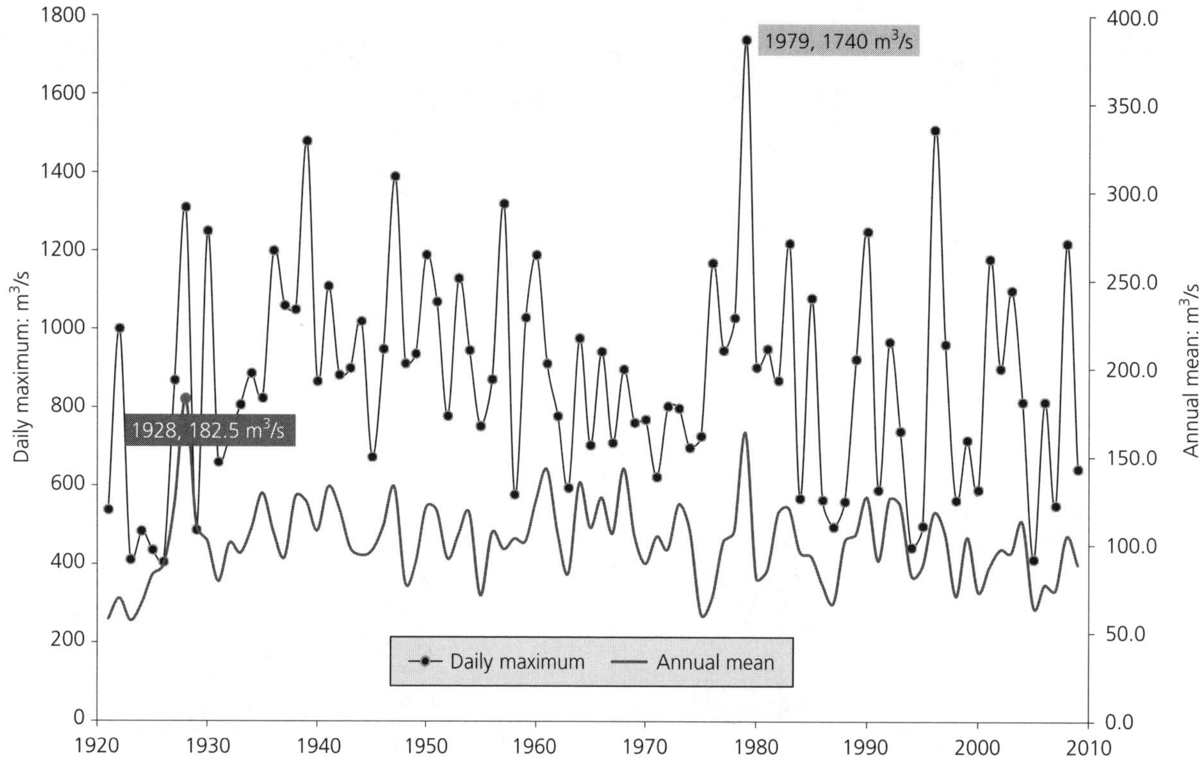


For an element, such as a reservoir, the balance of flow is such that the difference between the rates for the inflow and the outflow for a period of time is reflected in the change in storage for that period:

$$\frac{dS(t)}{dt} = I(t) - R(t) \tag{4.1}$$

where the storage $S(t)$, or more precisely active storage, is the volume of water stored at time $t$, with a maximum value imposed by the physical reservoir size or some other limit determined by the reservoir operating rules.

The inflow $I(t)$ is the sum of the cumulative inflows into the reservoir from time $(t - 1)$ to time $t$. It includes runoff from the local basin upstream from the reservoir, release from the upstream reservoir (if present) and flow that can be artificially conveyed from another basin through diversions or water transfers.

The release $R(t)$ is the actual amount of water released from the reservoir in order either to attempt to satisfy water or hydropower production demands or to prevent the reservoir from overtopping, for instance by adhering to an established operating rule. The latter case may take place when a reservoir is operated in the flood-management mode.

### 4.4.4 Hydraulics

A number of different waterways may exist for leading flow past, around or through a dam. Surface spillways and low-level outlets are designed to pass water safely from the reservoir to the river downstream, while intakes to hydropower stations provide a similar function as a by-product of the power generation.

An overview of the hydraulics of different kinds of waterways and conduits is discussed in detail in Chapter 5.

To pass flows in a safe and controlled way, control must be maintained over the rate of flow, the physical conveyance of the flow, and the kinetic and potential energies of the flow. These controls are usually met by different parts of a spillway system: Intakes, with or without gates, usually control the rate of flow, while chutes, canals and tunnels transport the flow physically, and energy dissipation systems control the kinetic energy of the flow.

Different performance considerations, and therefore different base parameters, are of interest for each of these parts of the spillway system, and thus the system states that are placed in the different parts (as considered below). One example is concern over cavitation damages in the high-velocity and energy-dissipation zones.

Cavitation is caused by the implosion of water vapour bubbles carried out of zones where the water pressure has fallen below the partial pressure of water vapour. Waterways with high flow velocities may experience cavitation wherever flow conditions create zones of very low pressure. The potential for cavitation is typically described by a cavitation index $\sigma$:

$$\sigma = \frac{p - p_v}{(\rho U^2)/2} \tag{4.2}$$

where $p$ is the mean pressure at a point, $p_v$ is the partial pressure of water vapour at the ambient temperature, $\rho$ is water density and $U$ is a reference velocity, which for spillway chutes is often taken as the potential flow velocity. The cavitation index is a dimensionless value describing the ratio of the mean pressure margin (the excess of pressure over vapour pressure at a certain point) to the dynamic pressure. The cavitation index can also be expressed as a ratio of the pressure head margin to vapour over the velocity head. The cavitation index is compared with an incipient cavitation index $\sigma_i$ at which cavitation has been found to start (usually in laboratory experiments) for the specific flow conditions at hand. The incipient cavitation index is accordingly an empirical value. Cavitation is predicted if $\sigma \leq \sigma_i$.

#### 4.4.4.1 Probability of cavitation damage

Cavitation does not necessarily cause damage to the structures of a waterway. Cavitation occurring close to fixed boundaries may, however, rapidly induce material damage to the structures. In undisturbed flow, damage tends to develop at cavitation indices (far) below the incipient index (i.e. $\sigma \ll \sigma_i$) provided it occurs close to a fixed boundary.

A variety of factors influences whether damage actually occurs. Vortices and intense turbulence may lead to increased damage rates, as do isolated large roughness elements. Air content, time of exposure and material properties of the surfaces attacked are important factors. The important properties are resilience and tensile strength, which are significantly influenced by the possible presence of micro-cracks in the surface. Once damage has started to occur, other factors may speed up or slow down the development of the damage. Thus, the extent of cavitation damage is an uncertain quantity, and consequently the fragility of the chute or channel to $\sigma$ follows an S-curve as do other reliability-based predictions of performance.

Just as the cavitation damage rate is uncertain, so too may the pressure variations and the predicted cavitation index be at a certain system state (Figure 4.10). Thus the reliability relationship involves the convolution of uncertainty in the prediction of the cavitation index at a system state with the conditional probability of cavitation damage for various cavitation indices:

$$P(\text{cavitation damage}) = \int f(\sigma) P(\text{cavitation damage} \mid \sigma)\, d\sigma \tag{4.3}$$

As the simulation proceeds, a set of basic parameters is tracked at each system state where cavitation may be a concern (Figure 4.11). This set of basic parameters includes, as a minimum, the flow depth (excluding air), pressures, flow velocities and air content.

Figure 4.10 Mean pressure along a simulated spillway chute

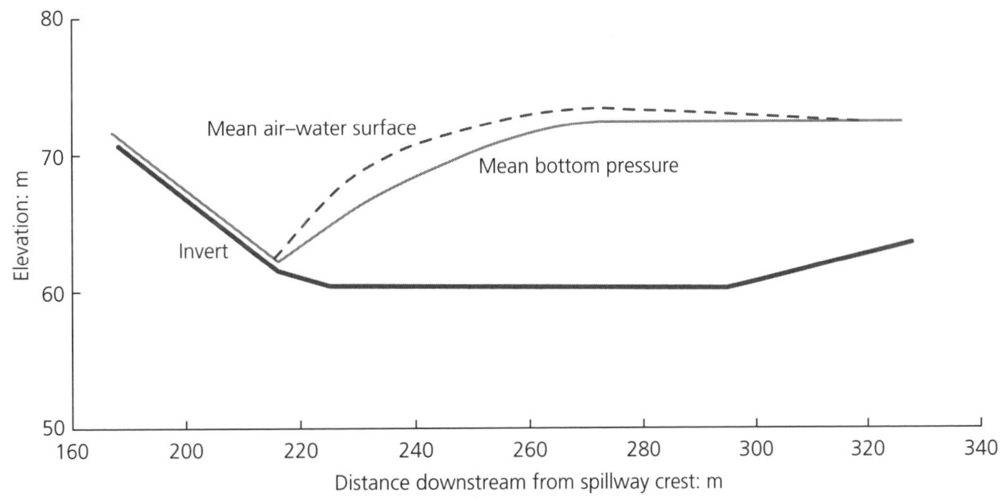

Figure 4.11 Cavitation potential index along simulated spillway chute

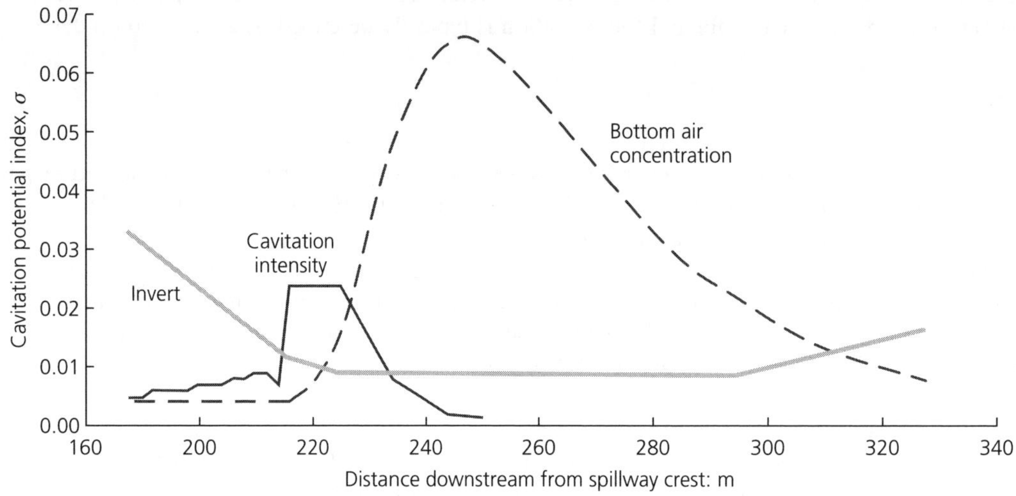


At any time in the simulation, the mean and variance of the cavitation index can be approximated by a simple first-order second-moment calculation of the moments:

$$E[\sigma] = \frac{E[p] - p_v}{(E[\rho]E[U]^2)/2} \tag{4.4}$$

$$\mathrm{Var}[\sigma] = \mathrm{Var}[p]\left(\frac{\partial\sigma}{\partial p}\right)^2 + \mathrm{Var}[\rho]\left(\frac{\partial\sigma}{\partial\rho}\right)^2 + \mathrm{Var}[U]\left(\frac{\partial\sigma}{\partial U}\right)^2 \tag{4.5}$$

where $E[\cdot]$ is the expected (mean) value and $\mathrm{Var}[\cdot]$ is the variance. $U$ might or might not be treated as uncertain. A distributional assumption such as Normal or logNormal is invoked to obtain $f(\sigma)$. The conditional probability of cavitation damage for a given value of $\sigma$ is found from a similar calculation or assessment of the probability distribution of $\sigma$, denoted by $f(\sigma)$. This assessment or calculation would typically be based on empirical correlations or expert judgement.

### 4.4.5 Disturbances

Sometimes in the course of reservoir operation an extraordinary event that is not hydrological in character occurs, such as an earthquake, an earth slide into the reservoir or conveyance works, a major inflow of debris, a major fire in the drainage (generating debris), or the like. Such disturbances, in principle, could include a broad variety of other unanticipated phenomena, both of natural origin (e.g. lightning strikes affecting power supplies or instrumentation) or of anthropogenic origin (e.g. the grid being unable to accept power or the transmission system being down, and thus the powerhouse waterway having diminished discharge capacity). Furthermore, some of these disturbances may not be probabilistically independent. For example, rainfall and debris may occur together, or reservoir landslides may cause debris.

Disturbances of the sort described are common in the long history of dam operations, although some of them may seem less probable for a specific dam or for a specific period of time, or to occur at the same time as some other type of disturbance. In the discussion to this point, extraordinary hydrological events such as extreme rainfall or snow melt have been treated as endogenous within the stochastic simulation for discharges in the river system, and have not been considered as disturbances in the sense of an exogenous event. The choice of where to draw this endogenous–exogenous boundary is a modelling issue.

A strength of the simulation approach to flow-control reliability is the relative ease with which disturbances such as seismic events, lightning strikes, grid unavailability or the like can be accounted for.

Many categories of disturbances can be included in the simulation approach. These can be simulated using, for example, a Poisson point process with a specified mean rate of occurrence, and overlaid on the underlying hydrological simulation (Figure 4.12):

$$p(n) = \frac{(\lambda t)^n n\, e^{-\lambda t}}{n!} \tag{4.6}$$

where $\lambda$ is the arrival rate per unit time, which might in turn be a function of other simulated variables such as discharge. Clearly, a disturbance such as a forest fires should be negatively correlated with rainfall and thus stream discharges. The

Figure 4.12 Stochastic model of 'disturbances' as Poisson point processes – here for seismic and fire events – overlain on a hydrological simulation of water flows

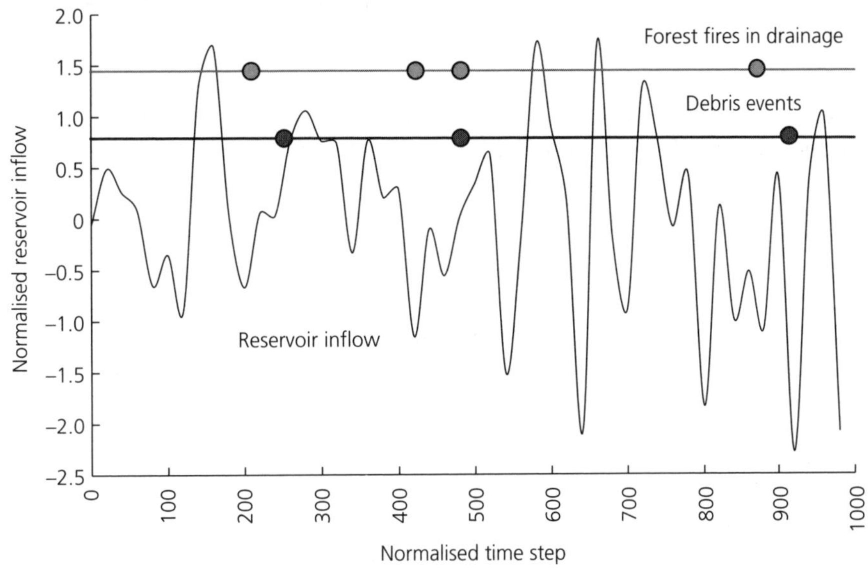


arrivals can also be generated including such a temporal correlation using, for example, a negative-binomial arrival model (Ripley, 2004).

The inclusion of disturbances within the simulation can affect systems performance in at least two ways. First, as in the upper part of Figure 4.13, the disturbance might affect the demand function on the flow-control system. For example, a seismic event could cause a slide into the reservoir, and consequently a wave load on the spillway outfall; or intense rainfall over a part of the catchment might increase the inflow of floating debris, overtax the debris boom and block the outflow openings. Second, as the in lower part of Figure 4.13, the disturbance might affect the capacity of some component of

Figure 4.13 Example of the effect of disturbances on flow-control reliability

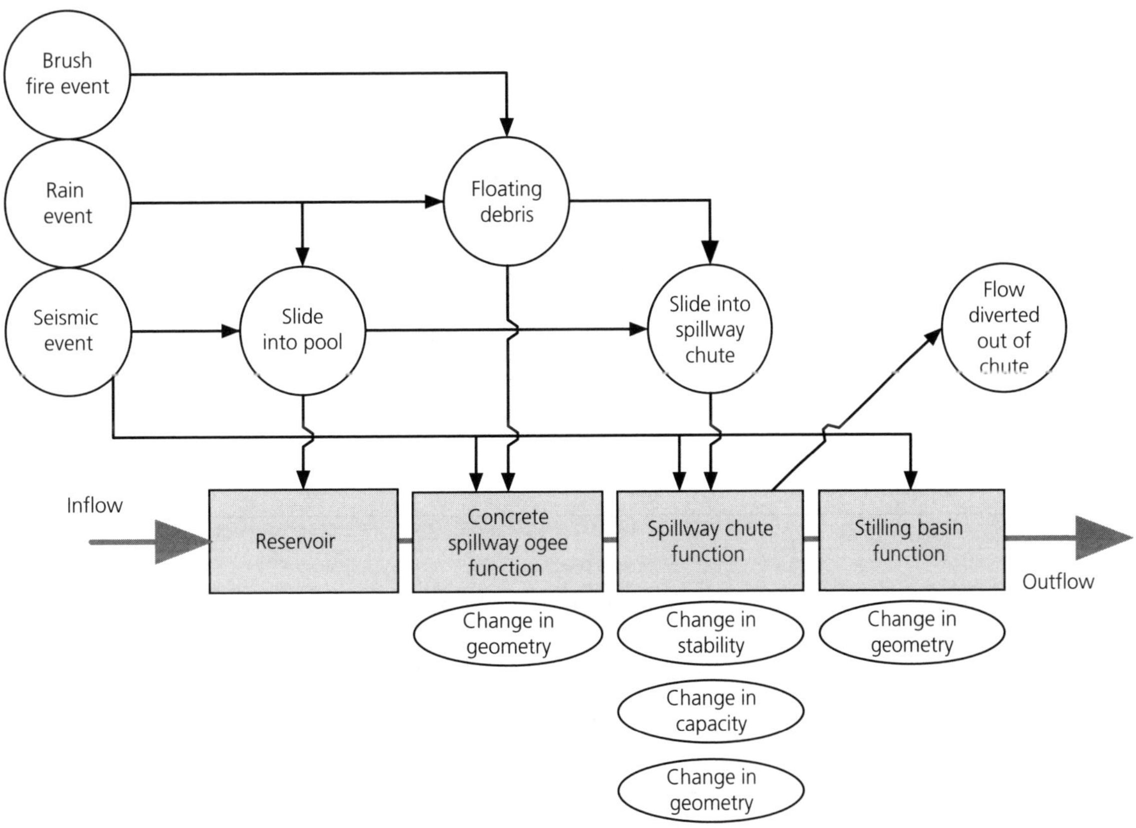

the flow-control system to withstand the demands placed upon it. For example, the seismic event might wrack a spillway gate, and consequently decrease the outflow capacity; or commercial factors or a disturbance to the power grid might mean that power cannot be generated, and thus the discharge capacity of the powerhouse waterway would be substantially reduced.

### 4.4.6 Sensors

The systems simulation model to this point deals with the reliability of the dam system and its flow-control aspects; that is, it targets the probability with which adverse behaviours occur, and what juxtaposition of behaviours is likely to lead to poor reliability. It deals with the natural, physical and mechanical character of the components of the flow-control system and their interactions. Other things, however, also affect the operational performance of a dam system, and these have to do with system monitoring and control.

We know from experience that the misfunctioning or misinterpretation of sensors is a not infrequent contributor to accidents and failures. The Taum Sauk failure is a recent example. A sensor might be located other than where the operator believes it to be, a sensor may return an inaccurate or biased signal, the operator may misread a sensor, or the reading may be misunderstood. There are many ways in which sensors can yield readings that lead the operator astray. Yet, this potential source of unreliable performance rarely appears in current dam safety risk analyses.

Because SCADA functioning has rarely been incorporated in current dam safety risk analysis, the modelling approach to this aspect of dam system performance is poorly explored. Of course, this is not true universally for other systems risk modelling, for example in application to chemical plant safety (AIChE, 2009) and nuclear power plant assurance (USNRC, 1999), so a starting point is available. Among the lessons learned in these other application areas is that the spectrum of errors that arise in sensor and SCADA systems is broad, and not all of these can be described and modelled in detail. As a result, analysts default to summary ways of modelling sensors by their frequency of erroneous readings, their possible inaccuracy (bias) and their possible imprecision (noise).

These summary measures intend to account for an essentially infinite number of specific ways that sensor errors arise, as, for example, in the case of Taum Sauk, yet without requiring that all those many specific errors be enumerated in detail. It is impossible to specifically list all the possibly sources of sensor error, first because their number is large, and second because they are, by definition, unanticipated. At Taum Sauk, the pool elevation sensor position had been displaced from where it was physically thought to be, and so the sensor generated water elevation measurements that were in error, unbeknown to the distant operator, due to the physical displacement of the device.

Analytically, the simple statistical model for sensor error is as depicted schematically in Figure 4.14. The error in a sensor measurement is characterised by a precision (usually as a mean square error, i.e. a standard deviation) and by a bias (i.e. an inaccuracy):

$$z = bx + \varepsilon \tag{4.7}$$

where $z$ is the measurement of some performance indicator $x$, $b$ is the bias is the measurement, and $\varepsilon \sim N(0, s)$ is a random zero-mean error term, usually taken to be Normally distributed with standard deviation $s$, although other similarly simple assumptions are sometimes made, for example, logNormal.

When the sensor is calibrated and operating correctly, the bias should be negligible, so the variation in the possible sensor readings is centred on the true value $x$. When a bias error occurs, with probability $p$, the variation in the sensor reading is offset by $b$. Thus, the sensor model is often taken to have the probability density function,

$$f(z) = \begin{cases} N(0, s), \Pr = (1 - p) \\ N(b, s), \Pr = p \end{cases} \tag{4.8}$$

Figure 4.14 Sensor measurement model

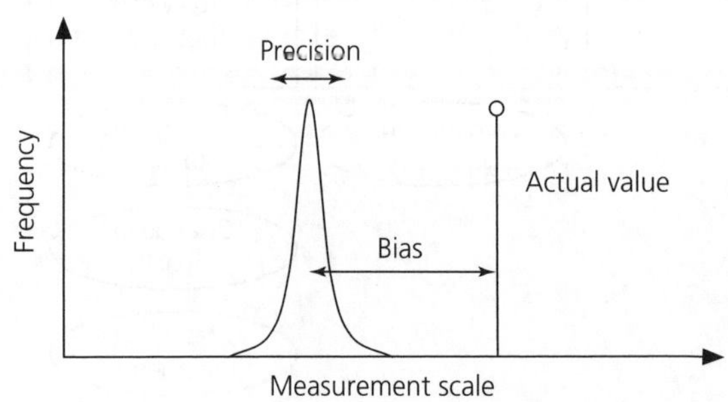

It is also possible that more than one bias might be anticipated, given a risk register of potential interferences with the sensor network, and thus Equation 4.8 might be multivariate rather than simply bivariate. While this model is simple, it captures a wide variety of possible sensor errors.

### 4.4.7 Communications and SCADA

In order that a human operator or an automated control system might act on the basis of a sensor reading, that reading must to be communicated to the operator, and presumably with little error. On site, a disruption may interfere with the SCADA system, for example through a computer error or failure. Electrical shorts or other mishaps might similarly affect the communication of sensor readings to an operator.

For non-locally operated facilities, the potential for breakdown in communications is potentially greater. Data must be communicated over distance via a grid or wireless link. This grid may be disrupted by electrical outages, weather events, landslides, accidents or many other disturbances. The default way of modelling breakdown in communication to a distant control centre is by a simple failure rate in time, as per Equation 4.6. This failure rate is correlated with other factors, time series or disturbances in the simulation, such as, for example, seasonality, rainfall or weather.

### 4.4.8 Control

Control functions in the context of systems simulation (Figure 4.2) are those feedbacks to the operating system from system monitoring data passed through operators or automated controllers. It is this loop of system monitoring to the control and back to the operating system that introduces strong non-linearity. A number of factors affect the performance and impact of controls on dam system performance, and many of these are not traditionally accounted for in reliability engineering. For example, among these are social and organisational factors in accidents, software errors, human operator reliability and adaptations over time.

In current risk analysis, event trees developed to explain an accident or failure usually concentrate on the proximate events immediately preceding the event. These will usually include factors such as those just mentioned, and a simulation will include probabilities or rates associated with these factors. Yet, the determinants of these factors may lie in decisions or policies or precursors laid long before the occurrence of the event. Figure 4.15 (ICOLD, 2010) suggests some of these factors in dam safety management programmes. For example, rates of human operator reliability surely depend on training, the adequacy of staffing, management policies and other factors that long pre-date the accident or failure. These may or may not be included in an analysis.

As reservoir systems become more complex and the monitoring and controls of those systems become more sophisticated and dependent on software, new threats to safety are emerging. Many of these are not represented in traditional reliability models like our flow-control simulation, and must be addressed in a separate or new manner. Examples of these new concerns include the following, although this is not an exhaustive list.

- *Increasing interactive complexity.* Reservoir and flow-control systems are becoming ever more complex, and the interactions between their many subsystems, instrumentation and controls is creating an interactive complexity that

Figure 4.15 Dam safety management system as proposed by ICOLD (2010)

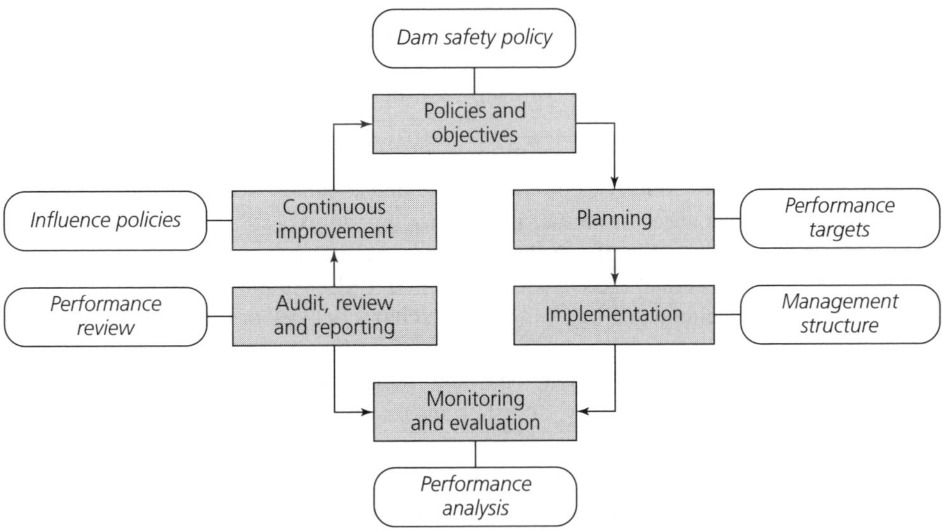

may be difficult for operators or even design engineers themselves to comprehend. As the number of subsystems increases, their interactions increase combinatorially. Automated control systems provide the operator with a way of dealing with this interactive complexity, but at the same time shield the operator from developing an intuitive understanding of the system functions.

- *Increasing dependence on information systems*. Computer and software systems are creating the potential for loss of information or incorrect information, which can lead to unacceptable physical, engineering or financial losses. This is coupled with the fact that engineered systems are becoming ever more complex, such that operators may no longer understand and be able to intuit the potential interactions between the system components, controls and software.
- *Complexity in the interactions between man and machine*. Automated control systems and SCADA systems are playing an increasingly important role in reservoir operations. Computer software is supporting human decision-making at ever higher and more complex levels. This is introducing new modes of maloperation (e.g. errors of omission versus errors of commission) (Sarter *et al.*, 1997).
- *Responding to the 'first one'* (Leveson, 2004). An extremely rare combination of natural events or circumstances is unlikely to have ever been experienced before in the operation of a particular reservoir system; this is true by definition. Learning lessons from a long operating history about how best to react to this particular rare combination of events is ineffective precisely because there is no such operating history. There needs to be an emphasis on mitigating the results of the first occurrence of rare combinations of events, which is not learnable just from the statistical record.

The hazards presented by these new conditions are not manifest in conventional reliability simulations and different ways are required for them to be analysed and understood.

## 4.5. Functional considerations

The simulation of stochastic reservoir inflows over time begins at time zero and proceeds initially in fixed time steps. The granularity of the steps (e.g. hourly, daily, monthly or annually) may vary from one simulation to another and be adjusted depending on the situation and the uses to be made of the results. The time step may also vary during the course of a single simulation, depending on the characteristic times of response of the flow-control system and in order to accurately capture the performance of the system. That is, the time steps within a single simulation might be nested.

### 4.5.1 Sampling the simulation

The trade-off consideration in setting simulation granularity is that one must balance short and explicit response modelling of the spillway and hydraulic structure, with its correspondingly intensive computation requirements, against long and implicit response modelling, with its corresponding loss of precision. For computational efficiency, the simulation might begin with daily or weekly increments of flow. For a 10 000-year time series, daily increments would generate slightly more than 3 650 000 time steps, which is computationally reasonable.

It is anticipated that, for most modelling situations, a nested sampling approach will be warranted. A nested sampling approach uses coarse granularity (say, daily or weekly time steps) as a backbone of the simulation, and switches to finer granularity (say, hourly or even smaller time steps) for critical periods, for instance when the gate position, and therefore the discharges, are changing quickly.

Daily or weekly time steps are inadequate during peak flows, when the system response at the spillway and outflow may depend on moment-by-moment operation. The nested approach uses coarse granularity in the simulation for intervals within ordinary limits of discharge, but converts to fine granularity for intervals during which some predefined threshold is exceeded (Figure 4.16). The concept of nested calculations is to provide efficient simulation most of the time, with greater precision but consequently less efficiency in those intervals during which adverse performances may occur.

Simulation continues until a total failure occurs in the model, regardless of how many years of simulation this takes. Design-life considerations do not enter the simulation (except when modelling assumed deterioration), as the goal is to estimate probabilities of specific sorts of performances in annual increments, not in expected life. The simulation is then restarted after each failure. By accumulating the number of specified performance failures or the number of physical failure events within a number of simulation years, the annual exceedance probabilities (AEP) of the demand function or of various descriptions of adverse performance can be accumulated and expressed as charts of AEP versus flow-control performance.

Repairable (partial) failures are modelled using a simple repair and replacement time during which the structure is either at decreased capacity or at a heightened state of fragility, and thus atypically prone to further damage, or both.

The principal flow that is tracked in the simulations is that of water. Nonetheless, other flows would need to be tracked, including floating debris and ice, other disturbances, sediments, human resources and information. These are considered further in Chapters 7 and 8.

Figure 4.16 Sampling the stochastic reservoir inflows

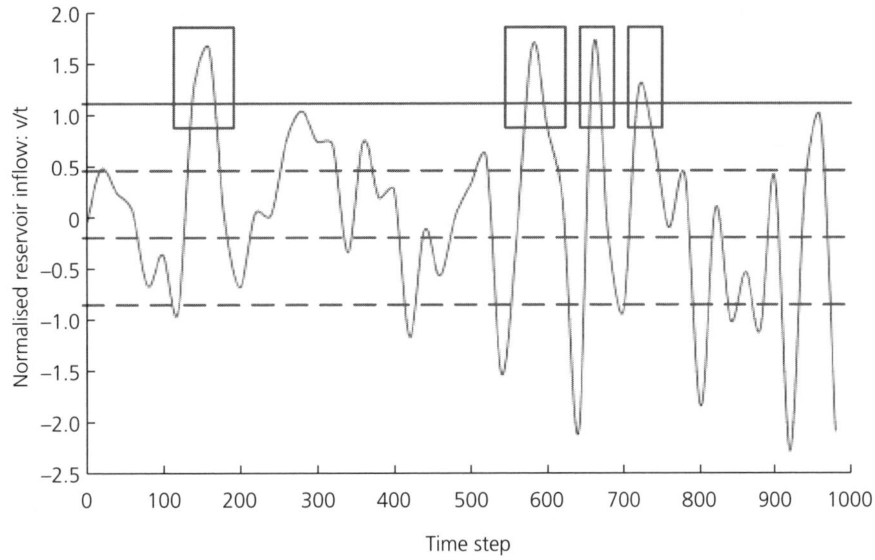


### 4.5.2 Combining engineering and economic simulation for power optimisation

The systems engineering approach to assessing flow-control reliability carries with it yet another potential benefit: Simulation modelling of multi-reservoir cascades has long been used to optimise hydropower generation (Joshi and Gupta, 2009), but usually without reference to engineering and operational reliability. The two aspects of reservoir operations and flow-control are, in fact, easily combined within a single simulation model to evaluate the impact of operational reliability on economic production.

Reservoir storage is used to regulate variable inflows to provide predictable and more uniform supplies for power generation, irrigation and navigation. Reservoir modelling (see e.g. Figure 4.16) is used to plan reservoir operating rules and plan management strategies (Wurbs, 2005). The inflows, outflows and storages of water in reservoir cascades are simulated presuming random time series of inflows, with the intent of maximising power production while satisfying the constraints created by dam safety requirements and other needs, for instance for water supply, navigation or recreation.

The development of simulation approaches to flow-control reliability analysis allows this aspect of operational analysis to be incorporated directly in the economic simulation analysis, expanding the evaluation of power and water supply reliability. In essentially all previous studies of reservoir systems reliability, the driver for availability (water or power on demand) has been limited to hydrological uncertainty. However, the many vagaries of engineering and flow-control operations also impact on availability, and yet have not been included due to the focus of reservoir cascade models only on inflows and outflows, with no consideration of engineering operations.

## 4.6. Tracking system states

The outputs of the simulated system are tracked at a set of specific and predefined locations where performance is judged to be important (Figure 4.17). These are called 'system states'. For example, the trash-racks upstream of power intakes could be identified as an important location to be tracked in the simulation, as debris or ice may accumulate there and affect the capacity in terms of both discharge and power. Similarly, critical points along the outflow conveyances could be identified as important locations to be tracked, as cavitation or erosion damages, or high or low water levels could cause structural failure or wear.

### 4.6.1 System states

These specific and predefined critical locations are the critical points – and the only points – at which systems performance is monitored in the simulation analysis. It is at these points that the output of the reliability model is tracked over time as the performance of the flow-control system is simulated. These system states satisfy a sort of '100% rule' as do work breakdown structures in project management: 100% of the performance of concern in the simulation modelling is contained in the system states, and if some performance is not represented at a system state it is not part of the simulation output.

The locations of system states from the outfall intake to the energy dissipation works for a simple free-surface overflow spillway is illustrated schematically in Figure 4.17. In a typical simulation analysis there could be a score or more system states, possibly hundreds, depending on the complexity of the system and the refinement of the analysis.

Figure 4.17 System-state concept of spillway function. $E_w(t)$, reservoir elevation at time $t$; $Z_g(t)$, opening of gate at time $t$; $E_o(t)$, tailwater elevation at time $t$

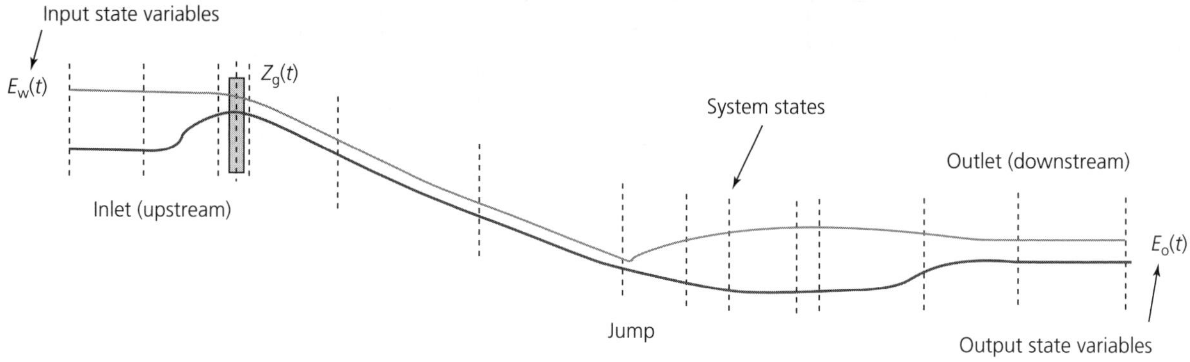


For each system state an FMEA is conducted to identify ways in which the flow-control performance at that point could be inadequate or adverse. For example, along the outflow conveyance the combination of flow velocity, water depth and bottom air content might be such to encourage cavitation. This location would be tagged as a system state, and a set of engineering parameters or metrics identified and tracked at this location.

The system states and failure modes listed in Table 4.3 are suggestive of the sorts of locations and metrics that might be tracked in the modelling. Obviously, the actual conditions to be tracked is a decision for the modelling effort. The base parameters summarise fundamental performance metrics (e.g. water depth and velocity). Other parameters may include environmentally important measures (e.g. tailwater discharges involving temperature and turbidity) or structurally important parameters (e.g. vibration parameters at the overflow crest).

### 4.6.2 Engineering performance

Engineering models are used to calculate and track the various base parameters at each defined system state. In principle, these engineering models are those commonly used in analysis and design for strength, deformation and flow predictions. Nothing precludes these from being complex engineering models, although from a computational point of view they will often be straightforward so that their outputs can be repeatedly calculated as the simulation proceeds.

The base parameters are tracked through each cycle and for each system state (Figure 4.18). This generates a matrix of parameter values for each system state, one dimension being time in the simulation and the other being the base parameter value. Overall, the output of the simulation is a three-dimensional matrix, the third dimension being the failure-state number. These time series of base parameter values are used in the component reliability analysis to determine whether a component is operating normally, adversely or has failed at a particular time.

Table 4.3 FMEA identification of failure modes for specific system states

| System state | Failure modes | Performance variables |
|---|---|---|
| Intake structure | Structural vibrations cause damage<br>Debris blocking<br>Ice blocking | Intensity of vibration<br>Characteristic frequencies<br>Debris loads<br>Ice loads |
| Conveyance structure (chute) | Cavitation of concrete lining<br>Uplift of concrete pad<br>Overtopping of conveyance sidewalls | Hydraulic pressures<br>Air contents<br>Uplift pressures<br>Water flow heights |
| Energy dissipation structure | Energy not dissipated<br>Structural damage to dissipation system | Hydraulic energies<br>Pressures<br>Water heights |
| Downstream outflow | Turbulent flow released downstream<br>Energetic flow causes downstream erosion | Energy in flows<br>Erosive potential<br>Water temperature<br>Sediment load |

Figure 4.18 Tracked system state metrics at each system state

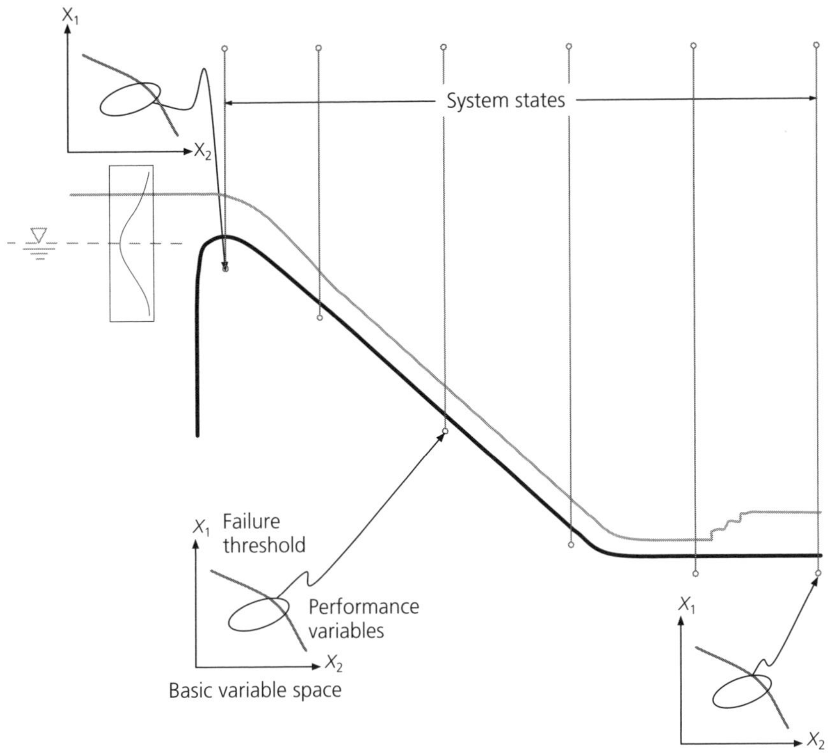


### 4.6.3 Reservoir outlets, spillways and gates

The outlet, spillways and gates have been traditionally viewed from the perspective of structural mechanical–electrical reliability considerations. They comprise, at a minimum, the following:

- structural subsystems
- mechanical subsystems
- electrical subsystems
- controls subsystems.

The controls subsystems include any SCADA technology that may be present. Reliability block diagram analysis and fault-tree methods have been the preferred approaches to qualifying the performance of these systems, whereas other methods of reliability modelling might be used to develop a fragility curve for hydraulic structures or geotechnical system states, and block diagram or fault-tree modelling might be used for the mechanical and electrical components of gate and control systems (Figure 4.19).

These systems are also those for which human factors are important. The human factors, in a broad sense, include the control room operations of people and equipment, operating rules and procedures, and the supervisory system (real-time interfaces, forecasts of reservoir inflows, etc.).

## 4.7. Component reliability

Performance variables are tracked at each system state. For example, at a system state defined at a point along the outflow conveyance these performance variables might be discharge, flow velocity, flow depth and bottom air content. From these, other indicator variables would be calculated and added to the database, for example changes in structural state due to wear or cavitation damages.

### 4.7.1 Defining and identifying adverse performance

The purpose and method of the modelling will determine what is to be considered as adverse performance or failure for each particular system state. A particular failure mode will be *loss of control*, in the sense that the performance or the conditions have passed beyond that which can be described and evaluated by the model used.

### 4.7.2 Limiting states

Limiting states are specified for each performance variable, or vector of performance variables, and at each system state. Typically, two limiting states may be considered: an ultimate limit state, and a serviceability limit state (Melchers, 1987).

Figure 4.19 Reliability analysis of electoral power support for an individual gate component. (Courtesy USACE)

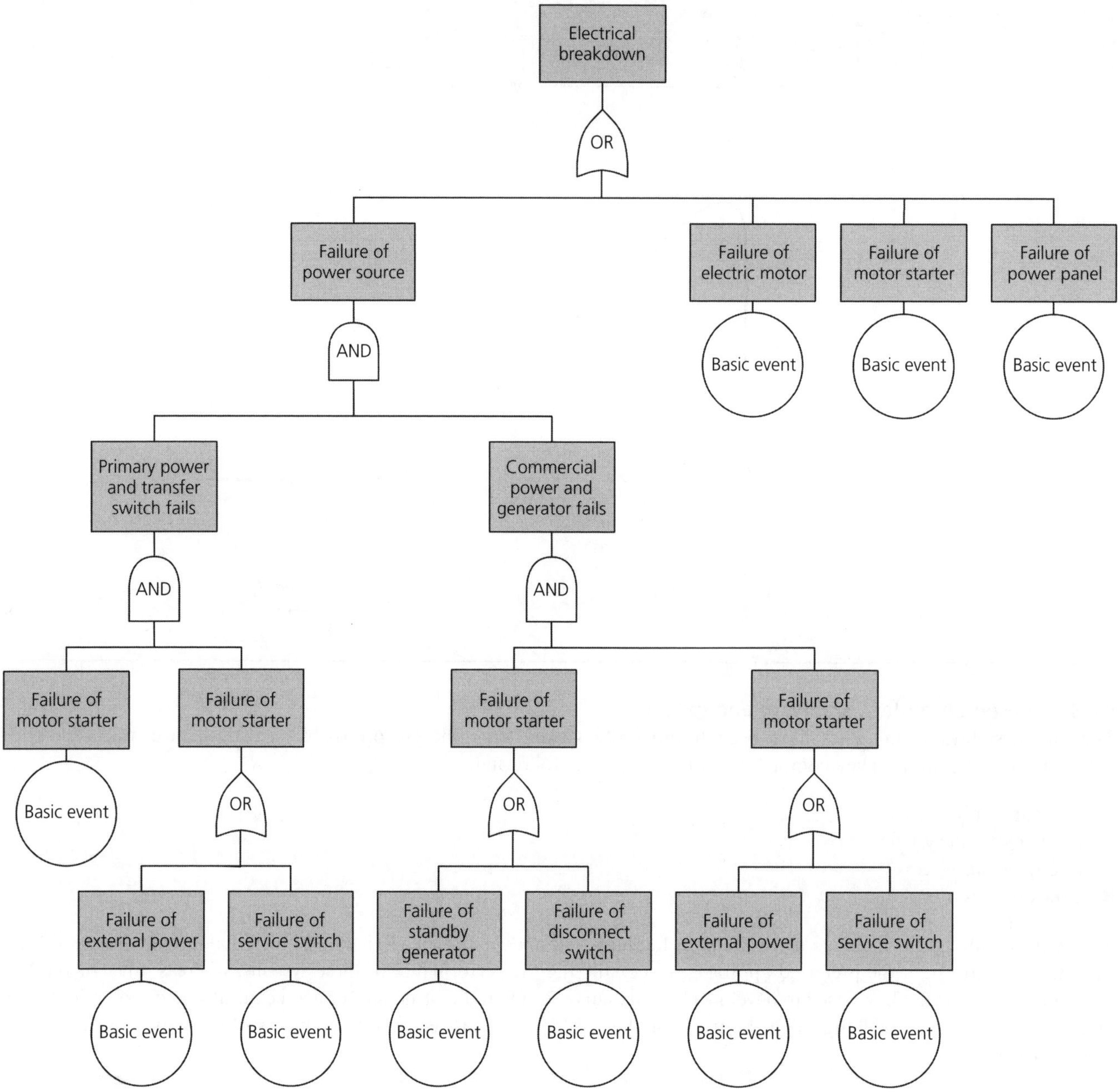


A limit state is a set of performance criteria (e.g. vibration levels, deflection, strength, stability, buckling, twisting and collapse) that must be met when a component system is subject to loads. The ultimate limit state is associated with failure (e.g. structural collapse), while the serviceability limit state is associated with routine performance (e.g. deformations).

To satisfy the ultimate limit state, the system must not be significantly damaged when subjected to simulated loads. A structure is deemed to satisfy the ultimate limit state criteria if all applied loads are below the resistance calculated for the system state under consideration.

Returning to the example of cavitation, the ultimate limit state could be defined as the condition that the cavitation index is less than some critical value $\sigma \leq \sigma_{i_i}$. The frequency with which this criterion is violated in the simulated results is an estimator of the probability (or return period) of adverse cavitation performance (damages).

To satisfy the serviceability limit state criteria, the system must remain functional for its intended use subject to routine loading. A system is deemed to satisfy the serviceability limit state when its components do not deflect by more than certain limits specified by the design requirements, vibrations remain within predetermined criteria and other possible requirements are met (Figure 4.20).

Figure 4.20 Limiting states for performance variables tracked at each system state. (a) Simulated realisations of the system state at the limiting state boundary indicate adverse performance or failure. (b) Evolution of the limiting state at one system state due to deterioration or other change over time, load history or other factor

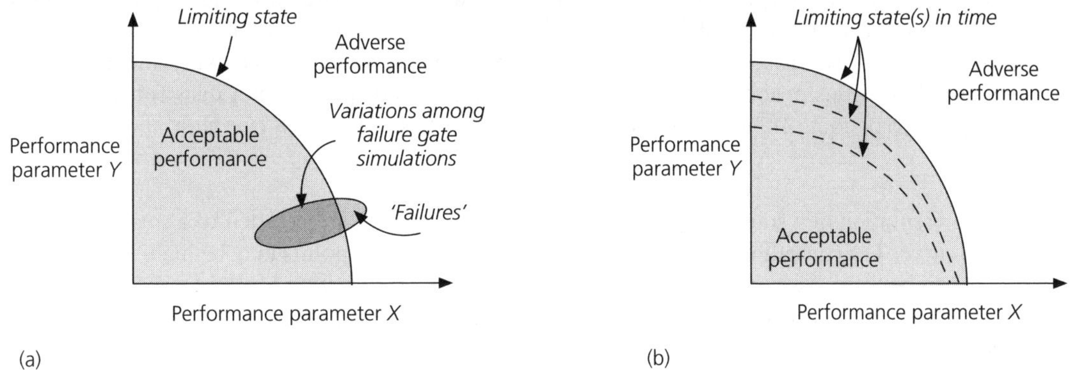


The frequency of simulated realisations lying outside the limiting-state boundary indicates the frequency of adverse performance or failure. Based on the number of simulated trials, the years of simulated performance and the frequency of excursions of the limiting state(s) within the simulations, statistical sampling theory is used to infer annual exceedance probabilities of 'failure' or adverse performance (Ditlevsen, 1981, 1996).

### 4.7.3 Probabilities of adverse performance

Reliability analysis deals with the balance of demands on a system and the capacity of the system to withstand those demands. Both the demands and the capacity may be uncertain, so the result of their interaction is also uncertain. This relationship is used to generate a fragility curve for a particular component. The fragility curve relates the demand function on the component to the probability of adverse performance (Figure 4.21).

Three computational approaches to calculating the probability of adverse performance or failure from the means, variances and covariances of the underlying demands and capacity are in common use: the error propagation (sometimes called the first-order second-moment) method (Cornell, 1969), the point estimate method (Rosenblueth, 1975), and the first-order reliability method (Hasofer and Lind, 1974). The present methodology may use any of these three calculation methods, and may mix them for convenience.

In essence, the simulations proceed by tracking indicator variables at each system state. For any interval of time, the values of these variables are compared with one or more limiting state functions for the system state. If the variable has an excursion outside the limiting-state boundary, an adverse performance or failure is recorded for that simulation at that time interval.

Another strength of the simulation approach is that it allows for the limiting state at any system state to change or evolve over time. For example, the concrete lining of an outflow conveyance may deteriorate over time, or it may be damaged by earlier peak discharges over the spillway.

Figure 4.21 Reliability analysis of an individual component based on upstream demand function

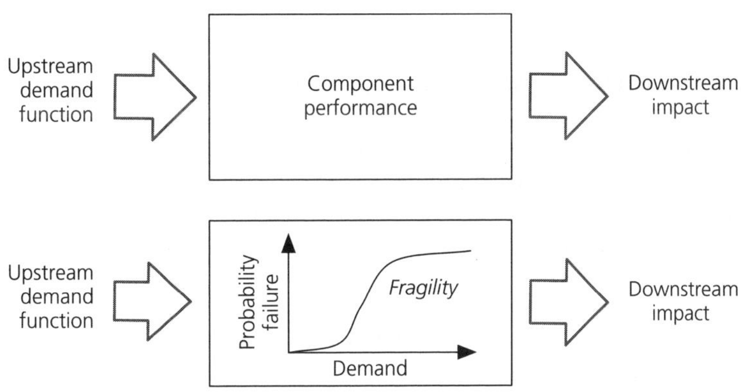

### 4.7.4 Replacing complex reliability calculations by response surfaces

Some of the component reliability analyses will become complicated and computationally intensive. This is true of many of the mechanical and structural calculations, and of at least some of the geotechnical and hydraulic calculations. Many of these reliability calculations may require the use of fault- or event-tree models, finite-element models or other detailed numerical models. Given the large number of simulated iterations required to infer low probabilities of systems failures, it will be computationally impractical to use these complicated and computationally intensive reliability models in the simulation itself. The practical solution to this problem is to pre-compute sample sets of parameter values for the reliability analysis, and to interpolate the corresponding fragility relations as response surfaces. This is a common approach in large systems simulations.

A response surface representation of a fragility curve is, in essence, a regression surface fitted to a small number of calculated points on the fragility curve. The computational cost of calculating each of these points may be high, but once they are calculated an interpolation can be fitted (i.e. regressed) to these points to create, in essence, a look-up table that can be quickly and efficiently used within a simulation involving repeated sampling. This dramatically reduces the computational burden on the simulation while sacrificing only a modest level of accuracy.

Response surface methodology (RSM) is a development of statistical practice, generally attributed to Box and Wilson (1951). The response surface approximates the relationship, typically between several 'independent' variables and one or more performance responses. For example, in the case of a vertical lift gate, the variables might have to do with the strength or performance characteristics of the mechanical and electrical components, while the response might be availability on demand. RSM uses a set of designed test cases or 'experiments' to generate an output, and then fits a mathematical surface, such as a second- or third-degree polynomial, to these data. The resulting surface is an approximation of the more complexly and expensively obtained performance data, which might only be modelled using, say, fault-tree analysis or finite-element models, and would therefore be too computationally intensive to include in a simulation involving hundreds of thousands of iterations.

### 4.7.5 Systems reliability

The reliability of the entire flow-control system, and its annual exceedance probabilities of failures in various modes, is estimated by cycling the simulation repeatedly, ten-thousand times or more, until adverse performance or failure is recorded at one or more system states. The frequencies of adverse performance or failures within the many simulated years are used as the basis for estimating annual probabilities of different sorts of performance.

The simulation commences with upstream flows entering the reservoir. As discussed above, these are simulated as a pure stochastic process, or through some at least partly physics-based hydrological model with random time-series inputs. For each time increment in the simulation, the (now deterministic) reservoir inflow is routed through the reservoir to the different waterways. This generates a demand function of water head and other variables on these waterways and other system components such as instrumentation and operations. For each system state in the system, all the associated base variables are updated and stored.

The fragility of each component is compared with the relevant demand function to generate a probability of adverse performance or failure. The performance (realisation) at the system state is calculated. This performance variable is compared against criteria to categorise it as being within operating requirements, adverse but repairable or a failure. A decision is made on whether or not a failure is of sufficient severity to terminate the simulation.

Failures that are not considered severe enough may reduce the capacity or change the function of the relevant component. This consideration is necessary for the modelling of sequential and common-mode failures.

## 4.8. Interpreting the results of simulation models

The reliability model focuses on the question of whether the flow-control system as implemented is fit for use. *Fitness-for-use* means that the design, implementation and the operational processes used for flow control are effective in delivering a well-functioning system that suits the operator's defined needs under specified operating conditions. A system that is fit for use may or may not be 'safe', depending on how it is operated and the functional or policy regime within which it is operated. In this context, 'safe' means that incidents, whether or not arising from the unreliability of the physical system, occur with low frequency.

The purpose of the simulation modelling is to improve our understanding of whether a dam system is both fit-for-use and safe. How is this done? Whether the system is fit-for-use is the somewhat easier of the two to investigate, because it depends on expected conditions. 'Safe' is the more difficult, because it depends on interactions between unusual conditions.

### 4.8.1 Fitness-for-use

The analysis of fitness-for-use involves simulating a dam system presuming that its design, implementation and operation adhere to a set of presumptions. That is, we take the configuration of the system as true, and the dam system to be operating in an expected way. Even though the process of monitoring and feedback through controls may introduce non-linearities in dam system performance, and even though disturbances may perturb that performance, still the simulation focuses on proximate events and interactions. This reduces the scope of the analysis, as well as the vagaries of possible causal chains leading to accidents and failures. The focus is on the rates of accidents and failures, not on the patterns of causal chains that lead to those accidents and failures.

As posited in Section 4.1, the function of a dam system is to retain water volumes for productive uses and to pass flows in a controlled way. Therefore, the question of fitness-for-use is whether the dam system reliably accomplishes these two things. One can define limiting states of the system in response to these two functions and test by simulation how frequently the dam system fails to achieve one or the other, or both. These limiting states need not be numerous. The limiting states for retained water volume derive from the set of productive uses intended for the dam system, and whether when the respective demands occur there is water volume to meet them (e.g. power generation, navigation and recreational use). The limiting states for passing flows in a controlled way derive, typically, from downstream damage criteria, and how frequently damages occur to dam structures (e.g. cavitation in the stilling basin) or downstream channels (e.g. failure to meet environmental flows).

Simulation provides a large number of possible realisations of the dam system performance over time. These can be thought of as a statistical sample of the universe of possible time streams of performance. Rates of the dam system exceeding the defined performance limiting states can be estimated using familiar statistical methods. These are discussed in Chapter 10.

### 4.8.2 Safety

The investigation of safety is more complicated than for fitness-for-use because it involves simulating the dam system to uncover adverse patterns of behaviour. The configuration of the system is not necessarily taken as true, and nor is the operation of the system presumed to be as expected. The simulation does not focus only on proximate events and interactions, but on chains of causal events leading to the accident or failure. These chains of events may depend on factors at the management or policy level, such as training or reducing operator numbers. The goal is to tease out and to understand the causal chains. This is more complicated than estimating simple rates. The focus is on patterns of causal chains that lead to accidents and failures.

Recognising incidents is the same as understanding fitness-for-use. By a defined set of limiting states, and probably the same set of limiting states. But now the goal is to recognise patterns and to identify recurring causal chains in a large number of simulations. This is a much different challenge. The chains are not all exactly the same, because while the patterns may be similar the details differ. The goal now is to recognise patterns of behaviour rather than probabilities of specific incidents or failures.

The probability of observing a specific causal chain of events leading to an accident or failure is small because the joint probability of any particular set of such events happening in the exact sequence needed to cause a particular incident or failure is small. This remains a challenge.

#### REFERENCES

AIChE (American Institute of Chemical Engineers) (2009) *Guidelines for Chemical Process Quantitative Risk Analysis*. AIChE, New York, NY, USA.

ANCOLD (Australian National Committee on Large Dams) (1994) *Guidelines on Risk Assessment 1994*. ANCOLD, Hobart, Tasmania, Australia.

ANCOLD (2003) *Guidelines on Risk Assessment*. ANCOLD, Hobart, Tasmania, Australia.

Bowles DS (2007) Dam safety governance and management. In *Dams – Securing Water for Our Future*. Proceedings of NZSOLD ANCOLD 2007 Conference, Queenstown, New Zealand. Institution of Professional Engineers New Zealand, Wellington, New Zealand.

Box G and Wilson K (1951) On the experimental attainment of optimum conditions. *Journal of the Royal Statistical Society, Series B – Statistical Methodology* **13(1)**: 1–45.

Brown T, Beyeler W and Barton D (2004) Assessing infrastructure interdependencies: the challenge of risk analysis for complex adaptive systems. *International Journal of Critical Infrastructures* **1(1)**: 108–117.

CDA (Canadian Dam Association) (2007) *Dam Safety Guidelines*, CDA, Montreal, Canada.

Cornell CA (1969) *Structural Safety Specifications based on Second-moment Reliability Analysis*. See http://www.e-periodica.ch/cntmng?pid = bse-re-001:1969:4324 (accessed 01/06/2016).

Defra (Department for Environment, Food and Rural Affairs) (2013a) *Guide to Risk Assessment for Reservoir Safety Management. Volume 1: Guide*. UK Environment Agency. London, UK.

Defra (2013b) *Guide to Risk Assessment for reservoir Safety Management. Volume 2: Methodology and Supporting Information.* UK Environment Agency. London, UK.
Ditlevsen O (1996) *Structural Reliability Methods.* Wiley, Chichester, UK.
Ditlevsen O (1981) *Uncertainty Modeling with Applications to Multidimensional Civil Engineering Systems.* McGraw-Hill International, New York, NY, USA.
EU (European Union) (2013) Directive of the European Parliament and of the Council concerning measures to ensure a high common level of network and information security across the Union. European Commission, Strasbourg, France.
Hartford DN and Zielinski PA (2013) Credibility and defensibility of risk assessment in dam safety. In *81st ICOLD Annual Meeting*, Seattle, WA, USA.
Hasofer AM and Lind N (1974) An exact and invariant first-order reliability format. *Journal of Engineering Mechanics, ASCE* **100(EM1)**: 111–121.
Hobbs G and Azavedo D (2000) Spillway gate reliability. ANCOLD Annual Meeting, Cairns, Queensland, Australia.
HSE (Health and Safety Executive) (2001) *Reducing Risks, Protecting People – HSE's Decision Making Process.* HMSO, London, UK.
ICOLD (International Commission on Large Dams) (2005) *Risk Assessment in Dam Safety Management. A Reconnaissance of Benefits. Methods and Current Applications.* ICOLD Bulletin 130. ICOLD, Paris, France.
ICOLD (2010) *Dam Safety Management: Operational Phase of the Dam Life Cycle.* ICOLD Bulletin 154. ICOLD, Paris, France.
Joshi GS and Gupta K (2009) A simulation model for the operation of multipurpose multireservoir system for River Narmada, India. *Journal of Hydro-Environment Research* **3(2)**: 96–108.
Leveson N (2004) A new accident model for engineering safer systems. *Safety Science* **42(4)**: 237–270.
Leveson N, Dulac N, Marais K and Carroll J (2009) Moving beyond normal accidents and high reliability organizations: a systems approach to safety in complex systems. *Organization Studies* **30(2–3)**: 227–249.
Melchers RE (1987) *Structural Reliability: Analysis and Prediction.* Wiley, Chichester, UK.
NCS (National Communications System) (2004) *Supervisory Control and Data Acquisition (SCADA) Systems.* NCS, Arlington, VA, USA.
Nelson P (2009) *Complex Infrastructure Systems.* New Jersey Institute of Technology, Newark, NJ, USA.
Norman DA (1981) Categorization of action slips. *Psychological Review* **88(1)**: 1–15.
OAG Canada (Office of the Auditor General of Canada) (2012) Protecting Canadian critical infrastructure against cyber threats. In *2012 Fall Report of the Auditor General of Canada.* OAG Canada, Ottawa, Canada.
Reason JT (1990) *Human Error*, 1st edn. Cambridge University Press, Cambridge, UK.
Reason JT (1997) *Managing the Risks of Organizational Accidents*, 1st edn. Ashgate, Farnham, UK.
Regan P (2010a) Dams and civil structures: An examination of dam failures vs. the age of dams. *Hydro Review*, 1 June. See http://www.hydroworld.com/articles/hr/print/volume-29/issue-4/articles/dams–civil-structures.html (accessed 01/06/2016).
Regan P (2010b) Dams as systems – a holistic approach to dam safety. In *30th Annual USSD Conference on Dams: Collaborative Management of Integrated Watersheds*, Sacramento, CA, USA.
Regan P (2012) Dam safety. In *First STAMP/STPA Workshop at MIT*, Cambridge, MA, USA. See http://csrl.scripts.mit.edu/home/stampstpa-workshop/materials (accessed 01/06/2016).
Regan P (2013) *Combined Events: A Systems Perspective.* US Federal Energy Regulatory Commission, Division of Dam Safety and Inspections. See http://www.nrc.gov/docs/ML1305/ML13059A397.pdf (accessed 01/06/2016).
Ripley BD (2004) *Spatial Statistics.* Wiley-Interscience, New York, NY, USA.
Rosenblueth E (1975) Point estimates for probability moments. *Proceedings of the National Academy of Sciences USA* **72(10)**: 3812–3814.
Ross SM (2006) *Simulation*, 4th edn. Academic Press, London, UK.
Sarter NN, Woods DD and Billings CE (1997) Automation surprises. In *Handbook of Human Factors and Ergonomics* (Salvendy G (ed.)). Wiley, New York, NY, USA, pp. 1926–1943.
USACE (US Army Corps of Engineers) (2001) *Reliability Analysis of Navigation Lock and Dam Mechanical and Electrical Equipment.* USACE, Washington, DC, USA.
USBR (US Bureau of Reclamation) (2003) *Guidelines for Achieving Public Protection in Dam Safety Decision Making.* USBR, Denver, CO, USA.
USNRC (US Nuclear Regulatory Commission) (1999) *Framework for Risk-informed Regulation in the Office of Nuclear Material Safety and Safeguards.* USNRC, Washington, DC, USA.
Wolfram S (2002) *A New Kind of Science*, 1st edn. Wolfram Media, Champaign, IL, USA.
Woods DD, Decker S, Cook R, Johannesen L and Sarter N (2010) *Behind Human Error*, 2nd edn. Ashgate, Farnham, UK.
Wurbs RA (2005) Modeling river/reservoir system management, water allocation, and supply reliability. *Journal of Hydrology* **300(1–4)**: 100–113.

# Part II

# Inputs and constraints

**Operational Safety of Dams and Reservoirs**
ISBN 978-0-7277-6121-7


http://dx.doi.org/10.1680/osdr.61217.107

# Chapter 5
# Hydrological modelling and river flow routing

Modelling of river systems has a long tradition in the water management and environmental management areas. A watershed or a catchment has always been at the centre of modelling attention due to its complexity in responding to external and internal inputs. Other components of a river system have always been perceived (with justification) as an easier modelling challenge. This chapter does not attempt to fully characterise specific modelling methodologies, rather the focus is mostly on the philosophical and scientific approaches to modelling of river systems and the implications of that philosophy on achieving the broader objective of characterising the reliability of a flow-control system – a small part of the dam system and an even smaller part of the entire river system.

As such, this chapter serves as a guide to selecting appropriate modelling approaches, and provides references to specific methods, techniques and examples of models currently used in applications. The selection of a particular modelling approach or a specific modelling method must depend on the modelling objective.

## 5.1. River systems modelling

There is no single definition of a river system, and definitions may change depending on the purpose of the study being undertaken. Dunne (2007) provides a comprehensive definition of a river system as seen by geomorphology. This follows the general principles below.

Rivers are complex systems, through which irregular fluxes of water and mobile terrestrial materials derived from the lithosphere, atmosphere, biosphere and technosphere are focused:

- The dynamics of the transport, storage and interactions of these materials creates channel and valley-floor environments with which the river continually interacts, creating certain functions and environmental conditions.
- The resulting functions and environments change both gradually and episodically due to both external forcings and internal dynamics.
- Ecological changes (including human exploitation) include both successional changes and perturbations of various intensities, which may re-set or replace the succession.
- The continual creation results in spatial and temporal complexity (rather than a single continuum of environments linked by transport).

In water resources management an equivalent to a river system is a water resources system, which can be defined as a collection of various elements that interact in a logical manner and are designed in response to various social needs in the development and improvement of existing water resources for the benefit of human use (Nandalal and Simonovic, 2003). The identification of a water resources system that follows this definition can become a complex endeavour. Nandalal and Simonovic point out that a system does not have to be related to any specific size, purpose or context but its essence can be better understood by the description of relevant aspects that need to be considered (e.g. political, administrative, legal, economic, social, environmental, financial, atmospheric and hydrological).

What is of interest in applications of the systems approach to regulated river networks are the models of river systems. A model is only a physical or mathematical representation of the system itself and of the interrelations and interactions between the system elements. As such, the models are only simplified and idealised abstractions of reality, and usually do not describe the entire reality. But this should not necessarily disqualify the model, as the purpose can be characterised as achieving only prescribed accuracy within predefined time and budget constraints.

In this chapter a river system model is developed along the following general principles:

- A system can be defined as a collection of interrelated elements purposely working toward achieving some common objective.
- Most natural systems or systems that include natural systems can be characterised as dynamic systems because the collection of elements constantly interacts over time.
- The elements of the system can be subdivided into three groups: components, attributes and relationships. These three groups can be characterised as:
  - *Components* or subsystems, which are the operating parts of a system and consist of input, process and output. Components can take the form of natural or technological artefacts. Each component may assume a variety of values defining the system state.
  - *Attributes*, which are the properties or discernible manifestations of the system, its components and the relationships between them.
  - *Relationships*, which are the links between components and attributes. The properties and behaviour of each component of the system influence the properties and behaviour of the system as a whole. At the same time, each component depends on the properties and behaviour of at least one other component in the system. Because of this interdependence, the components cannot be divided into independent subsets; the system is more than the sum of its parts. Also, if a component is removed from a system or if its characteristics change, the characteristics of the other artefacts in the system will alter accordingly, and the relationships between them may also change (Carlsson *et al.*, 2002).

River networks, whether in natural state or regulated (with man-made structures), can be understood as systems. Behaviour of the river system can be characterised by quantities that vary over time (water levels, flows, velocities at different places within the system, etc.). The components of a river system are:

- reservoirs providing storage behind the dams
- natural and artificial channels
- diversion and control structures (dams)
- confluence points
- local sub-basins (sub-watersheds).

It is important to understand at this point that such a definition of the model of a river system is not universal and may be inadequate for purposes that differ from the purpose of this book. It is obvious that if the river system under consideration includes dams, the system is a regulated river system and the purpose of the regulation is to satisfy the needs of various water uses. These uses may include hydropower generation, navigation support, water supply for agricultural and industrial or commercial use, drinking water supply and many others.

Depending on the objectives of the modelling endeavour the model may include or exclude certain elements of the system. For example, in the river systems for which the only objective of regulation is the support of navigational needs by flow augmentation, the elements of the system must include reservoirs and connecting river channels. Depending on the configuration of a particular river network it may also include confluence points (if there is more than one branch) and diversion or control dams. Depending on the way the hydrology of the system is undertaken it may also include local sub-basins supplying local inflows to the most upstream reservoirs, local inflow between reservoirs, confluence points and diversion or control dams. An example of such a river system is depicted in Figure 5.1.

If, on the other hand, the purpose of regulation is the allocation of available water resources to industrial, agricultural and municipal water users, additional elements of the river systems have to be included. These elements may include individual water users and intake and return points.

As the present objective is aligned closely with a particular subsystem of the component 'reservoir' (i.e. spillway system), the impact of water users on flow conditions within the entire river system is assumed to be negligible and is disregarded in further considerations. However, the general framework described in Chapter 4 can be easily modified to include those water users that impact flow conditions within the river system in a significant way.

The diagram in Figure 5.1 is a network model of the river system. In the real system the local sub-basins provide inflows to reservoirs. The reservoirs are thus among the nodes of the network. Other nodes include diversion or control dams, and those locations within the system where two or more sources of flow meet. In the diagram these points are denoted as confluence points. All nodes (with the exception of the most upstream reservoirs and control or diversion dams) have at least two arcs connected to them. The arcs characterising interconnections between components of the system represent physical fluxes (streamflow rates) between the nodes. As all components in the system (reservoirs, dams, channels and sub-basins) are physical artefacts, the continuity law has to be obeyed at all nodes.

Figure 5.1 Schematic diagram of a simple river system

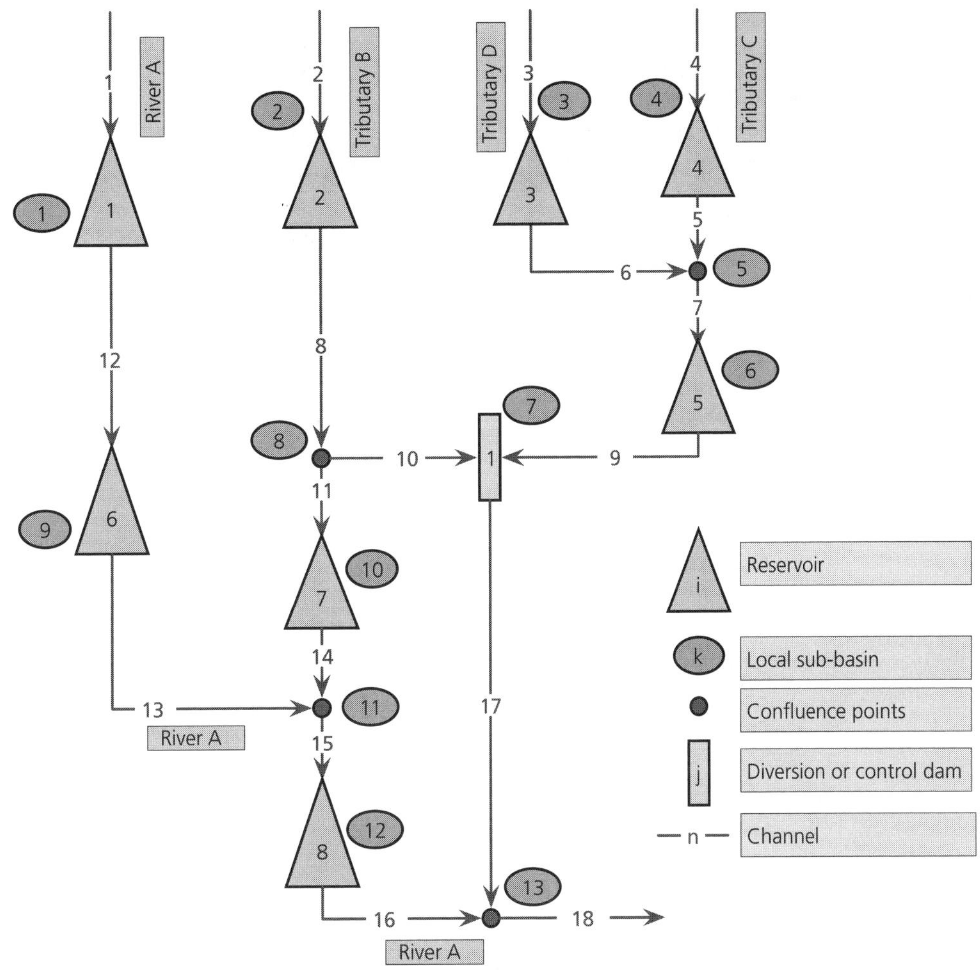


It is evident that any reservoir is interconnected with other components, a network of natural and artificial channels and other reservoirs. The collection of all these subsystems (or components) constitutes the river system of interest. Each component can be broken down into lower level components (or subsystems). For example, a reservoir can be further subdivided into reservoir rim, block dams and levees, and the main dam. The components of main dam, in turn, can include earth embankments, gravity walls, spillway system and powerhouse. Each of these components can be further subdivided into smaller and smaller subsystems. The river system analyst has to decide on the level to which the analysis is going to be limited, and for our purposes the level at which the reservoir constitutes the lowest level of disaggregation will be appropriate. Thus the schematic diagram plays the role of a template for any river system.

In some applications for which control and optimisation problems are being considered, as opposed to simulation of how the system works, additional nodes may be desirable (Figure 5.2). These nodes representing logical components of the system can be introduced in order to provide a means of modelling the flow of information through the system. Logical components describing the functions of the real system in an aggregated way are also connected with other components, but what flows through the arcs of the network now is information, and thus the continuity law does not apply.

Identification of all components of the river system together with the topology of interconnections is the first step in developing the model of a river system. Once this has been done, the identification of the component models begins.

### 5.1.1 Reservoir modelling

The 'reservoir' component is a complex subsystem in which a number of interacting processes take place. The attributes of a reservoir can include a large number of quantities characterising both physical properties and processes occurring within the component. Not all of the properties and the processes are relevant to the objectives of a particular modelling endeavour, and the system analyst has to make a decision which attributes should be included in the component model. The relevant attributes of the reservoir may include inflow, water level, discharge, chemical properties of water, bathymetry, water temperature and others. The first three of these are necessary in any model that is concerned with the flow of water through the river system.

Figure 5.2 Schematic diagram of a river system with water users

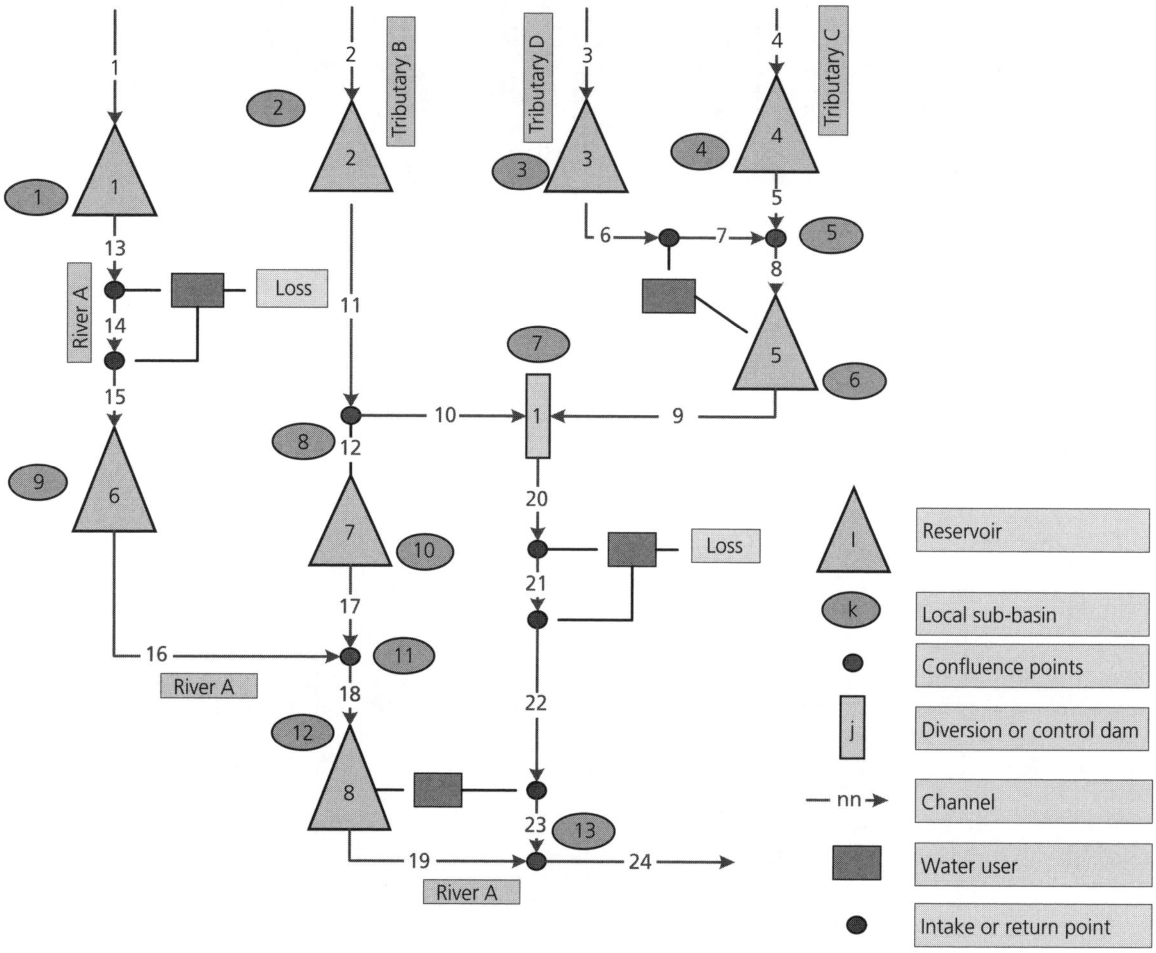


*Water level* (denoted by $h$) can be defined is the water level measured at a specific gauge and at a specific time. Often in the past, measurements at hydrometric stations were carried out daily at a specific hour (for example, 9:00 AM). Today, water level measurements are often continuous.

*Inflow* (denoted by $q$) can be defined as total inflow to the reservoirs from all sources. Streamflow, like many other quantities observed in natural systems, is always continuous in time. However, the sampling of not only inflows but all other attributes usually takes place at discrete time instants. For this reason, it is convenient to build the discrete-time model in which the time parameter $t$ represents sampling or decision time instants.

The inflow $q_{t+1}$ represents the amount of water entering the reservoir within the time interval $[t, t+1)$. Often it is convenient to separate the streamflow entering the reservoir into two parts: the flow released from the upstream reservoir (denoted by $u$) and the flow resulting from the runoff occurring in the local sub-watershed between the reservoirs ($w$).

Following this notation and accounting for continuity, inflow can be restated as the sum of discharge and storage.

*Discharge* (denoted by $d$) is the total flow released from the reservoir through all spillway facilities, turbines of the hydropower station (if present) and locks (if present). Similarly, discharge $d_{t+1}$ represents the amount of water released from the reservoir within the time interval $[t, t+1)$.

*Storage* (denoted by $s$) is the volume of water stored in the reservoir.

The general structure of the reservoir model can be developed following the methodology outlined in Soncini-Sessa *et al.* (2007). It begins with the construction of the *causal network* that explains the relationships between reservoir attributes. As these attributes are represented by physical quantities, they can also be called the 'model variables'. The causal network of a reservoir is shown in Figure 5.3.

**Figure 5.3** Causal network of a reservoir

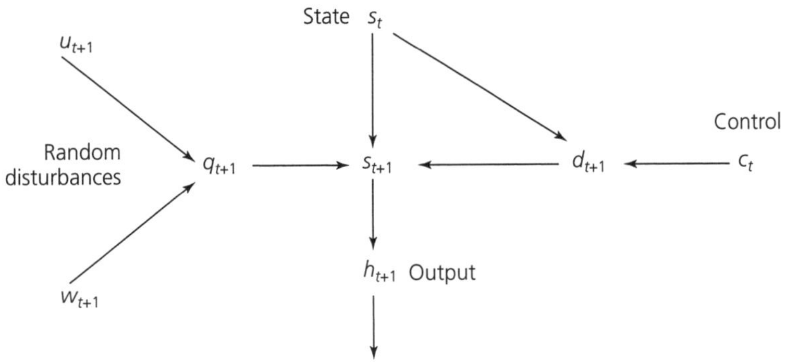


In developing the causal network, the notation will follow Soncini-Sessa *et al.* (2007), and subscripts referring to time for all variables denote the instant of time within the interval $[t, t+1)$ when the variable assumes a deterministic value. Thus water level will be denoted by $h_t$ because it is known (measured) at time $t$. Inflow within the interval $[t, t+1)$ is not known at time $t$ and can be known only when it has been realised at the end of the time interval, and for that reason will be denoted by $q_{t+1}$.

The variables depicted on the diagram of the causal network (see Figure 5.3) play different roles in modelling the behaviour of the reservoir and can be classified accordingly to their roles. These roles (with the exception of storage) can change depending on the purpose of the modelling exercise.

Storage has a unique property, as its value $s_{t+1}$ at time $t+1$ can be calculated from its value $s_t$ at time $t$ using the following formula:

$$s_{t+1} = s_t + q_{t+1} - d_{t+1} \tag{5.1}$$

When a variable at any instant of time depends on its value at the previous instant, it is called a *state variable*. It describes the state or, in other words, the present condition of the component. It is determined from its past history and is necessary in order to predict the future. Therefore, storage is a state variable in the reservoir model.

The water level $h_{t+1}$ at time $t+1$ can be determined from the storage $s_{t+1}$ using either the stage–storage relationship or from information about the bathymetry of the reservoir and topography of the reservoir rim. As the reservoir is regulated, decisions about the amount of the water to be released are also to be taken at discrete time intervals. These decisions are reflected in the causal network by a *control variable* $c_t$.

Inflow from the local sub-basin is not known at time $t$ and is denoted in the network by $w_{t+1}$. Similarly, the outflow from the upstream flow controlling structure (reservoir, diversion or control dam) is also unknown at instant $t$ and is therefore denoted by $u_{t+1}$. Such designation for $w$ and $u$ may appear in contradiction to the designation for the control variable $c$. However, although the decision about the amount of water to be released (variable $c$) is known at time $t$, the actual amount released during the interval $[t, t+1)$ may be different due to unavailability or reduced capacity of flow discharge facilities.

The control variable $c_t$ together with variables $u_{t+1}$ and $w_{t+1}$ do not depend on other system variables and have to be provided externally. For that reason they are called *exogenous variables* or *inputs*. The control variable $c_t$ is different from the other inputs because it is under direct control of the reservoir manager or regulator. The other two inputs are independent of the manager, and for this reason are called *disturbances*. In the casual network in Figure 5.3 both disturbances are non-deterministic because both are realised during the time interval $[t, t+1)$.

The total inflow to the reservoir $q_{t+1}$ and discharge from the reservoir $d_{t+1}$ are called *internal variables* because their role is only to retain values that are used for calculation of the state variable $s_t$.

Finally, the model of the reservoir constructed with the help of the causal network can be described by the following:

$$s_{t+1} = f(s_t, c_t, w_{t+1}, u_{t+1}) \tag{5.2}$$

$$h_t = g(s_t) \tag{5.3}$$

$$u_t \in U(s_t) \tag{5.4}$$

Figure 5.4 Causal network of a reservoir with a powerhouse and a bottom outlet

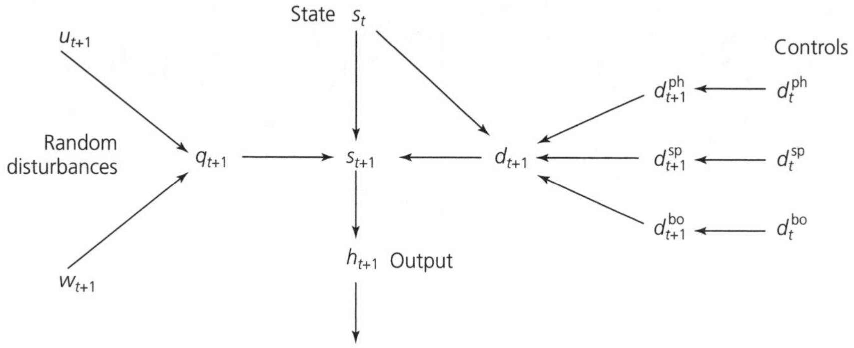


Equation 5.2 provides the relationship between the future state of the system and the present state and inputs, and is called a *state transition function*. Equation 5.3 expresses the output as a function of the states, and is called *output transformation function*. The constraint (Equation 5.4) defines the feasible controls as a function of the system state.

The causal network for a reservoir becomes more complicated if the dam has more than one discharge waterway. For example, the dam may have a powerhouse and a bottom outlet. If the amounts of water discharged through turbines, spillway and bottom outlet are denoted by $d^{ph}_{t+1}$, $d^{sp}_{t+1}$ and $d^{bo}_{t+1}$, respectively, then the diagram can be modified as shown in Figure 5.4.

### 5.1.2 Channels

The simplest model of a natural channel or an artificial canal can be built using the following causal network (Figure 5.5).

The inflow $p^{in}_{t+1}$ at the upstream end of the channel together with the inflow $q_{t+1}$ from the local watershed between the upstream and downstream end of the channel equals the outflow $p^{out}_{t+1}$ at the downstream end. This simple model can be described by the following equation:

$$p^{out}_{t+1} = p^{in}_{t+1} + q_{t+1} \tag{5.5}$$

If the time lag that accounts for the time $\tau$ necessary for a volume of water entering the channel to reach the downstream end of the channel is included, Equation 5.5 can be modified as:

$$p^{out}_{t+1} = p^{in}_{t-\tau+1} + q_{t+1} \tag{5.6}$$

More sophisticated techniques that account for the fact that time lags usually vary with the amount of water entering the channel, and which are capable of reflecting more realistically the dynamics of water flow through channels, are discussed in Section 5.4.

### 5.1.3 Diversion structures

The role of diversion dams in the river system is to redirect the flow of water by using the back water effect into a side channel, which is usually an artificial canal. The storage created by diversion dams is usually small and can be neglected in systems analysis. The regulation of the flow through the diversion dam is often passive and the amount of outflow is controlled by the head over the spillway crest. Sometimes the amount of flow through the dam is regulated with the help of flow discharge equipment. For the latter type of diversion dam the casual network can be described by the following system of formulae:

$$p^{div}_{t+1} = \min[u_t, (p^{in}_{t+1} - p^{env}_t)^+, p_{max}] \tag{5.7}$$

$$p^{out}_{t+1} = p^{in}_{t+1} - p^{div}_{t+1} \tag{5.8}$$

$$u_t \in U_t \tag{5.9}$$

Figure 5.5 Causal network of a channel

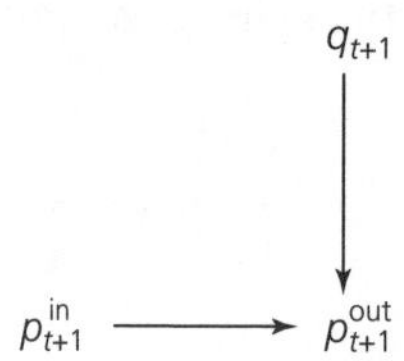

Figure 5.6 Causal network of a diversion dam

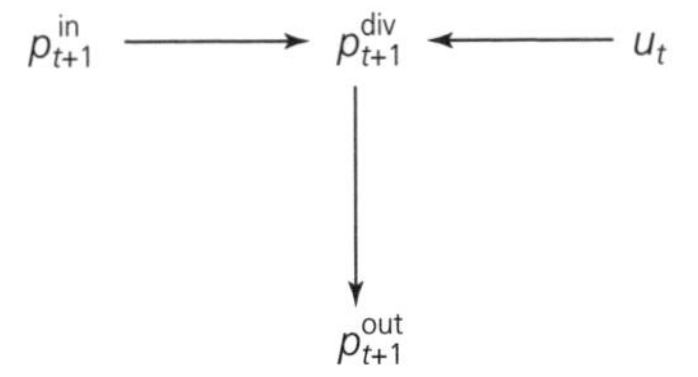


in which $p^{in}_{t+1}$ is the inflow to the diversion dam, $p^{div}_{t+1}$ is the diverted flow, $p^{out}_{t+1}$ is the outflow from the diversion dam, $u_t$ is the diversion decision, $p^{env}_t$is the minimum environmental flow and $p_{max}$ is the maximum flow that can be diverted. The operator $(\cdot)^+$ returns the argument for positive values and zero otherwise.

The causal network diagram for an unregulated diversion dam is different only in that the decision $u_t$ and the connecting arc are removed (Figure 5.6). The formulae simplify to:

$$p^{div}_{t+1} = \min[\,(p^{in}_{t+1} - p^{env}_t)^+, p_{max}] \tag{5.10}$$

$$p^{out}_{t+1} = p^{in}_{t+1} - p^{div}_{t+1} \tag{5.11}$$

### 5.1.4 Confluence points

The confluence point or confluence junction where $n$ channels meet is described by the following algebraic continuity equation:

$$p^{out}_{t+1} = \sum_{j=1}^{n} p^{in,j}_{t+1} \tag{5.12}$$

### 5.1.5 Local sub-basins

Local sub-basins are simple watersheds, but watersheds are the most complex elements of any river system. A watershed model is a specific kind of hydrological model that integrates component sub-models of individual hydrological processes over the entire watershed area. From the perspective of river system modelling, the quantity characterising a single local watershed is the streamflow that flows through the end of a cross-section of the river draining the watershed.

The amount of water flowing through the outlet of a watershed is only one of many quantities that are the outputs of watershed modelling. Although watershed models are capable of providing the answer required by a river system model, there are also other approaches that are characterised as synthetic streamflow generation and can be used as an alternative to watershed models. The current state of the art in watershed modelling is summarised in the following section, which provides brief descriptions of alternative ways of characterising local inflows to sub-basins.

## 5.2. Hydrological modelling

Hydrological models are now considered as indispensable tools in understanding, describing, simulating and predicting the dynamic behaviour of watersheds. The models can be designed for different purposes, which may include understanding how individual subprocesses act and interact, operational flow forecasting, synthetic streamflow generation for water resources planning, development and management, or advancement of hydrologic science.

The origins of mathematical modelling of watershed hydrology can be traced to the development of the Stanford Watershed Model (SWM) by Crawford and Linsley (1966). In the early 1960s the SWM was instrumental in introducing the civil engineering profession to the concept of mathematical hydrological modelling. In describing the motivation of these two pioneers of hydrological modelling, Crawford and Burges (2004) wrote:

> The catalysts in 1959 for the first steps toward hydrologic simulation modelling at Stanford were frustration and resignation. The frustration came from unsatisfactory pencil-and-paper predictions of peak flows on small watersheds that were based on watershed characteristics. Since the plan for predicting small watershed peak flows didn't work by trying co-axial correlations, Linsley suggested the last dregs of the grant money be spent to see if the 'digital computer' in the Electrical Engineering Department could be used.

First reactions to this novel approach to hydrology beyond Stanford were mixed, and the model was viewed as primarily a new research tool with limited opportunities for application. This situation changed quickly, and the following years witnessed rapidly growing interest in developing and applying computerised mathematical models of watershed hydrology. Over the

**Figure 5.7** Classification of models. (After Singh, 1995)

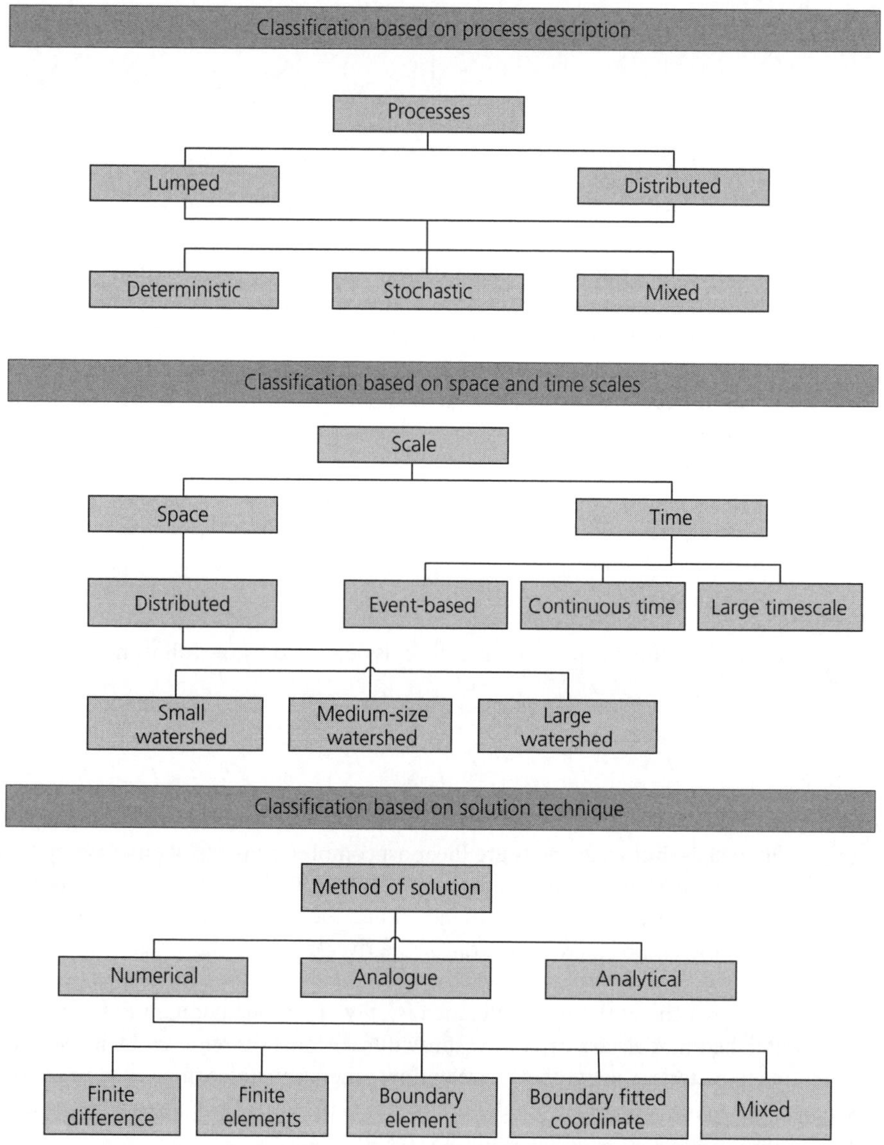


next few decades a large number of mathematical models for watershed hydrology were developed. A detailed characterisation of the most popular models can be found in Singh (1995) and Singh and Frevert (2006).

Different models were developed for different purposes and were based on different assumptions, leading to the classification by Singh (1995) shown in Figure 5.7. The classification is based on different criteria accounting for process description, scale and technique of solution.

Hydrological models are instrumental in improving the understanding of and in predicting and simulating the behaviour of hydrological systems. However, models only provide an approximation of a natural system. The processes occurring in the watershed are complex. The inputs forcing the response of the system are not completely known and are usually difficult to measure. The same applies to the characteristics of the system. Dynamic aspects of the system behaviour and the non-stationary nature of most of the processes complete this complicated picture.

### 5.2.1 Modelling philosophy

A hydrological system, similar to other natural systems, is a complex assembly of interacting physical processes and physical components, which can be characterised as a non-linear dynamic system. Historically (keeping in mind that this history is limited), the developments in hydrological modelling were following both deductive and inductive rules of scientific inference.

The first, the deductive rules, called by Young (2002) a hypothetico-deductive approach, are based on an *a priori* defined model structure established on the basis of the current hydrological paradigms and the perception of the modeller. The hypothetico-deductive approach has been popular since the 1970s, and is based on the preconception that the hydrological system can be described sufficiently well through a deterministic mathematical framework built on the first principles of physics. This point of view is not uncommon among modellers of natural systems because, as pointed out by Koussis (2008): 'The ability to describe a physical phenomenon deterministically fascinates us, perhaps, because such a mechanistic thinking has dominated the education in the natural sciences of many of us'.

On the contrary, the inductive approach does not use theoretical or personal preconceptions, but uses available data and information to infer the structure of the model. The interpretation of the model in physical terms, with or without agreement with the current science of hydrology, follows the identification phase. It is, perhaps, interesting to note that the inductive approach in hydrology precedes the hypothetico-deductive approach and can be traced to the development of the rational method in the middle of the 19th century (Mulvany, 1850), an example of empirical modelling based on observed data.

The hypothetico-deductive approach to hydrological modelling represents the deterministic reductionism philosophy, which presupposes that a watershed model can be built on the basis of detailed characterisation of various physical components and processes occurring within the watershed, and an *a priori* perception of how these components are linked and how processes interact. Sivapalan *et al.* (2003) indicate that the models developed using this approach are often called 'process based' or 'physically based', and that the development approach is characterised by using descriptors such as upward, mechanistic, bottom-up or reductionist. The essence of the approach is based on two principles:

- individual model components and processes are selected or omitted based on an *a priori* perception of which of the components and processes are relevant
- the structure of the interactions and interrelationships between the individual components and processes selected is also based on an *a priori* perception of the modeller.

Thus the focus of the mechanistic approach is on individual components and processes. The expectation is that if the individual segments of the model reflect the hydrological functioning of individual components and processes sufficiently well, the model will be a reliable tool characterising the response of the entire watershed to forcing inputs. Such an approach is in contrast to general systems theory, which focuses on the interactions, feedbacks and functional relationships between model components, and develops the individual components and processes only to the extent necessary to provide an accurate response of the system as a whole.

The systems approach was initially incorporated in hydrological modelling as early as the 1970s (Yevjevich, 1971), but it was restricted to defining input–output relationships from observed data, thus focusing on the behaviour of the analysed watershed as a whole. The approach initiated by Yevjevich gained popularity during the 1980s and 1990s, when a large number of complex statistical techniques commonly addressed as time-series analysis of hydrological data were developed.

The inductive approach to hydrological modelling, dominated for a long period of time by the deterministic reductionism approach, came back in the late 1980s in the form of a downward or top-down approach. Klemeš (1983) said that it 'starts with trying to find a distinct conceptual node directly at the level of interest (or higher) and then look for the steps that could have led to it from a lower level'. Thus the downward approach attempts to characterise the hydrological functioning of the watershed through the interpretation of the observed data. It can be contrasted with the upward approach by rephrasing the two principles governing the upward approach as follows:

- individual model components and processes are selected or omitted based on observed data for the watershed for which the model is built
- the structure of interactions and interrelationships between the individual components and processes selected is also based on the observed data.

Clearly, these two approaches use the data collected from observation differently. In general, the deductive approach is based on a preconceived structure of the model and uses data only for model calibration. The inductive approach uses observed data to infer the model structure that is to determine which components and processes to include and how these components and processes should interact with each other within the model.

### 5.2.2 Watershed modelling

The classification of hydrological models provides a general idea of the breadth of possibilities for model construction and development. A further discussion of the classification of models is developed in Wheater *et al.* (1993). Although the classification is for runoff–rainfall models, it can be extended to other classes of hydrological model. The five broad categories of watershed models are described below.

#### 5.2.2.1 Metric models

Metric models, often also called black-box models, are based on observations and characterise the system (watershed) response using available data. Wheater *et al.* (1993) indicate that the origin of metric models derives from the development of unit hydrograph theory for watershed-scale simulation, but this category also includes a wide variety of time-series models, which are described in more detail in Section 5.4.

Metric models are of inductive type; they seek to establish functional relationships approximating or fitting the available data. The simplest empirical models based on regression analysis can be significantly less complex than their physical and conceptual counterparts. More advanced empirical models that utilise artificial neural networks or synthetic regression through genetic algorithms are also available.

A detailed description of metric models – beginning with the time area concentration curve developed by Hawken and Ross (1921) and ending with geomorphological unit hydrograph and fuzzy systems based models – can be found in the World Meteorological Organization's *Manual on Flood Forecasting* (WMO, 2011).

#### 5.2.2.2 Conceptual models

Conceptual models represent the hypothetico-deductive philosophy of modelling. They attempt to describe hydrological processes with the help of simplified mathematical equations derived from physically based equations governing hydrological processes. Depending on the degree of simplification, a significant number of model variables and parameters may lose physical meaning.

Often, conceptual models have a mixture of physically based components for some of the modelled processes and empirical relationships describing the remaining processes. It is the presence of empirical relationships with parameters representing conglomerates of watershed properties that precludes parameters being determined from measurable physical properties.

The World Meteorological Organization characterises conceptual models as a network of interconnected stores, tanks or layers that are constantly recharged or depleted by various processes of the rainfall–runoff transformation system. The resulting runoff components are subsequently routed and combined to provide the estimated the outflow. A generic conceptual model usually starts with the pre-routing runoff production based on water balance, followed by the routing components responsible for slow- and fast-response routing.

Most conceptual models are of the *lumped parameter* form (Table 5.1), which means that their parameter values are spatially averaged (often the same spatial averaging approach applies also to inputs, such as precipitation or temperature). The *semi-lumped parameter* form of conceptual models involves disaggregation of the watershed into sub-basins, each with spatially averaged inputs and *semi-distributed* form, where spatial averaging is applied to model parameters for individual sub-basins (Table 5.2). Finally, if spatial disaggregation is extended further and each sub-basin has separately averaged inputs and parameters the model is of the *distributed parameter* type (Table 5.3).

The major difference between metric and conceptual models is that for conceptual models the structure is predefined in the form of major internal storage components (Eisenbies *et al.*, 2007).

Some examples of conceptual models from the report by Barrett *et al.* (2008) are described below.

The first example is the Sacramento Soil Moisture Accounting Model (Figure 5.8), a conceptual rainfall–runoff watershed water balance model working with daily, 6-hourly or 1-hourly time steps. The model contains two surface stores with surface evaporation, surface runoff and interflow, and three base flow stores representing soil evaporation and two stages of base flow. The original version of the Sacramento model had 16 parameters that can be estimated from soil moisture observations and soil-type data. The model is available in either lumped or distributed form.

The Australian Water Balance Model (Figure 5.9) is a conceptual rainfall–runoff water balance model working with an hourly or daily time step. The model calculates the moisture balance for three separate surface stores. At each time step, rainfall is added to each of the three surface moisture stores and evapotranspiration is subtracted. Surface flow occurs when the surface store is full. The model contains five stores comprising three surface stores, a base flow store and a surface runoff routing store.

The third example is the Soil Moisture Accounting Runoff (SMAR) model (Figure 5.10), a lumped conceptual hydrological water balance model. The model provides daily estimates of surface runoff, groundwater discharge, evapotranspiration and leakage from the soil profile for the catchment as a whole. The surface runoff component comprises overland flow, saturation excess runoff and saturated through-flow from perched groundwater. The SMAR model consists of two components in

**Table 5.1** Selected lumped-parameter watershed models

| | Model | | |
|---|---|---|---|
| | Identification of unit hydrographs and component flows from rainfalls, evaporation and streamflow data (IHACRES)[a] | Snowmelt runoff model (SRM)[b] | Water balance model (WATBAL)[c] |
| Temporal scale | Flexible (minute+) | Day | Day+ |
| Spatial scale | Flexible | Flexible | Flexible |
| Processes modelled: | | | |
| ■ Event simulation | Yes | No | No |
| ■ Continuous simulation | Yes | Yes | Yes |
| ■ Snow accumulation and melt | No | Yes | Yes |
| ■ Interception and infiltration | No | No | No |
| ■ Evapotranspiration | Yes | No | Yes |
| ■ Reservoir routing | No | No | No |
| Cost | Public domain | Public domain | Public domain |
| Set-up time (short/medium/long) | Short | Medium | Short |
| Expertise (low/medium/high) | Low | Medium | Low |
| Technical support | Possibly from the author<br>Workshops in the past | Message board | No support |
| Documentation (bad/medium/good) | Good | Good | Good |
| Ease of use (easy/medium/difficult) | Easy | Medium | Easy |
| Advantages | ■ Easy to use<br>■ Low data requirements<br>■ Both event and continuous simulations | Latest version allows modelling of the effect of climate change on the runoff regime for the entire hydrological year | ■ Easy to use<br>■ Developed for climate change impact studies<br>■ Some parameters can be optimised |
| Disadvantages | ■ Snowmelt not modelled<br>■ Evapotranspiration cannot be computed internally (temperature can also be used to estimate evapotranspiration effects) | ■ Developed for basins with dominant snowmelt runoff<br>■ No event simulations | ■ No event simulations<br>■ Empirical parameters |
| Reference | Littlewood *et al.* (1997) | USDA-ARS (1998) | Yates (1994) |

[a] Centre for Ecology and Hydrology, Wallingford, UK, and the Integrated Catchment Assessment and Management Centre (ICAM), Australian National University, Canberra, Australia.
[b] US Department of Agriculture, Agricultural Research Service (USDA-ARS), Hydrology Laboratory.
[c] International Institute for Applied Systems Analysis (IIASA), Laxenburg, Austria.

sequence, a water balance component and a routing component that simulates the associated lags between rainfall events and flow out of the watershed.

These three examples constitute only a sample of the conceptual models that have been developed and used in various applications over the past decades that illustrate a variety of concepts in modelling hydrological systems. Differences between the models result from the different ways of perceiving which components of the watershed need to be included and how various processes interact and affect each other.

**Table 5.2** Semi-distributed watershed models used by sponsoring organisations

| | Model | | | |
|---|---|---|---|---|
| | Hydrologiska byråns vattenbalansavdelning (HBV)[a] | Hydrologic Engineering Centre hydrologic modelling system (HEC-HMS)[b] | Streamflow synthesis and reservoir regulation model (SSARR)[c] | University of British Columbia watershed model (UBCWM)[d] |
| Temporal scale | Day | Flexible | Flexible | Flexible |
| Spatial scale | Flexible | Flexible | Flexible | Flexible |
| Processes modelled: | | | | |
| ■ Event simulation | Yes | Yes | Yes | Yes |
| ■ Continuous simulation | Yes | Yes | Yes | Yes |
| ■ Snow accumulation and melt | Yes | Yes | Yes | Yes |
| ■ Interception and infiltration | Yes | Yes | Yes | Yes |
| ■ Evapotranspiration | Yes | Yes | Yes | Yes |
| ■ Reservoir routing | Yes | Yes | Yes | Yes |
| Cost | Not known | Public domain | The SSARR model is no longer supported by USACE and is not in use as a standalone program. The AUTOREG model, however, has the SSARR model (all in FORTRAN) embedded within it | Public domain |
| Set-up time (short/medium/long) | Medium | Medium | Long | Short |
| Expertise (low/medium/high) | Medium | Medium | High | Medium |
| Technical support | SMHI International Consulting Services | Annual subscription service for USACE, fee-for-service from third-party vendors | SAR Consultants fee-for-service | Currently Michael Quick and Zoran Micovic; will transition to the Canadian Hydraulics Centre in the near future |
| Documentation (bad/medium/good) | Not known | Good | Good | User manual |
| Ease of use (easy/medium/difficult) | Medium | Medium | Difficult | Easy |
| Advantages | ■ All hydrological processes modelled<br>■ Fairly low data requirements | ■ State-of-the-art product in public domain<br>■ Input data in HEC format, compatible with other USACE software packages | ■ All hydrological processes modelled<br>■ Compatibility with USACE HEC Data Storage System | ■ All hydrological processes modelled<br>■ Low input data requirements<br>■ Good snow/glacial melt module with energy budget<br>■ Compatibility with stochastic modelling systems for extreme flood (e.g. SEFM and SCHADEX) |

**Table 5.2** Continued

| | Model | | | |
|---|---|---|---|---|
| | Hydrologiska byråns vattenbalansavdelning (HBV)[a] | Hydrologic Engineering Centre hydrologic modelling system (HEC-HMS)[b] | Streamflow synthesis and reservoir regulation model (SSARR)[c] | University of British Columbia watershed model (UBCWM)[d] |
| Disadvantages | ■ Only daily time step available<br>■ Simplified soil moisture dynamics<br>■ Limited information available | ■ Empirical modelling of snowmelt and evapotranspiration<br>■ Simplified river routing | ■ I/O operations<br>■ Extensive data demand<br>■ Limited available information | ■ Technical support not easy to obtain<br>■ Not fully physically based (i.e. needs to be calibrated); process representation conceptual |
| References | Lindström *et al.* (1997) | USACE (2000a,b, 2015) | USACE-NPD (1991) | Dr Zoran Micovic (personal communication, 2012) |

[a] Swedish Meteorological and Hydrological Institute (SMHI), Norrköping, Sweden.
[b] US Army Corps of Engineers (USACE), Washington, DC, USA
[c] US Army Corps of Engineers (USACE), Washington, DC, USA.
[d] University of British Columbia, Vancouver, Canada.

A detailed review of the most popular conceptual models can be found in Edsel *et al.* (2011), the Office of Hydrologic Development (OHD, 2007), Singh (1995) and Singh and Woolhiser (2002). The last four decades have witnessed a quest for development of better, more detailed and more comprehensive watershed response models, often tailored to specific applications and always based on the individual perceptions of the hydrologists and modellers involved in model building.

**Figure 5.8** Sacramento Soil Moisture Accounting Model. (Barrett *et al.*, 2008)

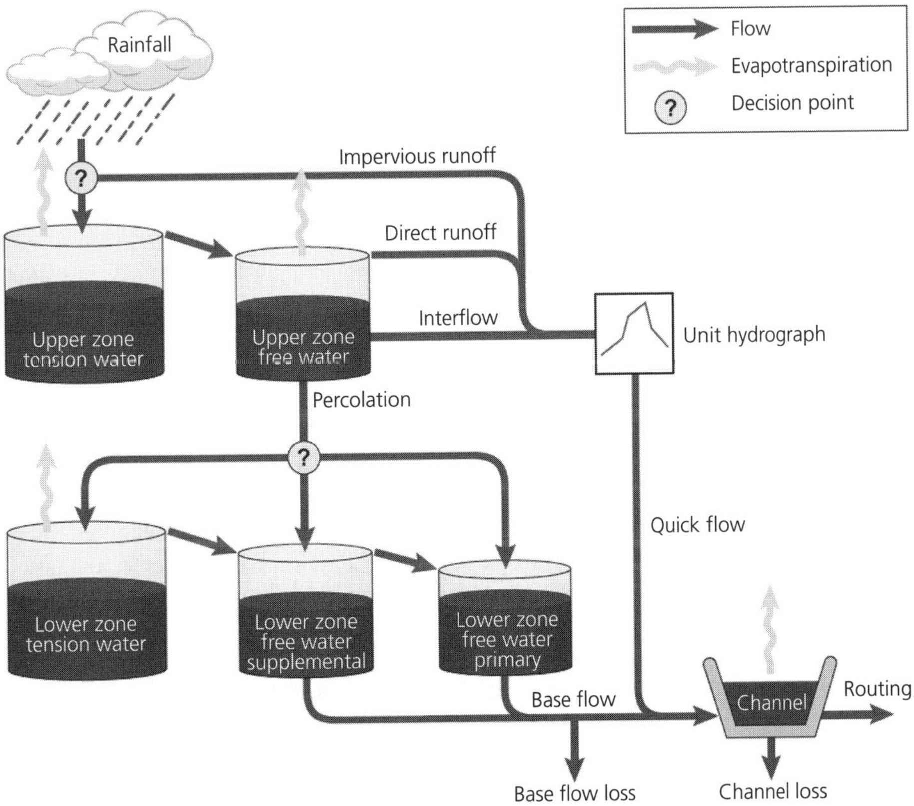

**Table 5.3** Selected distributed watershed models

| | Model | | |
|---|---|---|---|
| | CASC2D[a] | MIKE11/SHE[b] | WATFLOOD[c] |
| Temporal scale | Hour | Flexible | Flexible |
| Spatial scale | Flexible | Flexible | Flexible |
| Processes modelled: | | | |
| ■ Event simulation | Yes | Yes | Yes |
| ■ Continuous simulation | Yes | Yes | Yes |
| ■ Snow accumulation and melt | No | Yes | Yes |
| ■ Interception and infiltration | Yes | Yes | Yes |
| ■ Evapotranspiration | Yes | Yes | Yes |
| ■ Reservoir routing | No | Yes | Yes |
| Cost | Public domain | US $10 000 | CAD $2000 |
| Set-up time (short/medium/long) | Long | Long | Medium |
| Expertise (low/medium/high) | High | High | High |
| Technical support | No support for non-USACE users | ■ DHI software support centres<br>■ Training courses | From the author and model users |
| Documentation (bad/medium/good) | Good | Good | Good |
| Ease of use (easy/medium/difficult) | Difficult | Medium | Medium |
| Advantages | Free technical support for all licence holders of the Watershed Modelling System (WMS) graphical user interface (GUI) at the Engineering Computer Graphics Laboratory, Brigham Young University | ■ All hydrological processes modelled<br>■ State-of-the-art product<br>■ Highly flexible<br>■ Various add-ons available | ■ All hydrological processes modelled<br>■ Good support expected<br>■ Reasonable data demand |
| Disadvantages | ■ Extensive data requirements<br>■ For experienced users only<br>■ Inflexible time step | ■ Cost<br>■ Annual subscription<br>■ Technical support likely needed | ■ A step behind commercial products<br>■ Simplified channel routing |
| References | Julien *et al.* (1995), Ogden (1998) | DHI (2000a,b) | Kouwen (2001) |

[a] Julien *et al.* (1995), Ogden (1998)
[b] DHI (2000a,b)
[c] Kouwen (2001)

#### 5.2.2.3 Three selected models

The wide variety of modelling approaches and implementations of models, and available public-domain, research-oriented or commercial modelling systems and software allows users to select the most appropriate model for the task at hand. A detailed review of 18 watershed models commonly used by North American and European hydrologists is presented by Cunderlik (2003).

Based on Cunderlik's review, a brief characterisation is given here of three selected models:

- physically based models
- hybrid metric–conceptual models
- data-based mechanistic models.

Figure 5.9 Australian Water Balance Model. (Barrett *et al.*, 2008)

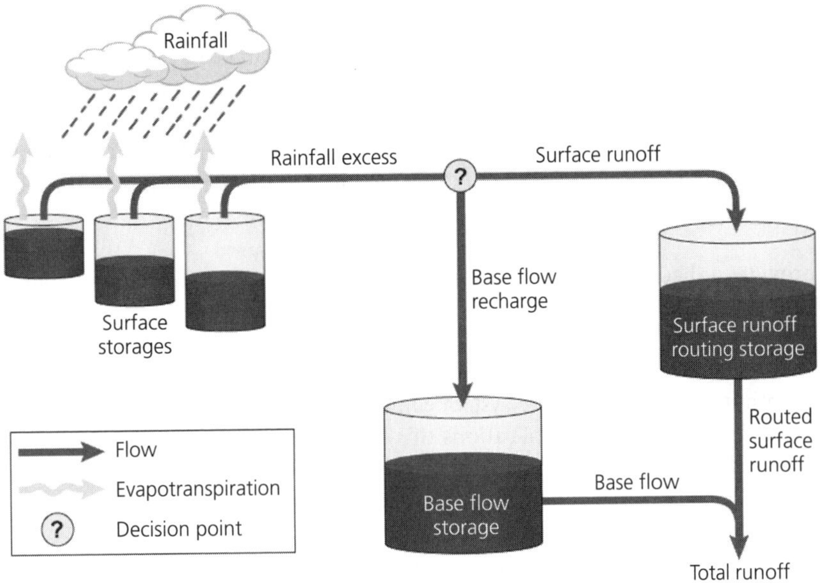


## Physically based models

Physically based models, often called 'mechanistic models', use the laws of physics as the basis for describing hydrological processes, components and their interactions within the watershed. The origins of physically based modelling go back to the 'blueprint for a physically based digitally simulated hydrologic response model' of Freeze and Harlan (1969). The blueprint proposed a mathematical framework consisting of a set of partial differential equations (Darcy–Richard equations) for the saturated and unsaturated subsurface portions of the watershed and another set of partial differential equations (St Venant equations) for the surface processes. The blueprint also specified necessary common boundary conditions accommodating the coupling of all components, which also included snowmelt and evapotranspiration.

The most often quoted advantage of physically based models over metric and conceptual models (the two previous categories) lies in the fact that, in principle, the parameters and state variables may be either directly measured or obtained from experimental work. For this reason physically based models have been considered a powerful tool in furthering the understanding

Figure 5.10 Soil Moisture Accounting Runoff Model. (Barrett *et al.*, 2008)

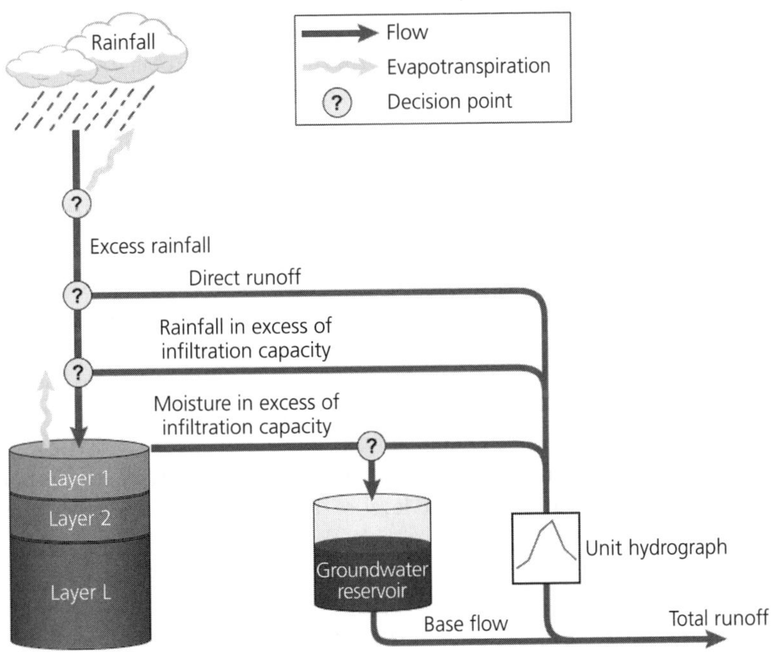

of how hydrological systems function. The disadvantage of these models is the requirement for complex numerical techniques and large amounts of input data.

Some hydrologists (Beven, 1989) disagree that true physically based models can be developed for applications because, even if the models are fully distributed and use exclusively parameters that are directly linked to physical characteristics of the watershed, the averaging of values of model variables and parameters at scales greater than the scales of processes modelled makes the model conceptual. Beven (2011) provides a more detailed discussion of this and the other limitations of the distributed physically based models based on the Freeze and Harlan (1969) blueprint. The main points of this discussion are summarised below.

Although the blueprint provides a theoretically consistent mathematical framework for the distributed hydrological response model, the following limitations are present in applications:

- The framework is based on a system of complex non-linear partial differential equations with arbitrarily changing boundary conditions, both in space in time. The system cannot be solved analytically, and numerical solutions require a spatial discretisation of the watershed. In applications this discretisation has to be larger than the point scales at which the flux equations hold. Lack of homogeneity at the sub-discretisation scales then causes gradients and fluxes to be improperly represented, leading to the conclusion that, although the processes are correctly modelled at the point scale, the same is not true for spatially discretised elements.
- Use of effective parameter values may alleviate some of the problems caused by lack of homogeneity, but for some of the parameters the effective values do not exist.
- Even if the effective values can be determined, it is impractical to define values for each element of the discretised space. Approximations have to be made because one cannot measure everything everywhere and some of the parameters cannot be measured at all.
- The implication is that, similar to conceptual models, distributed physically based models have to be calibrated (changing values of these parameters that for whatever reason cannot be measured until an acceptable fit between the observed and simulated watershed response values is achieved). The calibration process is complicated because the parameters, representing localised properties, are less identifiable. Thus, as Koren *et al.* (2004) point out, 'when parameter values are adjusted to calibrate a model, one must question if there is still physical significance to the parameter estimates or if at this point the physically based model has reverted to a type of elegant black box'.
- While the problem of calibration is relatively simple for conceptual models, this is not true for distributed models because the spatial discretisation of the watershed may lead to a huge number of parameters. The sheer number of parameters that need to be calibrated leads to the situation where many different combinations of parameter values give indistinguishable results.

The last topic has received considerable attention over the last 15 years, since Beven introduced the concept of *equifinality* (Beven, 1993, 1996) for hydrological modelling. Equifinality arises when more than one set of model parameters provide almost identical representation of the integrated response from the modelled watershed. This is a common problem for distributed physically based hydrological models. A thorough discussion of the implications of the presence of equifinality for hydrologic modelling can be found in the literature (Beven, 2001, 2002, 2006; Ebel and Loague, 2006; Hall *et al.*, 2007; Savenije, 2001; Sivakumar, 2008b).

Some of the most important conclusions from these papers are illustrated by the following quotations:

> Being fascinated by the complexity of hydrological processes, it is difficult to accept that at a higher level of aggregation this complexity may become irrelevant. This was not so disturbing a few decades ago, when people were forced to study the processes at larger levels of aggregation, merely because they didn't have the detailed knowledge or computational and mathematical tools to analyse the magnitude of data that detailed observations produce. With the rise of the modern computer, however, the problem has become apparent. Hydrologists who are blinded by the details of the processes (i.e. cannot see the forest through the trees) have attempted to use the powerful computers that are now available to simulate hydrological behaviour in highly fragmented distributed models. It required a visionary like Beven to see through this confusion. With the concept of equifinality he has made us realise that hydrological laws are purely laws of averaging, which only have a meaning at the particular scale at which they are applied. Even worse, the parameters in our hydrological laws are a function of the level of averaging (the size of the catchment, primarily).
>
> (Savenije, 2001)

> The development of more and more complex models that incorporate more and more detail about processes, but which introduce more and more parameters that must be calibrated, does not appear to be the future.
>
> (Beven, 2002)

> It took the discovery of equifinality to convince us that perhaps a new data-based 'downward' approach ... is necessary to counterbalance our faith in the distributed physically-based models. Equifinality may signal the end of the dead-end track of distributed modelling. However, it may be that it heralds the beginning of a new hydrology, which combines the old theories of linear and nonlinear reservoirs at the watershed scale with the computational powers of today.
>
> (Savenije, 2001)

> There is ... a growing infatuation with ever more complex models. It's gotten to the point where some models look as inscrutable as Nature itself. With numerous adjustable parameters, these models are generally unfalsifiable, so that the opportunity to learn from a wrong prediction is short-circuited. A Fermi approach based on models that capture the essence of the problem, but not all details, might get us farther.
>
> (Harte, 2002)

The second of the often quoted major significant disadvantages of physically based models (the requirement for much more input data than is required for the metric or conceptual models) appears to be of diminishing importance, at least for the surface components and processes. Increasing availability of powerful data management platforms such as geographical information systems (GIS), constantly improving capabilities of remote sensing technologies and the availability of digital terrain (DTM) and digital elevation (DEM) modelling tools are the primary reason why distributed physically distributed models are still considered to be valuable tools, especially for research purposes.

Examples of physically based models are not as numerous as of conceptual models. The most well-known models in this category include the SHE model developed by a collaborative effort of hydrology modellers from France, the UK and Denmark (Abbott *et al.*, 1986), its modified version in the UK known as SHETRAN (Ewen *et al.*, 2000), the physically based runoff production model TOPMODEL introduced by Beven and Kirkby (1979) and Beven (1995), and the Integrated Hydrology Model (inHM) (VanderKwaak, 1999, 2001).

### Hybrid metric–conceptual models

Hybrid metric–conceptual models use hypothetico-deductive thinking and are sometimes portrayed as a compromise between black-box (metric) models and complex distributed conceptual models. The approach is based on using available data (metric component) in order to test the hypotheses about the structure of the conceptual model. Thus the formulations of an *a priori* conceptual model structure based on the hydrological paradigms and the perceptions of the modeller are followed by the process of parameter optimisation supported by available observations.

A typical example of this modelling approach is the IHACRES model (Jakeman and Hornberger, 1993; Jakeman *et al.*, 1990). The modelling philosophy behind the model is simple: as most watersheds usually have inadequate data (streamflow, precipitation, temperature, etc.), the degree of complexity of the model should be appropriate to the data available and the modelling purpose. Thus the model includes only the most dominant processes lumped together at the watershed scale.

Some of the classification and categorisation systems of hydrological models include in this category data-based mechanistic (DBM) models. However, as the DBM models (unlike other hybrid metric–conceptual models) follow the inductive approach to scientific inference, the fundamentals of DBM hydrological models are described in a separate section below.

### Data-based mechanistic (DBM) models

The term 'data-based mechanistic modelling' was first used by Young and Lees (1993) but the general concepts of this approach date to the 1970s (Beck and Young, 1975). The DBM methodology was initially developed for environmental systems but it was quickly adopted for the modelling of any natural systems, including hydrological systems, for which time-series type data for real watersheds are available.

The evolution of the DBM modelling concept appears in a series of papers published since 2001 (Young 2001, 2002, 2003; Price *et al.*, 2000; Lees, 2000; Ratto *et al.*, 2000; Young *et al.*, 2007). At present, DBM modelling is an established approach to time-series modelling 'which seeks model structures and parameters that are both statistically optimal and consistent with plausible mechanistic interpretations of the system' (McIntyre *et al.*, 2011).

An outline of the DBM approach provided by Young *et al.* (2007) can be summarised in the following six steps:

1. Define the modelling objectives. Consider the modelling scale and the availability of data at this scale, and the type of model appropriate for the modelling objectives.
2. The initial phase of the modelling begins with the Monte Carlo simulation based on the stochastic form of the mass and energy conservation equations (uncertain parameters and inputs) derived from the deterministic modelling framework. Analysis of the sensitivity and uncertainty propagation in the model is used to identify the important parameters and the importance of the components of the model in explaining the dominant behavioural mechanisms.

3 Understanding of the model (in physical terms) gained in the previous step is enhanced by applying dominant mode analysis (DMA) to reduce the order of the model. The DMA is performed on the data generated by the simulation model with the objective of developing low-order approximations of dominant modes.
4 The appropriate model structure can now be identified using statistical inference from the time-series data. Initial identification is performed using generic class linear continuous or discrete-time transfer function (TF) models with time-varying parameters. Depending on this preliminary identification (linear or predominantly linear, linear with slowly varying parameters, linear with fast-varying parameters or non-linear) different advanced statistical methods can be used to estimate the model parameters.
5 In order to be accepted, the model has to be explained in physical terms. To demonstrate this, the modeller has to provide the interpretation of the final model structure that reflects the physical realities of the watershed. This is possibly the most important step in constructing DBM models, because if the explanation is missing or unavailable then the resulting model is not very different from the black-box models developed using traditional time-series analysis.
6 The last step is the model validation, which should at least include statistical evaluation of the model to confirm the satisfactory performance of the statistical diagnostics.

Until now in hydrology, the DBM modelling methodology has been use predominantly for rainfall–runoff models. Watershed DBM models that include snow accumulation and snow-melt processes have received considerably less attention, mostly due to the fact that snow-melt dynamics add a significant level of complexity to the identification of the model structure and the estimation of its parameters (Castelletti *et al.*, 2005; Young *et al.*, 2007).

### 5.2.3 Which model is the best?

This question comes up often when hydrology modellers meet, and equally often it remains unanswered when the meeting is over. It would be easier to come to a conclusion if the question were reformulated to: 'Which model is the best for my application?' The answer to this question will be driven now not only by the performance requirements and the quality and completeness of the model itself but also by realities of the case being considered, in other words by the availability of the data necessary to support the model application. This concept is illustrated in Figure 5.11.

## 5.3. Example of the Madawaska River watershed

The Madawaska River originates in the highland part of southern Ontario and flows east for 270 km, dropping 350 m of elevation by the time it reaches the confluence with the Ottawa River at Arnprior (Figure 5.12) (OPG, 2009). The total drainage area of the river is approximately 8500 km$^2$.

The main factor influencing water levels and flows is climate. Climate normals provided by Environment Canada for the period 1981–2010 for Combermere station, which is located in the centre of the watershed, are given in Table 5.4.

Most of the watershed is located within the Laurentian subregion of the Canadian Shield, which is characterised by numerous bedrock outcrops, ridges and predominantly shallow soils. Most of the watershed is covered by forests (85%), with open water accounting for over 10% and less than 3% used for agriculture. Ontario Power Generation (OPG) owns and operates two control dams and five generating stations on the main stem of the Madawaska River. Numerous small dams owned and operated by the Ontario Ministry of Natural Resources and Forests (MNRF) have a smaller impact on the variability of flows and levels than do the OPG dams.

Figure 5.11 Relationship between model performance, model complexity and data availability. (From Grayson and Blöschl, 2000)

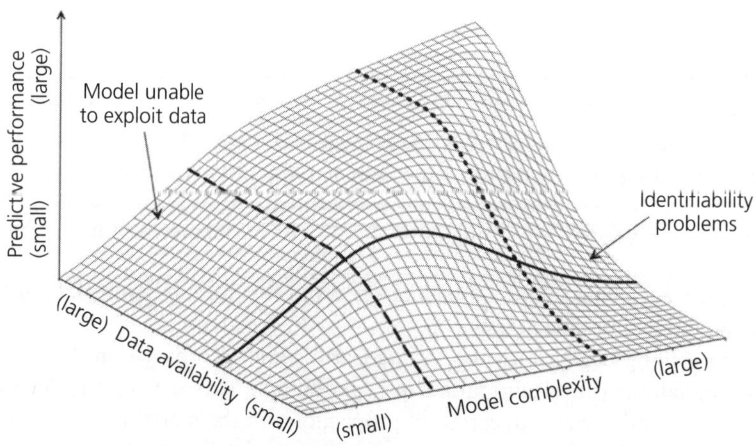

Figure 5.12 Madawaska River watershed

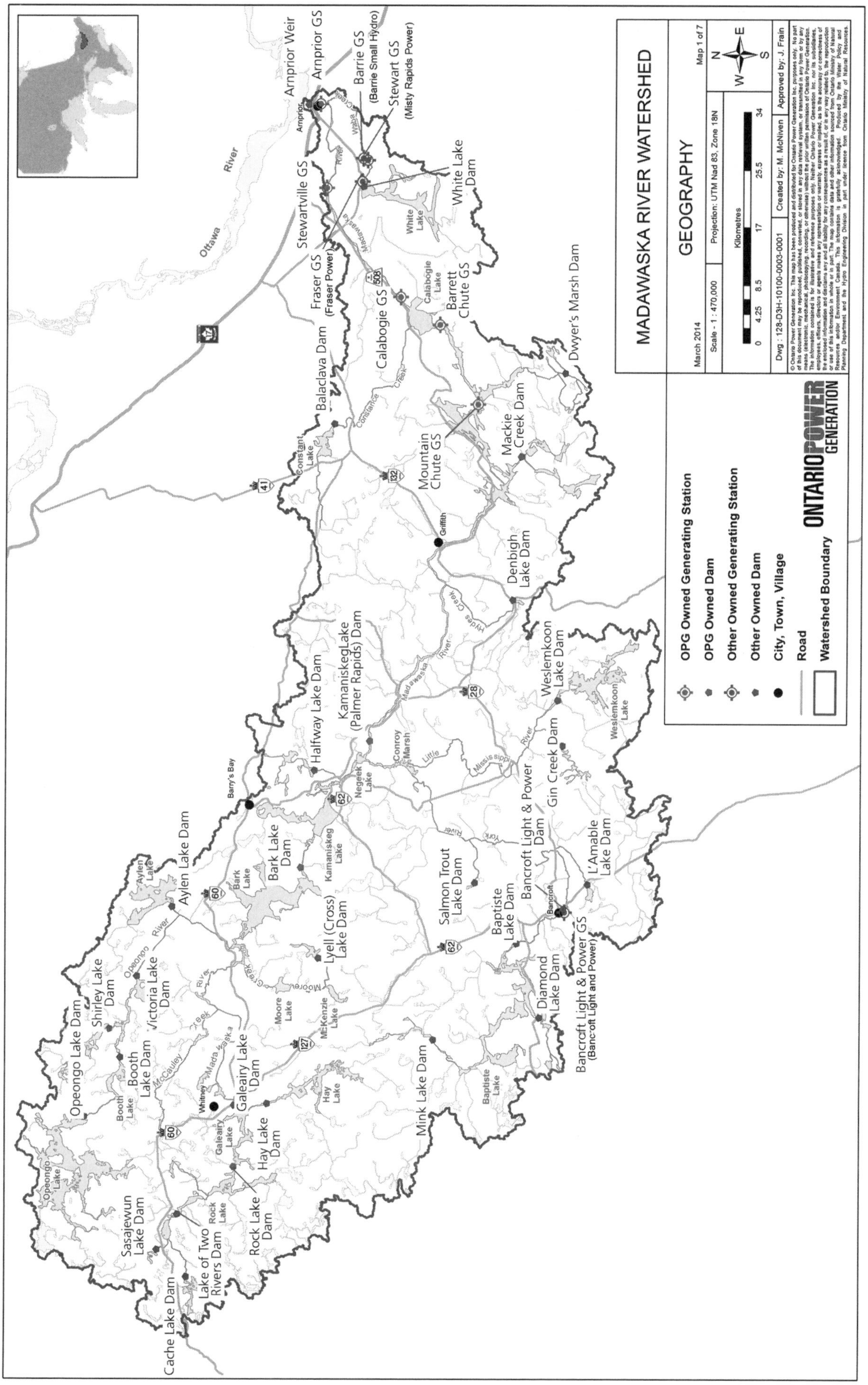

Table 5.4 Madawaska River hydrological data

| | Jan | Feb | Mar | Apr | May | Jun | Jul | Aug | Sep | Oct | Nov | Dec |
|---|---|---|---|---|---|---|---|---|---|---|---|---|
| **Temperature** | | | | | | | | | | | | |
| Daily average: °C | −11.6 | −9.4 | −3.8 | 4.4 | 11.5 | 16.8 | 19.4 | 18.1 | 13.4 | 6.7 | 0.4 | −6.8 |
| Daily maximum: °C | −5.5 | −2.9 | 2.4 | 10.4 | 17.9 | 23 | 25.5 | 24 | 19.2 | 11.8 | 4.6 | −2 |
| Daily minimum: °C | −17.6 | −15.9 | −9.9 | −1.6 | 5 | 10.5 | 13.3 | 12.1 | 7.6 | 1.6 | −3.8 | −11.5 |
| Extreme maximum: °C | 12 | 14 | 22.2 | 30 | 33 | 35 | 37.2 | 35 | 34.4 | 27 | 21.7 | 16 |
| Extreme minimum: °C | −40.0 | −37.8 | −32 | −23.9 | −10 | −3.9 | 2.2 | −0.6 | −5.6 | −10.6 | −23 | −35.6 |
| **Precipitation** | | | | | | | | | | | | |
| Rainfall: mm | 17.4 | 12.9 | 27.5 | 55.2 | 78 | 81 | 75.4 | 68.9 | 85.5 | 80.9 | 57.1 | 19.8 |
| Snowfall: cm | 45.9 | 36.6 | 31.4 | 11.6 | 0.2 | 0 | 0 | 0 | 0 | 4.2 | 21.9 | 45.7 |
| Precipitation: mm | 63.3 | 49.5 | 58.9 | 66.8 | 78.2 | 81 | 75.4 | 68.9 | 85.5 | 85 | 79 | 65.5 |
| Snow depth month-end: cm | 36 | 35 | 16 | 0 | 0 | 0 | 0 | 0 | 0 | 0 | 6 | 22 |
| Extreme daily rainfall: mm | 23.5 | 32.2 | 27.2 | 41 | 50 | 57.2 | 74.5 | 56 | 61 | 61.5 | 51.8 | 26.9 |
| Extreme daily snowfall: cm | 21 | 45.7 | 26 | 18 | 10.2 | 0 | 0 | 0 | 0 | 21 | 28 | 35.6 |
| Extreme daily precipitation: mm | 25 | 45.7 | 30 | 41 | 50 | 57.2 | 74.5 | 56 | 61 | 61.5 | 51.8 | 35.6 |
| Extreme snow depth: cm | 84 | 117 | 74 | 40 | 0 | 0 | 0 | 0 | 0 | 21 | 36 | 54 |

Two major storage sites within the system are Bark Lake Control Dam and Mountain Chute generating station, with the absolute volumes of 372 and 144 million $m^3$ (Figure 5.13). These two dams are operated on an annual cycle following the annual cycle of flows. The operating strategy of these two sites involves emptying the reservoirs before spring freshet in order to reduce peak flows in the river. Once the flows during the spring have peaked, Bark Lake flow is managed to balance the requirements to fill the lake to the summer operating range. Bark Lake is usually emptied by the end of February, and once its drawdown has been completed Mountain Chute generating station is emptied during March. Kamaniskeg Lake has little storage available and most of the flow reaching the lake has to be passed downstream.

The remaining four OPG hydroelectric facilities generate power on a daily cycle that involves storing water through the evening and early morning and releasing water during the day.

### 5.3.1 Dynamic simulation model of the Madawaska River system

The simulation model for the system was developed using the GoldSim modelling platform. GoldSim provides a flexible platform for visualising and dynamically simulating complex systems in engineering, science and business. It is user-friendly, highly graphical and object-oriented software, and can be easily applied for carrying out dynamic, probabilistic simulations to support management and decision-making.

Hydrological loadings for the Madawaska River system were developed using the stochastic weather-generating model MulGETS (Chen *et al.*, 2014). In general, stochastic weather generators are used to generate time series of climate variables that have statistical properties similar to those of observed data. MulGETS is a Matlab-based stochastic weather generator for generating multi-site precipitation and temperature, and it enables the user to generate time series of daily precipitation and maximum and minimum daily temperatures that preserve the spatial correlation between the observed data. The model uses a first-order two-state Markov chain method to generate precipitation occurrence, and multi-gamma or multi-exponential distribution for the precipitation amounts. Temperature is generated using the first-order autoregressive model.

The model was applied to generate the time series of daily precipitation and daily maximum and minimum temperatures for five climate stations with over 50 years of recorded data and located within or in the vicinity of the Madawaska River watershed (Figure 5.14). A sample of the generation results for 100 000 years of daily precipitation values is given in Figure 5.15.

The climate data generated were subsequently used to obtain synthetic time series of streamflow for the Madawaska River system with the use of the hydrological watershed modelling system HEC-HMS developed by the Hydrologic Engineering Center of the US Army Corps of Engineers (USACE, 2013). The structure of the model for the Madawaska River system is shown in Figure 5.16.

Time series of daily inflow were transferred into the GoldSim model of the Madawaska River system. The structure of the reservoir model, which is the main element of the system model, was further developed (Figure 5.17). Simulation of the system response to hydrological loading was carried out following the existing operating strategy briefly described above. A sample of results for Mountain Chute generating station is given in Figure 5.18.

Figure 5.13 Schematic diagram of the Madawaska River system

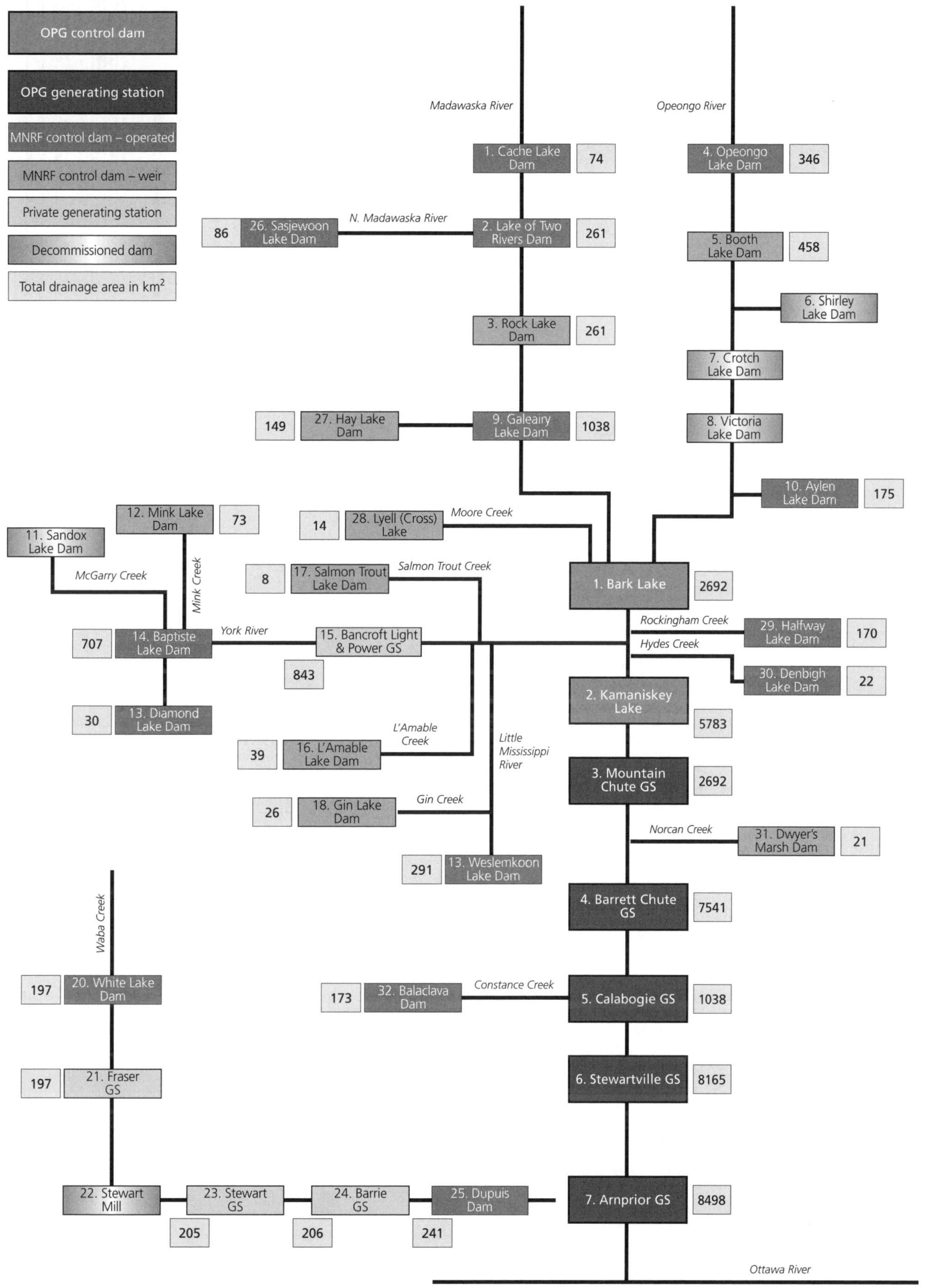

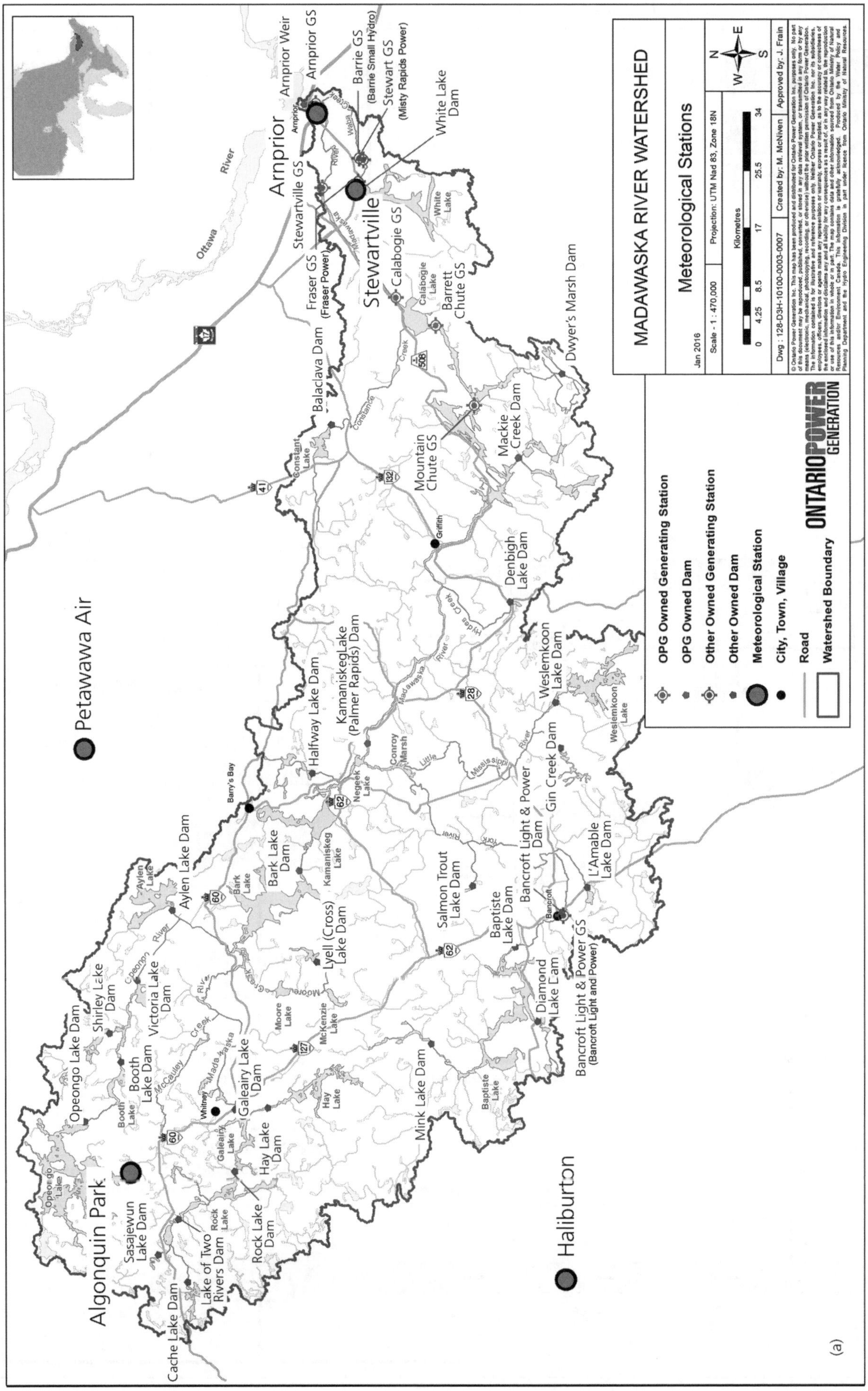


Figure 5.14 (a) Location of climate stations

Figure 5.14 (b) numbering of climate stations

(b)

Figure 5.15 Generation results for 100 000 years of daily precipitation: (a) maximum daily precipitation; (b) maximum 3-day spring precipitation

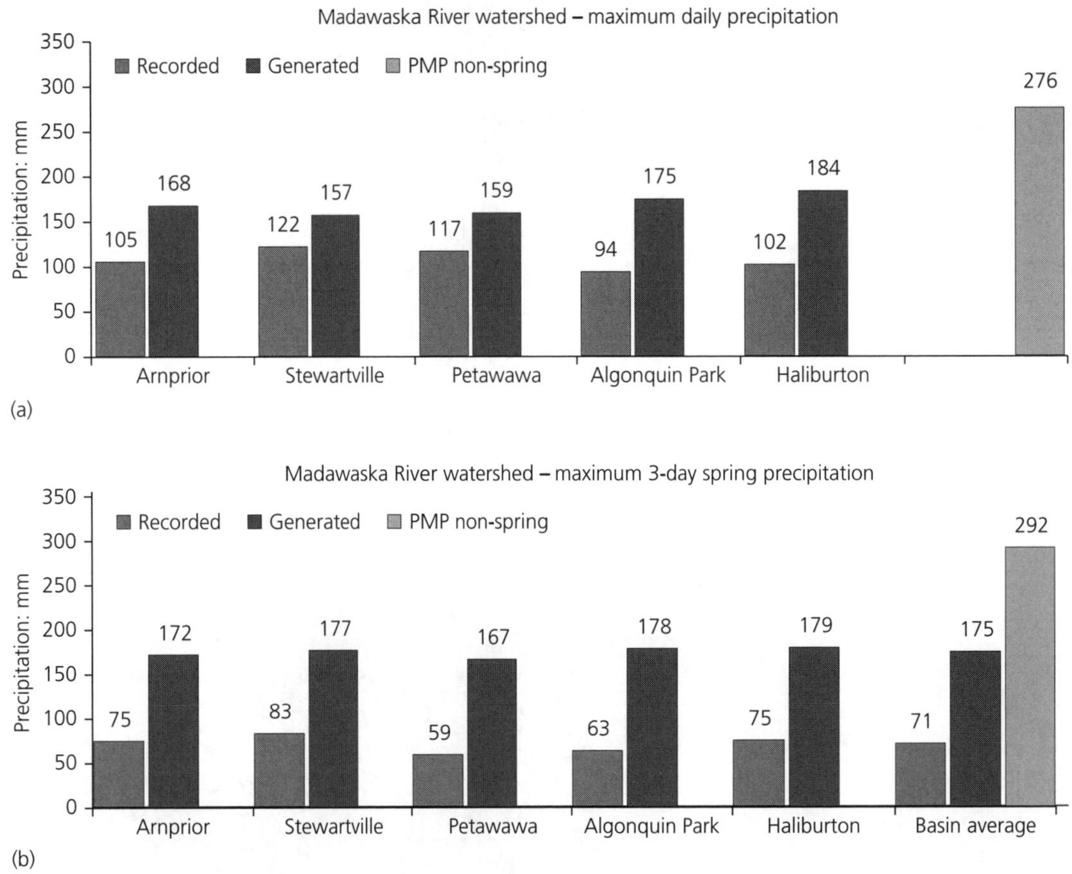


## 5.4. Time-series modelling of reservoir inflow

Time-series modelling has a long tradition in hydrological studies, based on the assumption that it is possible to develop a general model capable of representing most relevant statistical characteristics of historical streamflow series. The origins of the concept can be traced back to Hazen (1914), Sudler (1927) and Barnes (1954), who demonstrated the usefulness of statistics in the analysis of streamflow observations.

The approach received interest among modellers during the 1960s and 1970s. Salas *et al.* (1980) provides a comprehensive characterisation of the efforts of these hydrologists during this period. A variety of stochastic models was developed, including autoregressive (AR) models, fractional Gaussian noise (FGN) models, autoregressive moving-average (ARMA) models, broken-line (BL) models, shot-noise (SN) models, ARMA-Markov models, general mixture or Markov mixture models, and disaggregation models. The popularity of time-series modelling increased dramatically after the publication of the book *Time Series Analysis: Forecasting and Control* by Box and Jenkins (1976).

Time-series models belong to the family of stochastic models – mathematical models for describing systems the behaviour of which changes over time according to probabilistic laws. Climatological processes over the watershed and the resulting hydrological process, like most natural phenomena occurring in natural systems, behave, or at least appear to behave, in a random (probabilistic) manner. Cox and Miller (1965) categorise stochastic models on the basis of two attributes: time and space. Both attributes can be discrete or continuous, leading to the classification shown in Figure 5.19.

In general, a time series is understood as a series of observations that are arranged chronologically. This arrangement, which ensures that the order of occurrences is preserved, is mandatory. River flows data, unlike data for some climatological phenomena, are most often available at daily intervals and conform to the data requirements for most time-series models, which implicitly require that time series be observed and measured at equally spaced time intervals.

### 5.4.1 Time-series models for synthetic river flows

The major applications of time-series models in hydrology are inflow forecasting and flow simulation. Inflow forecasting is not of interest in this study, so the focus is on simulation. The aim of the simulation is to provide 'random' sequences of inflows of the required length that are consistent with the stochastic properties such as the historically observed inflows.

**Figure 5.16** Dynamic simulation model of the Madawaska system (GoldSim™)

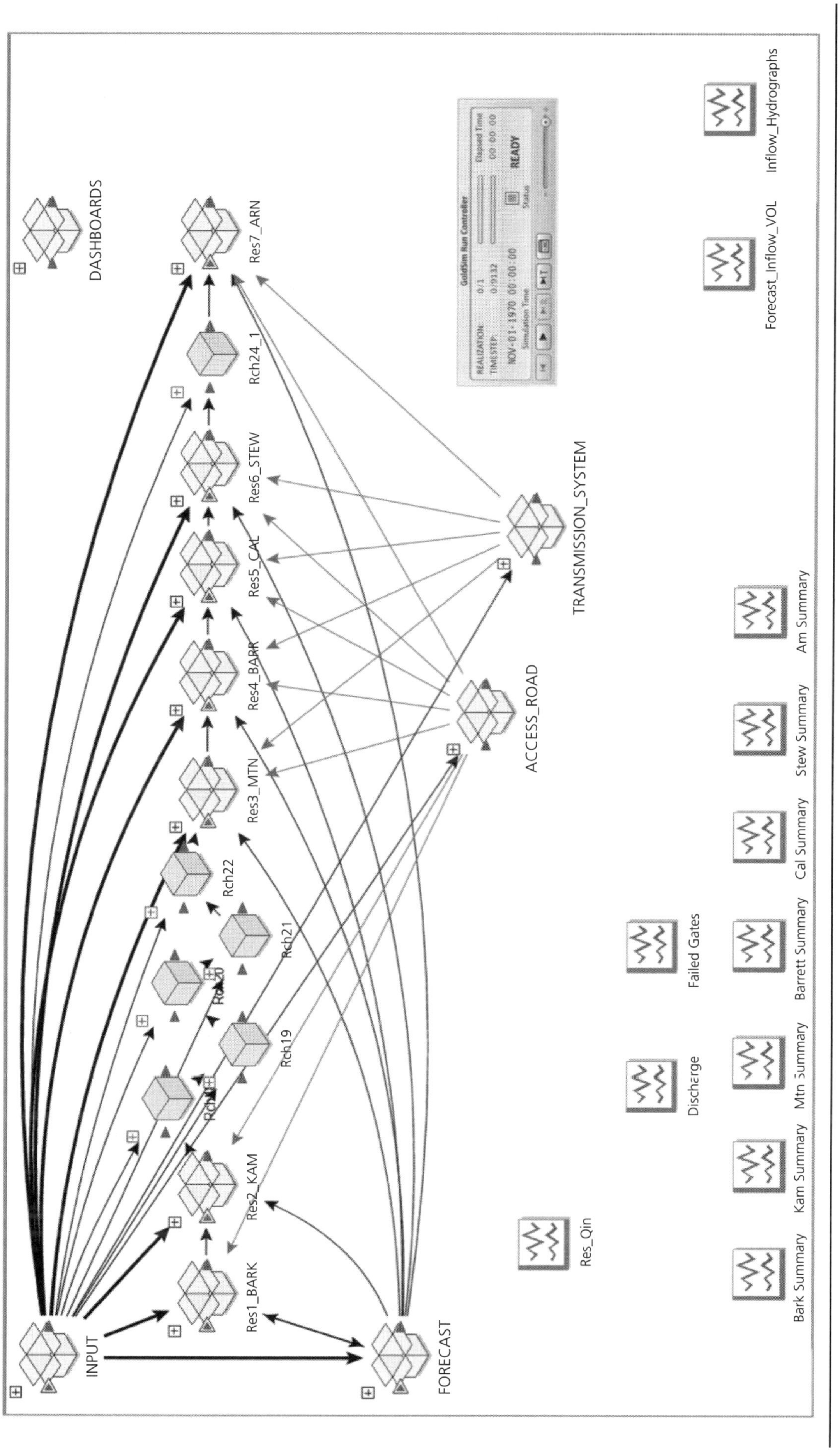

**Figure 5.17** Element Bark-GateFail (model of Bark Lane gates availability)

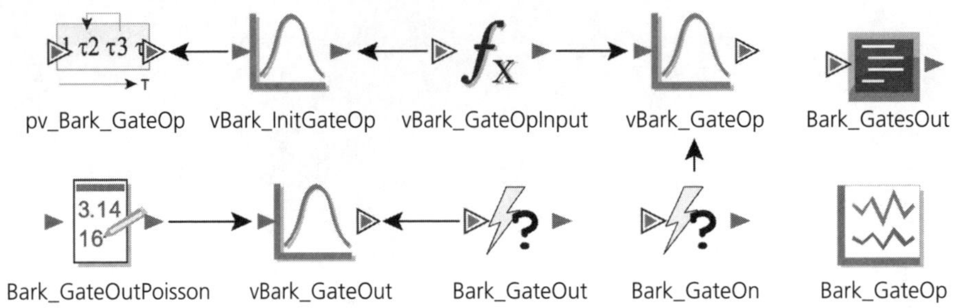


Several groups of models need to be considered. These differ with respect to the time interval used for simulation (annual, seasonal, monthly, weekly, daily). Due to the properties of the stochastic processes involved in modelling flows of different durations, the following groups of models are pertinent:

*Annual flows*
- Linear non-seasonal models of both stationary and non-stationary type
- Fractional Gaussian noise, fractional autoregressive moving-average models and broken-line models
- Generalised auto-covariance models.

*Annual and monthly flows*
- Multivariate split model
- Multivariate periodic autoregressive model with symmetric moving-average filter.

*Monthly flows*
- Seasonal autoregressive integrated moving-average models
- De-seasonalised models
- Periodic models
- Transfer-function models.

For practical applications that include stochastic modelling of time-series, details of these different modelling techniques can be found in Hipel and McLeod (1994). The list of the modelling alternatives described in this book includes:

- Linear non-seasonal models, both stationary and non-stationary of autoregressive type (ARMA)
- Derivatives of ARMA models:
  - long memory ARFIMA models (fractional ARMA models)
  - autoregressive integrated moving-average (ARIMA) models)
  - seasonal models (SARIMA – seasonal ARIMA) and periodic (PARIMA) models
  - threshold autoregressive models (TAR)
- Models with exogenous inputs:
  transfer function noise (TFN) models with single or multiple inputs
  - ARMA models with exogenous part (ARMAX).

Although the detailed background for time-series modelling together with the detailed description of various models is provided by Hipel and McLeod (1994), it is useful to review the mathematical foundation for the simplest class of models, namely the linear non-seasonal models. These models form the building blocks for the more complex models listed above.

### 5.4.2 Autoregressive and moving-average processes

An autoregressive process AR($p$) of order $p$ can be defined in terms of the following formula:

$$z_t - \mu = \phi_1(z_{t-1} - \mu) + \phi_2(z_{t-2} - \mu) + \cdots + \phi_p(z_{t-p} - \mu) + a_t \tag{5.13}$$

where $\phi_k$ is the $k$th *non-seasonal AR parameter*. By introducing the *non-seasonal AR operator* of the form

$$\Phi(B) = 1 - \phi_1 B - \phi_2 B^2 - \cdots - \phi_p B^p$$

Equation 5.13 can be simplified to

$$\Phi(B)(z_t - \mu) = a_t \tag{5.14}$$

**Figure 5.18** (a) Simulated headwater level; (b) simulated flows

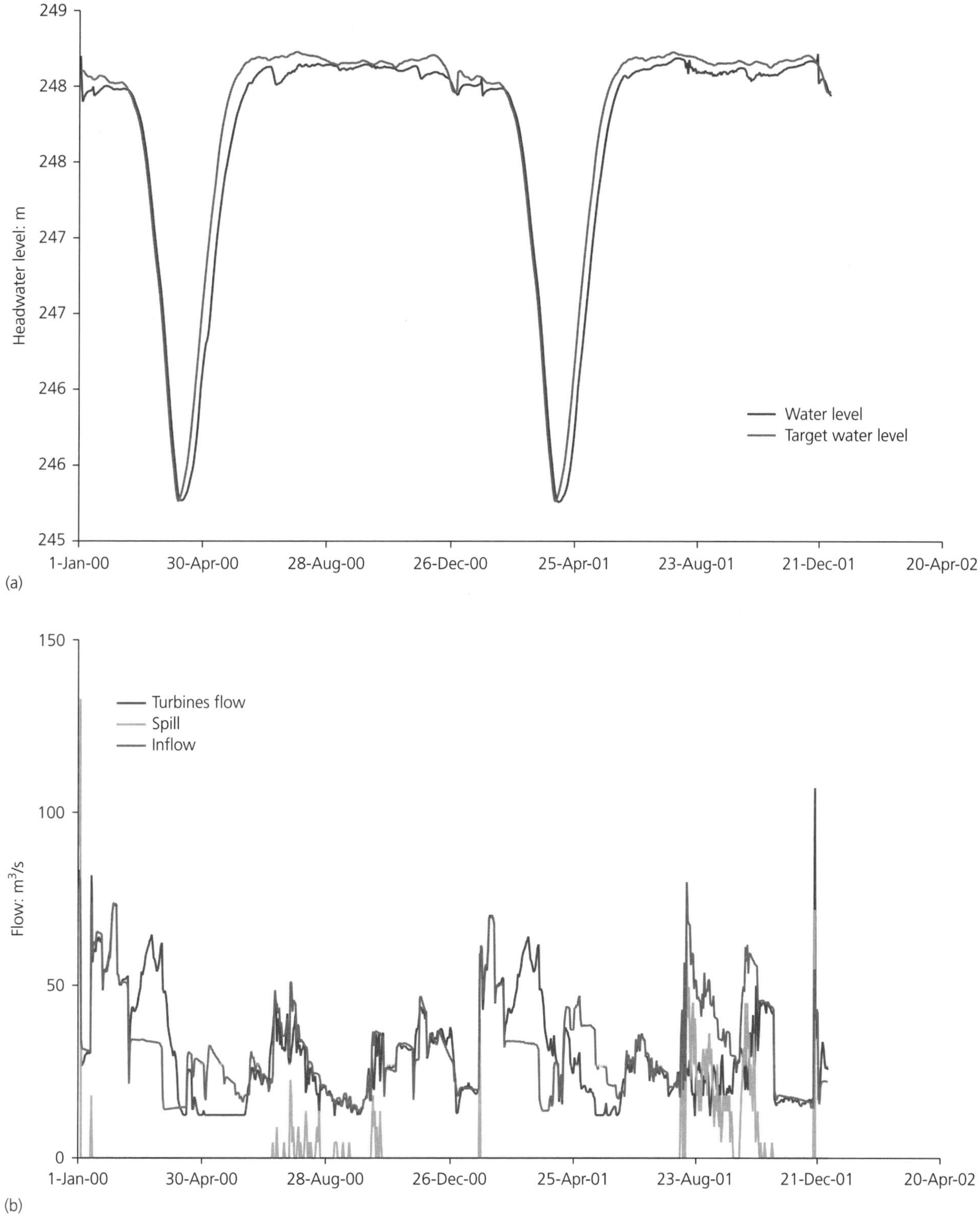


The AR(1) process is simply a Markov process. The Markov process has the property that the observation $z_t$ measured at time $t$ depends only on the time-series value at time $(t-1)$ and the value of a random shock $a_t$.

When the time-series value $z_t$ depends only on the value of the random shock at time $(t-1)$, $a_{t-1}$, and the value of the current shock $a_t$, then

$$z_t - \mu = a_t - \theta_1 a_{t-1} \tag{5.15}$$

**Figure 5.19** Stochastic models

<table>
<tr><td colspan="2" rowspan="2"></td><td colspan="2">STATE SPACE</td></tr>
<tr><td>Discrete</td><td>Continuous</td></tr>
<tr><td rowspan="2">Time space</td><td>Discrete</td><td>Markov chains</td><td>Time-series models</td></tr>
<tr><td>Continuous</td><td>Stochastic point processes</td><td>Stochastic differential equations</td></tr>
</table>

where $\theta_1$ is the *non-seasonal moving average (MA) parameter*. This process is called a moving average (MA) process of order 1, and is denoted by MA(1). It can also be written in the form of the operator $B$ as

$$z_t - \mu = (1 - \theta_1 B)a_t = \Theta(B)a_t \tag{5.16}$$

where $\theta(B) = (1 - \theta_1 B)$ is the *non-seasonal MA operator of order 1.*

The MA(1) process can be extended to a process with $q$ parameters providing the MA process of order $q$, MA($q$):

$$z_t - \mu = a_t - \theta_1 a_{t-1} - \theta_2 a_{t-2} - \cdots - \theta_q a_{t-q} \tag{5.17}$$

where $\theta_j$ is the $j$th *non-seasonal MA parameter*. Using the $B$ operator, Equation 5.16 takes the form:

$$z_t - \mu = \Theta(B)a_t \tag{5.18}$$

### 5.4.3 Disaggregation procedures

Seasonal, monthly values of synthetic (generated) flows can be converted into flows of shorter time intervals, such as days for example, with the help of disaggregation models.

The essence of staged disaggregation is the development of a staging framework where flow sequences are synthesised at a given level of aggregation and then disaggregated into component flows (e.g. from annual to seasonal, from seasonal to monthly). The higher level values can be derived using the models outlined in Section 5.4.1.

Disaggregation of higher level variables at one or more sites to lower level variables at those sites can be performed in more than one step. Modelling of the correlations between the lower level variables can also be restricted. Other variations can include disaggregation always into only two parts (e.g. annual into the first month and then the remaining 11 months).

The first disaggregation framework was developed in the early 1970s (Valencia and Schaake, 1973) and was based on a linear regression approach. The model can be written in the general form,

$$X = \boldsymbol{A}Y + \boldsymbol{B}w \tag{5.19}$$

where $X$ is the vector of disaggregated flows, $Y$ is the vector of aggregated flows, $\boldsymbol{A}$ and $\boldsymbol{B}$ are matrices with model parameters, and $w$ is the stochastic term.

This linear stochastic framework was modified over a number of years, with the most recent modifications being provided by Stedinger *et al.* (1985) and Santos and Salas (1992). All models developed using the original approach required a normalising transformation of the original data. As the models were fitted into the transformed space, both the additivity requirement (disaggregated flows adding to aggregated values) and the preservation of statistics requirement were often not met. Adjustments were subsequently proposed by Grygier and Stedinger (1988).

Other disaggregation schemes that avoided the problems caused by the transformation step were introduced by Koutsoyiannis (1992, 2001) and Koutsoyiannis and Manetas (1996). Although capable of preserving high-order statistics, all these methods are computationally intensive due to their iterative nature.

The 1990s witnessed a growing popularity of non-parametric methods in general stochastic hydrology. Applications of these methods followed synthetic flow generation (Lall and Sharma, 1996; Sharma *et al.*, 1997) and, later, flow disaggregation. Detailed descriptions of recently developed approaches to disaggregation to daily flows can be found in Kumar (2000), Prairie *et al.* (2007) and Tarboton *et al.* (1998). A simple heuristic method is provided by Nowak *et al.* (2010).

### 5.4.4 Time-series modelling of synthetic daily river flows

Direct development of time-series models for generating daily flows is complicated because of the high variability of the data. This problem has been studied by Bhattacharya (1997), who demonstrated the significance of heteroscedasticity using examples of observed time series of streamflow in the Midwestern USA. In analysing time series, a process is called homoscedastic if its variances are constant at all times. Heteroscedasticity has been recognised as a typical phenomenon in financial time series where the variance of the series changes over time. This makes the series non-stationary, which in turn makes the theory developed for stationary time series inadequate.

The problem of heteroscedasticity can be addressed only in a limited manner in traditional time-series modelling by considering season-dependent variances. A different approach, which directly addresses the problem of heteroscedasticity, was developed in the field of finance to study volatility clustering, in which large fluctuations follow large fluctuations and small changes tend to follow small changes (Bollerslev, 1986). The approach, which provides an approximate framework for developing time-series models with time-varying variances, is outlined in Andersen *et al.* (2009).

Autoregressive conditional heteroscedasticity (ARCH) models recognise the presence of successive periods of relative volatility and stability and can model volatility clustering. They do so by including previous variances when modelling future variances, and also account for fat tail behaviour. This feature of the time series, which exhibits autoregressive conditional heteroscedasticity, is called the ARCH effect. The ARCH model was first introduced by Engle (1982). A few years later, Bollerslev (1986) proposed the GARCH (generalised ARCH) framework for conditional variance specification, which allowed for a parsimonious parameterisation of the lag structure. Both models received interest in applications to financial time series, but their applications to hydrological modelling have been limited.

Applications of ARCH and GARCH to streamflow processes are discussed by Wang (2006). They begin with testing for the ARCH effect in daily streamflow time series and applying the ARCH test to classical ARMA and AR models. In this way, the question of why the de-seasonalised, seasonal or periodic models are incapable of efficiently modelling the ARCH effect for daily flows is addressed. Finally, a new ARMA-GARCH model was developed.

The model is a combination of an ARMA model, which is responsible for modelling the mean behaviour, and a GARCH model, which accounts for ARCH effects in the residuals from the ARMA model. The model has the form

$$\begin{aligned} &\Phi(B)(z_t) = \Theta(B)a_t \\ &a_t = \sigma_s x_t,\ x_t \sim N(0, h_t) \\ &h_t = \alpha_0 + \sum_{i=1}^{q} \alpha_i x_{t-i}^2 + \sum_{i=1}^{p} \beta_i h_{t-i} \end{aligned} \tag{5.20}$$

where $\sigma_s$ is the seasonal standard deviation of $a_t$ and $s$ is the season number (1 to 366) for daily time series. In the above set of equations the conditional variance $h_t$ of $a_t$ depends on the squared residuals in the previous $q$ time steps and the conditional variance in the previous $p$ time steps.

The model has been subsequently fitted to the 45-year-long sample of daily flows for the Yellow River, resulting in an ARMA-GARCH model composed of an ARMA(20,1) model and a GARCH(0,21) model. As the analysis of residual series obtained from the model indicated that significant autocorrelation remained in the standardised residuals, an extended version of the model (Equation 5.20) has been developed in the following form:

$$\begin{aligned} &\Phi(B)(z_t) = \Theta(B)a_t \\ &a_t = \sigma_s y_t \\ &\Phi'(B)(y_t) = \Theta'(B)x_t,\ x_t \sim N(0, h_t) \\ &h_t = \alpha_0 + \sum_{i=1}^{q} \alpha_i x_{t-i}^2 + \sum_{i=1}^{p} \beta_i h_{t-i} \end{aligned} \tag{5.21}$$

Subsequently, using the same set of observed daily flows, an AR(16) model was fitted to the seasonally standardised residuals from the ARMA(20, 1) model, and the ultimate model was composed of ARMA(20, 1), AR(16) and ARCH(21) sub-models. The addition of the AR sub-model removed the autocorrelation from the standardised residuals with the remaining autocorrelation of the squared residuals being negligible.

Although the results of Wang's research were promising (Wang, 2006), the wider applications of this methodology in hydrological modelling did not follow. For that reason, although ARCH and GARCH time-series modelling creates an interesting option for flows with daily and shorter time intervals, the approach has not been sufficiently examined in the context of hydrological time series.

### 5.4.5 Generalised framework for stochastic simulation of streamflow time series

Koutsoyiannis (2000) pointed out that, although the modelling based on Box and Jenkins' (1976) time-series analysis provided a strong and useful foundation for stochastic hydrology, it was incapable of accounting for the long-term persistence of hydrological processes. The property was first described by Hurst (1951), who investigated streamflow time series of the Nile. Hurst came to the conclusion that there was a tendency for annual average flows to stay above or below their mean value for prolonged periods of time (clustering of wet and dry years into drought and wet periods, respectively).

The unique dependence discovered by Hurst and observed at large timescales has significant consequences for hydrological modelling as it increases the uncertainty in the related processes. It is an important aspect of hydrological modelling because long-term persistence can be confused with natural or anthropogenically induced trends.

Even today, the importance of Hurst's discovery is not widely understood – or is ignored. Its nature is regarded as difficult to understand, and its reproduction in hydrological simulation is considered unnecessary (Koutsoyiannis, 2002). Although significant work has been done on this subject in the last decade, it remains ignored by the hydrological profession. Most recently, Koutsoyiannis (2011) developed a novel approach built on the premise that a change is not equivalent to non-stationarity, and even a significant change on multiple timescales can for practical purposes be satisfactorily described by a stationary model of Hurst–Kolmogorov stochastic dynamics.

Also recently, a unified theoretical framework for stochastic, physically based modelling was presented by Montanari and Koutsoyiannis (2011a,b,c). The framework allows for hydrological modelling and uncertainty assessment to be carried out jointly. The blueprint is built on the assumption that the use of probability is both an objective and an efficient approach to the assessment of uncertainty. The structural strength of the proposed approach is to some extent compromised in applications by large the computational effort required and the still unresolved issues about parameter, model and data uncertainties related to the physically based part of the approach.

### 5.4.6 Stochastic grey-box modelling

Jonsdottir (2006a) proposed still another way of modelling hydrological systems, where models are built on ordinary or partial differential equations describing the processes in deterministic terms (physical component), but containing a noise component. Thus, the physical knowledge about the system does not have to be complete. At the base of the modelling framework are stochastic differential equations resulting in semi-physical models. An example of the approach to parameter estimation for stochastic rainfall–runoff modelling (Jonsdottir, 2006b) builds on traditional conceptual hydrological modelling. The structure of the model is as presented in Figure 5.20.

Water is stored in reservoirs S1 and S2 and the outflows are routed with routing constants $k_1$ and $k_2$. The inflows to the reservoirs are derived from precipitation, as either direct rainfall or from snow melt. Thus the inflows are functions of precipitation $P(t)$ and temperature $T(t)$. The total precipitation is divided into rainfall and snowpack using a threshold function dependent on the air temperature:

$$\Phi[T(t)] = \frac{1}{1 + e^{[b_0 - b_1(T(t))]}} \tag{5.22}$$

Melting $M(t)$ is calculated using the positive degree constant $P_{\mathrm{dd}}$:

$$M(t) = P_{\mathrm{dd}} T(t) \Phi T(t) \tag{5.23}$$

In order to account for the fact that the melting water from the snowpack cannot be released until the threshold water content is reached, a low-pass filter is applied for the temperature:

$$\mathrm{d}T_s(t) = [-aT_s(t) + aT(t)]\,\mathrm{d}t + \mathrm{d}w_t \tag{5.24}$$

The work of Jonsdottir (2006a) was extended by Thordarson (2012), who generalised Jonsdottir's approach and provided the following concise framework for stochastic modelling of hydrologic systems by using stochastic differential equations (SDE).

Thordarson's approach derives from the premise that, as hydrological systems are complex with interacting processes evolving in time and space, the most appropriate approach to model the system is through a system of partial differential equations (PDEs). These describe physical processes taking place in the watershed.

**Figure 5.20** Model structure. (Jonsdottir, 2006b)

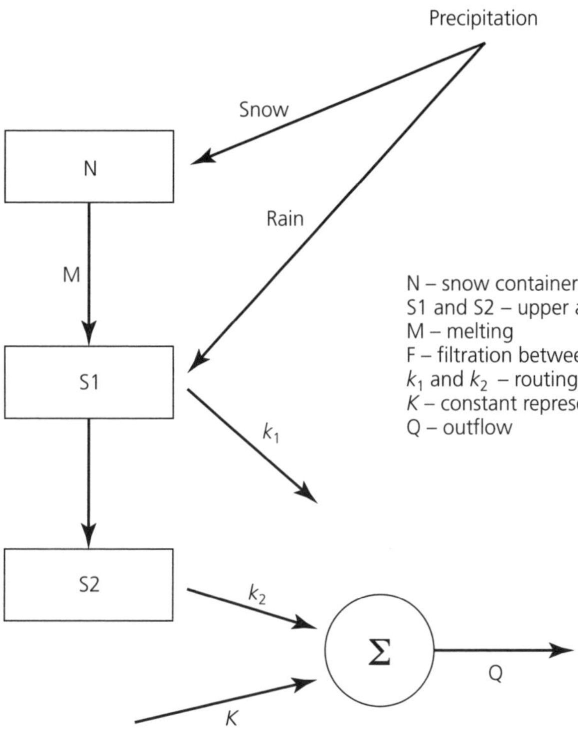


The PDEs, which are notoriously difficult to solve analytically, can be replaced with a set of ordinary differential equations (ODEs) representing a simplified, lumped-parameter model of the system. A single ODE of the form

$$\frac{\mathrm{d}X_t}{\mathrm{d}t} = f(X_t, t, \theta) \tag{5.25}$$

which represents the dynamic behaviour of $X_t$ in a deterministic way, can be replaced by the stochastic differential equation (SDE)

$$\frac{\mathrm{d}X_t}{\mathrm{d}t} = f(X_t, t, \theta) + \sigma(X_t, t, \theta)W_t \tag{5.26}$$

where $W_t$ is the Wiener process.

The solution can be found by the following integration:

$$X_t = X_0 + \int_0^t f(X_s, t, \theta)\,\mathrm{d}s + \int_0^t \sigma(X_s, t, \theta)\,\mathrm{d}w_s \tag{5.27}$$

where the second integral is the Ito integral.

A system to be modelled can thus be described by a system of SDEs related to discrete time measurements using the following state–space formulation:

$$\begin{aligned} \mathrm{d}\boldsymbol{X}_t &= f(\boldsymbol{X}_t, \boldsymbol{U}_t, t, \theta)\,\mathrm{d}t \\ \boldsymbol{Y}_k &= g(\boldsymbol{X}_k, \boldsymbol{U}_k, t, \theta) + \varepsilon_k \end{aligned} \tag{5.28}$$

The first of these two equations (system equation) describes the dynamics of the system physical states $\boldsymbol{X}_t$, while the second (observation equation) relates the observations to the states in discrete time (the index $k$ denotes discrete time instants of observations). $\boldsymbol{U}$ and $\boldsymbol{Y}$ are the vectors of input and output variables, respectively.

Thordarson (2012) indicates that this type (grey box) of modelling is better suited to accounting for the stochastic aspects of modelling of hydrological systems. The use of SDEs to describe the dynamics of the hydrological system allows for an increase in the complexity of the model by including all the necessary hydrological processes, and at the same time different characteristics of the process noise can be considered so that an adequate characterisation of the uncertainty in the model structure can be achieved.

## 5.5. Streamflow routing model

Streamflow (flood) routing is a process used to describe temporal and spatial variation of a (flood) hydrograph as it moves through the river channel. Maidment (1993) describes it as 'a mathematical procedure for predicting the changing magnitude, speed and shape of a flood wave as a function of time at one or more points along a river'. Thus the routing can be used to simulate the movement of water through a river reach, and to determine the magnitudes, volumes and temporal patterns of the flow as it moves downstream.

This tracing of the dynamic movement of water moving downstream through a river reach has a well-established theoretical background in computational fluid mechanics (ASCE 1996; Chow 1959; Maidment 1993; Singh 2004; Weinman and Laurenson 1979). Following the principles of conservation of mass and of momentum, the physics of the routing process one-dimensional unsteady flow can be completely described by the St Venant equations (Singh, 2004):

$$\text{Continuity:} \quad A\frac{\partial V}{\partial x} + VB\frac{\partial y}{\partial x} + B\frac{\partial V}{\partial t} = q \tag{5.29}$$

$$\text{Momentum:} \quad \frac{\partial V}{\partial t} + V\frac{\partial V}{\partial x} + g\frac{\partial y}{\partial x} - gS_0 + gS_f = 0 \tag{5.30}$$

where $A$ is the cross-sectional flow area, $V$ is the average velocity of water, $x$ is the distance along the river reach, $B$ is the water surface width, $y$ is the depth of water, $t$ is time, $q$ is lateral inflow, $S_f$ is the friction slope, $S_0$ is the channel bed slope and $g$ is gravitational acceleration.

The continuity equation describes the conservation of mass, and the momentum equation, derived from the Newton's second law of motion, preserves conservation of momentum. The five terms in the momentum equation are descriptions of local acceleration, convective acceleration, pressure force, gravity force and friction force, respectively.

It should be noted that the above model of flow routing is an idealised model based on a number of assumptions related to the properties of the river channel and the flowing water. More often than not, these assumptions are not fulfilled in applications.

The flow-routing methods based on Equations 5.29 and 5.30 or on their simplifications are usually called 'hydraulic routing', as opposed to the hydrological routing, methods. The latter are based on a different type of continuity relationship that uses relationships between storage within the channel reach and outflow. The movement of water through a river reach can vary significantly depending on the characteristics of the reach and the properties of the flow. These are usually the deciding factors in selecting the appropriate routing method (Figure 5.21).

### 5.5.1 Hydrological routing methods

Hydrological routing techniques model the downstream conditions within the river reach in response to a given flow hydrograph at the upstream end of the reach. The modelling uses only the continuity equation and analytical or empirical relationships between channel storage and discharge at the outlet. The continuity equation for the river reach can be conceptualised as:

$$\text{Continuous time:} \quad \frac{dS(t)}{dt} = I(t) - Q(t) \tag{5.31}$$

$$\text{Discrete time:} \quad I - Q = \frac{\Delta S}{\Delta t} \tag{5.32}$$

Figure 5.21 Flow conditions and flow-routing methods

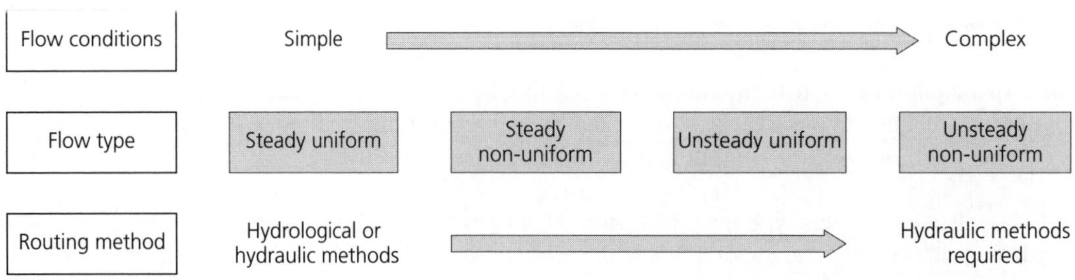

where for the discrete time formulation $I$ is the average inflow at the upstream end of the river reach during time interval $\Delta t$, $Q$ is the average discharge from the reach during time interval $\Delta t$, and $S$ is the reach storage.

The most commonly used methods applied for hydrological routing are:

- the linear reservoir(s) model
- the modified pulse method
- the Muskingum method
- the Muskingum–Cunge method.

#### 5.5.1.1 Linear reservoir model

The linear reservoir model uses the continuity equation (Equation 5.31) and the storage function relating the outflow from the reach $Q$, and the reach storage $S$ using a simple linear function in the form:

$$s = kQ \tag{5.33}$$

where $k$ is a storage constant that can be calibrated from observations.

Substituting the storage equation into the continuity equation leads to the analytic solution for $Q$:

$$\begin{aligned} Q(t) &= \int_0^t I(s)u(t-s)\,\mathrm{d}s \\ u(t-s) &= \frac{1}{k}e^{-t-s/k} \end{aligned} \tag{5.34}$$

However, as in river channels storage depends on both inflow and discharge, the single-storage model can be conceptually replaced with a series of $N$ identical linear models, which leads to the following formula for the outflow from the river reach:

$$Q(t) = \frac{1}{k\Gamma(N)}\left(\frac{t}{k}\right)^{(N-1)} e^{-t/k} \tag{5.35}$$

#### 5.5.1.2 Modified pulse method

The modified pulse method is one of the simplest routing techniques. It is based on the assumption that storage and discharge have a unique relationship that remains unchanged when downstream conditions vary. The method is also often used to route the flows through reservoirs with a fixed discharge or through natural lakes. The calculations are based on the modified form of the discrete continuity equation (Equation 5.32):

$$\frac{2S_{t+\Delta t}}{\Delta t} + Q_{t+\Delta t} = I_t + I_{t+\Delta t} + \left(\frac{2S_t}{\Delta t} - Q_t\right) \tag{5.36}$$

At any given time $t$, the right-hand side of Equation 5.32 is known and the left-hand side can be calculated, given that the relationship between storage and discharge is established. However, as in real river channels the storage is usually not a function of discharge alone, a modification of a simple scheme expressed by the equation above is necessary. This modification can be achieved by approximating the reach with a series of cascading reservoirs, each with a level pool (i.e. with a unique storage–discharge relationship). These relationships can be identified in a number of ways (USACE, 1994).

#### 5.5.1.3 Muskingum method

The Muskingum method is the most widely used method for hydrological flow routing. It is based on representing storage by a series of containers shaped as prisms (regularly shaped volume with identical inflows and outflows) or wedges (representing the positive or negative storage occurring when rate of flow varies, for example with the passage of flood wave). The method is based on the continuity equation and a linear relationship between channel storage and both inflow and outflow, called the 'storage equation':

$$S = K[cI + (1 - X)Q] \tag{5.37}$$

where $K$ is the storage constant and $X$ is the weighting factor. Both these constants are dependent on channel characteristics. USACE (1994) provides a detailed discussion of approaches for the identification of the constants $K$ and $c$.

#### 5.5.1.4 Muskingum–Cunge method

The Muskingum–Cunge method is based on the same continuity equation as the Muskingum model but it accounts for the diffusion form of the momentum equation. As the diffusion term is included, the solution of the equations requires the use of at least finite difference methods, and it is for this reason that some classify this method as belonging to the hydraulic routing class of models. USACE states that this method has its parameters more closely linked with physical properties, and that models based on this methodology perform quite well compared with full unsteady flow models. With the exception of this method, all the other techniques are implemented in spreadsheets.

### 5.5.2 Hydraulic routing methods

The system of non-linear equations (Equations 5.29 and 5.30) cannot be solved analytically. If the complete dynamic model is desired, numerical methods can be applied. Although Equations 5.25 and 5.26 provide a complete mathematical description of the physical process, the complexity of the physical properties of natural river channels requires extensive averaging of data and measurements that reduces the model to a conceptual type.

Most of the flow-routing models use the continuity equation in a similar manner, but it is often possible in applications that some terms in the momentum equation are small compared with the bed slope, and the equation can be simplified. USACE (1994) provides the following guidance on selecting the most appropriate simplification scheme, depending on the properties of the channel reach and the flow (Figure 5.22). Guidance on the selection of the most appropriate method is provided in Table 5.5.

#### 5.5.2.1 Kinematic wave

If gravitational and frictional forces are close to achieving equilibrium, the changes in depth and velocity are small in magnitude compared with the bed slope of the reach. This means that the first three terms in Equation 5.25 can be neglected and the momentum of flow can be approximated by Manning's equation for steady uniform flow. As the Manning's formula can establish a relationship between the discharge and the cross-sectional flow area, the movement of water can be described solely by the continuity equation.

#### 5.5.2.2 Diffusion wave

The diffusion wave approximation includes the pressure force term, resulting in the following momentum equation for attenuation effects:

$$S_f = S_0 - \frac{\partial y}{\partial x} \tag{5.38}$$

The equation can also be used for backwater effects by specifying appropriate downstream boundary conditions.

#### 5.5.2.3 Quasi-steady dynamic wave

This approximation includes both pressure force and convective acceleration, which accounts for changes in velocity along the channel.

$$S_f = S_0 - \frac{\partial y}{\partial x} - \frac{V}{g}\frac{\partial V}{\partial x} \tag{5.39}$$

Figure 5.22 Approximation of the full momentum equation. (USACE, 1994)

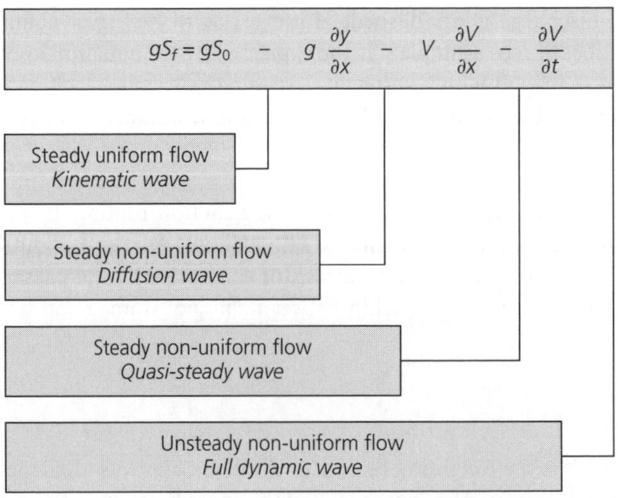

Table 5.5 Selecting channel routing technique

| Factors to consider | Methods that are appropriate | Methods not recommended |
|---|---|---|
| No observed hydrograph data available | Full dynamic routing<br>Diffusion wave<br>Kinematic wave<br>Muskingum–Cunge | Modified pulse<br>Muskingum |
| Presence of significant backwater effects influencing discharge hydrograph | Full dynamic routing<br>Diffusion wave<br>Modified pulse | Kinematic wave<br>Muskingum<br>Muskingum–Cunge |
| Flow outside of the main channel and into the floodplain | All hydraulic and hydrological methods that calculate hydraulic properties of main channel and overbanks separately | Muskingum |
| Channel slope >0.002, $TS_0v_0/d_0 \geq 171$ | All methods | None |
| Channel slope 0.002–0.0004, $TS_0v_0/d_0 < 171$ | Full dynamic routing<br>Diffusion wave<br>Muskingum–Cunge<br>Modified pulse<br>Muskingum | Kinematic wave |
| Channel slope <0.0004, $TS_0(g/d_0)^{1/2} \geq 30$ | Full dynamic routing<br>Diffusion wave<br>Muskingum–Cunge | Kinematic wave<br>Modified pulse<br>Muskingum |
| Channel slope <0.0004, $TS_0(g/d_0)^{1/2} < 30$ | Full dynamic routing | All others |

The convective and local acceleration terms in the full momentum equation are often of opposite sign for fast-changing water flows (flood routing), and in such cases the error introduced by simplification is greater than for diffusion waves. For that reason, this approach is commonly used for steady flow-water surface profile computations when the convective acceleration term may still be significant but the changes in velocity with respect to time (local acceleration) are negligible.

### 5.5.3 Reservoir routing

Flows that pass through a reservoir are delayed and attenuated as they enter the storage and spread over the reservoir surface. Water that is stored is then released through either a controlled or uncontrolled outlet. The routing of inflows through a reservoir (including natural lakes) is based on the same general principles described by the continuity and momentum equations. Thus, Equations 5.29 and 5.30 still apply to reservoir routing with free overflow spillways but a number of simplifications are made, as compared with channel routing which effectively transforms the problem into kinematic routing applied to a single reach (the reservoir) and backwater effects are neglected.

The modified pulse method is traditionally the method of choice for routing flow through reservoirs. It has been described above when discussing routing techniques for river channels. Fenton (1992) proposed a convenient modification of the original technique based on solving Equation 5.36. He pointed out that at each time step several iterations evaluating the right-hand side of the equation are necessary. An alternative formulation is based on reservoir surface area instead of on the storage, and is derived from the observation that any change in storage can be expressed as

$$\frac{dS}{dh} = A(h) \tag{5.40}$$

where $h$ is the surface water level and $A$ is the water surface area at the elevation $h$.

Substituting Equation 5.40 into the continuity equation (Equation 5.31) leads to the following:

$$\frac{dh}{dt} = \frac{I(t) - Q(t, h)}{A(h)} \tag{5.41}$$

The advantage gained by using this formulation lies in the fact that the storage does not need to be calculated and the outflow is explicitly expressed as a function of the water elevation, which is always the case whether the discharge from the reservoir is through a free overflow spillway or through controlled flow discharge facilities (gates, stoplogs, valves, etc.). The solution to Equation 5.41 can be expressed as

$$h(t+\Delta t) = h(t) + \Delta t \frac{I(t) - Q(t, h)}{A[h(t)]} + O(\Delta t^2) \tag{5.42}$$

where $O(\Delta t^2)$ is the Landau order symbol indicating that the omitted terms vary as the second or higher powers of $\Delta t$.

When the outflow from the reservoir is controlled using flow discharge equipment (gates, stoplogs, valves, turbines etc.), the continuity equations still apply but the outflow $Q$ can become defined by a complex relationship that depends not only on the storage and inflow but also on a number of other factors that take precedence over the natural relationship between the inflow, the outflow and the storage. These factors are usually related to the purposes for which the dam and reservoir were built – flood management, flow augmentation, water supply, irrigation, hydropower generation and others.

### 5.5.4 Simulating Mattagami River inflows

The Mattagami River watershed is located in the northern part of Ontario. It has a drainage area of almost 37 000 km$^2$ at the Little Long Generating Station and drains into Hudson Bay. Little Long Generating Station is a part of a cascade system of four hydroelectric generating stations located in the lower part of the Mattagami River system. The Little Long reservoir–dam system is located within a modified continental climatic zone. During the winter, cold polar air masses produce dry, clear, cold weather, and in the summer successions of cyclonic storms sweep the area and warm humid air masses from the south alternate with cooler drier air from the north. The average mean daily temperatures for January and July are approximately −19°C and +17°C, respectively. In a normal year, the area is frost free from mid-May to early September. Average annual total precipitation is about 86 cm (water equivalent mean). Rainfall accounts for 63% of the total precipitation, with the maximum occurring in the summer months. Snow cover is present for about 160 days/year, reaching maximum depth on the ground in February (average depth 61 cm).

As a consequence, strong seasonal patterns can be identified for the historical Little Long reservoir inflow data. Freeze-up usually occurs by late November or early December. The inflow volume into the reservoir reduces gradually. By mid-December, ice cover is complete, except in tailraces and rapids, which stay open all winter. The inflows stay small but positive.

During the winter peaking operation, ice hinges form along the shoreline, allowing the central ice sheet in each reservoir to move with the changing water levels without breaking. In late winter, the central ice sheet subsides, and as the inshore ice settles to the substrate the central floating ice sheet breaks from the inshore ice and can be pushed downstream. The ice breaks and the snow melts quickly during the spring freshet in mid-March.

The historical inflow data for Little Long comprises 18 394 records, corresponding to the average daily inflow rate for the 50 years from 1 August 1963 to 9 December 2013. In order to simplify the analysis, the daily data ranging from 1 January 1964 to 31 December 2013 were selected, giving 18 250 values over a 50-year time period, and these were completed by means of the 2D interpolation technique (365 days were counted for each year, extra days for leap years were deleted). The mean, median and range of daily inflows over a year are shown in Figure 5.23.

The preliminary simulation results of reservoir inflows were generated using the following process.

1 *Obtain the logarithm of inflow data* (Figure 5.24). The logarithm is a non-linear transformation that reduces positive skewness and compresses the upper tail of the distribution, while stretching out the lower tail. There are three reasons why the data were transformed by taking the logarithm: (i) the statistical techniques used work best with symmetrical distributions, (ii) the techniques work best when the data are homoscedastic, and (iii) the relationship between variables becomes approximately linear.
2 *Fourier decomposition.* In order to find the seasonal pattern within each year, an eight-term Fourier model was fitted to the logarithms of the historical inflow data. Figure 5.25 shows the corresponding data over the 4 years 2011–2014 as an example.
3 *Residuals.* The residuals of logarithm inflow minus the value of fitted Fourier model were calculated (Figure 5.26).
4 *Differencing and seasonal differencing* (Figure 5.27). A seasonal difference is the difference between an observation and the corresponding observation from the previous year. Time series with trends, or with seasonality, are not stationary, as the trend and seasonality will affect the value of the time series at different times. In general, a stationary time series will have no predictable patterns in the long term. As a result, differencing and the seasonal differencing were conducted to make the time series stationary. Lags were picked as 1, 7 and 365 days.
5 *Adjust variance of differencing data.*

Figure 5.23 Inflow hydrograph for the Little Long dam–reservoir system (1964–2013)

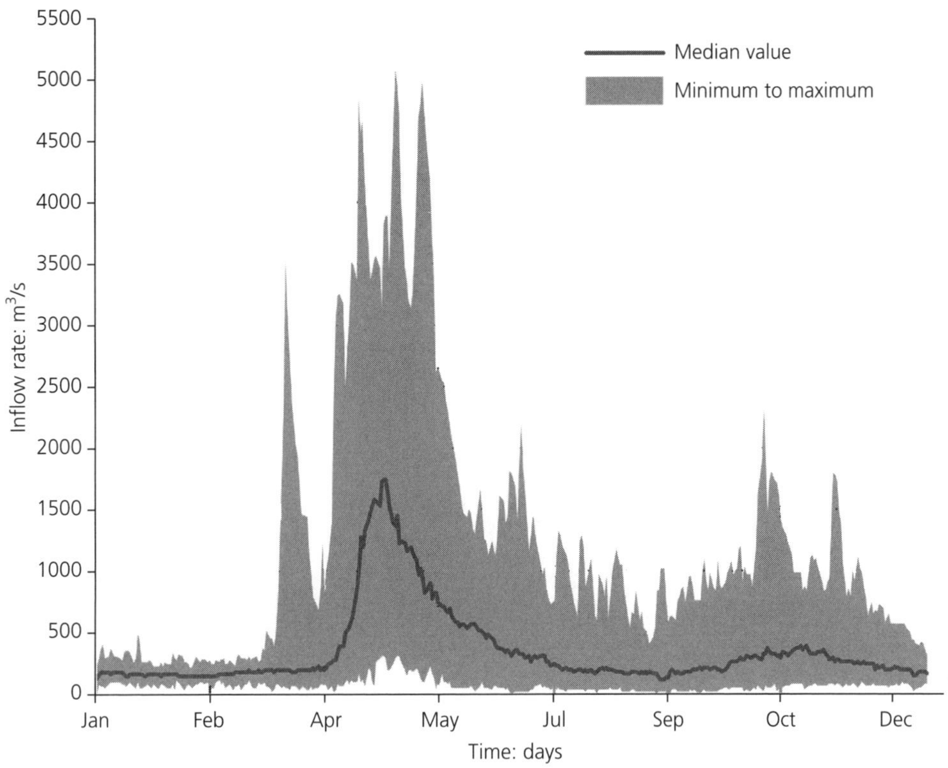


6 *Modelling choice and fitting.* A seasonal ARIMA process was selected. Sample autocorrelation functions are shown in Figure 5.28.
7 *Simulation results.* The sample path of the Monte Carlo simulation based on the seasonal ARIMA model is shown in Figure 5.29.
8 *Simulation results for future inflows.* The 50 years for historical data and the 50 years for simulated data are shown in Figure 5.30.

## 5.6. Uncertainty in watershed and streamflow modelling

The importance of uncertainty assessment in model predictions is hard to overestimate when one realises that the models are used for practical purposes in supporting decision-making. From that perspective the most important aspect of uncertainty assessment is the estimation of the model-output uncertainty arising from various sources during the simulation. Depending on the overall purpose of the model development, other aspects may be important and may include assessment of uncertainty caused by individual sources (model parameters, model structure and measurement errors) and apportioning of the total output uncertainty to the various sources.

The nature of this uncertainty is a crucial question when investigating watershed response to the climatic and weather forcings. Although there is acknowledgment within the hydrological modelling community that both the aleatory and epistemic uncertainties are present in hydrological modelling, the implementation of this distinction in modelling practice is inconsistent. The intrinsic character of uncertainty in hydrology should be indisputable, but practice is still focused on deterministic paradigms.

A closer look at the methodology proposed by Freeze and Harlan (1969) and the further developments of their modelling philosophy reveals that the underlying theory, while applicable at point or small scales, is inadequate for larger scales (e.g. watersheds). In addition, the presence of spatial heterogeneity of watersheds forces the introduction of 'effective parameters', which invalidates the concept of a physical basis.

This general approach (followed also, with some modifications, in conceptual modelling) reflects a deterministic view of the world in which uncertainty is a subjective element that should be either eliminated or at least reduced as far as possible. Such a position is not unexpected, as the approach presumes that an understanding of the causality mechanisms is available.

At present, the predominant position of the hydrology profession is that models based on principles of physics are 'a must' if one wants to make reliable predictions beyond the range of available records. However, the problem of implicit upscaling from

Figure 5.24 Historical data of Little Long inflows and the corresponding logarithm value

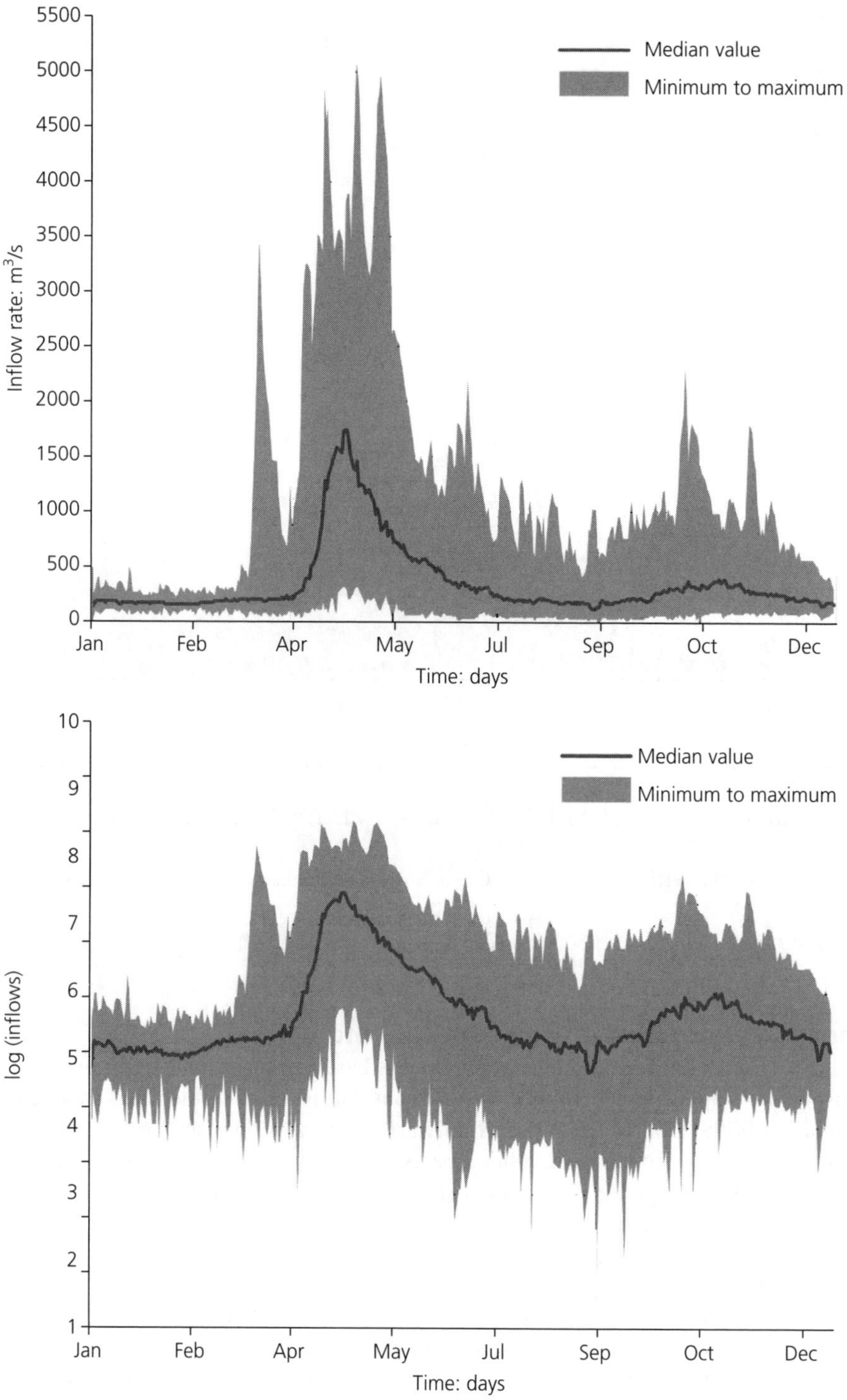


micro- to macro-physics, and the necessity for averaging model parameters both in time and space creates doubts about the degree of reliability of these predictions. This has considerable impact on our ability to provide credible uncertainty estimates for outputs from hydrological models.

### 5.6.1 On uncertainty and its assessment

The problem of assessing uncertainty in hydrological predictions is complex. Often, 'uncertainty analyses' are performed without an attempt to provide a scientific definition of their meaning. The problem of ambiguity in understanding or interpreting what *uncertainty* means is unresolved in applications of hydrological modelling.

An in-depth discussion of this fundamental issue for hydrological modelling can be found in Beven (2006), Montanari (2007) and Todini and Mantovan (2007), but it is worth giving a brief summary here. Beven points out that, although there are at

Figure 5.25 An example of fitting an eight-term Fourier model to the logarithm of the inflow data (2011–2014)

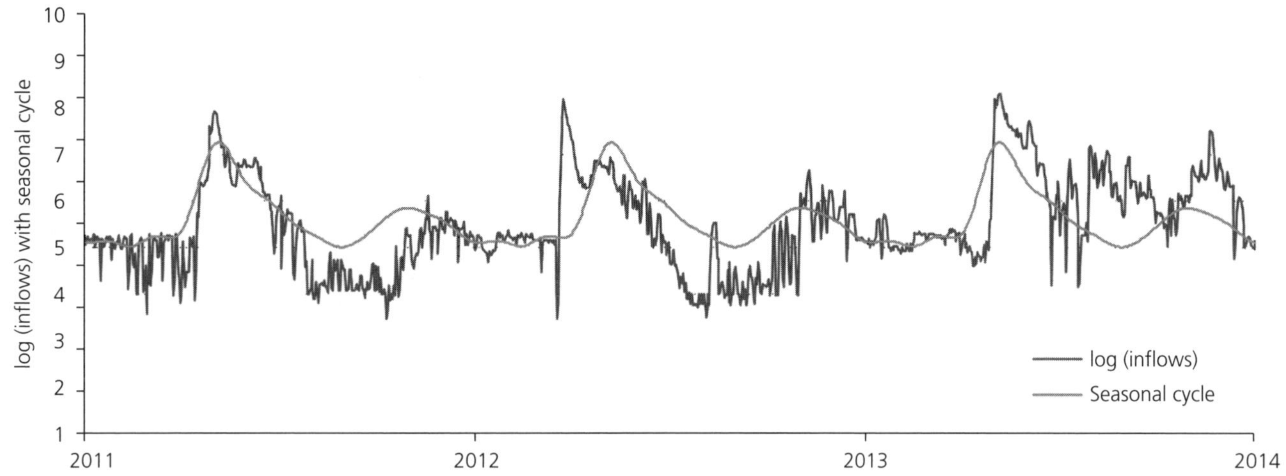


present different philosophies about how to estimate uncertainties, it is not possible to recommend any particular method of uncertainty estimation. On the question of whether the uncertainties are currently being under- or overestimated, his answer is that we do not actually know. Todini and Mantovan argue that in most, if not all, uncertainty analyses the uncertainty in model output (prediction) is confused with the uncertainty of the 'real' predicted value, which is conditional on the prediction. Montanari's main point is that the present lack of coherent terminology and a systematic approach makes it difficult, if not impossible, to assess the adequacy of currently used individual methods to quantify the reliability of hydrological simulations, design variables and predictions. Thus, from his point of view, the difficulties in uncertainty estimations can be attributed strictly to inferior or incomplete methods available for uncertainty estimation and the lack or low quality of available data. This point of view should not be surprising, because by following the deterministic approach the complete characterisation of any natural process is possible, and even if it is not achievable today due to the constraints in scientific knowledge and data-acquisition technology, it will be possible one day – we just need to work harder.

Thus, the general, and almost never questioned, approach to the assessment of uncertainty in hydrological modelling is to concentrate most of the effort on epistemic uncertainties related to measurements (data), model structure and model parameters errors. The literature on the methods and practice of assessing the epistemic uncertainty in hydrological modelling is abundant (see Pechlivanidis *et al.*, 2011), nevertheless we are far from a successful resolution of the problem.

Beven (2006) posed the question as to whether the uncertainties in hydrologic models are being overestimated. His answer to this question was that we do not actually know because currently available methods of uncertainty estimation are inadequate. He pointed out that in any application of uncertainty estimation methodologies, a number of subjective decisions must be made, from the choice of parameter distributions to be considered to assumptions about observational errors or residual error structures. Although, in principle, many of these assumptions can be checked as part of the analysis, this is usually not done and the posterior analyses of such assumptions are not available. Later, Beven (2006) provided the following succinct explanation.

> However, although epistemic errors are endemic to environmental models, there is no formal theory for dealing with them. Effectively, there can be no formal theory for dealing with epistemic error because we do not have adequate knowledge of the nature of the errors. If we had adequate knowledge, we would have a better idea of how to deal with them and they would no longer be epistemic. This is the core dilemma in modelling catchment systems for decision-making.

The reader interested in a more detailed characterisation of the shortcomings of the present state of the art in the area of epistemic uncertainty assessment is referred to the discussion initiated by Beven (2006) and followed by Andreassian *et al.* (2007), Hall *et al.* (2007), Montanari (2007), Todini and Mantovan (2007) and Sivakumar (2008a,b).

The aleatory part, the uncertainty caused by natural variability, is usually treated in a more or less perfunctory manner, which is not surprising as the origins of this source remain outside of the hydrological science domain and hydrologists are, in that respect, no different from other scientists believing that only what they do is important. Aleatory uncertainty (sometimes also called stochastic uncertainty) is the property of the hydroclimatic system and cannot be reduced or eliminated. In his commentary, Sivakumar (2008a) argues that it is 'unpredictability of Nature' that is behind the aleatory uncertainty, and as the models we develop are always fitted only to the data, not to the Nature itself, they will always be only crude approximations.

Figure 5.26 Residuals of the logarithm of the inflow minus the value of the fitted Fourier model

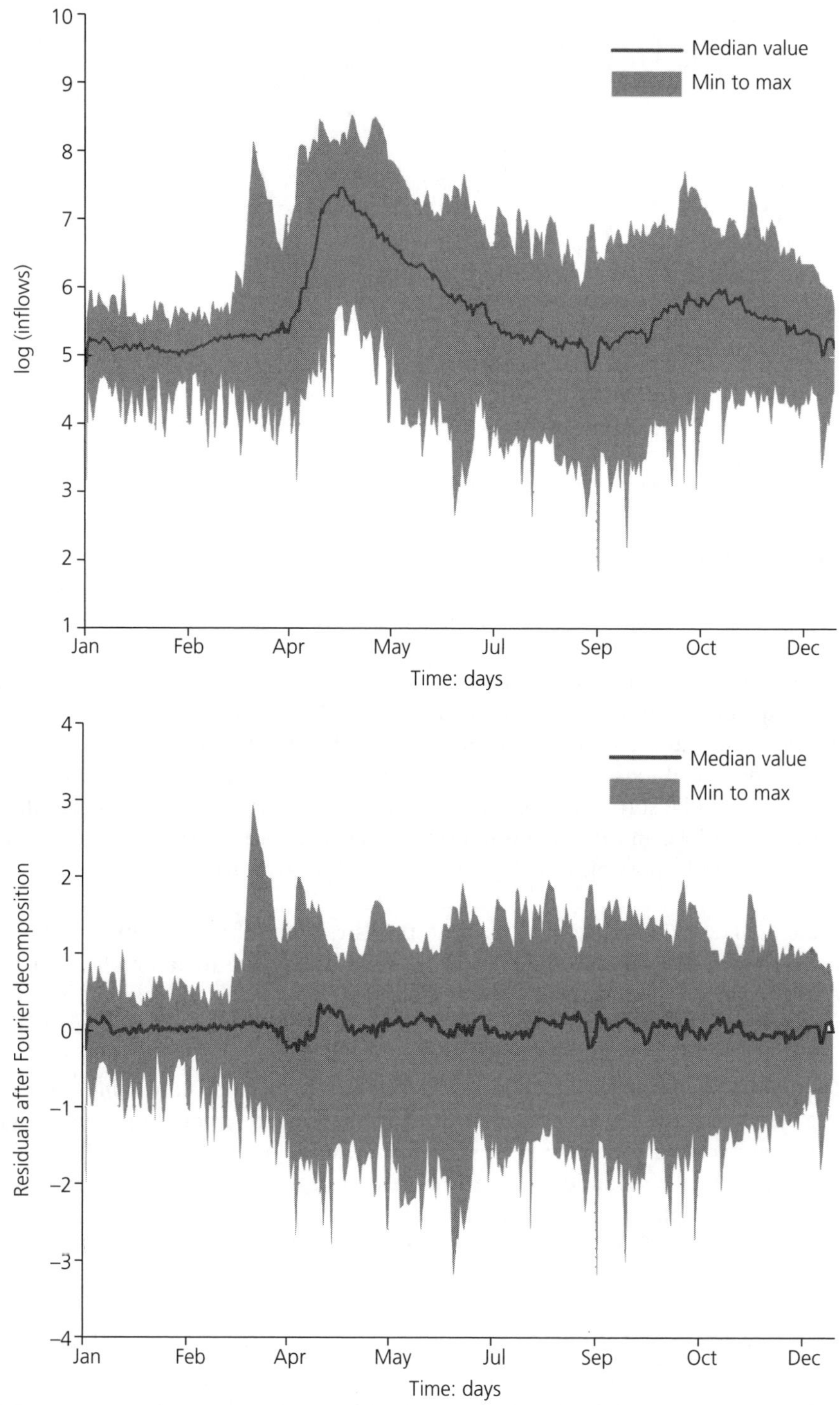


It is interesting to explore in more detail the nature of the problem of uncertainty in hydrological modelling and the way it is being handled at the present. Figure 5.31 illustrates the complexity of the problem that needs to be solved when following a deterministic deductive approach.

The climate drives the weather, which in turn provides the input to the watershed (hydrological system). The watershed transforms the inputs into runoff, and the watershed drainage systems then routes the streamflow to locations within the watershed river system. At some of these locations the river can be dammed, resulting in a presence of a reservoir that can modify the temporal characteristic of the streamflow process. This is, of course, a simplified picture in that that there are other links and feedbacks between these components, and a complete deterministic approach should consider all these components as belonging to a larger hydroclimatic system. In a simplified hydroclimatic system the feedback loops are absent and the driving forces move only from the top to the bottom.

Figure 5.27 Residuals after differencing and seasonal differencing

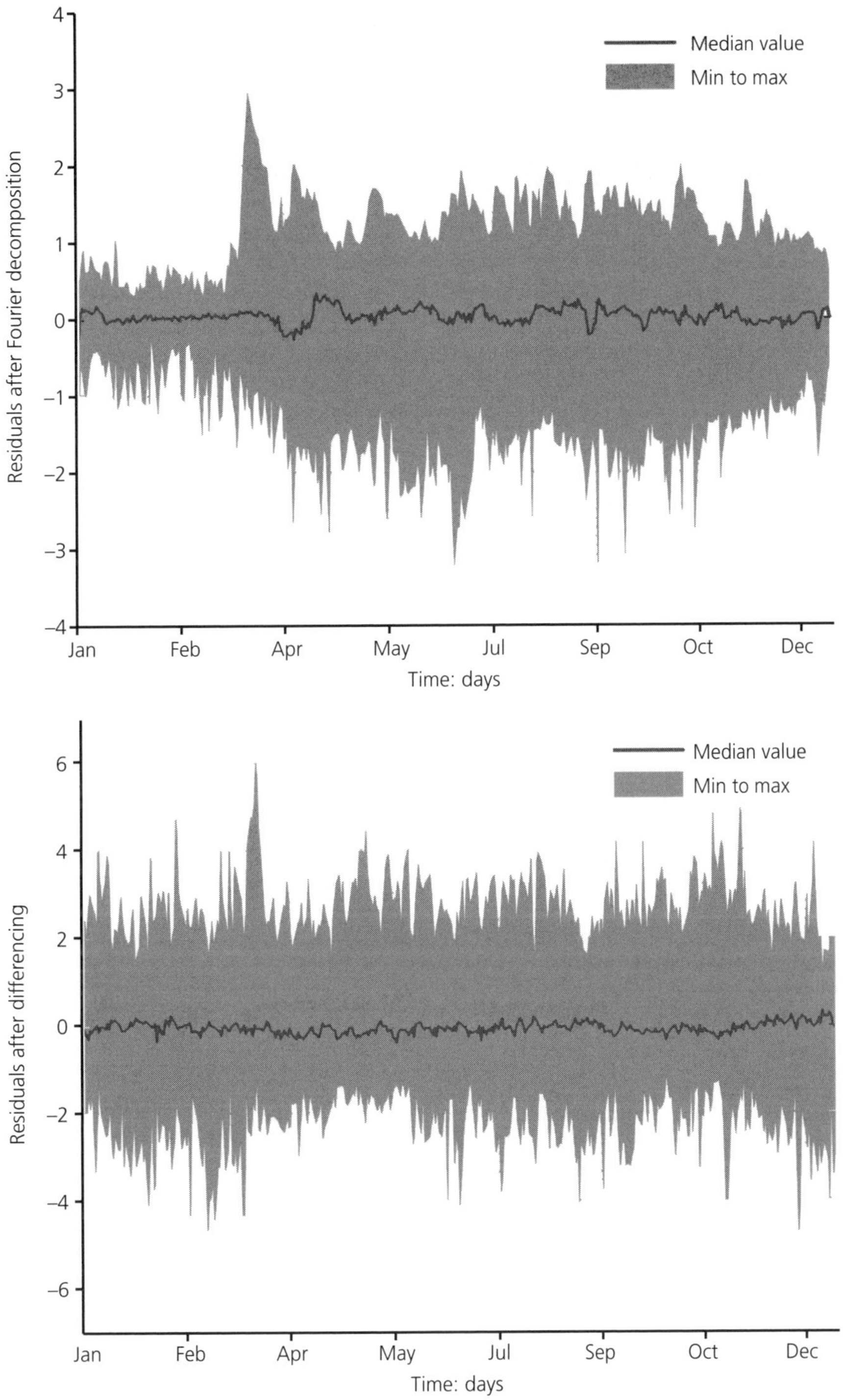


A further application of a reductionist philosophy allows detachment of the watershed from the other components, making it the only component of the hydroclimatic system to be modelled, with the inputs from the other components (climate and weather) being determined separately. It should be pointed out that the modelling of the entire hydroclimatic system would be simply impractical, making some degree of separation of subsystems necessary. The complexity of physically based models of the atmosphere that would be required to model the entire global atmosphere and of physically based models responsible for regional-scale climate and weather is orders of magnitude larger and more complex than the complexity of a hydrological system, both in terms of mathematical formulation and of the computational effort required in applications. Thus modelling of the climate and weather as a deterministic systems provides a huge challenge in applications of hydrological modelling, especially when the issue of the uncertainty in the modelling results is considered.

Figure 5.28 The autocorrelation function and partial autocorrelation function of residuals after variance adjustment

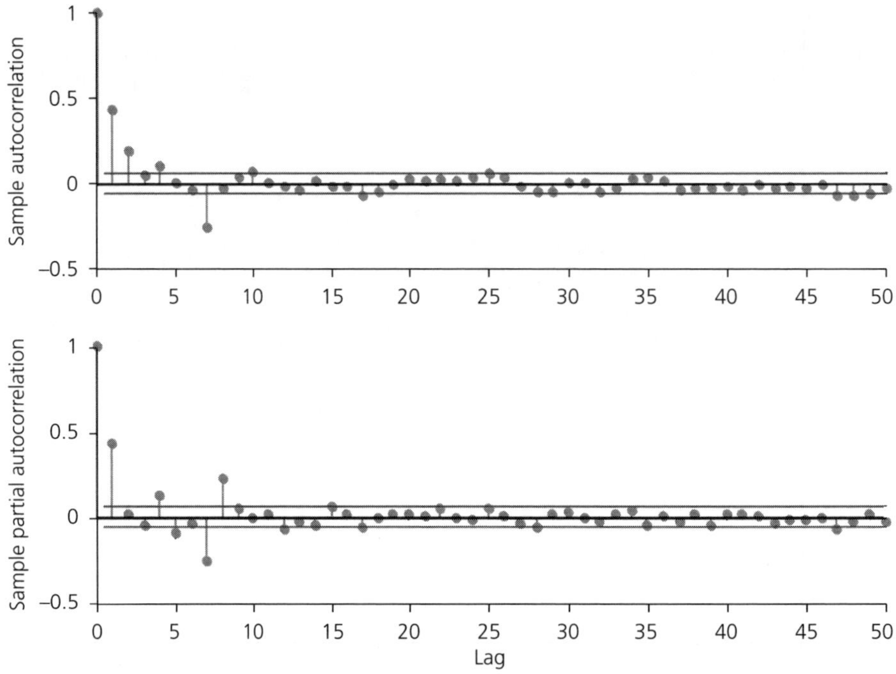


The challenge originates from the positioning of the hydrological system as a recipient of aleatory uncertainty inputs from the other components of the hydroclimatic system as modelled following the deterministic paradigm. These inputs to the hydrological model (most often these are precipitation and temperature, but other climatic variables can also be present) are the only sources of the aleatory uncertainty and represent the total natural variability affecting the watershed. However, the current physically based modelling of climate and weather remains incapable of providing a sound and credible stochastic characterisation of these outputs.

It is interesting to note that atmospheric modelling, which still remains predominantly deterministic, will have to evolve to the state of efficiently dealing with the problem of aleatory uncertainty. The general and unrealistic expectation is that the problem of epistemic uncertainty will be solved in time in a manner similar to hydrological modelling, where following the principles of physics in building physically based models allows the inference of the stochastic properties of outputs from the model, for which all parameters are measurable and have known probabilistic characteristics. However, for climate and weather modelling the problems associated with accounting for aleatory uncertainty are quite different because the natural variability of climatic processes has both internal and external sources.

Figure 5.29 Sample path of the simulated residuals after variance adjustment based on seasonal ARIMA

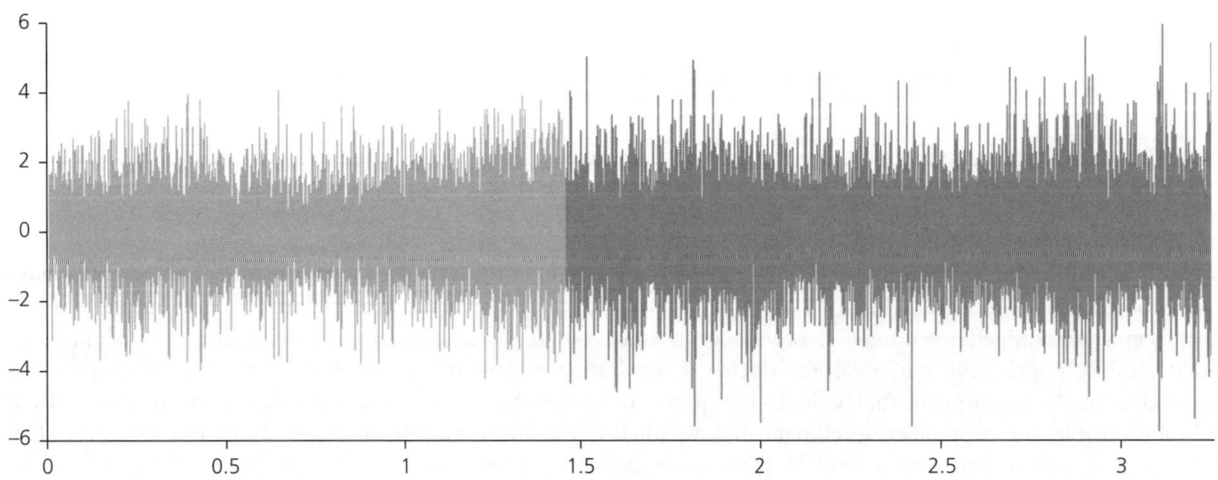

Figure 5.30 Historical and simulated Little Long inflow hydrograph

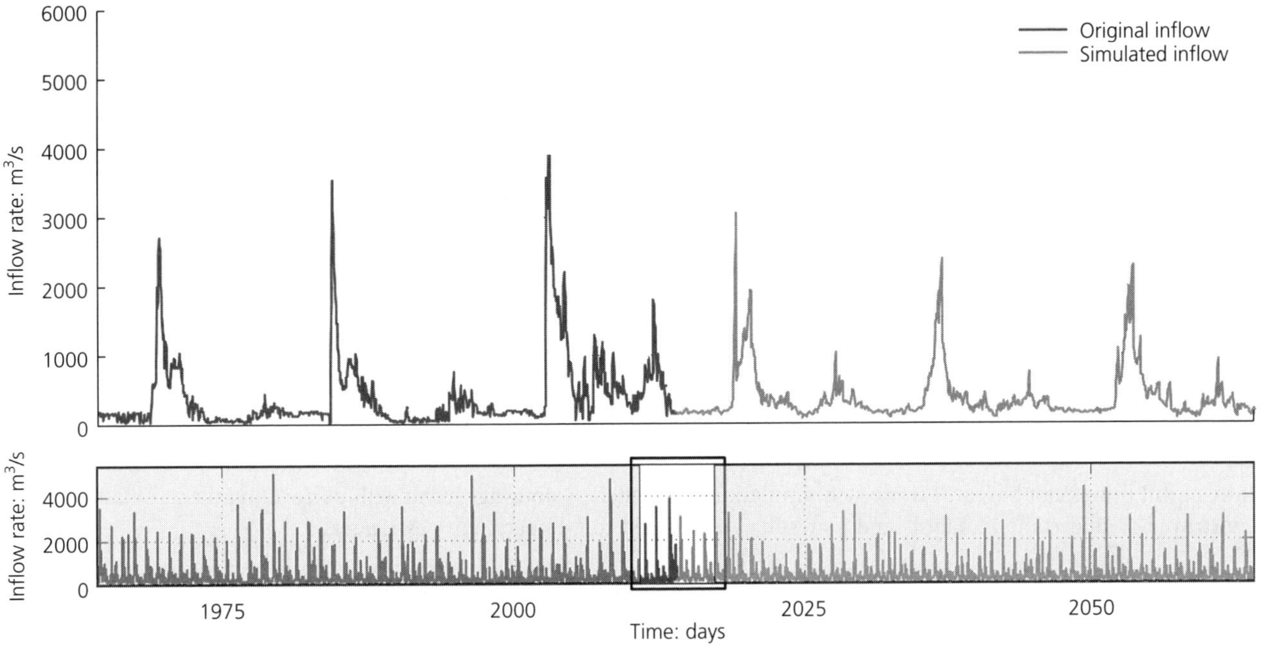


Where does all this leave the hydrologists and their models in terms of assessing uncertainty? The epistemic part (the watershed model itself) can be dealt with efficiently, but not necessarily correctly, using the procedures mentioned above. The aleatory input has to be addressed differently, and this is usually done using various forms of statistical analysis of climatic records. As the analytical solutions for hydrological models are not available, the analysis of aleatory (climate inputs) and epistemic (the watershed model) uncertainty is carried out by means of Monte Carlo techniques, generating series of watershed model outputs responding to stochastic time series of inputs and randomly sampled model parameters. This form of modelling is often called 'hybrid' or 'stochastic–deterministic' modelling, but it seems that the term 'randomised' deterministic modelling better reflects the nature of the process.

The general routine is to take an available watershed model (this can range from an empirical to a fully distributed physically based model) and attach probability distributions to anything that is uncertain (parameters, relationships, etc.). Additional terms are also used to represent uncertainty in the model structure. If the distributions are also attached to inputs, there is also some accounting for aleatory uncertainty (it is puzzling that we, as hydrologists, do not hesitate to generate time series of inputs – temperature and precipitation – that have a much more complex stochastic structure than flows, but we consistently refuse to perform the same operation on flows, arguing that the recorded information is insufficient to provide accurate stochastic characterisation of flows).

Figure 5.31 Physically based modelling – deductive approach

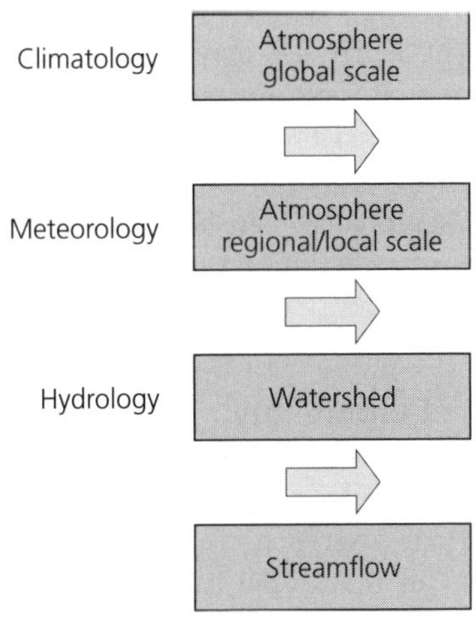

Repeated simulations with random sampling from these distributions will, of course, provide a sample of results that can be subsequently used to numerically characterise 'uncertainty bounds'. What is wrong with this approach?

- If the problem is not deterministic (and we acknowledge that by using Monte Carlo simulation) then, in general, the approach should begin with establishing a mathematical framework consisting of stochastic equations describing the physical processes. Most often, if not always, these equations are of differential type. Thus the problem at hand is described by a system of stochastic differential equations. The solution of such a system is a stochastic process. The properties of this process are, in general, different from the properties of randomised solutions of the system of deterministic equations.
- Parameters are, in general, unknown, and in practice are not measurable (see the discussion above), thus eliminating the advantage of using a physically based model. The practice is to attach more or less arbitrarily selected probability distributions to the parameters of the model, with distribution parameters selected in a similar manner. As the distributions and their parameters are unknown, the resulting Monte Carlo simulations are, in fact, providing something that more resembles a sensitivity analysis than an uncertainty analysis. (In fact, the sensitivity analysis can be defined as an uncertainty analysis with uniform distribution attached to anything that is uncertain in the model.) Thus, what we in fact learn from the entire process is mostly 'what may happen if' but not how likely it is.

Many useful things can be learnt from such simulations, and there is nothing wrong with performing them, unless the process is portrayed as uncertainty analysis and its results are claimed to be uncertainty estimates.

It is common in hydrological modelling practice to state that this 'stochastic physically based simulation approach' captures both aleatory and epistemic uncertainties, while in reality it just attempts to capture both uncertainties without any actual confirmation that it is successful in doing so.

### 5.6.2 Impacts of non-stationarity

The non-stationarity aspects in hydrological modelling (especially when the extremes are concerned) were recognised a long time ago, and then immediately abandoned. However, this is now gaining more attention due to the recently increased general interest related to climate change and the means of characterising its impact on the availability of water resources. There is also a great deal of confusion and imprecision in using the term 'non-stationarity', and for this reason it may be beneficial to state more precisely what the term means in the context of hydrological modelling. The issue was raised following the paper by Milly *et al.* (2008) and from the reaction following its publication it is quite clear that there are different interpretations of the terms 'stationarity' and 'non-stationarity'. In the interest of a balance between scientific precision and practical interpretation the following explanation can be suggested.

The scientific term 'stationarity' does not necessarily mean constancy of variables. What it does mean is constancy of laws and patterns. A stationary stochastic process, by definition, has the property that its probability distribution does not change with time. That is, if parameters such as mean and variance exist, they are constant. In contrast, statistical properties of non-stationary processes vary over time, for example, featuring abrupt or gradual changes in the mean, variance, higher moments, or characteristics of extremes (Kundzewicz, 2011).

It is worth pointing out that the definition refers to a stochastic process, which in hydrological modelling is typically understood to be a family of random variables indexed by time. Recorded data (time series) associated with the process is a specific realisation of the process.

The notions of stationarity and non-stationarity in physics may have entirely different meanings, and these meanings may differ from case to case. This creates room for confusion and misunderstanding, as the notion of stationarity in hydrology is associated with the model but often it is attributed to the real-world system that is being modelled. It is an important issue because it affects both the deterministic branch of hydrological modelling (can a deterministic model account for non-stationarity?) and the way physical non-stationarity can be modelled using the framework of stochastic calculus. A more complete explanation of the problem with the definition of non-stationarity as applied to natural systems; a discussion of its implications for the characterisation of uncertainty can be found in Koutsoyiannis (2006).

### 5.6.3 Methods for the characterisation of uncertainty

The traditional approach to uncertainty assessment in hydrological modelling is linked to the model calibration process, and it is usually assumed that the uncertainty in the model output is attributable primarily to uncertainty associated with the values of the parameters within the model. This approach effectively neglects uncertainty in forcings, and assumes that uncertainty caused by structural inadequacies in the model can be accounted for using relatively simple additive error structures.

In recent years there has been an increasing interest in developing uncertainty assessment frameworks which explicitly recognise that the total uncertainty in the model output depends on the uncertainty characteristics of all three components of the model (forcings, parameters and model structure).

The characterisation of the methods and techniques that are currently used for the purpose of the characterisation of uncertainty in hydrological modelling can be found in Blasone (2007) and Pechlivanidis *et al.* (2011), and a brief summary of these sources is given below. The methods can be roughly categorised into two groups: methods based on statistical inference and Bayesian methods.

#### 5.6.3.1 Methods based on statistical inference

These methods were the first to find application in hydrological modelling and were founded on traditional statistical theory. The applications in hydrology followed either various techniques based on Taylor-series expansion or the point-estimation method. Both methodologies are rarely used at present due to their limited characterisation of uncertainty (restricted to the statistical moments only) and the requirement for assumptions that are not justifiable for hydrological systems.

Another group of methods that are still in limited use is based on non-linear regression analysis. The regression is performed simultaneously on forcings (inputs), and model parameters provide the user with the joint confidence intervals for the parameter set.

#### 5.6.3.2 Bayesian methods

These methods received increased attention from hydrology modellers beginning in the 1990s. Three of the methods that have been most commonly used since their inception in hydrological modelling are briefly described below.

##### The Generalised Likelihood Uncertainty Estimation (GLUE) procedure

The GLUE procedure (Beven and Binley, 1992) was one of the first of the robust and comprehensive frameworks developed for the purpose of assessing uncertainty in hydrological modelling, and it is still the most often used approach. As GLUE uses an informal (subjectively selected) likelihood measure, it is quite often characterised as an informal Bayesian method. Behind the idea of GLUE was the belief that the formal Bayesian approach cannot successfully deal with the equifinality problem, and thus leads to overconditioning of the likelihood surface. The most often used likelihood measure is the inverse of the error variance.

Although the method is theoretically simple and easy to implement, it has often been criticised as being not coherent and too subjective. Besides the use of the subjective likelihood measure, which leads to inconsistency with classical Bayesian statistical inference, the other most often quoted shortcoming of the method is the dependency of the estimated uncertainty on a subjectively defined threshold separating acceptable and unacceptable parameter sets.

##### Multi-model methods

The Bayesian model averaging (BMA) (Duan *et al.*, 2007; Hoeting *et al.*, 1999) approach provides the joint assessment of model structure and parameter uncertainties by deriving the posterior distribution of the predicted output from a weighted sum of the posterior distributions from a number of different models. The multi-model ensemble (MME) approach (Hagedorn *et al.*, 2005) samples from the output distribution of several models.

Both methods are computationally demanding to apply and, although multiple models may compensate for errors in different structures, there is no guarantee that the range of chosen models will adequately cover the 'true' model.

##### Bayesian total error analysis (BATEA)

BATEA is the approach based in Bayesian hierarchical methods (Kavetski *et al.*, 2006a,b) and in the use of explicit uncertainty models for input/output data and structural uncertainty. It provides the opportunity to directly address the sources of uncertainty.

#### 5.6.3.3 More recent developments

As mentioned previously, the progress in developing new methods and techniques capable of addressing the total uncertainty of model outputs has been quite rapid. New methods and techniques are being developed every year, and in the last decade significant progress has been made toward the development of a systematic framework for the assessment of uncertainty in hydrological modelling. The developments in adaptive parameter estimation techniques, Bayesian analyses and Monte Carlo Markov chain techniques offer new possibilities for the simultaneous characterisation of all sources of uncertainty in hydrological models. The recent work in this area is briefly characterised below, and examples of proposed methods are provided.

### State space filtering

Vrugt *et al.* (2005) developed a simultaneous parameter optimisation and data assimilation (SODA) method. The method puts together the Monte Carlo Markov chain (MCMC) scheme, called the Shuffled Complex Evolution Metropolis (SCEM) algorithm, and the ensemble Kalman filter, providing a general framework for explicit treatment of input, output and model-structure uncertainty.

Moradkhani *et al.* (2005) proposed a dual-state-parameter estimation approach based on the ensemble Kalman filter (EnKF) for sequential estimation of the parameters and states of a hydrological model. The approach allows all three sources of uncertainty (input, output and parameters) to be addressed.

Moradkhani and Hsu (2005) used a sequential Monte Carlo method, also known as a particle filter, for the assessment of the uncertainty of both the parameters and the states of a conceptual hydrological model.

Vrugt *et al.* (2006) demonstrated a stochastic ensemble implementation of the Sacramento model used routinely by the National Weather Service for deterministic streamflow forecasting. The Simultaneous Optimisation and Data Assimilation (SODA) method uses an ensemble Kalman filter for recursive state estimation, which enables treatment of the streamflow data error, model structural error and parameter uncertainty, while enabling implementation of the Sacramento model without major modification to its current structural form. Model parameters are estimated in batches using the shuffled complex evolution metropolis stochastic ensemble optimisation approach (SCEM-UA).

### Multi-model averaging

Butts *et al.* (2004) developed a modelling framework to explore different hydrological model structures in both the rainfall–runoff and channel-routing components of their hydrological models. This framework was then used in a systematic analysis of the performance of a number of different model structures and a split sample test was carried out. An automatic multiple objective calibration procedure was used to determine an 'optimal' parameter set and to define the Pareto fronts for each structure. The absolute value of the average error and the root mean square error were used as the calibration criteria. These 'optimal' models for each model structure were used to create model structure ensembles to evaluate model-structure uncertainty and to evaluate the performance of combinations of different model structures.

Georgakakos *et al.* (2004) developed and applied sample-path methods to characterise streamflow-simulation uncertainty by diverse distributed hydrological models, with the focus of the approach being on the model-parameter and model-structure uncertainty. Multi-model ensembles were analysed for six application catchments in the central USA to characterise model-structure uncertainty within the sample of models. Ensembles from single distributed and lumped models were also used for one of the catchments to provide a basis for characterising the impact of parametric uncertainty versus model-structure uncertainty in flow-simulation statistics.

Vrugt *et al.* (2006) presented a multi-criteria optimisation approach to the calibration of a forecast ensemble using Bayesian model averaging (BMA). The proposed framework implemented various diagnostic measures to account for different but complementary metrics of forecast skill. It used a genetically adaptive evolutionary algorithm to solve for the Pareto set of solutions. Two case studies demonstrated that the proposed multi-objective calibration approach provided efficient guidance on the selection of the appropriate values of the BMA weights and variances.

Marshall *et al.* (2007) applied Bayesian model averaging within a hierarchical mixtures of experts (HME) framework. The framework provided additional flexibility, compared with simple combinations of models, by allowing the way in which model predictions are combined to depend on predictor variables. The capability of the proposed approach was illustrated by example application on ten selected Australian catchments.

Ajami *et al.* (2006) developed the Integrated Bayesian Uncertainty Estimator (IBUNE) for explicit accounting of the major uncertainties in hydrological rainfall–runoff predictions. The uncertainties in the input (forcing) data and from the model parameters are reduced through the use of the SCEM algorithm, which was extended to include a precipitation-error model. A Bayesian Model Averaging (BMA) scheme was then employed to improve the prediction skill and uncertainty estimation using multiple model outputs. A series of ten case studies using three rainfall–runoff models to predict the streamflow in the Leaf River basin, Mississippi, were used to examine the usefulness of the technique. The results suggested that ignoring either the input forcings error or the model structural uncertainty leads to unrealistic model simulations and their associated uncertainty bounds.

Vrugt and Robinson (2007) compared ensemble Kalman filtering, based on a single model, and the ensemble BMA for probabilistic streamflow forecasting. The results demonstrated that in its current form BMA could not achieve a performance matching that of the EnKF method for the best model in the ensemble, in terms of either forecast accuracy or precision. The authors observed that, as both methods rely on normal or gamma probability distributions, they may be inappropriate for accurately characterising input and model uncertainty. The overall conclusion was that more research is needed on the development of appropriate error structures for errors in rainfall, discharge, evaporation, model structure and parameters forecasting.

### Other approaches based on Bayesian methods

Kavetski *et al.* (2006a) developed a Bayesian total error analysis (BAETA) methodology for hydrological models that require the modeller to directly and transparently incorporate, test and refine existing understanding of all sources of data uncertainty. The input uncertainty was assumed to be multiplicative Gaussian and independent for each storm, but the general framework allows for alternative uncertainty models. The general methodology is computationally intensive, but powerful modern numerical methods (Monte Carlo analysis combined with fast Newton-type optimisation methods and Hessian-based covariance analysis) can be employed to obtain practical solutions.

Kavetski *et al.* (2006b) applied the BATEA methodology to directly address both input and output errors in the hydrological modelling of two North American watersheds (French Broad River and Potomac River). The study assessed the performance of a conceptual watershed model with and without accounting for input (precipitation) uncertainty. The results demonstrated considerable effects of precipitation errors on the predicted hydrographs (especially the prediction limits) and on the calibrated parameters. While BATEA allows direct treatment of input uncertainty and yields some limited insight into model errors, it requires the specification of valid error models, which are currently poorly understood and require further work.

Kuczera *et al.* (2006) began with the argument that conceptual rainfall–runoff (CRR) is intrinsically stochastic. A hypothesis that the CRR model error can be characterised by storm-dependent random variation of one or more of the CRR model parameters formed the foundation for the study, and a Bayesian hierarchical model was then formulated to explicitly differentiate between forcing, response and model error. A case study calibrating a six-parameter CRR model against daily data from the Abercrombie catchment (Australia) demonstrated the considerable potential of this approach. One of the important findings of the study was clear indication that total uncertainty was dominated by the uncertainty in the rainfall data.

Vrugt *et al.* (2008) introduced a novel Markov chain Monte Carlo (MCMC) algorithm, called Differential Evolution Adaptive Metropolis (DREAM), which was designed to efficiently estimate the posterior probability density function of hydrological model parameters in complex, high-dimensional sampling problems. They also demonstrated how DREAM can be used to analyse forcing-data error during watershed model calibration using a five-parameter rainfall–runoff model with streamflow data from two different catchments. Explicit treatment of the precipitation error during calibration of the hydrological model resulted in prediction-uncertainty bounds that were more appropriate but which also significantly altered the posterior distribution of the watershed model parameters.

## 5.6.4 Stochastic physically based modelling

Section 5.3 provided a general characterisation of methods attempting to address complex issues of uncertainty estimation in hydrological modelling. The focus of the hydrological modelling has for several decades been on the physically based or conceptual models, which aim to pursue a deterministic representation of all the processes involved. Consequently, the efforts in advancing the knowledge about uncertainty assessment of the outcomes of such models have concentrated on these methods which, while accepting the deterministic concept of model structure (or structures for multi-model methods), attempt to characterise the model-output uncertainty through propagation of known or assumed uncertainties in forcings, parameters and deviations from the 'true' structure.

Montanari and Koutsoyiannis (2011a,b,c) have pointed out that the fascination of the hydrology community with the deterministic modelling of hydrological systems has been the primary reason why other classical concepts of stochastic physically based modelling have been forgotten by hydrologists. They argued that stochastic physically based modelling is capable of providing the tools to effectively synthesise the knowledge and understanding of natural systems. There is sufficient knowledge to develop a framework that can take into account the always present and inherent uncertainty, and which can synthesise hydrological modelling and uncertainty assessment.

The entire framework is described in detail in Montanari and Koutsoyiannis (2011a), and the foundation of the approach is described briefly below. Beginning with deterministic hydrological model formulation in the form

$$Q_{\mathrm{p}} = S(\boldsymbol{\varepsilon}, \boldsymbol{I}) \tag{5.43}$$

where $Q_{\mathrm{p}}$ is the model prediction, which in a deterministic framework is implicitly assumed to be equal to the true value of the variable to be predicted. The relationship $S$ represents the model structure, and $\boldsymbol{I}$ and $\boldsymbol{\varepsilon}$ are vectors of the input data and the parameters.

In the stochastic framework, the equivalent of Equation 5.38 can be written as

$$f_{Q_p}(Q_p) = Kf_{\boldsymbol{\varepsilon},\boldsymbol{I}}(\boldsymbol{\varepsilon}, \boldsymbol{I}) \tag{5.44}$$

where $f$ is the probability density function and $K$ is a transfer operator that depends on the deterministic model $S$.

Equation 5.39 can be written in the full form as

$$f_{Q_p}(Q_p) = \int_{\boldsymbol{\varepsilon}}\int_{\boldsymbol{I}} f_{\boldsymbol{\varepsilon},\boldsymbol{I}}(\boldsymbol{\varepsilon}, \boldsymbol{I})[Q_p - S(\boldsymbol{\varepsilon}, \boldsymbol{I})/\boldsymbol{\varepsilon}, \boldsymbol{I}], f_{\boldsymbol{\varepsilon}}(\boldsymbol{\varepsilon}), f_{\boldsymbol{I}}(\boldsymbol{I})\,\mathrm{d}\boldsymbol{\varepsilon}\,\mathrm{d}\boldsymbol{I} \tag{5.45}$$

where $f_{Q_p}(Q_p)$ is the probability density function of the true value of the model output, and $S(\boldsymbol{\varepsilon}, \boldsymbol{I})$ is the deterministic model, $f_{\boldsymbol{\varepsilon},\boldsymbol{I}}(\boldsymbol{\varepsilon}, \boldsymbol{I})[Q_p - S(\boldsymbol{\varepsilon}, \boldsymbol{I})/\boldsymbol{\varepsilon}, \boldsymbol{I}]$ is the conditional probability density function of the global model error.

There are no restrictions on the type of model $S$ – it can be a physically based or any other kind of deterministic model. If a physically based model is used, the framework provides additional opportunity to improve the understanding of how the model works.

The difficulties in the implementation of the framework, as seen by the authors, are three-fold:

- the numerical integration of Equation 5.45 can be computationally intensive and the cost may be prohibitive
- the problem of achieving a reliable estimation of the global model error is still not satisfactorily resolved, and further research may be required.
- the estimation of parameter uncertainty is also a challenge, and like the numerical integration of Equation 5.45 can be also computationally intensive.

The method characterised in Section 5.6.4 can be used to estimate the global model error and the parameter uncertainty.

It seems that the theoretical framework described above has the potential to open up new perspectives for modelling of uncertain hydrological systems. The knowledge gained through statistical analysis of uncertainty and predictability can lead to an improved understanding of real systems and a better understanding of their changing or shifting behaviours.

## REFERENCES


Abbott MB, Bathurst JC, Cunge JA, O'Connell PE and Rasmussen J (1986) An introduction to the European hydrological system – Systeme Hydrologique Europeen, SHE 2: structure of a physically-based, distributed modelling system. *Journal of Hydrology* **87(1)**: 61–77.

Ajami NK, Duan Q and Sorooshian S (2006) *An Integrated Hydrologic Bayesian Multi-Model Combination Framework: Confronting Input, Parameter and Model Structural Uncertainty in Hydrologic Prediction.* Report No. UCRL-JRNL-221204. Lawrence Livermore National Laboratory, Livermore, CA, USA.

Andersen TG, Davis RA, Kreiß J-P and Mikosch T (2009) *Handbook of Financial Time Series.* Springer-Verlag, Heidelberg, Germany.

Andreassian V, Lerat J, Loumagne C *et al.* (2007) What is really undermining hydrologic science today? *Hydrological Processes* **21(20)**: 2819–2822.

ASCE (American Society of Civil Engineers) (1996) *Handbook of Hydrology.* ASCE Manuals and Reports on Engineering Practice No. 28. ASCE, New York, NY, USA.

Barnes FB (1954) Storage required for a city water supply. *Journal of the Institute of Engineers, Australia* **26**: 198–203.

Barrett DJ, Kuzmin VA, Walker JP, McVicar TR and Draper C (2008) *Improving Stream Flow Forecasting by Integrating Satellite Observations, in situ Data and Catchment Models using Model-Data Assimilation Methods.* Technical Report. Water Cooperative Research Centre, Bruce, Australia.

Beck MB and Young PC (1975) A dynamic model for BOD-DO relationship in a non-tidal stream. *Water Research* **9**: 769–776.

Beven KJ (1989) Changing ideas in hydrology – the case of physically based models. *Journal of Hydrology* **105(1–2)**: 157–172.

Beven KJ (1993) Prophesy, reality and uncertainty in distributed hydrological modelling. *Advances in Water Resources* **16**: 41–51.

Beven KJ (1995) TOPMODEL. In *Computer Models of Watershed Hydrology* (Singh VP (ed.)). Water Resources Publications, Highlands Ranch, CO, USA: 627–668.

Beven KJ (1996) Equifinality and uncertainty in geomorphological modelling. In *Scientific Nature of Geomorphology* (Rhoads BL and Thorn CE (eds)). Wiley, Chichester, UK: 289–314.

Beven KJ (2001) On modelling as collective intelligence. *Hydrological Processes* **15**: 2205–2207.

Beven KJ (2002b) Uncertainty and the detection of structural change in models of environmental systems. In *Environmental Foresight and Models: A Manifesto* (Beck MB (ed.)). Elsevier Science, Amsterdam, the Netherlands: 227–250.

Beven KJ (2006) On undermining the science? *Hydrological Processes* **20(14)**: 3141–3146.

Beven KJ (2011) Distributed models and uncertainty in flood risk management. In *Flood Risk Science and Management* (Pender G and Faulkner H (eds)). Wiley-Blackwell, Oxford, UK: 291–312.

Beven KJ and Binley AM (1992) The future of distributed models: model calibration and uncertainty prediction. *Hydrological Processes* **6**: 279–298.

Beven KJ and Kirkby MJ (1979) A physically-based variable contributing area model of basin hydrology. *Hydrological Sciences Bull*etin **24(1)**: 43–69.

Bhattacharya D (1997) *Memory, Periodicity and Heteroskedasticity in Hydrologic Time Series*. PhD thesis, Purdue University, West Lafayette, IN, USA.

Blasone R-S (2007) *Parameter Estimation and Uncertainty Assessment in Hydrological Modelling*. PhD thesis, Institute of Environment & Resources, Technical University of Denmark, Lyngby, Denmark.

Bollerslev T (1986) Generalized autoregressive conditional heteroskedasticity. *Journal of Econometrics* **31**: 307–327.

Box GEP and Jenkins GM (1976) *Time Series Analysis: Forecasting and Control*. Holden-Day, Oakland, CA, USA.

Butts MB, Payne JT, Kristensen M and Madsen H (2004) An evaluation of the impact of model structure on hydrological modelling uncertainty for streamflow simulation. *Journal of Hydrology* **298**: 242–266.

Carlsson B, Jacobsson S, Holménb M and Rickne A (2002) Innovation systems: analytical and methodological issues. *Research Policy* **31**: 233–245.

Castelletti A, Pianosi F, Soncini-Sessa R and Young PC (2005) Data-based mechanistic modelling of a snow affected basin. *Proceedings of the 16th IFAC World Congress*, Prague, Czech Republic.

Chen J, Brissette FP and Zhang XJ (2014) A multi-site stochastic weather generator for daily precipitation and temperature. *Transactions of the ASABE* **57(5)**: 1375–1391.

Chow VT (1959) *Open-channel Hydraulics*. McGraw-Hill, New York, NY, USA.

Cox DR and Miller HD (1965) *The Theory of Stochastic Processes*. Chapman and Hall, London UK.

Crawford NH and Burges SJ (2004) History of the Stanford Watershed Model. *Water Resources Impact* **6(2)**: 3–5.

Crawford NH and Linsley RS (1966) *Digital simulation in hydrology: The Stanford Watershed Model IV*. Technical Report 39. Department of Civil Engineering, Stanford University, Palo Alto, CA, USA.

Cunderlik J (2003) *Hydrologic Model Selection for the CFCAS Project: Assessment of Water Resources Risk and Vulnerability to Changing Climatic Condition*. Report 046. Department off Civil and Environmental Engineering, University of Western Ontario, Ontario, Canada.

DHI (Danish Hydraulic Institute) (2000a) *MIKE 11: A Modelling System for Rivers and Channels*. See https://www.tu-braunschweig.de/Medien-DB/geooekologie/mike-11-short-introduction-tutorial.pdf (accessed 01/06/2016).

DHI (2000b) *MIKE SHE Flow Modules, Technical Reference*. Danish Hydraulic Institute, Water and Environment.

Duan Q, Ajami NK, Gao X and Sorooshian S (2007) Multi-model ensemble hydrologic prediction using Bayesian model averaging. *Advances in Water Resources* **30(5)**: 1371–1386.

Dunne T (2007) *Introduction to Lectures on River Systems*. Bren School of Environmental Science & Management, University of California, Santa Barbara, CA, USA.

Ebel BA and Loague K (2006) Physics-based hydrologic response simulation: seeing through the fog of equifinality. *Hydrological Processes*. **20**: 2887–2900.

Edsel BD, Camp JV, LeBoeuf EJ, Penrod JR, Dobbins JP and Abkowitz MD (2011) Watershed modelling and its applications: a state-of-the-art review. *The Open Hydrology Journal* **5**: 26–50.

Eisenbies MH, Aust WM, Burger JA and Adams MB (2007) Forest operations, extreme flooding events and considerations for hydrologic modelling in the Appalachians – a review. *Forest Ecology and Management* **242**: 77–98.

Engle RF (1982) Autoregressive conditional heteroskedasticity with estimates of the variance of United Kingdom inflation. *Econometrica* **50**: 987–1007.

Ewen J, Parkin G and O'Connell PE (2000) SHETRAN: distributed river basin flow and transport modelling system. *Journal of Hydrological Eng*ineering **5(3)**: 250–258.

Fenton JD (1992) Reservoir routing. *Hydrological Sciences Journal* **37**: 233–246.

Freeze RA and Harlan RL (1969) Blueprint for a physically based digitally simulated hydrologic response model. *Journal of Hydrology* **9**: 237–258.

Georgakakos KP, Seo D-J, Gupta H, Schaake J and Butts MB (2004) Towards the characterization of streamflow simulation uncertainty through multimodel ensembles. *Journal of Hydrology* **291**: 297–318.

Grayson R and Blöschl G (2000) *Spatial Patterns in Catchment Hydrology*. Cambridge University Press, Cambridge, UK.

Grygier JC and Stedinger JR (1988) Condensed disaggregation procedures and conservation corrections for stochastic hydrology. *Water Resources Research* **24(10)**: 1574–1584.

Hagedorn R, Doblas-Reyes FJ and Palmer TN (2005) The rationale behind the success of multi-model ensemble in seasonal forecasting: I. Basic concept. *Tellus, Series A* **57**: 219–233.

Hall J, O'Connel E and Ewen J (2007) On not undermining the science: coherence, validation and expertise. Discussion of invited commentary by Keith Beven. *Hydrological Processes* **21**: 985–988.

Harte J (2002) Toward a synthesis of the Newtonian and Darwinian worldviews. *Physics Today*, October.

Hawken WH and Ross CN (1921) The calculation of flood discharges by use of a time-contour plan. *Journal of the Institute of Engineers, Australia* **2**: 85–92.

Hazen A (1914) Storage to be provided in impounding reservoirs for municipal water supply. *Transactions of the American Society of Civil Engineers* **77**: 1539–1569.

Hipel KW and McLeod AI (1994) *Time Series Modelling of Water Resources and Environmental Systems*. Elsevier, Amsterdam, the Netherlands.

Hoeting JA, Madigan D, Raftery AE and Volinsky CT (1999) Bayesian model averaging: a tutorial. *Statistical Science* **14(4)**: 382–417.

Hurst HE (1951) Long-term storage capacity of reservoirs. *Transactions of the American Society of Civil Engineers* **116**: 776–808.

Jakeman AJ and Hornberger GM (1993) How much complexity is warranted in a rainfall-runoff model? *Water Resources Research* **29(8)**: 2637–2649.

Jakeman AJ, Littlewood IG and Whitehead PG (1990) Computation of the instantaneous unit hydrograph and identifiable component flows with application to two small upland catchments. *Journal of Hydrology* **117(1–4)**: 275–300.

Jonsdottir H (2006a) *Stochastic Modelling of Hydrologic Systems*. PhD thesis, Informatics and Mathematical Modelling, Technical University of Denmark, Lyngby, Denmark.

Jonsdottir H (2006b) Parameter estimation in stochastic rainfall-runoff models. *Journal of Hydrology* **326**: 379–393.

Julien PY, Saghafian B and Ogden FL (1995) Raster-based hydrologic modelling of spatially-varied surface runoff. *Water Resources Bulletin* **31**: 523–536.

Kavetski D, Kuczera G and Franks SW (2006a) Bayesian analysis of input uncertainty in hydrological modelling: 1. Theory. *Water Resources Research* **42**: W03407. doi:10.1029/2005WR004368.

Kavetski D, Kuczera G and Franks SW (2006b) Bayesian analysis of input uncertainty in hydrological modelling: 2. Application. *Water Resources Research* **42**: W03408. doi:10.1029/2005WR004376.

Klemeš V (1983) Conceptualization and scale in hydrology. *Journal of Hydrology* **65**: 1–23.

Koussis A (2008) Appropriate model and uncertainty. Interactive comment on 'HESS opinions – climate, hydrology, energy, water: recognizing uncertainty and seeking sustainability' by D. Koutsoyiannis *et al. Hydrology and Earth System Sciences Discussions* **5**: S2196–S2201.

Koutsoyiannis D (1992) A nonlinear disaggregation method with a reduced parameter set for simulation of hydrologic series. *Water Resources Research* **28(12)**: 3175–3191.

Koutsoyiannis D (2000) A generalized mathematical framework for stochastic simulation and forecast of hydrologic time series. *Water Resources Research* **36(6)**: 1519–1533.

Koutsoyiannis D (2001) Coupling stochastic models of different time scales. *Water Resources Research* **37(2)**: 379–391.

Koutsoyiannis D (2002) The Hurst phenomenon and fractional Gaussian noise made easy. *Hydrological Sciences Journal* **47(4)**: 573–595.

Koutsoyiannis D (2006) Nonstationarity versus scaling in hydrology. *Journal of Hydrology* **324**: 239–254.

Koutsoyiannis D (2011) Hurst–Kolmogorov dynamics and uncertainty. *Journal of the American Water Resources Association* **47(3)**: 481–495.

Koutsoyiannis D and Manetas A (1996) Simple disaggregation by accurate adjusting procedures. *Water Resources Research* **32(7)**: 2105–2117.

Kouwen N (2001) *WATFLOOD/SPL9 Hydrological Model & Flood Forecasting System*. University of Waterloo, Waterloo, Canada.

Koren V, Reed S, Smith M, Zhang Z and Seo D-J (2004) Hydrology laboratory research modeling system (HL-RMS) of the US national weather service. *Journal of Hydrology* **291**: 297–318.

Kuczera G, Kavetski D, Franks S and Thyer M (2006) Towards a Bayesian total error analysis of conceptual rainfall-runoff models: characterising model error using storm-dependent parameters. *Journal of Hydrology* **331**: 161–177.

Kumar DN (2000) Multisite disaggregation of monthly to daily streamflow. *Water Resources Research* **36(7)**: 1823–1833.

Kundzewicz ZW (2011) Nonstationarity in water resources – Central Europe perspective. *Journal of the American Water Resources Association* **47(3)**: 550–562.

Lall U and Sharma A (1996) A nearest neighbour bootstrap for resampling of hydrologic time series. *Water Resources Research* **32(3)**: 679–693.

Lees MJ (2000) Data-based mechanistic modelling and forecasting of hydrological systems. *Journal of Hydroinformatics* **2(1)**: 15–34.

Lindström G, Johansson B, Persson M, Gardelin M and Bergström S (1997) Development and test of the distributed HBV-96 hydrological model. *Journal of Hydrology* **201**: 272–288.

Littlewood IG, Down K, Parker JR and Post DA (1997) *The PC Version of IHACRES for Catchment-scale Rainfall–Streamflow Modelling. Version 1.0. User Guide*. Institute of Hydrology, Wallingford, UK.

Maidment DR (1993) *Handbook of Hydrology*. McGraw-Hill, New York, NY, USA.

Marshall LA, Nott DJ and Sharma A (2007) Towards dynamic catchment modelling: a Bayesian hierarchical mixtures of experts framework. *Hydrological Processes* **21**: 847–861.

McIntyre N, Young PC, Orellana B, Marshall M, Reynolds B and Wheater H (2011) Identification of nonlinearity in rainfall-flow response using data-based mechanistic modelling. *Water Resources Research* **47(3)**.

Milly PCD, Betancourt J, Falkenmark M *et al.* (2008) Stationarity is dead: whither water management? *Science* **319**: 573–574.

Montanari A (2007) What do we mean by 'uncertainty'? The need for a consistent wording about uncertainty assessment in hydrology. *Hydrological Processes* **21**: 841–845.

Montanari A and Koutsoyiannis D (2011a) *A Blueprint for Physically-based Modelling of Uncertain Hydrological Systems.* Unpublished draft. See http://distart119.ing.unibo.it/albertold/files/draft-paper-stochasticmodeling.pdf (accessed 01/06/2016).

Montanari A and Koutsoyiannis D (2011b) *Uncertainty Estimation in Hydrology: Incorporating Physical Knowledge in Stochastic Modelling of Uncertain Systems*. Invited Seminar, University of Uppsala, Uppsala, Sweden.

Montanari A and Koutsoyiannis D (2011c) Stochastic physically-based modelling in hydrology: towards a synthesis of different approaches for a new target. European Geosciences Union General Assembly 2011, Vienna, Austria. *Geophysical Research Abstracts* **13**: EGU2011-11775.

Moradkhani H and Hsu K-L (2005) Uncertainty assessment of hydrologic model states and parameters: sequential data assimilation using the particle filter. *Water Resources Research* **41**: W05012. doi:10.1029/2004WR003604.

Moradkhani H, Sorooshian S, Gupta HV and Houser PR (2005) Dual-state parameter estimation of hydrological models using ensemble Kalman filter. *Advances in Water Resources* **28**: 135–147.

Mulvany TJ (1850) On the use of self-registering rain and flood gauges. *Proceedings of the ICE* **4(2)**: 1–8.

Nandalal KDW and Simonovic SP (2003) *State-of-the-Art Report on Systems Analysis Methods for Resolution of Conflicts in Water Resources Management*. Report SC-2003/WS/60. International Hydrological Programme, UNESCO, Paris, France.

Nowak K, Prairie J, Rajagopalan B and Lall U (2010) A nonparametric stochastic approach for multisite disaggregation of annual to daily streamflow. *Water Resources Research* **46(8)**: WO8529.

OHD (Office of Hydrologic Development) (2007) *Hydrology Laboratory Strategic Science Plan*. National Weather Service, National Oceanographic and Atmospheric Administration (NOAA), Washington, DC, USA.

Ogden FL (1998) *CASC2D Reference Manual*. Department of Civil and Environmental Engineering, University of Connecticut, Storrs, CT, USA.

OPG (Ontario Power Generation) (2009) *Madawaska River Water Management Plan*. OPG, Toronto, Canada.

Pechlivanidis IG, Jackson BM, McIntyre NR and Wheater HS (2011) Catchment scale hydrological modelling: a review of model types, calibration approaches and uncertainty analysis methods in the context of recent developments in technology and applications. *Global NEST Journal* **13(3)**: 193–214.

Prairie J, Rajagopalan B, Lall U and Fulp T (2007) A stochastic nonparametric technique for space–time disaggregation of streamflow. *Water Resources Research* **43(3)**.

Price LA, Goodwill P, Young PC and Rowan JS (2000) A data-based mechanistic (DBM) approach to understanding dynamic sediment transmission through Wyresdale Park Reservoir, Lancashire, UK. *Hydrological Processes* **14**: 63–78.

Ratto M, Young PC, Romanowicz R, Pappenberger F, Saltelli A and Pagano A (2007) Uncertainty, sensitivity analysis and the role of data based mechanistic modelling in hydrology. *Hydrology and Earth Systems Science* **11**: 1249–1266.

Salas JD, Delleur JW, Yevjevich V and Lane WL (1980) *Applied Modeling of Hydrologic Time Series*. Water Resources Publications, Littleton, CO, USA.

Santos EG and Salas JD (1992) Stepwise disaggregation scheme for synthetic hydrology. *Journal of Hydraulic Engineering* **118(5)**: 765–784.

Savenije HHG (2001) Equifinality, a blessing in disguise? *Hydrological Processes* **15**: 2835–2838.

Sharma A, Tarboton DG and Lall U (1997) Streamflow simulation: a nonparametric approach. *Water Resources Research* **33(2)**: 291–308.

Singh VP (1995) *Computer Models of Watershed Hydrology*. Water Resources Publications, Highlands Ranch, CO, USA.

Singh VP (2004) Flow routing in open channels: some recent advances. *Proceedings of the Second International Conference on Fluvial Hydraulics: River Flow 2004*, Naples, Italy.

Singh VP and Frevert DK (2006) *Watershed Models*. Taylor & Francis, London, UK.

Singh VP and Woolhiser DA (2002) Mathematical modelling of watershed hydrology. *Journal of Hydrological Engineering* **7(4)**: 270–292.

Sivakumar B (2008a) Undermining the science or undermining Nature? *Hydrological Processes* **22**: 893–897.

Sivakumar B (2008b) The more things change, the more they stay the same: the state of hydrologic modelling. *Hydrological Processes* **22**: 4333–4337.

Sivapalan M, Blöschl G, Zhang L and Vertessy R (2003) Downward approach to hydrologic prediction. *Hydrological Processes* **17**: 2101–2111.

Soncini-Sessa R, Castelletti A and Weber E (2007) *Integrated and Participatory Water Resources Management: Theory*, vol. 1A. Elsevier, New York, NY, USA.

Stedinger JR, Pei D and Cohn TA (1985) A condensed disaggregation model for incorporating parameter uncertainty into monthly reservoir simulation. *Water Resources Research* **21(5)**: 665–675.

Sudler CE (1927) Storage required for the regulation of streamflow. *Transactions of the American Society of Civil Engineers* **91**: 622–660.

Tarboton DG, Sharma A and Lall U (1998) Disaggregation procedures for stochastic hydrology based on nonparametric density estimation. *Water Resources Research* **34(1)**: 107–119.

Thordarson FÖ (2012) *Grey Box Modelling of Hydrological Systems with Focus on Uncertainties*. PhD thesis, Informatics and Mathematical Modelling, Technical University of Denmark, Lyngby, Denmark.

Todini E and Mantovan P (2007) Comment on: 'On undermining the science?' by Keith Beven. *Hydrological Processes* **21**: 1633–1638.

USACE (US Army Corps of Engineers) (1994) *Engineering and Design – Flood-Runoff Analysis*. Engineering Manual EM 1110-2-1417. USACE, Washington, DC, USA.

USACE (2000a) *Hydrologic Modelling System HEC-HMS. Release Notes, Version 3.5*. USACE, Hydrologic Engineering Centre, Washington, DC, USA.

USACE (2000b) *Hydrologic Modelling System HEC-HMS. Technical Reference Manual*. USACE, Hydrologic Engineering Centre, Washington, DC, USA.

USACE (2013) *Hydrologic Modeling System HEC-HMS. User's Manual, Version 4.0*. USACE, Washington, DC, USA.

USACE (2015) *Hydrologic Modelling System HEC-HMS. User's Manual*. USACE, Hydrologic Engineering Centre, Washington, DC, USA.

USACE-NPD (USACE North Pacific Division) (1991) *SSARR Model Streamflow Synthesis and Reservoir Regulation*. USACE North Pacific Division, Portland, OR, USA.

USDA-ARS (United States Department of Agriculture, Agricultural Research Service) (1998) *Snowmelt Runoff Model (SRM) User's Manual. Updated Edition, Version 4.0*. USDA-ARS, Washington, DC, USA.

Valencia RD and Schaake IC (1973) Disaggregation processes in stochastic hydrology. *Water Resources Research* **9(3)**: 580–585.

VanderKwaak JE (1999) *Numerical Simulation of Flow and Chemical Transport in Integrated Surface–Subsurface Hydrologic Systems*. PhD dissertation, University of Waterloo, Waterloo, Ontario, Canada.

VanderKwaak JE (2001) Hydrologic response simulation for the R-5 catchment with a comprehensive physics-based model. *Water Resources Research* **37(4)**: 999–1013.

Vrugt JA and Robinson BA (2007) Treatment of uncertainty using ensemble methods: comparison of sequential data assimilation and Bayesian model averaging. *Water Resources Research* **43**: W01411. doi:10.1029/2005WR004838.

Vrugt JA, Diks CGH, Gupta HV and Bouten W (2005) Improved treatment of uncertainty in hydrologic modelling: combining the strength of global optimization and data assimilation. *Water Resources Research* **41**: W01017. doi:10.1029/2004WR003059.

Vrugt JA, Clark MP, Diks CGH, Duan Q and Robinson BA (2006) Multi-objective calibration of forecast ensembles using Bayesian model averaging. *Geophysics Research Letters* **33**: 1–6.

Vrugt JA, ter Braak CJF, Clark MP, Hyman JM and Robinson BA (2008) Treatment of input uncertainty in hydrologic modelling: doing hydrology backward with Markov chain Monte Carlo simulation. *Water Resources Research* **44**: W00B09. doi:10.1029/2007WR006720.

Wang W (2006) *Stochasticity, Nonlinearity and Forecasting of Streamflow Processes*. IOS Press, Amsterdam, the Netherlands.

Weinman PE and Laurenson EM (1979) Approximate flood routing methods: a review. ASCE *Journal – Hydraulics Division* **105(HY12)**: 1521–1535.

Wheater HS, Jakeman AJ, Beven KJ (1993) Progress and directions in rainfall–runoff modelling. In *Modelling Change in Environmental Systems* (Jakeman AJ, Beck MB and McAllen MJ (eds)). John Wiley, Chichester, UK, pp. 101–132.

WMO (World Meteorological Organization) (2011) *Manual on Flood Forecasting*. Draft report. Commission for Hydrology, World Meteorological Organization, Geneva, Switzerland.

Yates D (1994) *Watbal – An Integrated Water Balance Model for Climate Impact Assessment of River Basin Runoff*. WP-94-64. International Institute for Applied Systems Analysis (IIASA), Laxenburg, Austria.

Yevjevich V (1971) *Systems Approach to Hydrology: Proceedings, Bilateral US–Japan Seminar on Hydrology*, University of Hawaii. Water Resources Publications, Fort Collins, CO, USA.

Young PC (2001) Data-based mechanistic modelling of environmental systems. Keynote paper presented at the *IFAC Workshop on Environmental Systems*, Tokyo, Japan.

Young PC (2002) Advances in real-time flood forecasting. *Philosophical Transactions of the Royal Society London, Series A* **360**: 1433–1450.

Young PC (2003) Top-down and data-based mechanistic modelling of rainfall-flow dynamics at the catchment scale. *Hydrological Processes* **17**: 2195–2217.

Young PC and Lees MJ (1993) The active mixing volume: a new concept in modelling of environmental systems. In *Statistics for the Environment* (Barnett V and Turkman KF (eds)). John Wiley, Chichester, UK: 3–44.

Young PC, Castelletti A and Pianosi F (2007) The data-based mechanistic approach to hydrological Modelling. In *Topics on System Analysis and Integrated Water Resource Management* (Castelletti A and Soncini-Sessa R (eds)). Elsevier, Amsterdam, the Netherlands: 27–48.

**Operational Safety of Dams and Reservoirs**
ISBN 978-0-7277-6121-7


http://dx.doi.org/10.1680/osdr.61217.161

# Chapter 6
# Reservoir and outflow control

While the focus of this book is broader than the traditional view of the safety of reservoir and river operations and utilisation, this chapter briefly summarises established reservoir and river system operational practices.

Notwithstanding the reliance on established practices, the capacity of the physical and operational outflow control processes to respond to an adverse condition at a dam/reservoir at an instant in time is influenced, and most usually limited, by prior operational conditions and past decisions that apparently have a significant bearing on real-time dam safety or operational control.

As introduced in Chapter 2, in analytical terms, the question 'What do we mean by system operation?' can be addressed in terms of the following two problem statements:

1 Given the overall objectives of the system, a mechanism is needed to decide how much water needs to be released through every piece of flow control/discharge equipment at all times.
2 Given the decisions in (1) have been determined, what needs to be done (and how) in order to implement the decisions at all times?

Statement 1 presents a problem of decision-making under uncertainty. The uncertainty arises from: (i) uncertainty in natural processes that influence both the supply (available water) and demand (production and other objectives) sides, and (ii) uncertainty about the prospects of successful (partially successful, partially unsuccessful or unsuccessful) implementation of the decision.

Statement 2 presents the problem of conditioning the system such that the system can successfully respond to the various and often conflicting demands within the system. Conditioning the system in the absence of any internal conflicting demands necessarily involves a certain degree of technologically based operational uncertainty in the result. The introduction of internal conflicting demands and the need to make trade-offs between them increases the overall uncertainty, and may even increase the technologically based operational uncertainty. Typically, in practice, treatment of part (i) of problem 1 usually involves the development of some more or less detailed inflow/outflow management rules. More sophisticated approaches, such as the decision-support systems (DSS) that emerged in the 1980s, are available and, while not widely utilised in their full form, may be used to inform the more traditional 'rules-based' approach to inflow/outflow management.

Part (ii) of problem 1 is typically not addressed in general operational practice, although it could be addressed in terms of a sophisticated DSS that goes beyond *traditional practice, which does not explicitly address the uncertainty implications.* In this regard, problem 2 provides the basis for the inputs to part (ii) of problem 1. How, and the extent to which, problems 1 and 2 are addressed in the operation of a dam–reservoir system are organisational matters. These matters of 'how' and 'extent to which' are complex matters that frequently reflect a wide range of organisational positions and views over considerable periods of time. For example, financial decisions that influence the design stage of an asset may eventually emerge as reliability issues in operational practice many years in the future. Furthermore, political considerations concerning public perceptions and expectations when they influence operational decisions may provide the short-term objectives while concealing longer term latent safety hazards (e.g. holding water behind a dam to alleviate downstream flooding may put the system in an operational state that renders it incapable of withstanding the effects of the, as yet not appreciated, imminent larger inflow). Similarly, maintenance and operational decisions made long in the past, and often forgotten, influence present-day assumptions and actions concerning the operation of equipment in ways that can have unexpected outcomes now and in the future.

## 6.1. Operating concepts

It is quite obvious that the operation of a single-purpose reservoir is much simpler than the operation of a multi-purpose one. Let us consider first an example of a reservoir dedicated strictly to flood control. The most effective use of such reservoir is to

**Figure 6.1** Simple operation for flood control

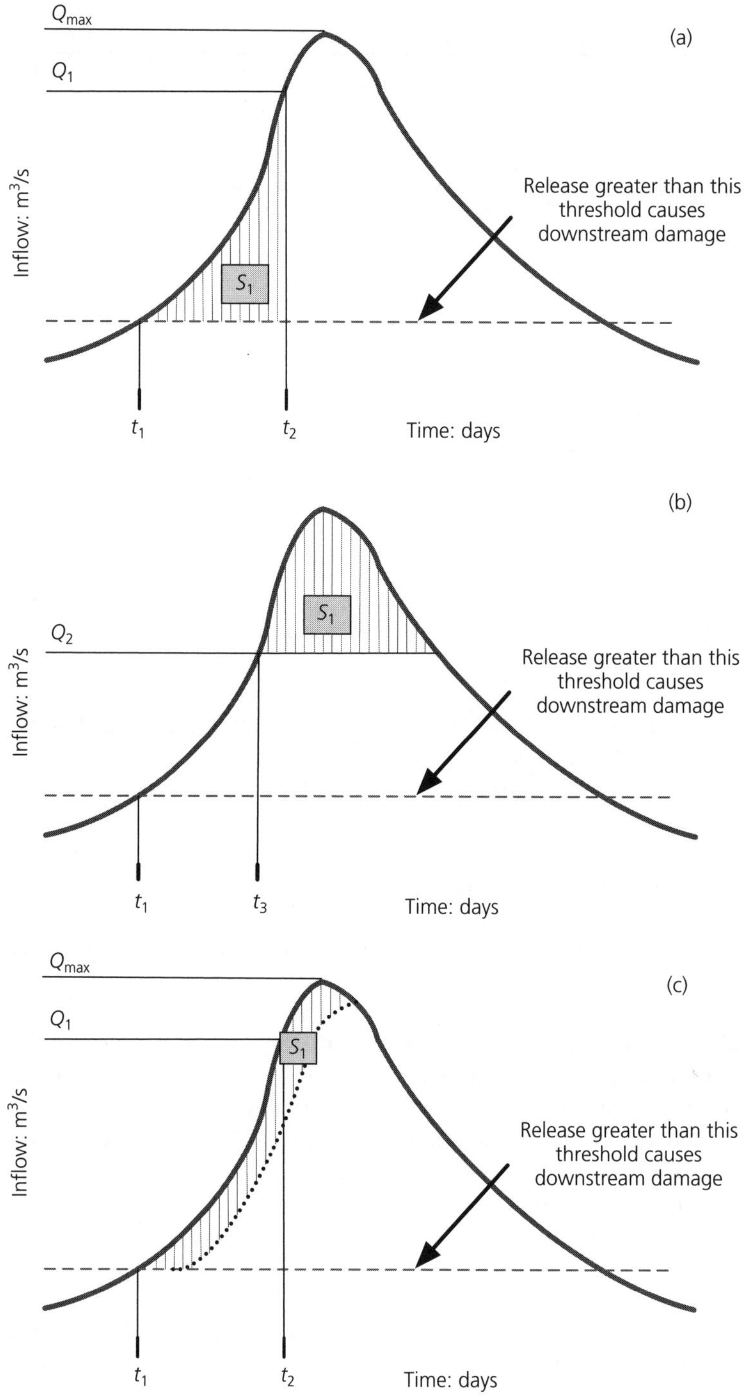


keep it essentially empty at all times, with the exception of periods when the inflow volume, if released in its entirety, would cause adverse impacts downstream. Thus, if the inflow exceeds a certain threshold level, the excess inflow can be stored in the reservoir and only the amount equal to the threshold would be released. However, even such a simple case gets more complicated in practice, because every reservoir has a finite storage, the inflows are uncertain and following the simple rule explained above can lead to damages that would be avoidable if the operating rule was constructed in such a way that the purpose of the operation would have been to capture only the highest inflows. This concept is illustrated on Figure 6.1.

On the upper chart (a), as long as the inflow (solid line) is below the threshold release (dashed line) above which the outflow from the reservoir causes downstream damage, the entire inflow can be released or part of it can be stored. At time $t_1$ the operation changes and the excess of inflow above the threshold has to be stored, resulting in releases being equal to the threshold. This operation continues as the inflows are rising, until time $t_2$ when the entire available storage (denoted by $S_1$) is filled. At that point the releases have to match the inflows, and the most damaging inflows greater than $Q_1$ cannot be captured and stored.

It is quite obvious that if the inflows for the entire period of flood were known (a perfect forecast) then the best operating policy would have followed the one illustrated in (b). Storing of inflows does not happen until time $t_3$. The releases between times $t_1$ and $t_3$ would just cause some damage downstream, but the most damaging inflows (greater than $Q_2$) would be reduced and no release greater than $Q_2$ would occur.

The scenario presented in (b) is, of course, unrealistic because perfect forecasts of inflows for prolonged periods of time are unavailable. A more realistic procedure, as depicted in (c), would follow first the operation in (a) with periodic modifications increasing the amount of releases when the available forecasts indicate that the available storage would be filled before the peak inflows arrive.

### 6.1.1 Development of operating rules

The term 'operating rules' can have a very broad meaning, ranging from long-term policies spanning months or even years to real-time decision-making rules governing releases from reservoirs at very short timescales, which may vary from days to hours depending on the characteristics of the reservoir and the upstream river system.

The long-term policies usually address multi-year or annual reservoir operation, and can be derived from strategic operations studies oriented toward providing general guidance for system operations (USACE, 1997). Such guidance often provides specific targets for operations, usually in the form of targets for reservoir storage, and is commonly presented as so-called 'rule curves'.

### 6.1.2 Rule curves

The reservoir rule curve describes the ideal or target reservoir pool elevation, and thus storage volume, as a function of season (Figure 6.2). Deviations from the rule curve provide a basis for making decisions relative to releases. In addition to the rule curve(s), operating rules also specify quantities of water to be released throughout the year and in relation to pool elevation,

**Figure 6.2** Rule curve for Lake Simcoe, on the Trent-Severn Waterway, Ontario, Canada. The solid line shows rule curve; the dashed line shows the 50-year average elevations. (Courtesy of Parks Canada)

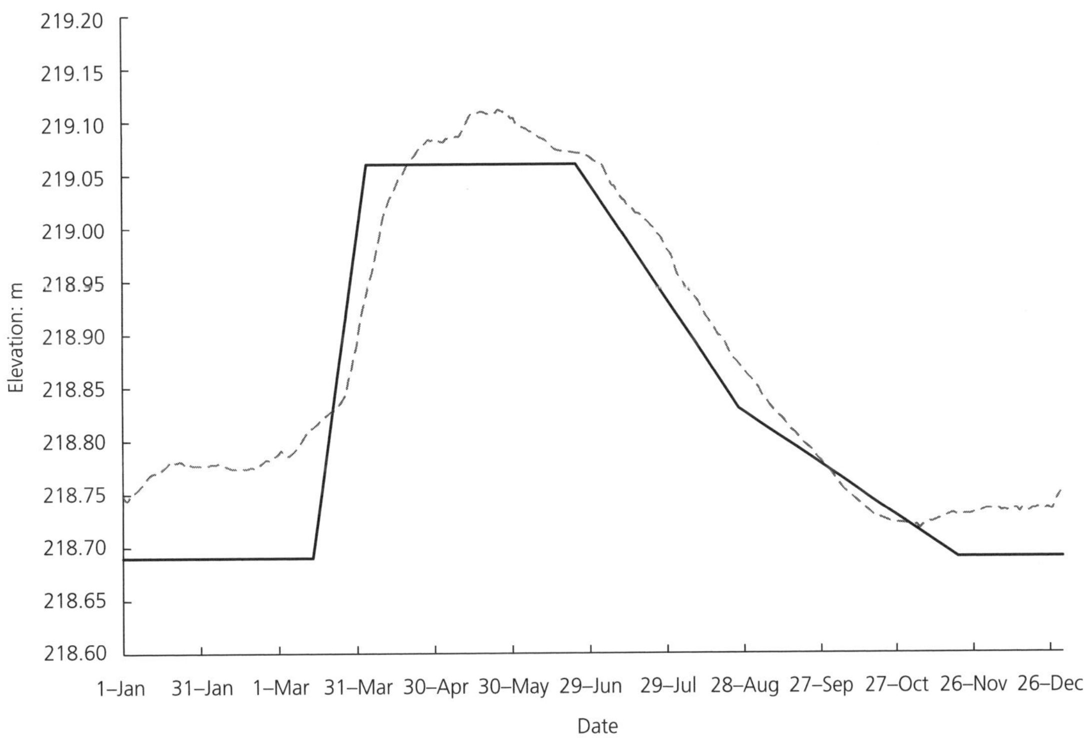

Figure 6.3 Rule curve and results of operation in 2012 for Bark Lake Control Dam on the Madawaska River, Ontario, Canada

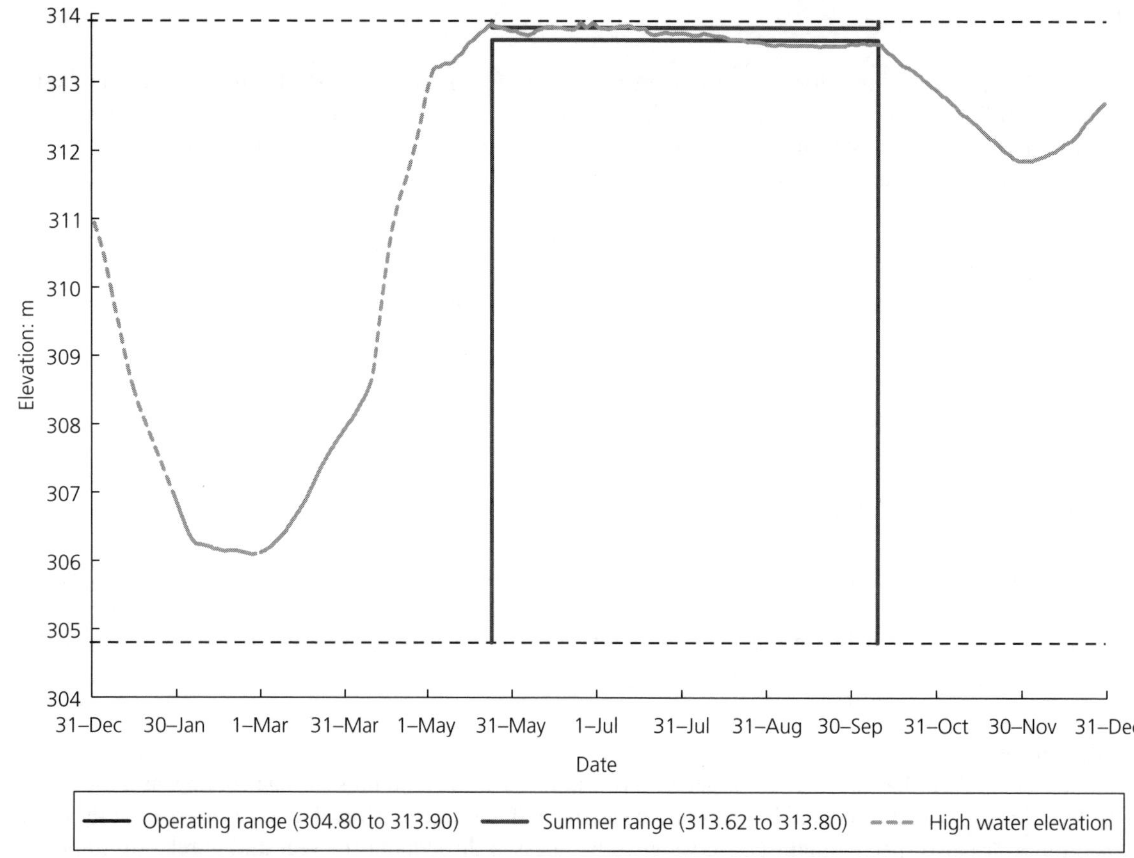


quantities of power generated and flood releases to be made. These rules reflect contracted and non-contracted services provided to users, but also the required environmental flows.

A modification of a rule-curve approach replaces a single target line with the bounds (upper and lower) for the targeted variable. An example of the bounds for reservoir elevations is presented in Figure 6.3. The operation of Bark Lake Control Dam is quite tightly restricted during summer but still leaving the operator with the task of establishing daily releases to ensure that the elevation in the reservoir remains within the bounds. Winter time provides the operator with considerable leeway when preparing the storage for floods, which always occur during the freshet period.

Both of the above examples refer to so-called 'fixed storage rules', as the target is either the storage or its equivalent elevation. Yet another approach that utilises the set of rule curves for flood control (USACE, 1991) is based on defining release as a function of the pool elevation and some other variable characterising inflow either directly or indirectly (Figure 6.4). Such rules are often classified as 'fixed release rules'.

The development of rule curves is usually based on assumed requirements under normal hydrological conditions for contracted services such as power, navigation and water supply, or on presumed levels of non-contracted services such as flood protection and recreation. Criteria for decreasing services under abnormal hydrological conditions are common. Rule curves under conservation conditions are usually based on normal power generation and water supply requirements, but alternative criteria for decreased services are needed for periods in which reservoir levels are crucially low.

#### 6.1.2.1 Standard operating policy and other related concepts

USACE (1996b) provides a detailed characterisation of operating rules that can be applied to single reservoirs which are operated independently from other reservoirs within the system. These rules are straightforward and of a limited use in actual operation but can be useful in general operational studies for planning purposes and can also serve as a starting point for developing operational policies and strategies.

The simplest in this group is the classic standard operating policy (SOP), which can be applied to reservoirs supplying downstream water needs, either for single purpose or multiple purposes. By following the SOP the decisions made about releases from the reservoir can be determined from the curve presented in Figure 6.5.

Figure 6.4 Family of rule curves. (USACE, 1996a)

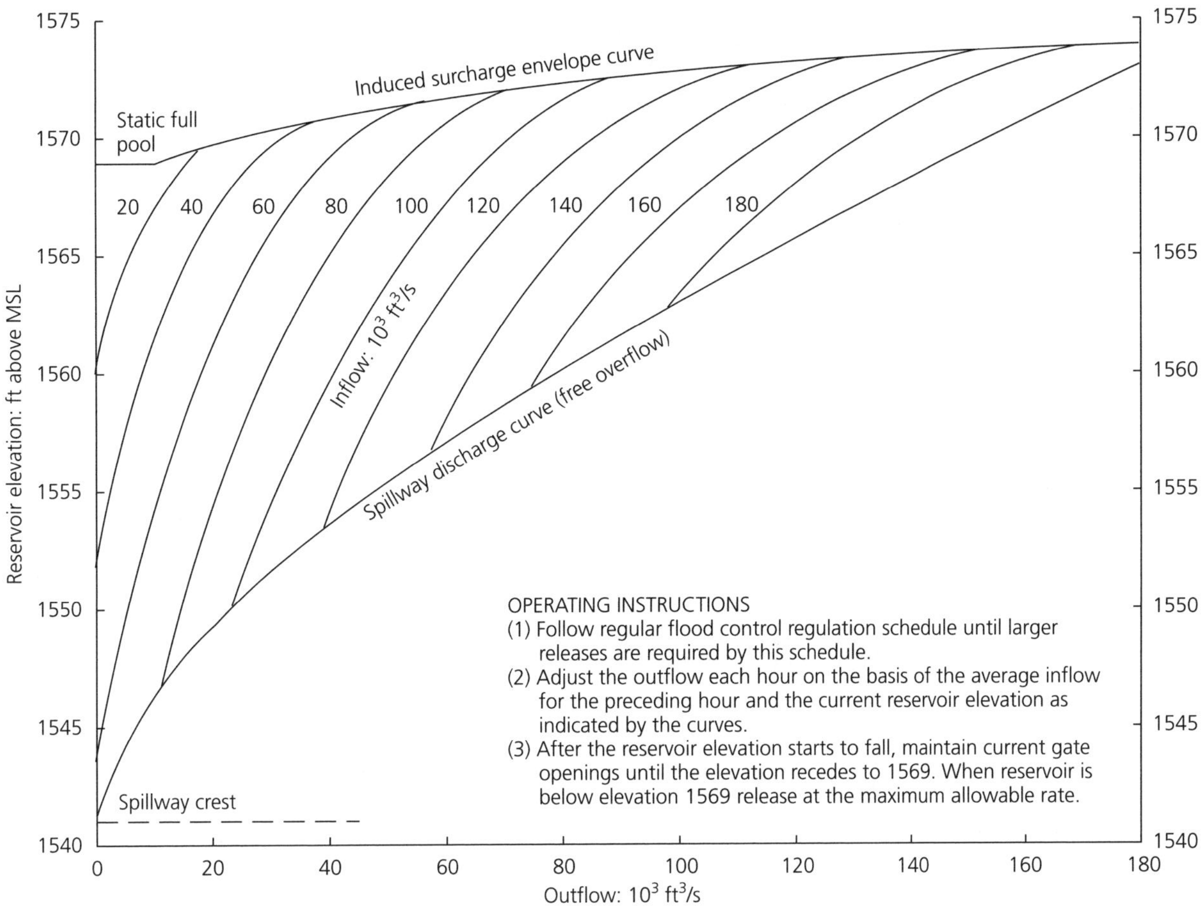


The SOP is closer to a general concept than a rule of actual operation because it does not discriminate in times of available water shortages and excesses. The policy can be modified in a variety of ways using the hedging approach to water shortages (as illustrated in Figure 6.6) or so-called 'pack rules' for better utilisation of excess water. Details on the development of these two policies can be found in USACE (1996a).

Figure 6.5 Standard operating policy

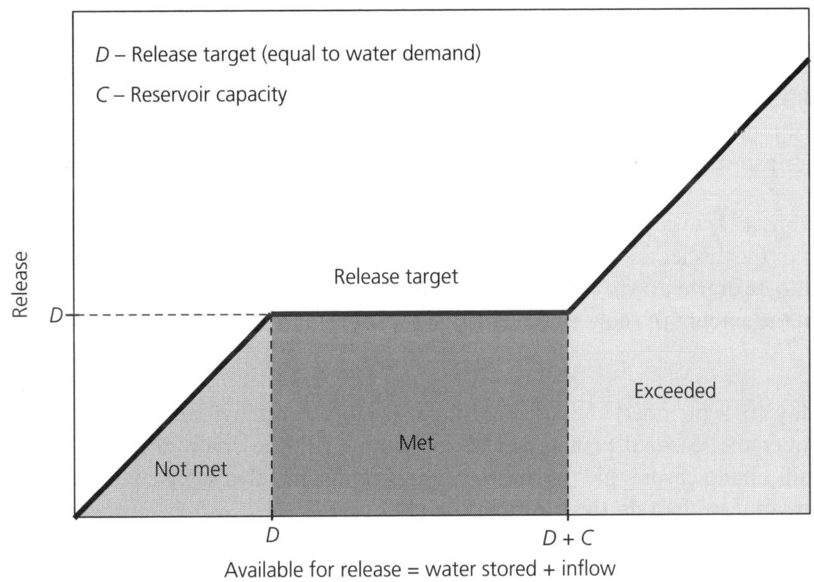

Figure 6.6 Hedging policies

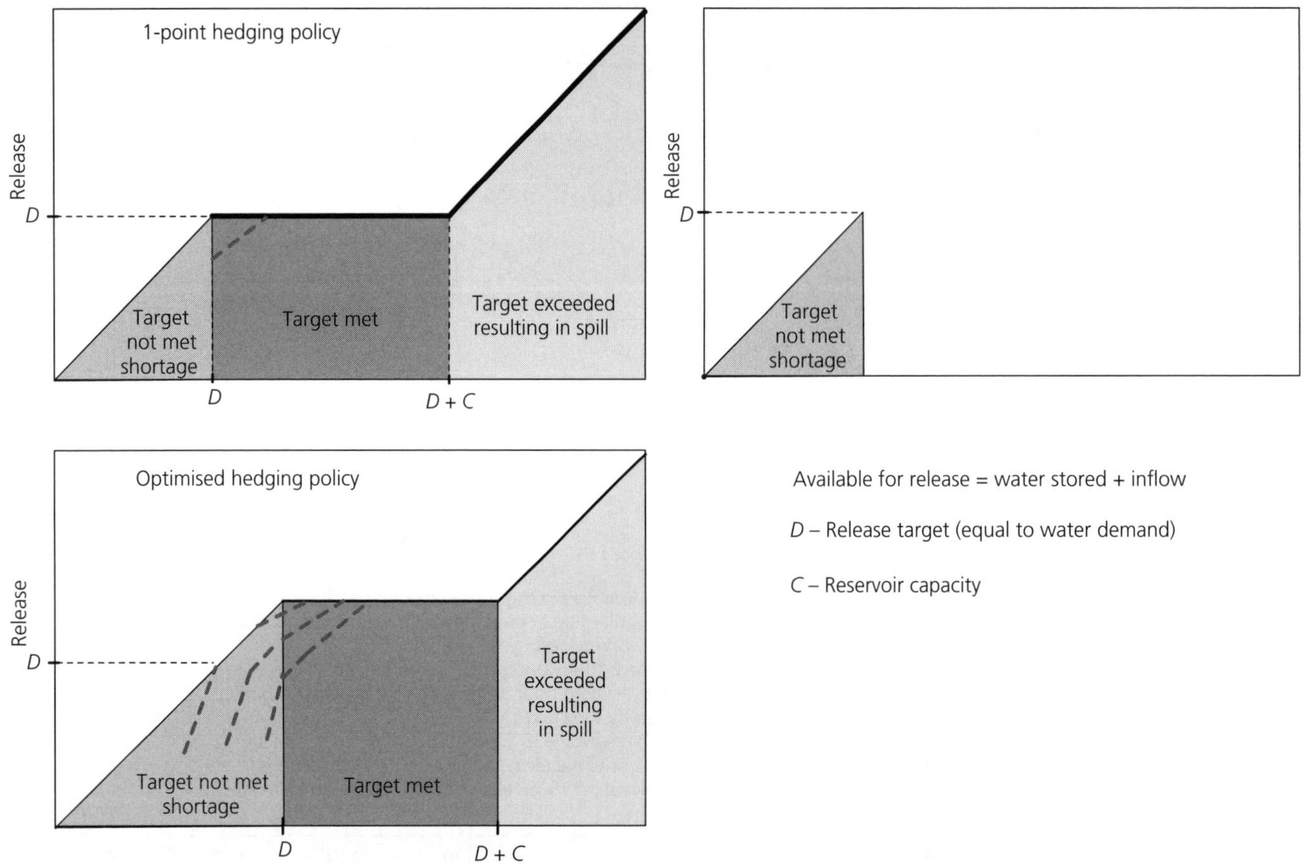


USACE also provides a detailed description of specific control rules for floods (basic single purpose control rule, hedging pre-release rules, inflow following filling rules, storage evacuation rules and zone-based control rules) and for systems of reservoirs either in series (cascades) or in parallel. As the report is available on the USACE website and can be downloaded at no cost, no further details on the construction of operating rules are provided here.

### 6.1.3 Time horizon

The operation of a reservoir or a system of reservoirs typically follows long-term policies established on the basis of the system purpose, the nature and magnitude of demand for water, licenses, regulatory conditions and restriction, and societal expectations. All of these elements may change over time, and when changes occur the strategy of how to operate the system may change as well. The strategies usually reflect priorities in the operation and address time horizons spanning multiple years or decades.

Another aspect of operational strategies relates to the natural variability in the hydrological cycle and the seasonality of conditions that result in inflows and demands changing with the season. Thus each of the seasons may require a separate operational plan. The rule curve shown in Figure 6.7 is a reflection of a seasonal strategy of operation of Bark Lake Control Dam – the most upstream reservoir, with a significant storage, in the Madawaska River System (Ontario, Canada).

The main purpose of Bark Lake is to capture freshet floods and then use the stored water to augment flows downstream during the summer. Thus, prior to the freshet flood, which in the past occurred usually between the mid-March and the end of May, the reservoir is supposed to be drawn down. The target elevation of water in the reservoir at the end of drawdown is not stated explicitly and depends on the amount of snow accumulated in the watershed during the winter and the general condition of the watershed.

During the summer (May to September) the seasonal plan is more restrictive and provides tight boundaries for water elevations. After September the seasonal plan leaves the operator of the reservoir with the freedom to adjust releases to meet downstream demands and to ensure that the autumn floods can be handled safely. The autumn floods have significantly smaller peak inflows and volumes than do the snowmelt floods.

Figures 6.7 and 6.8 illustrate further the seasonal aspects of Bark Lake reservoir operation within a much shorter time frame. Decisions based on specific current conditions, such as current water levels and recorded and anticipated inflows, are carried

Figure 6.7 Bark Lake operation – year 1

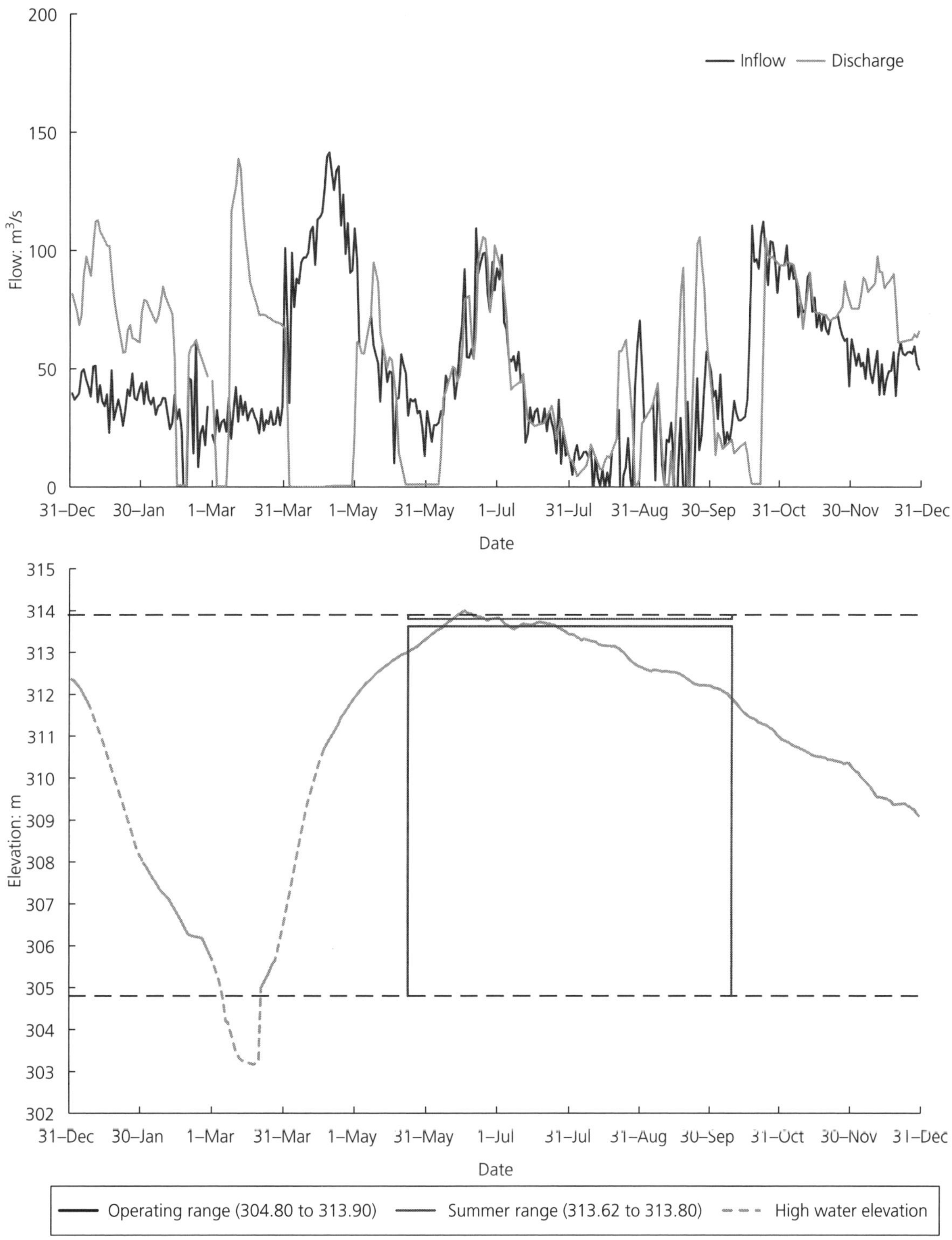


out on a daily basis. However, there are no exact rules supporting the operator in making decisions about specific daily releases from the reservoir. Such decisions are generated by the operator in a heuristic manner in which the following factors are accounted for with different degrees of formality:

- minimum and maximum downstream flow requirements and restrictions
- general watershed conditions that include information on snowpack, snow conditions, soil moisture, etc.
- qualitative and/or quantitative forecasts of inflows to the reservoir, and inflows to other sites with a capability to store water and regulate flows
- effects of routing outflows through the system
- demand for water downstream.

Figure 6.8 Bark Lake operation – year 2

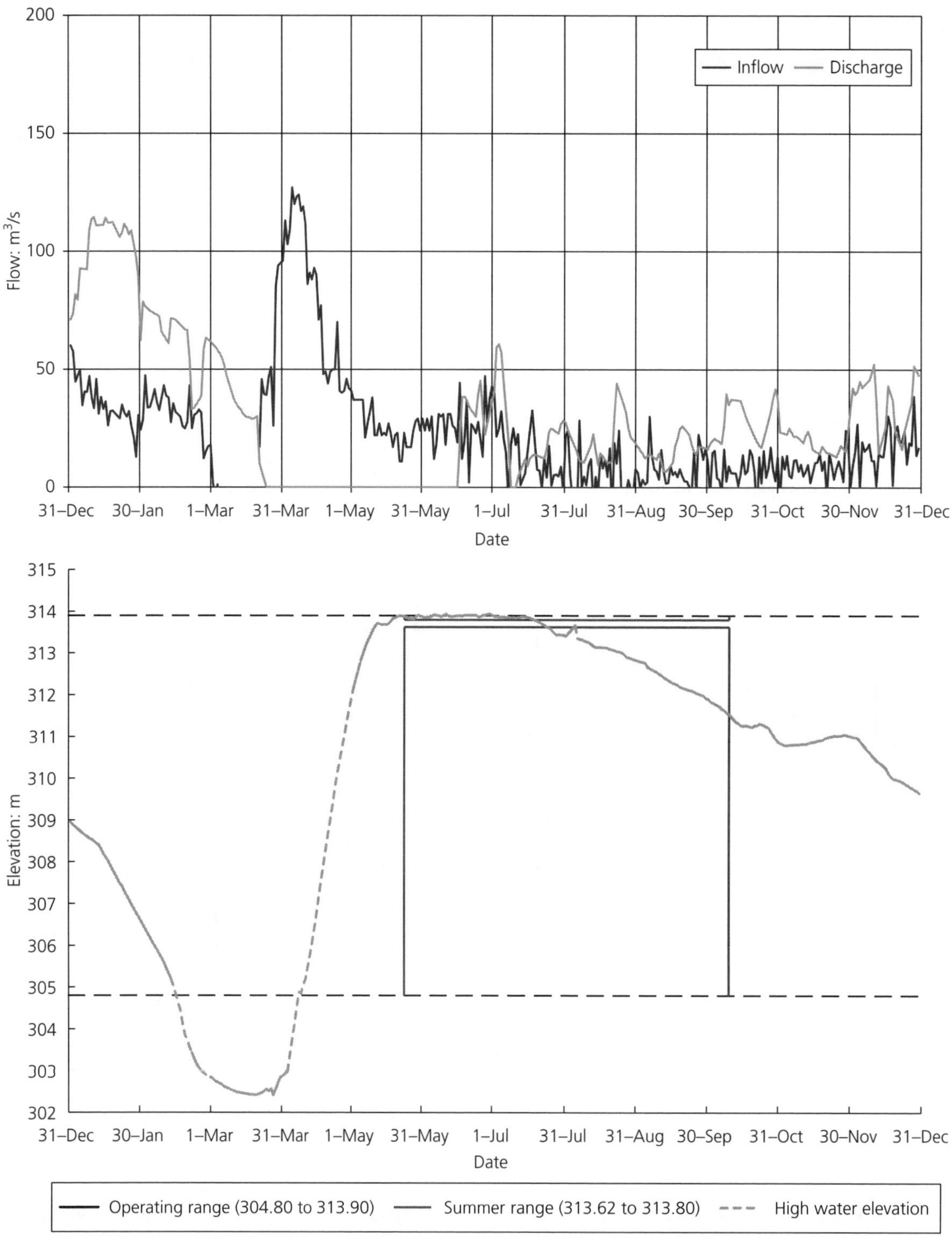


In general, the type of information used for the purpose of short-term operation and the manner in which this information is used for decision-making differs from organisation to organisation, and may also vary within a single organisation. For example, the family of rule curves depicted in Figure 6.8 leaves the operator with an exact release decision if the curves are followed to the letter. One may question whether such simple rules can be further improved by considering not just the current inflow but also future anticipated inflows.

## 6.2. Role of simulation and optimisation in system operation

Both simulation and optimisation techniques can be used for the development of strategies for long-term and seasonal operation and for short-term and real-time operation of a single reservoir or of the entire river system. Both techniques can provide an important insight into operational issues for single reservoirs and thus are capable of providing a very effective support when developing both strategies and rules.

An early understanding that rivers with multiple reservoirs on the main stream as well as on the tributaries require a systems approach to resolving very complex operational issues promoted extensive use of both techniques in the planning and development phases of water resources systems (Buras 1972; Hall 1970; Hufschmidt 1966; Maass 1962). These developments were subsequently followed in the next two decades by increasing interest in the application of both techniques in developing operational strategies and rules for multiple-reservoir river systems. The reader can find an extensive bibliography on this topic in USACE (1991).

While at the beginning of this period the emphasis was on simulation approach, developments in the area of applied operations research and rapidly increasing computer power quickly shifted the attention to optimisation methods and focused on optimisation of reservoir system and river systems operations. However, as USACE pointed out at that time:

> A large gap exists between research studies and innovative applications reported in the literature and the more traditional proven practices followed by the agencies responsible for the actual planning, construction, and operation of reservoir projects.

Although the interest of the research community in the optimisation aspect of river systems operations is still continuing and has already progressed to the stage of development of complex DSS, the reality for many dam owners is that the actual operations are still following either the rules providing upper and lower boundaries of targets or more specific rule curves that provide unique release decisions for any combination of factors taken into account in constructing the curve. The approaches and methods for developing DSS range from relatively simple models for a single reservoir (Georgakakos, 2007) to a sophisticated decision framework for multi-purpose multi-reservoir river systems (Soncini Sessa, 2007).

The simulation of the behaviour of the river system, or more precisely the simulation of the response of the system to various forcings (natural and induced by humans), that forms the scientific foundation of this book requires that at each time interval a unique decision is made with respect to the amount of water to be released from each reservoir. It is quite clear that if the river system under consideration has a fully developed DSS available, the DSS can be incorporated in the overall simulation model and thus can provide the information required to continue the simulation process.

On the other hand, a rule curve that defines only the guiding boundaries is insufficient for simulation purposes: firstly, because no unique decision can be derived from such curve; and, secondly, because it is doubtful if not impossible to construct the curves that can reasonably address the operational issue of any given reservoir in separation from other reservoirs in the system. Thus in such situations the general approach proposed in this book requires that a model of a simplified simulator of existing practice of operating the system needs to be developed and incorporated in the overall simulation model. A relatively straightforward solution to developing such a simulator can be found in Rinaldi *et al.* (1982), where a simple and heuristic method based on the actual operation of a multi-purpose reservoir is proposed, tested and implemented for an existing real case.

## 6.3. Decision-support systems in operation

The development and applications of computerised DSS in water resources began in the early 1970s following the increasing power and availability of computers. Initially the applications were related mostly to general problems of water resources planning and development. Beginning in the late 1980s the applications moved into the area of reservoir management and operation, and later into the management and operation of the entire river systems.

The term 'decision-support system' itself is not uniquely defined and understood, but presently applications in hydrology and water resources follow the definition provided by Adelman (1992): 'interactive computer programs that utilise analytical methods, such as decision analysis, optimisation algorithms, program scheduling routines, and so on, for developing models to help decision-makers formulate alternatives, analyse their impacts, and interpret and select appropriate options for implementation'. Within water resources management the DSS concept has evolved through enhancements of existing mathematical models by adding a graphical user interface and a set of post-processing tools. Advanced DSS may have a sophisticated information technology framework that is tailored to the specific requirements of the decision-making process. A DSS typically encompasses additional capabilities beyond the modelling framework, including advanced tools for data and information management and performance evaluation tools.

Present developments of DSS in support of operational decision-making follow the broad principles of integrated water resources management (IWRM) and integrated river basin management (IRBM). IWRM is 'a process that promotes coordinated development and management of water, land and related resources in order to maximise the resultant economic and social welfare in an equitable manner without compromising the sustainability of vital ecosystems' (GWP, 2000). IRBM recognises that a watershed is a system of interconnected parts and that effective watershed management has to address the governance of watershed as a multiple dam owner, multiple user and multiple stakeholder system.

Although the concepts of IWRM and IRBM were formally introduced only recently, there are examples of DSS that were developed in line with these concepts in the early 1990s. In 1992 the Tennessee Valley Authority (TVA) and the Electric Power Research Institute (EPRI) initiated a project aimed at the development of a suite of software tools to assist in the coordinating scheduling and planning of power generation and reservoirs operation, balancing water quality, water supply and multiple social needs of system stakeholders (Miller *et al.*, 1993; Lindquist *et al.*, 1996). The DSS named TERRA (TVA-EPRI River Resource Aid) uses a geographic information system background map of the Tennessee River Valley system of hydroelectric, nuclear and thermal plants. Management of real-time and historical system status states (pool levels, turbine discharges, minimum flow, water temperatures, dissolved oxygen levels, etc.) and management of system constraints was implemented with the help of an ORACLE-based database. The system is capable of tracking non-compliances and alerting the system operator to potential developing problems.

At almost the same time, the US Bureau of Reclamation, the TVA and the Center for Advanced Decision Support for Water and Environment Systems at the University of Colorado in Boulder began developing a modelling tool that could be applied to any river system for either planning or operational applications (Zagona *et al.*, 1998). In terms of operational application the tool was aimed at supporting long- and mid-term planning, operational forecasting and daily scheduling. At present it represents a complex DSS known as RiverWare, which provides a platform for operational decision-making, forecasting, operational policy evaluation, system optimisation, water accounting, water rights administration and long-term resource planning. It has the capability to model:

- the hydrology and hydrological processes of reservoirs, river reaches, diversions, distribution canals, consumptive uses, groundwater interaction and conjunctive use
- water quality, including total dissolved solids, dissolved oxygen and temperature
- hydropower production and energy uses
- water rights, water ownership and water accounting transactions.

RiverWare's object-oriented, data-centred approach enables the modeller to represent site-specific conditions by creating a network of simulation objects, linking them together to form the river–reservoir network, populating each with data, and selecting physical process algorithms on each object that are appropriate to the purposes of the object and its representation in the overall model. For example, there are numerous methods for routing flow in river reaches, for evaporation calculations and for computing the hydropower generated (CADS, 2016).

A general approach to designing a DSS for planning and managing river systems can follow a process proposed by Castelletti *et al.* (2002). The modified approach introduces three different levels of decision-making in the dam-owning organisation, namely strategic planning, operational control and implementation, as depicted in Figure 6.9.

The DSS at the 'planning level' (DSS-P) focuses on strategic goals of the operation and the ways to achieve them. The outcome of the considerations at this level is a general strategy and operating policy that can be expressed as a set of rules. The rules can be developed through offline analyses that typically will include simulation–optimisation modelling. Alternatively, the rules can be developed within the DSS with the aid of optimal control schemes and simulations. The rules reflect the management goals for the long and medium term.

The 'operation level' (DSS-O) takes the general direction from the planning-level rules, and determines reservoir releases for the entire system in a way that maximises the utilisation of available water resources. The frequency of decision-making at this level depends on the characteristics of the system. For some, such decisions are being made at the beginning of each day, while for others it may once every few days. The frequency will also depend on the flow conditions. During flood events corrective decisions may be taken more frequently than during normal flow conditions.

To reach a release decision the DSS-O can use the same models as those applied at the previous level, but with a higher degree of resolution. The simulations can be carried out with shorter time steps, and the probabilistic characterisation of system components and inputs can be more detailed. Similarly, the optimisation schemes selected at this level can accommodate dynamic aspects of system behaviour.

The 'implementation level' (DSS-I) may seem to be quite simple, because in fully automated systems (where all sluiceway gates are remotely operated) the only decision required from the system operator is to push the button that activates the gate or turns the generators on or off. It may be more complicated when the sites are not remotely operated and require work crews to reach

Figure 6.9 Multi-level decision process

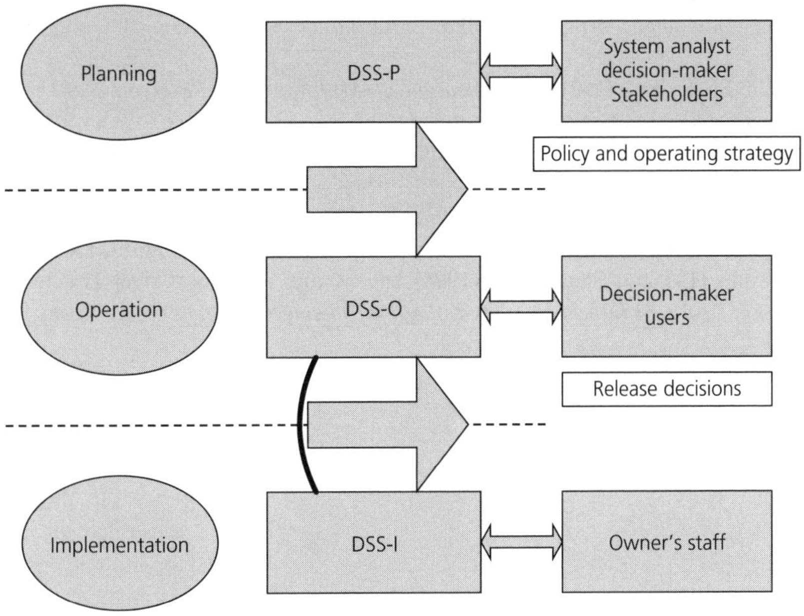


the site and operate equipment manually. The implementation level needs also to address the issues of equipment failures, as indicated by a feedback loop to the operation level in Figure 6.9.

It should be pointed out that in the actual management of a river system the release decisions are in fact recommendations for the system operator, and it is up to the system operator to accept the suggested releases or prompt the DSS for more possible scenarios and additional suggestions. This caveat is easy to understand if one bears in mind that the outcomes of simulation–optimisation analyses are carried out on a simplified model of the real world and that the simplifications are unavoidable in order to make the decision-making problem mathematically and computationally tractable.

## REFERENCES

Adelman N (1992) *Evaluating Decision Support and Expert Systems*. Wiley, New York, NY, USA.

Buras N (1972) *Scientific Allocation of Water Resources: Water Resources Development and Utilization – A Rational Approach*. Elsevier, New York, NY, USA.

CADS (Center for Advanced Decision Support for Water And Environmental Systems) (2016) RiverWare demonstration tutorial. CADS, University of Colorado, Boulder, CO, USA. See http://www.riverware.org/riverware/training/informational/Intro/InformationalTutorial_1_Intro.pdf (accessed 24/06/2016).

Castelletti A, Soncini-Sessa R and Weber E (2002) A two-level DSS for participatory planning of water reservoir systems. *Proceedings of the International Conference on Policies and Tools for Sustainable Water Management in the EU*, Venice International University, Venice, Italy.

Georgakakos A (2007) Decision support systems for integrated water resources management with an application to the Nile basin. In *Topics on System Analysis and Integrated Water Resources Management* (Castelletti A and Soncini-Sessa R (eds)). Elsevier, Amsterdam, the Netherlands, ch. 5.

GWP (Global Water Partnership) (2000) *Integrated Water Resources Management*. GWP Technical Advisory Committee Background Paper No. 4. GWP, Stockholm, Sweden. See http://www.gwp.org/Global/ToolBox/Publications/Background%20papers/04%20Integrated%20Water%20Resources%20Management%20(2000)%20English.pdf (accessed 24/06/2016).

Hall WA (1970) *Water Resources Systems Engineering*. McGraw-Hill, New York, NY, USA.

Hufschmidt MM (1966) *Simulation Techniques for Design of Water-resource Systems*. Harvard University Press, Cambridge, MA, USA.

Lindquist K, McGee M and Cole L (1996) TVA-EPRI river resource aid (TERRA) reservoir and power operations decision support system. *Water, Air and Soil Pollution* **90(1-2)**: 143–150.

Maass A (1962) *Design of Water-resource Systems: New Techniques for Relating Economic Objectives, Engineering Analysis, and Governmental Planning*. Harvard University Press, Cambridge, MA, USA.

Miller BA, Alavian V, Giles JE and Parsly JA (1993) Requirements for an Integrated modeling system to support water resources and power operations. Proceedings of the Hydraulic Engineering '93 Conference, San Francisco, CA, USA, vol. 1-2, pp. 1499–1512.

Rinaldi S, Guariso G and Zielinski P (1982) *The Value of Information in Reservoir Management*. International Institute for Applied Systems Analysis, Laxenburg, Austria.

Soncini Sessa R (2007) *Integrated and Participatory Water Resources Management*, 1st edn. Elsevier, Amsterdam, the Netherlands.

USACE (US Army Corps of Engineers) (1991) *Optimization of Multi-purpose Reservoir System Operations: A Review of Modeling and Analysis Approaches*. USACE, Washington, DC, USA.

USACE (1996a) *Developing Seasonal and Long-term Reservoir System Operation Plans using HEC-PRM*. USACE, Washington, DC, USA.

USACE (1996b) *Risk-based Analysis for Flood Damage Reduction Studies*. USACE, Washington, DC, USA.

USACE (1997) *Hydrologic Engineering Requirements for Reservoirs*. USACE, Washington, DC, USA.

Zagona EA, Fulp TJ, Goranflo HM and Shane RM (1998) RiverWare: A general river and reservoir modeling environment. *Proceedings of the First Federal Interagency Hydrologic Modeling Conference*, Las Vegas, NV, USA, pp. 113–120.

**Operational Safety of Dams and Reservoirs**
ISBN 978-0-7277-6121-7


http://dx.doi.org/10.1680/osdr.61217.173

# Chapter 7
# Disturbances

This chapter introduces a methodology for understanding the effect of various discrete-time disturbances or disruptions to the operation of dam systems. Disruptions can be caused by natural hazards (e.g. earthquakes, landslides and meteorological events, such as severe winds, extreme temperatures and intense rainfall) or any number of other things, such as geomagnetic storms, lightning strikes and fires. Here these are collectively referred to as 'disturbances', regardless of their cause or character. Human factors (e.g. errors of operating rules, misreadings of gauges and maloperation) are treated separately in Chapter 8. Disturbances may be external to dam systems or internal, depending on how the system boundaries are defined.

The *consequences* of disturbances to dam syste'ms may be sudden imposition of loads, blockage, access problems, increased lead times, or sudden loss of component or system functionality. Consequences usually have (much) longer duration than the disturbances that cause them. Consequences of disturbances may be both external (indirect) and internal (direct). An example of the former would be changes in the runoff characteristics of a catchment appearing as the result of forest clear cutting or forest fires, while an example of the latter would be the loss of generation and discharge capability due to a lightning strike or fire in a critical transformer or cable.

## 7.1. Disturbances defined

Sometimes in the course of the operation of a dam system an extraordinary event occurs, such as an earthquake, an earth or rock slide into the reservoir or conveyance works, or a major fire in the drainage area, at a time that cannot be anticipated, but that may affect dam operations or even dam safety. A systems engineering study of BC Hydro's Campbell River generating stations identified the set of disturbances shown in Figure 7.1.

**Figure 7.1** Dam safety considerations in the Campbell River system. (BMT Isis, 2010)

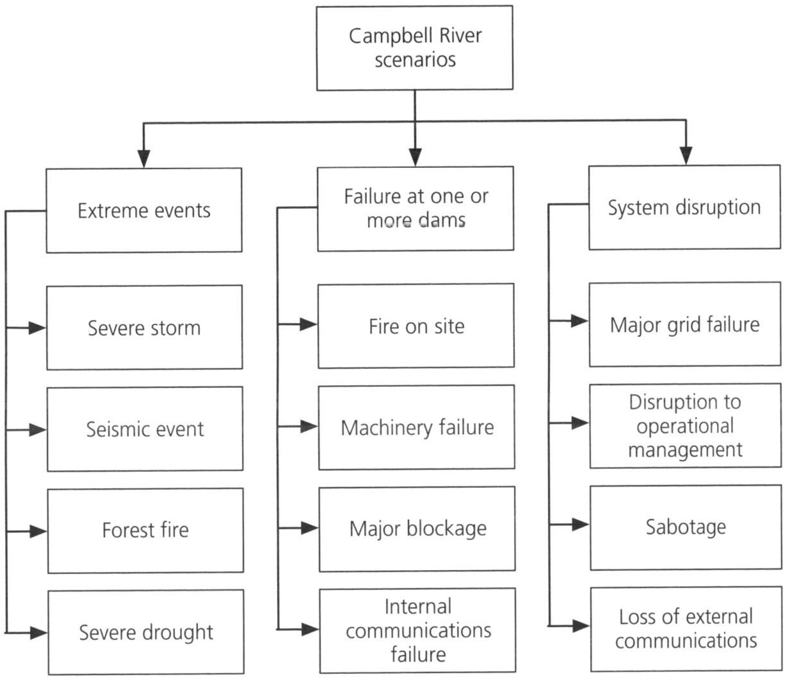

**Table 7.1** Possible disturbances to flow-control systems

| Disturbance | Possible effect |
|---|---|
| Lightning | Effect on power supply<br>Effect on controls<br>Puts powerhouse out by hitting wiring, and thus discharge capacity (common mode failure of production–discharge system) |
| Seismic event | Seiche, leading to oscillating pool elevation<br>Slide, leading to wave, blockage at spillway, loss of storage or debris<br>Damages concrete structure, leading to destruction of dam or blockage<br>Basin changes that modify reservoir inflow<br>Damages to gate<br>Damages to instruments, power supply or controls<br>Inhibits access<br>Puts powerhouse out, and thus discharge capacity (common mode failure of production–discharge system) |
| Debris | Depends on storm events<br>Geomorphology of reservoir slopes or slides<br>Climate inputs: intense rainfall in catchment<br>Sediment loading is a 'futures' issue<br>Debris arrival, type (density, time of year, species, length from outflow, trunks or brush, lumped) and rate of arrival<br>Basin changes that modify reservoir inflow<br>Changes spillway capacity<br>Causes physical load on the concrete structure<br>Changes the hydraulic loads on the structure<br>Prevents gate opening or closing, or damage to gate<br>Puts powerhouse out, and thus discharge capacity (clogging cooling water intakes or trash racks, if any) |
| Icing at spillway | Reduces free opening<br>Changes tailwater conditions<br>Gates stick and de-icing fails to function<br>Puts powerhouse out and thus discharge capacity (icing at intakes) |
| Ice | Reduces free opening<br>Changes reservoir inflows<br>Damages gate |
| Wind waves | Builds ice on gates<br>Overtops gates<br>Overloads gates |
| Slides | Impact waves<br>Increase debris<br>Decrease storage if big enough<br>Modify reservoir inflow |
| Communications | Telephone/radio/internet outages<br>Roads out of service<br>Air transport unavailable<br>Controls, supervisory control and data acquisition (SCADA) system |
| Grid disturbance | Affects powerhouse availability |
| Power supply | Puts powerhouse out and reduces discharge capacity<br>Backup failure or unavailability<br>Instrumentation malfunction |

Disturbances can, in principle, include a broad variety of phenomena, both of natural origin, such as lightning strikes affecting power supplies or instrumentation, and of anthropogenic origin, such as the grid being unable to accept power and thus the powerhouse waterway having diminished discharge capacity, or operational incidents or accidents such as powerhouse fires. The exact definition is a matter of modelling convenience; severe floods or droughts may, for instance, be considered as disturbances or, alternatively, just as aspects of the stochastically modelled catchment hydrology.

A significant strength of the simulation approach to flow-control reliability is the relative ease with which disturbances such as seismic events, lightning strikes or grid unavailability can be accommodated. A variety of disturbances can, in principle, be included within a simulation approach.

Disturbances are of two basic types: external and internal. *External disturbances* are those disruptive events that arise outside the boundaries of the system being modelled (e.g. earthquakes, forest fires or flows of floating debris). *Internal disturbances* are those disruptive events within the boundaries of the system for which there are no identified external causes (e.g. plant fires, forced outages or the unforeseen mismanagement of procedural factors).

Examples of external disturbances are given in Table 7.1. These are things that could occur outside the boundary of the system being modelled, either in space or in the management hierarchy (e.g. an unanticipated disrupting directive from political authorities or from higher levels in an organisation not included in the modelled system).

When a disturbance occurs, for example an earthquake or a landslide into the reservoir, it may cause damages to parts of the flow-control system, or it may trigger other disturbances or consequences, which in turn cause damage to the dam system. It may simultaneously damage more than one component of the system; for example, seismic ground-shaking might damage mechanical equipment in one or more spillway gates while at the same time damaging the concrete conveyance carrying the outflow from the gates. The combinations of damages can exaggerate the consequences by limiting the redundancies in the way the flow-control system operates.

An important factor in analysing disturbances is their causal interactions. For example, both earthquakes and reservoir slides could be individually modelled as disturbances occurring randomly in time with some frequency and severity. However, a reservoir slide might also be caused by an earthquake, and thus not be 'random' at all, except in the sense that it is triggered by another natural hazard which is considered random (Figure 7.2).

The inter-arrival time between occurrences of the same or different types may itself become a concern. For example, in traditional dam safety evaluations the joint occurrence of an earthquake and an extreme hydrological event is ignored; because the two events are phenomenologically independent and each is of low probability, their joint probability within a

Figure 7.2 Effect of disturbances on flow-control reliability

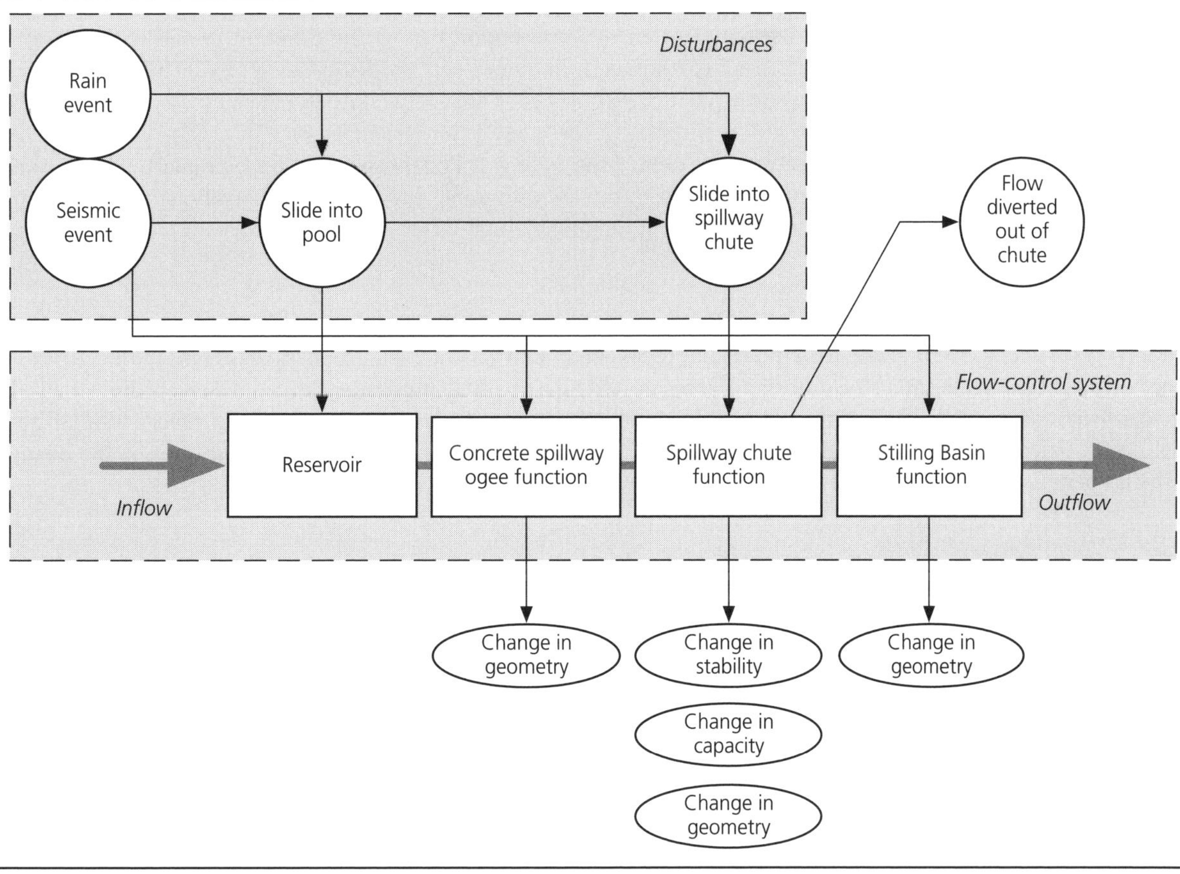

Figure 7.3 Common modelling protocol for natural hazard risk analyses showing annual exceedance probabilities (AEP) of hazard and consequences and probability of failure (Pf) with respect to hazard load

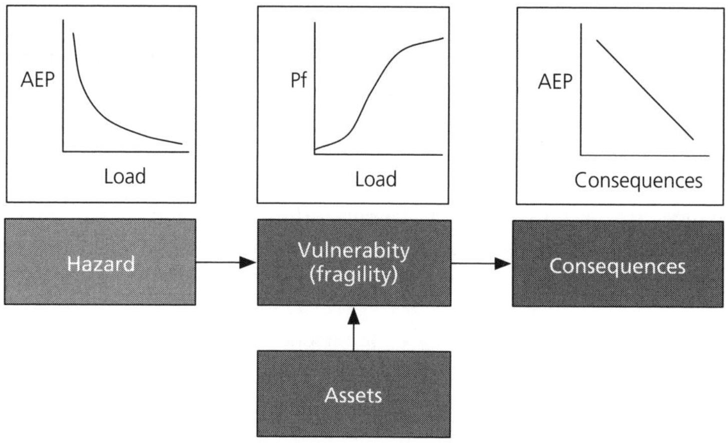


small period of time is vanishingly small. In practice, however, should an earthquake occur and cause damage to the flow-control system that cannot be repaired within an impending flood season, a subsequent large, but not extreme, hydrological event may be uncontrollable due to the pre-existing spillway damage. Thus, while the disturbance events are probabilistically independent, their consequence durations may intersect.

## 7.2. Probabilistic model of disturbances

Disturbances are generally modelled – as are most natural hazards – as:

- a probability of occurrence of the hazard
- an identification of the assets that could be damaged
- the fragility of those assets (i.e. the conditional probability of damage given the hazard occurs)
- an assessment of the economic and other consequences (Figure 7.3).

The occurrence of the hazard is characterised by a time at which it initiates, a severity and a duration (Figure 7.4). The duration may be that of the disturbance itself, but often of more importance is the duration of the consequence of the disruption (e.g. the repair of damages due to an earthquake). Should a second disturbance (e.g. a high reservoir inflow) occur before the damages are repaired, the subsequent consequences of that second disturbance may be exacerbated.

The time of initiation can be modelled as a random point process (e.g. a Poisson or negative binomial), while severity and duration are typically modelled as continuous random variables with some probability density functions. For simplicity, severity is sometimes described in discrete increments (e.g. small, medium, large; or Mercalli intensity) with a corresponding probability mass function.

The probabilistic approach to simulating disturbances borrows from actuarial risk theory and models of property and casualty insurance (Daykin *et al.*, 1993). These involve a point process in time to generate disturbances (or claims in the actuarial case), combined with an uncertain intensity process to generate the size of the disturbance (or claim loss in the actuarial case). Natural hazards risk modelling, for example, involving seismic risk or hurricane risk, has long used a similar approach (Cornell, 1968; Friedman, 1984). The disturbance model adds one more factor, which is the duration of the disturbance, from which operational interdependencies may arise within the flow-control system.

Figure 7.4 Simple disturbance model

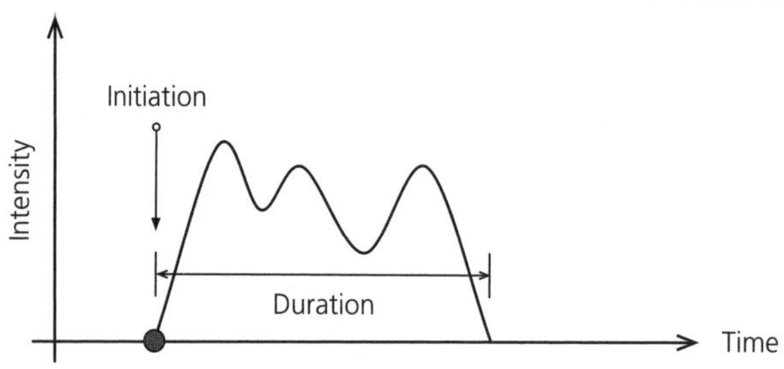

The inclusion of disturbances within the simulation can affect systems performance in at least two ways. First, as in the upper part of Figure 7.2, the disturbance might affect the demand function on the flow-control system. For example, a seismic event could cause a slide into the reservoir and consequently a wave load on the spillway outfall; or a fire in the basin might increase the debris load in the basin and consequently a debris pile up on the outflow opening. Second, as the in lower part of Figure 7.2, the disturbance might affect the capacity of some component of the flow-control system to withstand the demands placed upon it. For example, the seismic event might twist a spillway gate and consequently decrease the outflow capacity; or a disturbance to the power grid might mean that power cannot be generated, and thus the discharge capacity of the powerhouse waterway would be substantially reduced.

### 7.2.1 Occurrence model in time

If we assume that a disturbance is initiated at an instant in time, and if we ignore the duration of the disturbance, then we may represent the initiation as a point in time. For example, earthquake occurrences are modelled in this way, as are forest fire initiations, reservoir landslides, many types of human error and grid outages. Point process models are also common in the earth and environmental sciences (Gaetan and Guyon, 2009).

#### 7.2.1.1 Initiation of a disturbance

If the initiation of disturbances is modelled as a random point process in time, then there are a random number of occurrences within any interval of time, and the times between occurrences are also a random process (Figure 7.5). Point processes in time are one-dimensional and thus inherently ordered. Every individual occurrence has predecessors and successors. This ordering is not present in higher dimensional processes such as spatially random point processes.

A temporal random point process can be described in various ways. The most obvious is to describe the *arrival times* of the occurrences $\{T_1, T_2, \ldots, T_n\}$, but the probabilistic description of these is made difficult by their order dependences, in that $T_1 < T_2, \ldots, T_{n-1} < T_n$. Often, it is simpler to describe the *inter-arrival times* $\{T_2 - T_1, \ldots, T_n - T_{n-1}\}$, in that for some random process, such as Poisson and Renewal processes, these times are independent and identically distributed. An alternative way to describe the random arrivals is to count the cumulative number $N_t$ up to a point in time. This is called a *counting process*. Counts of arrivals within an interval of time are simply the differences between the counting process for the two times. In spatial point processes the concept of inter-arrivals is less easily defined than in the one dimension of time, and thus for these spatial or spatio-temporal processes the metric of counts in more common than arrivals or inter-arrivals (Lewis, 1972).

#### 7.2.1.2 Stationary point processes

Presuming that the arrivals of disturbances are random in time with a constant rate $p$ per unit time, and presuming that the individual disturbances are probabilistically independent of one another and that the probability of more than one disturbance in a unit of time is negligible, then the number of initiations $k$ within $n$ intervals of time follows a simple binomial probability mass function:

$$p(k) = C_k^n p^k (1-p)^{(n-k)} \tag{7.1}$$

in which $C_k^n$ is the number of combinations of $n$ things taken $k$ at a time, and $p$ is the probability of one disturbance in a unit time interval. The process is said to be *stationary* if the rate of arrival is constant in time: Somewhat more formally, a process is said to be stationary if the probability distribution of the number of points in some interval $t_0 + \Delta t$ does not depend on $t_0$. For large values of $np$ the discrete binomial can be approximated by a continuous *normal* distribution, but this is seldom useful in modelling the particular disturbances related to reservoirs and flow-control because they are usually rare events. Were a disturbance frequent enough to satisfy the above condition, it would most likely be managed operationally rather than as a disturbance.

#### Poisson point process

For rare events, as the interval of time becomes ever narrower, while the probability per time interval decreases in proportion, the number of disturbances in time $t$ can be approximated asymptotically as $t \to 0$ by the *Poisson* probability mass function:

$$p(n) = \frac{(\lambda t)^n n e^{-\lambda t}}{n!} \tag{7.2}$$

**Figure 7.5** Temporal point process

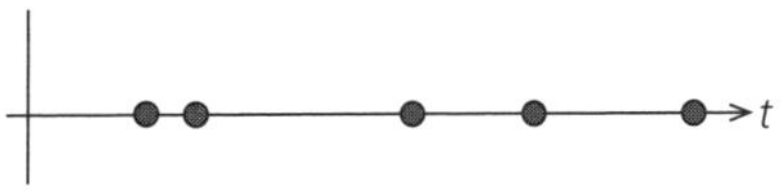

in which $p(n)$ is the probability mass function of the number of arrivals $n$ in time $t$, and $\lambda$ is the arrival rate per unit time. This is arguably the most common probability model used to represent random disturbances. It assumes stationarity of arrivals (i.e. the rate of occurrences in time is constant) and that any two arrivals are probabilistically independent of one another. For many natural processes and operational disturbances these assumptions of stationarity and independence are reasonable.

The probability density function of inter-arrival times for the Poisson process is

$$f(t) = (\lambda t)e^{-\lambda t} \tag{7.3}$$

in which $t$ is the time between the initiations of disturbances. The probability density function of the $k$th arrival time is Gamma distributed according to

$$f(t) = \frac{\lambda^k}{\Gamma(k)} t^{k-1} e^{-\lambda t} \tag{7.4}$$

for $t > 0$, and in which $\Gamma(k)$ is the *gamma function* of $k$, which for integer values of $k$ equals $\Gamma(k) = (k - 1)!$. For integer values of $k$ this is sometimes called the *Erlang distribution*.

An important assumption of the Poisson distribution is that all occurrences are independent of one another. Thus, the probability of occurrence of a new disturbance does not depend on other recent occurrences. Regardless of the start point in time, whether at the occurrence of a disturbance or at an arbitrary time, the probability distribution of the arrival time of the next disturbance is always the same – disturbances are random in time and unrelated to one another.

#### Cox process

The Cox process is among the simplest extensions of the stationary Poisson process. Sometimes called a doubly stochastic, inhomogeneous, or mixed Poisson process, this is a Poisson process for which the rate parameter is itself a stochastic function of time, $\lambda(t)$ (Cox, 1955). For intervals of time in which $\lambda$ is high, more disturbances occur and thus the disturbances are clumped. For intervals of time in which $\lambda$ is low, fewer disturbances occur and thus the disturbances are dispersed. A Cox process is stationary if $\lambda(t)$ is stationary.

#### Renewal process

Renewal processes are similar to Poisson processes (the disturbances are independent and identically distributed, stationary, etc.) but the probability density functions of inter-arrival times are generalised from the simple exponential (Cox, 1962). For example, the inter-arrivals might be normally or log Normal distributed, or distributed according to a power law, or something else, but they remain independent and identically distributed. The name derives from the individual inter-arrival times $\{T_2 - T_1, \ldots, T_n - T_{n-1}\}$, which are called *renewal times*. On the other hand, the future time of an arrival is not independent of the time of the last arrival as in a Poisson process with exponential inter-arrivals. For an exponential distribution $f(t) = \Lambda e^{-\Lambda t}$, the conditional probability density function, given that a time $t$ has already elapsed (i.e. $t \geq T$), is unchanged from the marginal probability density function $f(t \mid t \geq T) = \lambda e^{-\lambda t}$.

Renewal processes are typically applied to mechanical and electrical equipment, rather than to natural hazards, because the former are periodically maintained or repaired, and thus made 'new' at each inspection time. Typically, renewal processes are not used to model natural hazards.

### 7.2.1.3 Non-stationary disturbances

*Temporal stationarity* is the property that the statistical model and parameters of the point process are constant over time. *Non-stationarity* is the opposite property, that the statistical model or its parameter changes over time, and the rate term $\lambda$ is the rate in time and space (e.g. number per year-km$^2$).

#### Non-homogeneous Poisson process

The non-homogeneous Poisson process is among the simplest extensions of the stationary Poisson process to non-stationarity. This is simply a Poisson process for which the rate parameter $\lambda$ varies deterministically with time, and is thus related to the Cox process. For intervals of time in which $\lambda$ is high, more disturbances occur and thus the disturbances are clumped. For intervals of time in which $\lambda$ is low, fewer disturbances occur and thus the disturbances are dispersed (Figure 7.6).

The variation in the rate of occurrence may be seasonal in that, for example, extreme stream flows presumably occur more frequently in spring than in autumn at most temperate climate sites. Cox and Isham (1980) cite the case of coal-mining fatalities varying over time with employment in the mines. For operation and maintenance issues the rate of equipment breakdowns may slowly increase between times of maintenance. Similarly, the rate of human errors may increase with worker fatigue during the daily shift.

Figure 7.6 Temporal point process with a time-varying arrival rate (Cox process)

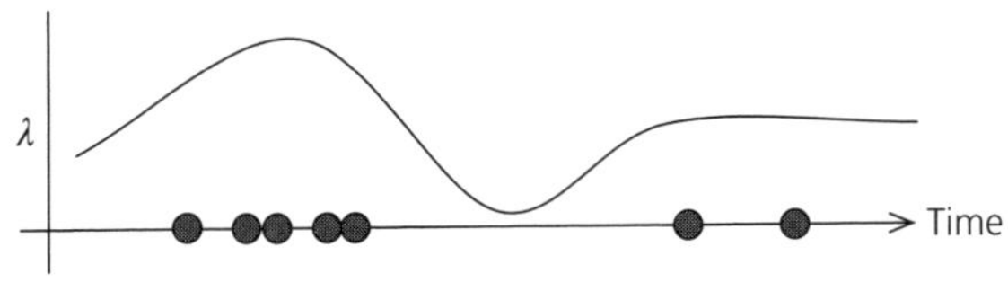


### Conditional intensity function

Some point processes have history-dependent rates through which future rates of occurrence depend on the history of previous occurrences. As a generalisation, any point process in time can be described by a *conditional intensity function* $\lambda(t \mid H_t)$, which is the probability of a point occurrence within an infinitesimal interval of time given the history of occurrences $H_t$ up to that time:

$$\lambda(t \mid H_t) = \lim_{\Delta t \to 0} \frac{\Pr(\Delta N = 1 \mid H_t)}{\Delta t} \tag{7.5}$$

in which $\Delta N$ is the event in which a point occurs (i.e. the count increments), and $\Pr(\Delta N = 1 \mid H_t)$ is the instantaneous probability of a point. In the limit as $\Delta t \to 0$, the probability of more than one point within a single time interval $\Delta t$ is zero. The conditional intensity function provides a general way of expressing dependence in the point process. A point process described by a conditional intensity function is usually called a *doubly stochastic point process.*

As the occurrence or non-occurrence of a point within a given time interval $\Delta t$ is a simple Bernoulli event, the inter-arrival probability density function for a doubly stochastic point process depending only on the time of the last occurrence is

$$f(t \mid H_t) = \lambda(t \mid H_t) \exp\left\{-\int_{t_{i-1}}^{t} \lambda(t \mid H_t)\,\mathrm{d}t\right\} \tag{7.6}$$

That is, the inter-arrival probability density function is a simple exponential distribution with variable $\lambda$ depending on the time since the last occurrence. This probability density function is easily generalised for the case of dependence on multiple precursors (Eden, 2005).

Among the simplest conditional intensity functions, and among the easiest to statistically estimate, is that related to an auto-regressive dependence:

$$\lambda_k = \exp\left\{\alpha_0 + \sum_{i=1}^{n} \alpha_i \Delta N_{k-i}\right\} \tag{7.7}$$

This is dependent on the number occurrences (1 to $n$) in the most recent $n$ time periods. Depending on the values chosen for the parameters $\alpha_i$, the recent occurrences may either heighten or lessen the $(n + 1)$ probability.

A similar conditional dependence can be specified with respect to the intervals $\{X_j = T_j - T_{j-1}\}$ between occurrences, as for a renewal process. For the special but common case in which the time to the next occurrence depends only on the duration of the preceding interval, the intervals are as for a Markov chain. This special case is sometimes called a *Wold process* (Wold, 1948).

#### 7.2.1.4 Cross-related initiations of disturbances

Clearly, a disturbance such as *forest fires in the basin* should be negatively correlated with rainfall, and thus stream discharges (Figure 7.7(a)). This can be accomplished by defining a Cox model with the rate parameter $\lambda$ as a function of rainfall within some prior window of time $\Delta$:

$$\lambda(t \mid D_t) = \alpha_0 + \alpha_i \int_{i=t-\Delta}^{t} D_i\,\mathrm{d}t \tag{7.8}$$

in which $D_t$ is the depth of rain at time interval $t$, and $\alpha_0$ and $\alpha_i$ are empirical parameters. The arrivals could also be generated to include such a temporal autocorrelation using, for example, a negative-binomial arrival model (Ripley, 2004).

For the case of cross-dependent point processes – for example, *forest fires statistically related to increased debris load* – a common and convenient approach to modelling cross-correlation between two or more point processes is via the conditional intensity function.

Figure 7.7 (a) Stochastic models of: 'disturbances' as Cox point processes dependent on another variable – here for fire – overlain on the hydrological simulation of water flows; (b) dependent 'disturbances' as point processes – here for fire and debris

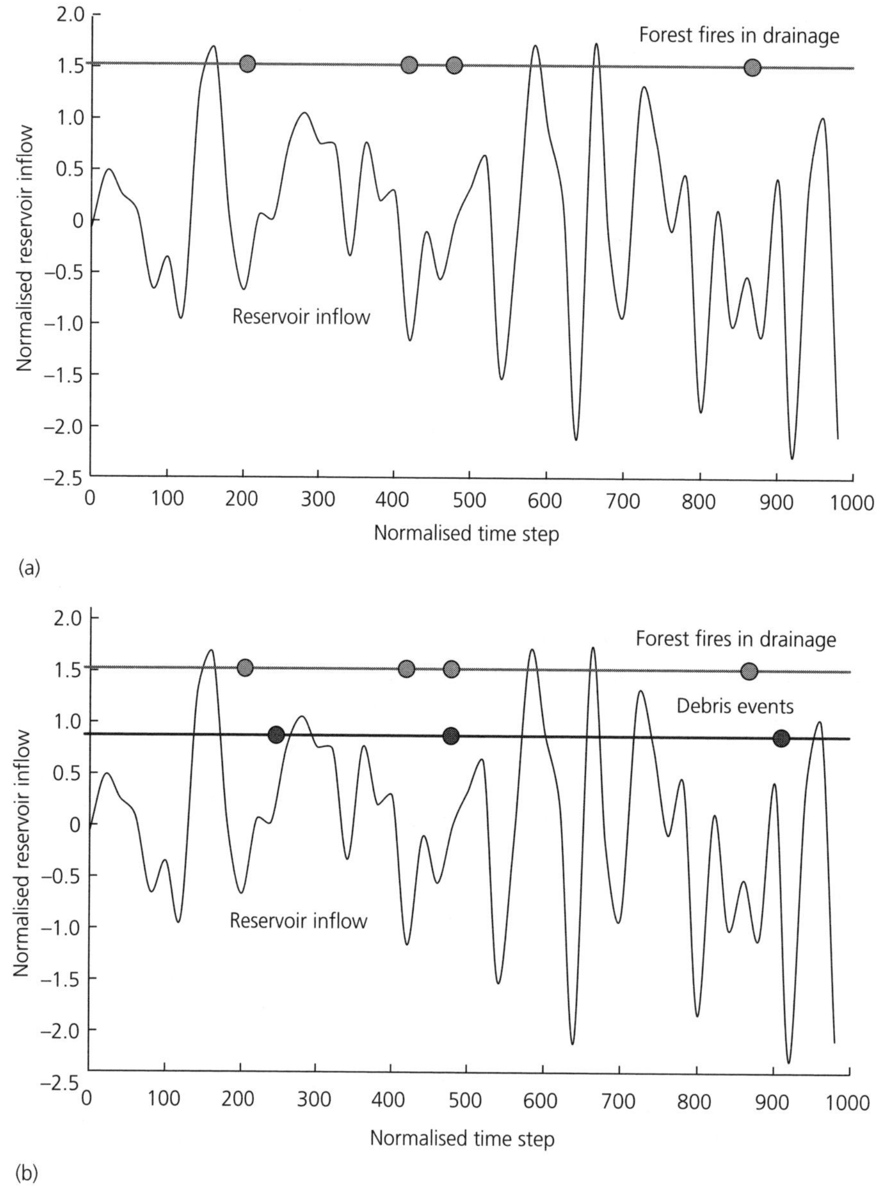


If multiple point processes are modelled, a conditional intensity function can be defined using the occurrences of a separate point process as arguments (e.g. the arrivals described by the right-hand side of Equation 7.7 could be those of a separate disturbance). For example, the probability of the occurrence of debris loading events could be conditioned on the number of occurrences of forest fire events in the previous $n$ time periods. In the simplest case, just the most recent time step ($n = 1$).

#### 7.2.1.5 Causal interaction of disturbances

From the diagram in Figure 7.2 it is clear that disturbances might be causally related as well as statistically correlated; that is, the occurrence of, say, a seismic event might physically cause another disturbance such as a reservoir land slide. Disturbances might even interplay together to physically cause a third disturbance. For example, if a seismic event occurs immediately after a large rainfall event the probability of inducing reservoir slides may even be greater.

While it is possible to develop engineering reliability models from which to calculate the probability of an induced disturbance given the occurrence of a disturbance of another type, in practice this is complicated, data-intensive and expensive (Hartford and Baecher, 2004). A much simpler and less-expensive approach is to treat the casual relationship in a statistical way using a conditional intensity function.

### 7.2.2 Occurrence model in space

Most natural hazard disturbances such as earthquakes, forest fires and lightning strikes occur not only in time but also in space – they are spatio-temporal processes. For example, data for the times and spatial location along the Parkfield, California, fault of micro-seismic events between 1988 and 1996 have both time and spatial location, and have been widely studied in geography, the earth sciences and meteorology (Cliff and Ord, 1981; Cressie, 1993; Lewis, 1972).

#### 7.2.2.1 Spatial Poisson process

As in the case of the purely temporal disturbances, the simplest model of spatio-temporal disturbances is the Poisson process. For the spatial Poisson process every bounded area $A$ contains a Poisson-distributed number of disturbances, and disjoint areas have numbers of disturbances that are probabilistically independent. The probabilities of disturbances occurring in time and space are uniform over those dimensions. As in temporal Poisson processes, the distribution of closest-neighbour points for the spatial Poisson process is exponential, but adjusted for the dimension of the space: in $R^2$ the probability density function of nearest-neighbour spacing decays in proportion to $f(r \mid \lambda) \propto \exp(-\pi\lambda r^2)$, in which $r$ is the nearest-neighbour distance and $\lambda$ is the spatial density of points. As in the case of the Cox process in time, in the spatio-temporal process $\lambda$ can be allowed to vary in space, yielding clumps of points in some regions and dispersions in others.

#### 7.2.2.2 Clustering: the negative binomial process

Some natural disturbances (e.g., tornados) exhibit spatial correlation or clustering. That is, if one tornado occurs the probability of additional nearby tornados increases. These disturbances are said to be 'contagious', and the Poisson model tends not to provide a good fit. The spatial distribution of particular species of trees suggests clustering (Wold, 1948).

One way to detect clustering in disturbance occurrence data is by using cell counts. Dividing the region of occurrence into a grid or a set of cells allows for a counting of disturbance occurrences by cell. For a Poisson distribution of points the mean number of counts per cell and the variance of the counts should be equal. Thus, the ratio of the variance to the mean provides an index of clustering:

$$I = \frac{s^2}{\bar{x}} - 1 \qquad (7.9)$$

where $s^2$ is the variance of the point counts per cell, and $\bar{x}$ is the mean or average number (Ripley, 2004). For a Poisson distribution of points this ratio should be 1.0. If the data points are clustered, this index should be positive; if the data points are overly dispersed, this index should be negative. Other similar indices are also available for testing empirical distributions of disturbances (Rogers, 1974).

In such cases a better model is often a Cox process, or its special case the *negative binomial*:

$$p(k) = \frac{(k+r-1)!}{k!(r-1)!} p^k (1-p)^r \qquad (7.10)$$

where $k$ is the number of occurrences and $r$ is the number of non-occurrences. The negative binomial results when the probability density function of the partial rate $\lambda$ of the Cox process is itself gamma distributed over space. Equation 7.5 is also called the *Pólya distribution* when $r$ is real valued. For discrete $r$, Equation 7.5 is sometimes called the *Pascal distribution*, especially in engineering and climatology. Note, the negative binomial is similar to the binomial but with clustering of disturbances, and is usually applied to spatial distributions. For large values of $r$ the negative binomial approaches the Poisson distribution asymptotically.

### 7.2.3 Severity (size) and duration models

In addition to occurring randomly in time and space, most disturbances also have an associated severity: some disturbances are small and some are large. The magnitude of an earthquake can be minor or major, the number of barrels of oil spilled in an accident can be negligible or immense, the number of hectares covered by a forest fire can be restricted or expansive. The severity of a disturbance is itself uncertain before the disturbance occurs, and so the severity is usually also modelled as having a random size. Mathematically, this is exactly what the insurance industry does when setting premiums and forecasting losses using actuarial models (Hogg and Klugman, 1984; Kleiber and Kotz, 2003). For the particular disturbances mentioned above, Cornell (1968) used actuarial models to predict the magnitude of earthquakes for seismic risk assessments of power plant sites, Eschenbach *et al.* (2010) used actuarial models to forecast oil spills for risk analyses of offshore platforms, and Preisler *et al.* (2011) used actuarial models for the sizes of wildfires to forecast fire-fighting costs (Table 7.2).

#### 7.2.3.1 Exponential distribution

The exponential distribution is among the simplest of continuous probability density functions. It arises as the inter-arrivals of the Poisson point process, but is also used to describe the distribution of sizes in a variety of applications.

**Table 7.2** Common severity (size) probability models

| Name | Probability density function | Support | Mean | Variance |
|---|---|---|---|---|
| Exponential | $f(x) = \lambda \exp[-\lambda x]$ | $x \geq 0$ | $1/\lambda$ | $1/\lambda^2$ |
| Normal | $f(x) = \frac{1}{\sigma\sqrt{2\pi}} \exp\left[-\frac{1}{2}\left(\frac{x-\mu}{\sigma}\right)^2\right]$ | $-\infty \leq x \leq \infty$ | $\mu$ | $\sigma^2$ |
| log Normal | $f(x) = \frac{1}{x\sigma\sqrt{2\pi}} \exp\left[-\frac{1}{2}\left(\frac{\ln(x)-\mu}{\sigma}\right)^2\right]$ | $x \geq 0$ | $e^{\mu+1/2\sigma^2}$ | $(e^{\sigma^2}-1)e^{2\mu+\sigma^2}$ |
| Laplace | $f(x) = \frac{1}{2\sigma} \exp\left[-\left(\frac{\lvert x-\mu\rvert}{\sigma}\right)\right]$ | $-\infty \leq x \leq \infty$ | $\mu$ | $2\sigma^2$ |
| Power law | $f(x) = ak^a x^{-(a+1)}$ | $x \geq 0$ | $\frac{ak}{a-1}$ | $\frac{ak^2}{(a-2)(a-1)^2}$ |
| Weibull | $f(x) = \frac{k}{\lambda}\left(\frac{x}{\lambda}\right)^{k-1} \exp\left[-\left(\frac{x}{\lambda}\right)^k\right]$ | $x \geq 0$ | $\lambda\Gamma(1 + 1/k)$ | $\lambda^2[\Gamma(1 + 2/k) - \Gamma^2(1 + 2/k)]$ |

It is the maximum entropy distribution of all probability density functions with support $[0, \infty)$ and a finite mean (Jaynes, 1996). The minimum of $n$ exponential variates with corresponding parameters $\lambda_i$ is also exponentially distributed, but the maximum is not (Gumbel, 1958).

#### 7.2.3.2 Normal distribution

Among the simplest size models for disturbances is the *normal or Gaussian distribution.* The normal probability density function is symmetrical and decays rapidly both above and below the mean. The normal distribution arises naturally from the central limit theorem in the case where a random size is generated as a sum of a large number of factors, $x = \sum_i z_i$, in which the $z_i$ are random and independent variables. The mean is also the mode, or most likely value, both properties of which distinguish the normal from most other commonly used size distributions, and limit its use. For most natural hazards the upper tail of the normal decays too quickly to be a good model of empirical data. Experience suggests that the upper distribution of disturbance size decays more slowly (i.e. the upper tail is heavier) than the normal.

#### 7.2.3.3 log Normal distribution

The log Normal distribution is widely used in the earth sciences to model size behaviour, for example, of the trace lengths of faults (Cruden, 1977), and pool sizes in oil and gas exploration (Kaufman and Barouch 1976). The log Normal arises by applying a normal distribution to the logarithm of the variable, and as such arises naturally from the central limit theorem in the case where a random size is generated as a multiplication of a large number of factors, $x = \prod_i z_i$, in which the $z_i$ are random and independent variables. The mean is typically greater than the mode, while the median and mode are the same. This gives the log Normal distribution a thick upper tail, which has been proven empirically to be a good fit for many naturally occurring size distributions in the natural sciences, and thus it is widely used in geology, geography and wildlife biology.

#### 7.2.3.4 Power law distribution

The power law distribution is another simple-size distribution depending only on the power of size. Thus, it is a complementary cumulative distribution function (i.e. annual exceedance probability curve) and is linear on a log–log grid. Power law distributions are commonly observed in naturally occurring size data, such as earthquake magnitudes, stream lengths and the sizes of cities. They are also common in social science and economic data, such as income distributions. The *Pareto distribution* of income is a power law. Various explanations have been put forward for this empirical phenomenon (Reed and McKelvey, 2002). If $X$ is a random variable and $X \sim$ power law, then $\log X \sim$ exponential.

#### 7.2.3.5 Laplace distribution

The Laplace distribution, which is also referred to as the 'double exponential' distribution, comprises two simple exponential distributions joined together at the distribution mean (Kotz *et al.*, 2001). In the standard case, this is at the origin. The distribution is defined by two parameters, a mean and a scale parameter. Thus, it is similar to the normal distribution except that its decay is expressed in linear deviations from the mean rather than as squared-deviations as is the case of the normal distribution. The log(Laplace) variant of this distribution is similar in concept to the log Normal distribution in that it uses log(deviations) for the decay of probability away from the mean, and in consequence has thicker tails than the common Laplace distribution (Kozubowski and Podgorski, 2002). The log(Laplace) is also called the *double power law.*

#### 7.2.3.6 Weibull (gamma) distribution

The Weibull distribution is similar to the exponential distribution except for there being a double exponent in the exponential term. This distribution is commonly attributed to the Swedish physicist W. Weibull (1887–1979), although other provenances are claimed. Russian literature sometimes refers to it as the 'Gnedenko distribution', while the French literature sometimes refers to it as the 'Fréchet distribution'. The double exponent parameter $k$ is referred to as the 'shape parameter', while $\lambda$ is referred to as the 'location parameter'. The distribution displays a thick upper tail for small values of the shape parameter ($k < 1$). For $k < 1$ the Weibull distribution becomes the exponential distribution. The distribution arises in extreme-value theory, which deals with the maximum or minimum number of multiple observations (Gumbel, 1958). As a result, the distribution is widely used in reliability engineering studies of the length of life of mechanical and electrical equipment, and in actuarial models for the distributions of insured losses in property and personal injury markets.

#### 7.2.3.7 Duration of the consequences of disturbances

Many disturbances, including those that are perhaps most of interest in flow control, have a short physical duration. For example, an earthquake event may last but tens of seconds. A forest fire may last days, but even this can be considered short with respect to the practical performance of a dam. On the other hand, the duration of the consequences of a disturbance may not be short. If an earthquake occurs and damages a gate structure, it may take months or years to repair the damage. That period of repair may overlap with seasons of high reservoir inflow, and thus the probability of a high spillway demand at a time when the gates are impaired due to earthquake loading could be non-negligible. The consequences of earthquake and flood loading in this way are not mutually exclusive, as is sometimes assumed in design.

In simulating flow-control performance, not only is the duration of the physical disturbance possibly of interest, but so too may be the time to repair the consequences of the disturbance. Of course, this time to repair may itself be uncertain, and could be specified in a probability distribution, although for practical purposes that may not be necessary and the duration might be identified just by an expected value.

The way in which the simulation works is that a disturbance occurs in a particular cycle. This disturbance has some associated random magnitude. The disturbance and its magnitude are generated in the normal way using pseudo-random numbers. All the failure gates are polled to check whether a system state has been changed and a flag tripped. A change in a system state (i.e. the occurrence of a fault or a failure) in turn depends on the fragility curve for that failure gate, which gives the probability of a fault or failure at that failure gate as a function of disturbance magnitude. In the event that a fault or failure occurs, the corresponding failure-gate flag is tripped, and that flag remains active for a period of time in the simulation (found from a look-up table). Should a subsequent disturbance, for example a flood or a debris load, occur on a later simulation cycle, the impact on this failure gate would possibly be altered by the impaired system state at the failure gate.

## 7.3. Seismic hazard

A probabilistic seismic hazard analysis (PSHA) is conducted to quantify the probability of exceeding a given earthquake characteristic (e.g. peak ground acceleration) at a site given all the possible earthquakes that may influence the site (Field, 2001).

Earthquakes cause damage in four ways:

- vibratory ground motion causing damage to natural or artificial structures, or ground failures
- rupture of the fault at the ground surface or below causing damage to structures in the immediate area
- upheaval or subsidence in the fault region causing damage related to ocean or river waters
- tsunamis or seiches causing inundation of coastal or lake-shore regions.

Various ground-motion parameters may be of interest at a particular site, such as the peak displacement, velocity or acceleration, which can lead to damage of structures at the site. Most analyses examine the peak ground acceleration (PGA) or the response spectral acceleration to quantify ground motion in the PSHA. The spectral acceleration is defined as the maximum acceleration experienced by a damped single-degree-of-freedom oscillator, which is a simplified representation of building response. Various spectral periods may be analysed depending on the natural period of the structure.

The hazard computation involves calculating the probability that a specified ground-motion level will be exceeded in a specific period of time, typically based on the following methodology:

1. Potential seismic sources (faults and area sources) within a specific distance (e.g. at least 200 km) of the study site are identified by geological and seismological studies.
2. For each source, a magnitude–recurrence relationship (magnitude probability density function) is developed that describes the probabilities that earthquakes of different magnitudes will be exceeded within a particular time period. This is based on the historical earthquake record for a given source.

3 The probability that the earthquake source is a specific distance from the site is determined (distance probability density function).
4 One or more ground-motion prediction models applicable to the region are selected.
5 The probability that the ground motion at the site will exceed a specific value is computed by integrating over all sources and all possible earthquake magnitude–distance combinations and their probabilities, also incorporating the prediction model uncertainty $\varepsilon$.
6 The computation is repeated for a number of specific ground-motion values to develop a hazard curve showing the (annual) exceedance frequencies for a range of ground motions (PGA or spectral acceleration at specific vibration periods).

This methodology is often referred to as the *Cornell method* (Cornell, 1968), and can be summarised in the general form

$$P(Y \geq y) = \iiint_{M\,\min,\,0,\,-2.5\sigma}^{M\,\max,\,D\max,\,+2.5\sigma} f_M(m) f_D(d) f_\varepsilon(\varepsilon) P(Y \geq y \mid M, D, \varepsilon)\, \mathrm{d}m\, \mathrm{d}d\, \mathrm{d}\varepsilon \tag{7.11}$$

where $P(Y \geq y)$ is the probability that the ground motion $Y$ exceeds a specific value of $y$, the magnitude probability density function, the distance probability density function, and $f_\varepsilon(\varepsilon)$ the attenuation randomness density function (an aleatory uncertainty).

This computation is based on the assumption that the occurrence of earthquakes is a Poisson process, with no memory of or dependency on previous earthquakes. Thus the chance of an earthquake occurring in or on a specific source during an upcoming time period is independent of when the last earthquake occurred.

### 7.3.1 Seismic-hazard source model

The seismic-hazard source model describes the magnitude, location and timing of potential earthquakes above a certain threshold magnitude within a certain distance of the site (e.g. 200 km). The magnitude could represent the single characteristic magnitude of a specific fault, or it might represent a discrete value sampled from a fault or region that has a continuous distribution of events (Field, 2001). The location is usually given as a point, line or rectangular surface. The timing represents the annual rate of the earthquakes. In general, the assumption is made that earthquake occurrences follow a time-independent Poisson process. Time-dependent models can be implemented by converting the conditional probability into an equivalent Poisson, time-independent rate.

BC Hydro carried out comprehensive seismic hazard analyses for its dams and power system facilities starting in late 2007. The project adopted a 'SSHAC Level 3' analysis, following the methodology developed by the US Nuclear Regulatory Commission (USNRC, 2012). The project team comprised more than 20 individuals organised into a seismic-source characterisation team and a ground-motion modelling team, each led by a small group of technical integrators.

The identification of potential seismic sources is shown in Figure 7.8. For each seismic source a magnitude–recurrence relationship was developed. It is preferable to identify and characterise the activity rate of individual faults, but in many parts of the world there very few identified active faults so this is difficult to achieve. In such cases, area sources are defined, based on an evaluation of regional geology, tectonic conditions and observed seismicity. It is generally assumed for modelling purposes that conditions are homogeneous across an area source such that the calculated rate of seismicity is uniform over time and area within that area.

### 7.3.2 Ground-motion model

A ground-motion model is a probabilistic model for predicting the level of ground shaking that an earthquake may produce at a particular site given its source of occurrence. This is commonly referred to as an *attenuation relationship*. Several such models have been developed by fitting an analytical expression to observed data. The simplest models relate ground motion to source distance and magnitude, with additional complexity added if accounting for parameters such as site characteristics. The attenuation shows the relationship between the PGA at a source and the PGA attenuated at distance. The relationship may change depending on the magnitude of the event.

Given a site, the seismic-hazard source model describes earthquake scenarios each of which has an associated magnitude $m_n$, location $L_n$ and rate $r_n$. Knowing the site location, the distance from the source $D_n$ can be calculated. Knowing $m_n$ and $D_n$, a log Normal attenuation relationship provides the distribution of possible ground-motion levels:

$$p_n(\ln PGA) = \frac{1}{\sigma_n \sqrt{2\pi}} \exp\{-[\ln PGA - g(m_n, D_n)]^2 / 2\sigma_n)\} \tag{7.12}$$

Figure 7.8 Seismic source model and location of 42 dam sites. (Courtesy of BC Hydro)

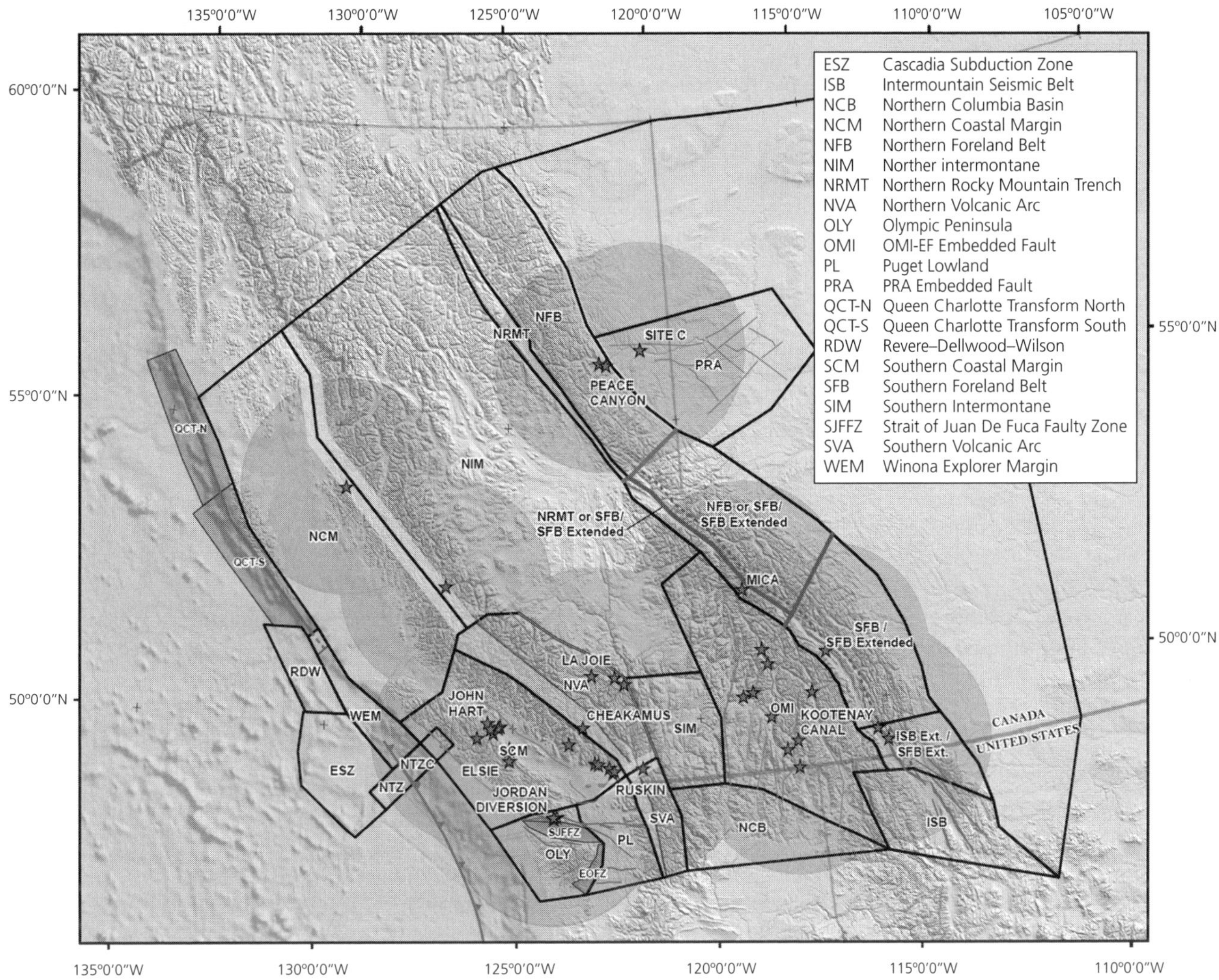


where $g(m_n, D_n)$ and $\sigma_n$ are the mean and standard deviation of ln $PGA$, respectively. The conditional probability of exceeding a particular ln $PGA$ is

$$P_n(\geq \ln PGA) = \frac{1}{\sigma_n\sqrt{2\pi}} \int_{\ln PGA}^{\infty} \exp\{-[\ln PGA - g(m_n, D_n)]^2/2\sigma_n\} \, \mathrm{d}\ln PGA \qquad (7.13)$$

This is the annual exceedance probability. Multiplying by the rate $r_n$ gives the annual rate $R_n$ at which any ln$PGA$ is exceeded in this scenario:

$$R_n(\geq \ln PGA) = r_n P_n(\geq \ln PGA) \qquad (7.14)$$

By summing over all the earthquake scenarios $N$ in the source model, the total annual rate of exceeding each ln $PGA$ is

$$R_{\mathrm{tot}}(\geq \ln PGA) = \sum_{n=1}^{N} R_n(\geq \ln PGA) = \sum_{n=1}^{N} r_n P(\geq \ln PGA) \qquad (7.15)$$

Assuming that the occurrence of damaging earthquakes can be represented as a Poisson process, the probability of exceeding each ground-motion level in $t$ years is

$$P_{\mathrm{pois}}(\geq \ln PGA, T) = 1 - \exp\{R_{\mathrm{tot}} T\} \qquad (7.16)$$

which can then be plotted as a hazard curve.

### 7.3.3 Results

Two key tasks were to develop a composite model of the seismic sources and a set of ground-motion prediction models that would be fundamental inputs to the PSHA. The key deliverables were the seismic-source and ground-motion prediction models, software tools capable of computing seismic hazard, and computed site-specific seismic hazard parameters for each BC Hydro dam. Such parameters typically include peak horizontal and vertical ground motions for a range of annual exceedance frequencies (AEFs), the uncertainties in the computed hazards, uniform hazard response spectra for a range of damping ratios and AEFs, and de-aggregations of computed seismic hazards.

In addition to this technical work, it was intended that a policy would be developed to define how the results of the seismic hazard analysis were to be implemented in the organisation's dam safety programme.

The PSHA calculation computes the annual frequency of exceeding specific levels of strong ground shaking at a site due to the occurrence of potential future earthquakes in the region surrounding the site. The mathematical formulation assumes that the occurrence of damaging earthquakes can be represented as a Poisson process. Under this assumption, the probability that a ground motion parameter $Z$ will exceed a specified value $z$ in time period $t$ is given by

$$P(Z \geq z \mid t) = 1 - \exp[-v(z)t] \leq v(z)t \qquad (7.17)$$

where $v(z)$ is the average annual frequency during time period $t$ at which the level of ground-motion parameter $Z$ exceeds value $z$ at the site from all earthquakes on all sources in the region. Equation 7.17 is valid provided that $v(n)$ is the appropriate average value for the time period $t$.

The frequency of exceedance $v(z)$ is a function of the frequency of earthquake occurrence, the randomness in the size and location of future earthquakes, and the randomness in the level of ground motion they may produce at the site:

$$v(z) = \sum_n \alpha_n(m^0) \int_{m^0}^{m^u} f_n(m) \left[ \int_0^{\infty} f_n(r \mid m) P(Z \geq z \mid m, r)\, \mathrm{d}r \right] \mathrm{d}m \qquad (7.18)$$

in which $\alpha_n(m^0)$ is the annual frequency of earthquakes from source $n$ above a minimum magnitude $m^0$, $f_n(m)$ is the probability density of earthquake size for source $n$ between $m^0$ and a maximum earthquake the source can produce $m^u$, $f_n(r \mid m)$ is the probability density function for the distance to an earthquake of magnitude $m$ occurring at source $n$, and $P(Z > z \mid m, r)$ is the probability that, given an earthquake of magnitude $m$ at distance $r$ from the site, the peak ground motion will exceed $z$. The frequency of earthquake occurrence and the size distribution of earthquakes are defined by the earthquake recurrence relationships developed in the seismic source models. The distributions for the distance between the earthquake rupture and the site are defined by the geometry of the seismic sources and the spatial distribution of earthquakes defined in the seismic source models. The conditional probability of exceedance is defined using the selected ground-motion models.

The PSHA calculations were implemented in summation form:

$$v(z) = \sum_n \alpha_n(m^0) \sum_{m_i - m^0}^{m_i = m^u} P_n(m = m_i) \sum_{r_j = 0}^{r_j = r^{max}} P_n(r = r_j \mid m_i) \times P(Z \geq z \mid m, r) \qquad (7.19)$$

PSHA calculations were conducted for PGA and pseudo-spectral acceleration at spectral periods of 0.03, 0.05, 0.075, 0.1, 0.15, 0.2, 0.3, 0.5, 1.0, 2.0, 3.0 and 5.0 s. Calculations were done using logic trees (epistemic uncertainties) defined for each source. All calculations were performed for a reference site condition. All the calculations were done using single values of aleatory variability in the peak ground motions. The sources used in the calculation for each site are indicated in the results and were limited to those sources within 300 km of the site. Composite fractiles of total hazard were computed using composite logic trees for groups of seismic sources to preserve dependencies between input parameters across sources. The magnitude probability mass function was obtained by discretising the earthquake recurrence relationships for the source using 0.1 magnitude unit increments centred on nominal magnitude values.

The distance probability mass function was computed numerically by incrementally moving earthquake ruptures over the region representing the seismic source. Fault sources and fault-like sources (e.g. embedded faults) were modelled as segmented planar surfaces. Crustal earthquake ruptures were modelled as quadrilateral shapes.

The conditional probability of exceedance $P(Z > z \mid m, r)$ is computed using a numerical approximation to the cumulative normal distribution. The calculations were performed using without truncation of the ground-motion distribution (Figure 7.9).

Figure 7.9 Seismic hazard curves for PGA

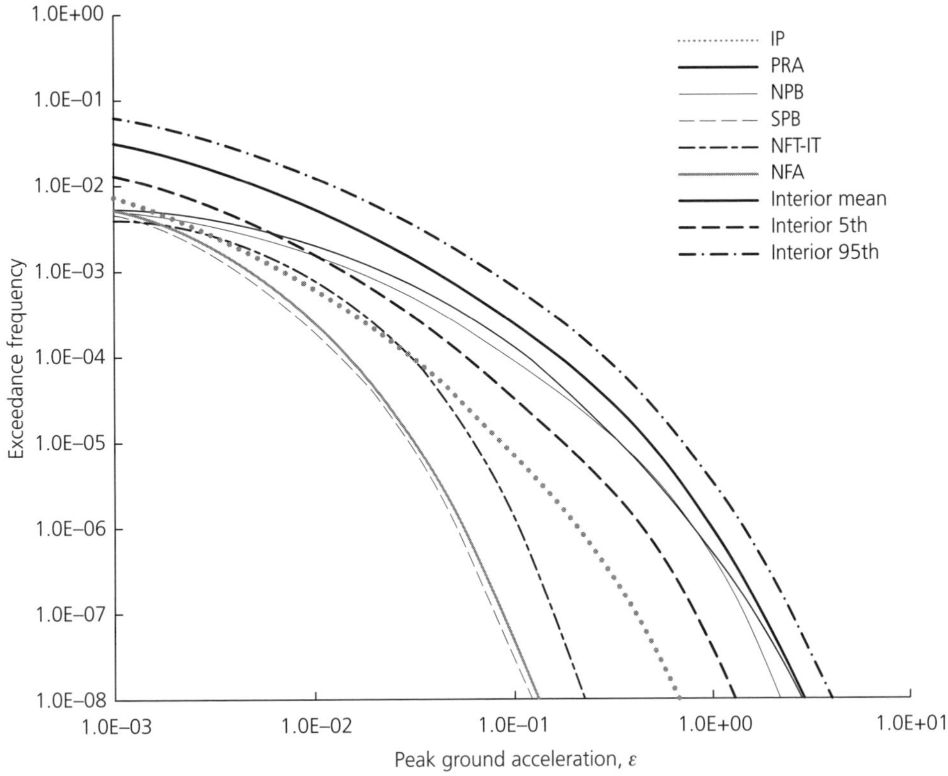


## 7.4. Floating debris

Floating debris – trees, brush and other vegetation, or even houses, caravans and boats – can block spillways or intakes, sometimes with disastrous consequences (Figure 7.10). Even if not blocked entirely, the capacity of outlet works can be significantly and seriously reduced by debris, leading to inadequate control of reservoir pool elevation, and in extreme cases to overtopping. Hydropower stations regularly have to close down for cleaning during significant inflows of floating debris, as trash racks and cooling water intakes become blocked (Figure 7.11). Large debris such as trees can also damage gates, spillway conveyances and energy-dissipation structures when the debris passes through a waterway. Debris can also damage the upstream face of earth or rockfill dams and displace riprap.

The sorts of damages that debris can cause to dams and other in-river structures include, among other things (ISAB Dammar & Kraft, 2005):

- reduction in discharge capacity
- damage to the upstream face of a dam
- erosion due to increased stream velocities
- erosion due to disturbed flow patters
- surges and unsteady flow caused by debris being washed away.

In essence, the dam operator has only a limited number of options for protecting a structure from debris, which include reducing the yield of debris entering the reservoir, intercepting the debris once it is in the reservoir (e.g. with debris booms), facilitating the passage of debris through outlet works and physical removal.

Experience suggests that an authoritative understanding of the source of the debris, the process of its generation and transport to the reservoir, the loads it places on spillways and other outlet structures, and techniques for its capture or management are necessary for maintaining reliable spillways and dam safety. Thus, a fundamental understanding of debris processes is worth pursuing.

The 'normal' annual debris yield in a catchment may contain little information about the potential yield under extreme hydrological conditions. Many forests in the northern hemisphere are, in effect, plantations, where generations of trees are genetically close. In this case, root systems may not be 'well designed' for extreme conditions, and an intense rainstorm over even a moderate extent of the catchment may produce extreme amounts of debris. Furthermore, under normal

Figure 7.10 Marseilles Lock and Dam (Illinois, USA) – barge impacts, April 2013. (Courtesy of USACE)

hydrological conditions the capacity for transport of debris to the main river in a catchment may not exist, while under more extreme hydrological conditions tributaries may be wide enough to facilitate transport of debris from a distance. Also, under extreme conditions groundwater levels may be high, and slides may be initiated in slopes adjoining rivers and streams so that even more areas become active in debris production.

The result is that the debris yield during major floods may be many times higher than and have little in common with 'normal yields'. The determination of design debris yields needs to be based on the potential for debris production under extreme events rather than on normal yields, although the latter may be pertinent to planning and budgeting of normal activities.

### 7.4.1 Debris incidents at dams

Galeati (2009) cites the example of Palagnedra Dam (Switzerland) at which a large flood in August 1978 led to debris blockage of the free overflow spillway, pressure tunnel intake and two bottom outlets (Bruschin *et al.*, 1982). The original spillway, which was later redesigned with a wider unobstructed crest, used a free overflow ogee crest with 13 openings, each 5 m wide. These were completely blocked by debris. Figure 7.12 shows the redesigned crest.

As reported by Wallerstein *et al.* (1997), on 7 August an extreme storm carried 25 000 m$^3$ of wood and nearly 2 million m$^3$ of sediment into the reservoir. Debris partially obstructed the spillway openings and caused the dam to overtop. Peak discharge was estimated at a 2000 cm/s. The main dam suffered structural damage, and 50 000 m$^3$ of glacial material was eroded from the downstream side, endangering the dam's stability.

Figure 7.11 A field crew tries to dislodge debris on the spillway gates at Elko Dam, Elk River, BC. (Courtesy of BC Hydro)

Galeati (2009) describes a similar example of an event at the Sa Teula Dam (Sardinia) in 2004. On 6 December the masonry gravity dam was affected by a flood event that had an estimated peak of 340 cm/s, the return period for which was evaluated to be 500 years. The dam had a spillway equipped with an automatic 12 × 3 m gate that was opened by the lifting movement of two floats, pushed by water, situated in appropriate galleries created in the pier walls. In addition, there was a free overflow

Figure 7.12 Redesigned wider crest of Palagnedra Arch Dam after the August 1978 flood

spillway measuring 45.0 m and consisting of 15 openings, each with a width of 3.0 m, and a 239.5 m high sill located 1.5 m from the intrados of the footbridge situated at the top of the dam.

The dam, empty due to maintenance work being done on the penstock of the hydroelectric plant, was affected by a considerable transport of floating debris, including tree trunks of a wide dimension. The floating debris strongly slowed down and reduced the flow of water to the galleries containing the floats, and consequently prevented the gate from opening. At the same time, the restricted width of the free overflow spillways (5 m), together with the reduced distance of the intrados of the footbridge, impeded the release of the floating debris. As a consequence there was an almost complete blockage of the openings, leading to the dam overtopping. The considerable debris load that affected the gate, together with some tree trunks trapped between the edge of the gate itself and the service footbridge, finally determined the failure of the gate and its complete detachment.

Dam accidents caused by debris are listed in Table 7.3. In an evaluation published in 1983, the US National Research Council concluded that about 30% of spillway malfunctions are caused by spillway blockage due to floating debris (NRC, 1983), and the National Performance of Dams Program reported 23 accidents at US dams involving debris blockage of spillways (NPDP, 2015) (Table 7.4). Furthermore, it should be noted that the contribution of floating debris to overtopping failures may not always be possible to discern after the fact.

Floating debris continues to be a problem with modern dams, especially those equipped with gates. For example, it has been an expensive nuisance at the newly built Three Gorges Dam in China (Government of China News Agency, 2006) where 200 000 $m^3$ of debris piles behind the dam in a normal year. The debris consists of 'plant roots, straw, rotten leaves, branches, and household and industrial refuse'. The current treatment is to collect the debris and dump it on land higher than the final reservoir pool elevation of 175 m, where it is burned or buried.

Table 7.3 Dam accidents caused by debris (by date of occurrence)

| Dam | Date | Description | Reference |
|---|---|---|---|
| Nacimiento Dam, CA, USA | 1969 | 'After several intense storms, the high-level outlet slide gate clogged with trash and failed.' | USCOLD (1975) |
| Pilar Rock Dam, WA, USA | 1970 | 'Logging road fill culvert blocked by debris, overtopped and failed, caused 25 foot high concrete gravity dam to overtop and fail. Three homes and fish cannery destroyed, 80 000 evacuated.' | ASDSO (2012) |
| Palagnedra Dam, Switzerland | 1978 | '... uncontrolled ogee crest on top of the dam and a steep chute terminated by a ski jump. The crest was divided into 13 openings 5 meters wide, the boundary walls of which functioned as piers for an overpassing road, which were completely occlused during the event.' | Larivière *et al.* (1997) |
| Smiths Pond, MA, USA | 2004 | 'Dam overtopped; spillway clogged by debris. Divers from the Leominster (MA) EMA and crane operators worked to clear the spillway.' | ASDSO (2012) |
| Sa Teula, Sardinia | 2004 | 'The dam, empty due to maintenance work ... was affected by a considerable transport of floating debris ... As a consequent, there was an almost complete blockage of the openings creating the dam overtopping.' | Galeati, (2009) |
| Testalinden dam, British Columbia, Canada | 2010 | 'Testalinden dam on Testalinden Creek, eight kilometres south of Oliver, British Columbia, failed on 13 June, causing an enormous debris and mud torrent that impacted a number of homes and an agricultural area.' | WPDC (2012) |
| Kerckhoff Dam, CA, USA | 1997 | Kerckhoff Dam (PG&E) outside Fresno above Millerton on San Joaquin. Kerckhoff Dam (PG&E) above Millerton on the San Joaquin River. Debris caused serious impediment to spillway function, but this did not lead to dam failure. | Patrick Regan, Federal Energy Regulatory Commission, personal communication |

Table 7.4 National Performance of Dams database of 23 incidents or failures involving debris build up (NPDP, 2015)

| Dam | Date | Description | Dam failure |
|---|---|---|---|
| La Blonde | 7/1992 | Inflow flood; debris | Yes |
| Unnamed dam | 7/14/1994 | Debris reservoir | No |
| Holcombe Dam (North Lake) | 6/27/1994 | Debris; animal attack | No |
| Hutton Dam | 4/26/1994 | Debris reservoir | No |
| Scomp (Former Brewer) | 2/13/1995 | Debris reservoir | No |
| Stafford Pond | 3/21/1995 | Debris reservoir | No |
| Chowan College Dam | 5/17/1995 | Debris reservoir | No |
| 102 River Tributaries Watershed Structure Lt-36 | 12/4/1996 | Debris reservoir | No |
| Apple Valley Lake Dam | 4/16/1997 | Debris; piping | Yes |
| Leonard Phillips Dam | 1/21/1998 | Debris reservoir | No |
| Jackson City Dam | 10/7/1999 | Debris reservoir | No |
| SCS-Broken Leg Creek Site-2 | 9/16/1999 | Debris reservoir | No |
| Hobbs Lake Dam | 10/25/2000 | Debris reservoir | No |
| Joe English Acres Pond | 5/2/2000 | Debris reservoir | No |
| Unnamed dam (NCS00038) | 11/21/2000 | Debris reservoir | No |
| McDougal Lake | 5/17/2001 | Debris reservoir | No |
| Gaston Shoals Middle | 9/5/2001 | Debris reservoir | No |
| Pinecrest High School Dam | 1/17/2001 | Debris reservoir | No |
| Thomas Lake Dam #2 | 1/19/2001 | Debris reservoir | No |
| Low Gap Wildlife Club Dam | 12/17/2002 | Debris reservoir | No |
| Old Mill No. 1 Upper | 7/11/2002 | Debris reservoir | No |
| N. State Orchard Dam #1 | 11/2002 | Debris reservoir | No |
| Anderson (Powell) | 4/16/2004 | Debris reservoir | Yes |

Neither of these is a suitable long-term solution, however, as the debris contains little flammable content and pollutes the ground. The 2010 flood year produced a debris mat covering 50 000 $m^2$ of the reservoir surface to depths averaging 60 cm.

### 7.4.2 Debris yield

From an engineering perspective it would be useful to be able to predict debris loads, their times of concentration and their occurrence in relation to other events such as storms and hydrological inflows. The research done in this area is not extensive and is somewhat inconclusive.

#### 7.4.2.1 Bank erosion

In the present context 'debris production' means the development in the basin of felled timber and brush of a character that could become floating debris in the reservoir (Bruschin *et al.*, 1982). Research suggests that bank erosion in streams is the main cause of whole trees washing into streams and thus into the reservoir under normal hydrological conditions. McFadden and Stallion (1976) and Ott *et al.* (2001) studied 141 miles of the Connecticut River in the USA and concluded that bank erosion due to flowing water, pool fluctuations and the forces of waves, ice and seepage were the major causes of whole trees in floating debris (Figure 7.13). In mountainous catchments, fallen trees can become stuck in small tributaries, collecting a large mass of debris behind them, which is washed out during large storms (Perham, 1988). Forestry and logging practices can also contribute to debris production (Box 7.1 and Figure 7.14).

Thus, debris is principally produced by bank erosion or landslides, and as fallen trees due to a storm event of wind, rain and gully wash. This production can be measured, typically, either as a weight per hectare or in numbers of significant trees. For analytical purposes these measures might be expressed as rates or densities (i.e. as tons per hectare or number of felled trees per hectare). From an analytical view, debris production is the first step in generating debris accumulation at the outlet works and subsequently in passing debris through the outlet waterways.

Guidance from the Norwegian Water Resources and Energy Administration presumes that debris is principally generated by the 'severity of landsliding' in the basin. Bruschin *et al.* (1982), in the context of Palagnedra Dam, concluded that the important factors to be considered in the design of spillways for debris are:

- the hydrological and meteorological environment of the basin
- the potential for extreme events

Figure 7.13 Whole-tree debris along the Tanana River, Alaska. (Ott *et al.*, 2001)

- high-yield sediment and debris sources within the basin
- the stability of slopes in upstream forests.

Wallerstein and Thorne (2004), in a study of debris processes and channel evolution in the Yazoo Basin, Mississippi, USA, concluded:

> It has been found that, while spatially unpredictable debris input processes ... cause a small percentage of jams, the majority of woody material enters these unstable channels through bank erosion, much of which results from mass failure in degraded reaches. ... Thus, for management purposes, the spatial density of debris jams in degrading rivers may be qualitatively predicted on the basis of the location and migration of head cuts and degradational processes.

This is similar to the conclusion drawn by McFadden and Stallion (1976). Bank erosion and slip can be exacerbated by wind and current, ice break up, and other processes in the reservoir (Simons *et al.*, 1979). Man-made features and shore structures are prone to flooding or floating away, including, for example, docks, boats, caravans and other portable items.

Given that the dominant generating mechanism for floating debris is mass movements (i.e. slope failures) in the upstream basin or bank failures, and these mass movements occur primarily as a result of intense storm events (i.e. rainfall, wind and waves), this suggests that there may be different possible approaches to characterising the amount of debris resulting from storms. One way would be to use generic geospatial properties of the basin in some form of empirical index, along the lines that surface erosion is estimated. Another way would be to predict the specific number of slope failures along tributaries or the reservoir boundary. The first approach might be called a 'debris-production equation', while the second might be called a 'landslide forecasting model'.

On water bodies with significant fetch (e.g. reservoirs, lakes and large rivers), waves can erode the shoreline, causing slope instabilities and trees to topple. Structures such as docks, buildings or impermanent items near the water can be loosened from the land. The flotsam generated by this wave action ends up in the reservoir. Wind and wave action can also cause the removal of debris from natural impoundments such as coves or the mouths of streams. Wind is a major source of debris from forested areas.

The state of the art for predicting the yield of floating (i.e. vegetative or woody) debris at a reservoir is not far advanced. There are four general approaches:

- empirical formulae (Rickenmann 1997; Rundqvist, 2005)
- theoretical models (Vansickle and Gregory, 1990)

Box 7.1 Catchment factors that affect debris production (FHWA 2005)

**Flooding** increases the production of debris as the associated discharges serve as a means to produce and deliver debris to a site. Higher discharges are more likely to cause erosive forces on bank and floodplains. The inundation of the flooding event affects more of the floodplain area, facilitating transport of debris into the main channel.

**Fires** can decrease the amount of floating debris introduced into the stream system. However, fires increase the magnitude of runoff from the burned area, increase the erodability of soils and increase the probability of catastrophic events such as debris flows and landslides, resulting in a significant increase in sediment yield from the affected area. This increase could cause an increase in fine and coarse detritus to be transported to and deposited at a culvert or bridge structure.

**Urbanisation**, over time, causes the opposite effect on the yield from a watershed from that of fires. Initially, the yield can be significantly increased during the construction phase of development, due to the removal of existing vegetation and the disturbance of the soil. However, over time the sediment yield decreases as the developed land becomes restabilised and the land surface area exposed to the erosive effects of rainfall and runoff is reduced as a result of the increase in impervious area, such as roads, structures and car parks. Hydrological effects from urbanisation include an increase in runoff volume, higher peak flows and longer durations.

**Logging** has been identified as a source of floating debris, and different logging practices cause substantial differences in the loads of floating debris. Practices that reduce the quantities of floating debris include directional felling uphill with a tree-pulling system and providing a buffer strip of undisturbed vegetation along the streams.

**Land clearing** associated with logging, grazing or agricultural practices could cause the same effects as are associated with fires. However, the magnitude of these effects would most likely not be as severe. Also, grazing allowed near a stream can result in a significant increase in bank erosion.

**Conservation practices** have the opposite effects from those associated with clearing of land. The implementation of a different conservation practice can reduce the amount of both erosion and runoff from the land.

**Channel improvements** or modifications to the channel geometry and/or vegetation clearing from the channel can influence the quantities of both floating debris and fine/coarse sediment. Improper design of such improvements can cause significant instabilities to develop within the system, including increased bank erosion, increased degradation and/or aggradation of the stream bed, and/or significant changes in the plan form, which could increase the generation and delivery of floating debris to a structure site.

**Growth of riparian forest buffer strips** has been recommended and encouraged by the US Environmental Protection Agency for their water quality, ecological and bank erosion benefits, especially in agricultural areas. These forested buffer strips adjacent to stream channels are now common and have maturing trees, especially in heavily agricultural areas.

**Extreme events**, such as ice storms, debris flows, forest fires and insect infestations can drastically increase the debris load at some point in the life of the structure.

Figure 7.14 Debris accumulation and loading process

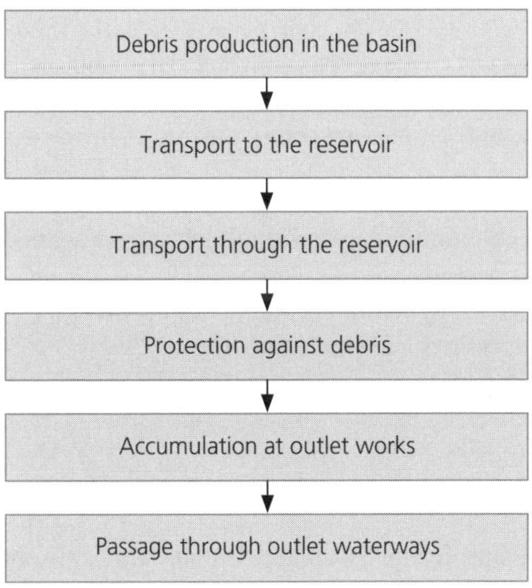

- historical analysis of past events in the same or similar drainage
- detailed (geographical information system (GIS) remote sensing) mapping of the drainage (R2 Resource Consultants, 2003).

CEATI International Inc., on behalf of the Dam Safety Interest Group, commissioned a three-part study of debris management in spillways and waterways during floods (Comfort and Abdelnour, 2009). The study investigated debris-yield management, as well as loads on and failure modes of debris booms, including model tests. The principal conclusions of the survey of debris-yield estimation methods were as follows (Comfort and Abdelnour, 2010):

> *Regulations and Guidelines regarding Debris Yield Estimation* – No regulations or widely-accepted guidelines were found that provide detailed, specific direction or guidance regarding yield estimation for woody or man-made debris. A standard, widely-accepted method for estimating the yield of woody or man-made debris is not available. The methods used regarding debris estimation are specific to the site and to the requirements of the investigation. Broadly, they can be categorised as: (a) historical, (b) comparative, and (c) survey-based.
> *Debris Yield Estimation* – The current state-of-the-art for estimating the yield of woody or man-made debris is very rudimentary. This is due to a number of factors including the following. (a) Debris yield estimation is a very complex problem. Many types of debris can occur. Furthermore, conditions vary greatly among the possible cases (e.g. remote northern locations versus ones in populated areas). In our opinion, it would be unrealistic to expect that a single debris yield estimator could be developed that would be applicable to all classes of debris yield estimation problem. (b) There is a general lack of quantitative data on which reliable numerical modelling or debris yield estimation can be based.
> [*Way forward* –] The following are recommended as a way forward. (a) Debris yield estimation cases should be categorised into various classes of problems (e.g. debris yield estimation for northern remote locations). This would limit the extent of the overall estimation problem (e.g. number of factors and processes to be included). Further work could then be focused on debris yield estimation approaches for each class of problem. (b) Efforts should be made to collect and archive data regarding debris-related events, including high floods, intense rainfall and catchment topography and geology, with the intent of eventually building up a sufficiently large database that would allow reliable debris yield predictors to be developed. In essence, this is the approach that was used for developing the USACE's model (Gatwood *et al.*, 2000), which was developed based on regression analyses for a large number of events.

BC Hydro undertook a major study in the early 1990s to understand debris production in its reservoirs (BC Hydro, 1992). The results of this study apply specifically to the spruce forests of the Pacific coast of Canada, but the principles learned in the study may have broader applicability. The study was in part instigated by the accident at Alouette, Stave Lake and Ruskin Dams in 1955. The purpose was better to understand and to estimate the quantity of floating (organic) debris entering reservoirs due to river bank erosion, debris torrents, slides and like mechanisms during major storms.

Alouette Dam lays 50 km east of Vancouver on a tributary of the Frasier River. Originally constructed in 1926 at a time of little local development, by 1955 the dam was surrounded by residential properties. Breach of the dam would affect many hundreds of residential and commercial establishments and cause significant loss of life. The dam was originally 21 m high and 288 m long, and had a concrete overflow intake weir with three vertical lift gates and 12 small under sluices. These fed into a concrete-lined conveyance on the right-hand abutment.

In November 1955, a large flood caused about 1.5 m of overflow of the weir and an accumulation of large tree trunks on top of the concrete panels. This damaged the panels and is suspected to have led to seepage under the panels. This seepage eroded the stiff silty clay foundation of the weir and led to a 25 m breach. Fortunately, the failure happened at a time of decreasing pool elevation, so a loss-of-containment failure was averted. The weir was later replaced by a concrete gravity section.

To estimate the quantity of debris that could enter a reservoir during a storm, Brown and Nielson (1992) considered the flowing variables:

- forest cover, including forest cover types, amounts of residual debris and areas of forest killed by insects or fire
- terrain, including slope angle and soil cover
- transport mechanism, including landslides, debris flows and shoreline erosion
- hydrology, including weather, snow pack and snowmelt flooding
- other factors, including geology, logging and access-road construction, wind speed and direction, and debris storage on tributaries.

These variables were quantified as well as possible in order to estimate the amount of debris anticipated in a major storm.

Rickenmann (1997) studied a number of sites in the Swiss and Italian Alps and compared the results to those reported by Ishikawa (1990) in his PhD dissertation on driftwood in Japanese mountain streams, finding them similar.

Figure 7.15 Summary Swiss and Italian Alpine (including the Palagnedra Dam discussed above) and Japanese data showing the expected mass of driftwood transported downstream in high-water events as a function of drainage area. (Adapted from Rickenmann, 1997)

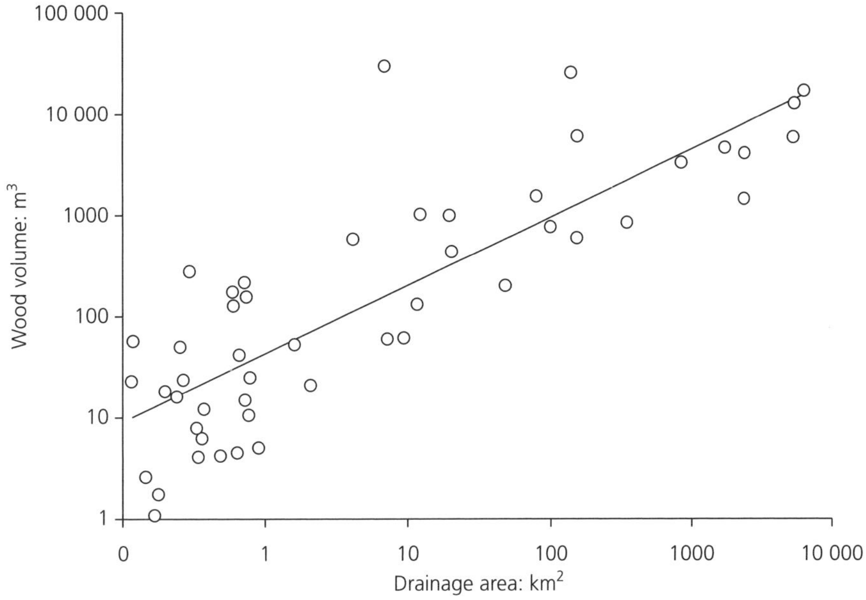


Figure 7.15 shows summary results of wood volume as a function of drainage area. The driftwood volume was found also to depend on the wooded fraction of the drainage area and the total water volume. All these empirical relationships contain a great deal of scatter, and depend on forest type.

A subsequent study by Rimböck (2002) summarised a number of empirical formulae for predicting debris yield, including Rickenmann's (Table 7.5).

Consecru (1966) among others have introduced theoretical models of debris generation in forests. They develop a probabilistic model, which predicts means and variances of the number and volume of large woody debris falling into a stream per unit time. The inputs to the model are the number of trees per acre, tree size distribution, and tree-fall probability adjacent to the reach. Distributions of volume, length, and orientation of delivered debris pieces are the output. Predictions of the model are in reasonable agreement with observed debris volumes, although the model has not been applied to hydropower cases (Figure 7.16).

### 7.4.3 Gate loads

Procedures for estimating the forces transmitted from a floating debris mat to a spillway are presented in CEA (1982) on the basis of Kennedy's (1962) experimental and field work. More recent results from experimental studies are presented in Godtland and Tesaker (1994), who propose the following equation based on model tests:

$$F^{\mathrm{D}} = C_{\mathrm{d}} b_{\mathrm{D}} (30 t_{\mathrm{s}} + L^{\mathrm{D}}) \rho_{\mathrm{w}} \tfrac{1}{2} v_{\mathrm{s}}^{2} \qquad (7.20)$$

where $F^{\mathrm{D}}$ is the load transmitted from the debris mat to the spillway, $C_{\mathrm{d}}$ is the drag coefficient, $L^{\mathrm{D}}$ is the length of the debris mat parallel to flow, $b_{\mathrm{D}}$ is the width of the debris mat perpendicular to the flow, $\rho_{\mathrm{w}}$ is the density of water, $_{\mathrm{s}}$ is the surface water velocity and $t_{\mathrm{s}}$ is the submerged depth. Léger *et al.* (1996) and Larivière *et al.* (1997) present a more detailed analysis, as illustrated in Figure 7.17. Suggested values of the hydraulic drag coefficient are reproduced in Table 7.6.

## 7.5. Ice and icing effects on flow control

Ice interferes with dam and spillway operations in a variety of ways, from changing flood-stage frequency relationships, to creating ice-sheet loads on dam faces and other structures, to interfering with mechanical equipment by accumulating on gates and other equipment. Dam gates and other moving equipment are especially vulnerable to icing, which may block movement or weigh down lifting equipment such as hoisting chains or gate trunnion arms, or increase hydraulic head losses at the spillway (Figure 7.18). Ice floes passing over or through discharge structures may block water passage, increase river bank erosion or foster ice jams downstream. The present discussion focuses only on the loads placed on spillway structures by ice moving against them or adhering to them.

Table 7.5 Probable debris yield as a function of basin characteristics. (Rimböck, 2002)

| Source | Formula | Base data | Comments |
|---|---|---|---|
| Probable debris yield | | | |
| Ishikawa (1990) | $V_D = 500A_{DR}$ | 27 cases in Japan | Upper limit of yield |
| Zollinger (1983) | $V_D = 10A_{DR}$ to $180A_{DR}$ | 2 cases in the Alps | Observed values |
| Ishikawa (1990) | $V_D = 0.02V_S$ | 27 cases in Japan | Upper limit. Establishment of actual sediment transport difficult |
| Rickenmann (1997) | $V_D = 545A_{DR}^{2/3}$ | 50 cases in Japan and the Alps | Mean value of debris volume |
| Uchiogi *et al.* (1996) | $V_D = 4V_w^{0.4}$ | 18 cases in Switzerland | Mean value of debris volume |
| Maximum debris yield | | | |
| Ishikawa (1990)<br>Rickenmann (1997) | $V_{DO} = CA_{DR}$<br>$C = 1000$, upper limit for coniferous forests<br>$C = 100$, upper limit for floriferous forests<br>$C = 10$, lower limit | 9 cases in Switzerland, 88 cases in the USA | Upper and lower limits |
| Rickenmann (1997) | $V_{DP} = 90A_{FOR}$ | 9 cases in Switzerland | Mean value of debris potential |
| Consecru (1966) | $V_{DP} = 40A_{FOR}{}^2$ | 9 cases in Switzerland | Mean value of debris potential |
| Rickenmann (1997) | $V_{DP} = 90CL_{DR}$<br>$C = 1.2$ to $36\ m^3/m$ | 4 cases in Switzerland | Estimated values |

$A_{DR}$, area of drainage basin ($km^2$)
$A_{FOR}$, forested area of drainage basin ($km^2$)
$V_D$, volume of debris ($m^3$)
$V_{DP}$, potential volume of debris ($m^3$)
$V_S$, volume of sediment ($m^3$)
$V_w$, volume of water ($m^3$)
$L_{DR}$, length of drainage channel (km)
$L_{FOR}$, length of forested area (km)

Ice also has a variety of effects on the hydraulics of a basin and of channels. These are mostly ignored here because the present focus is on spillways, other waterways and discharge controls. Nonetheless, some of the channel effects may affect flow controls at the dam. For example, floating ice approximately doubles the wetted perimeter of a flat channel, and influences the distribution of velocities in the flow and the discharge that can be passed. Ice may cause bed and shoreline scour, and affect sediment transport. Ice jams may cause flooding, and ice breakup in spring can cause rapid pulses of flow into a reservoir. Ice dams may reach heights of many metres. Ice dams are generally created by frazil adhering to the river bottom and banks, where these are strong enough to stabilise an ice dam. Ice floes may damage floating booms and

Figure 7.16 Kinbasket reservoir (Mica Dam) obscured by floating debris. (Courtesy of BC Hydro)

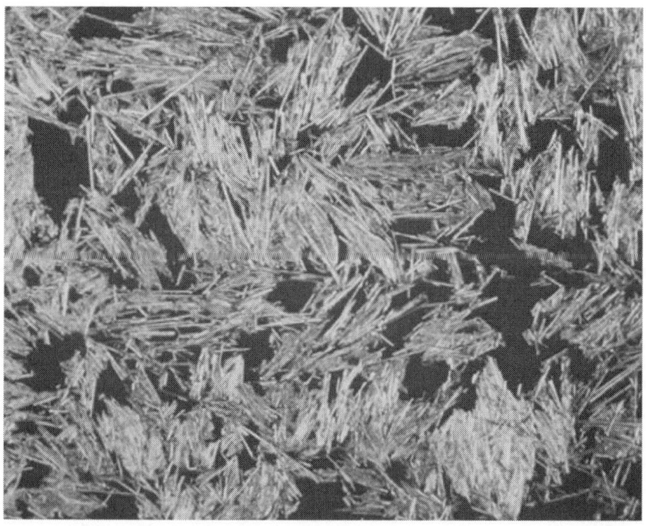

Figure 7.17 Forces exerted by floating debris: (a) plan view, (b) elevation. (Larivière *et al.*, 1997)

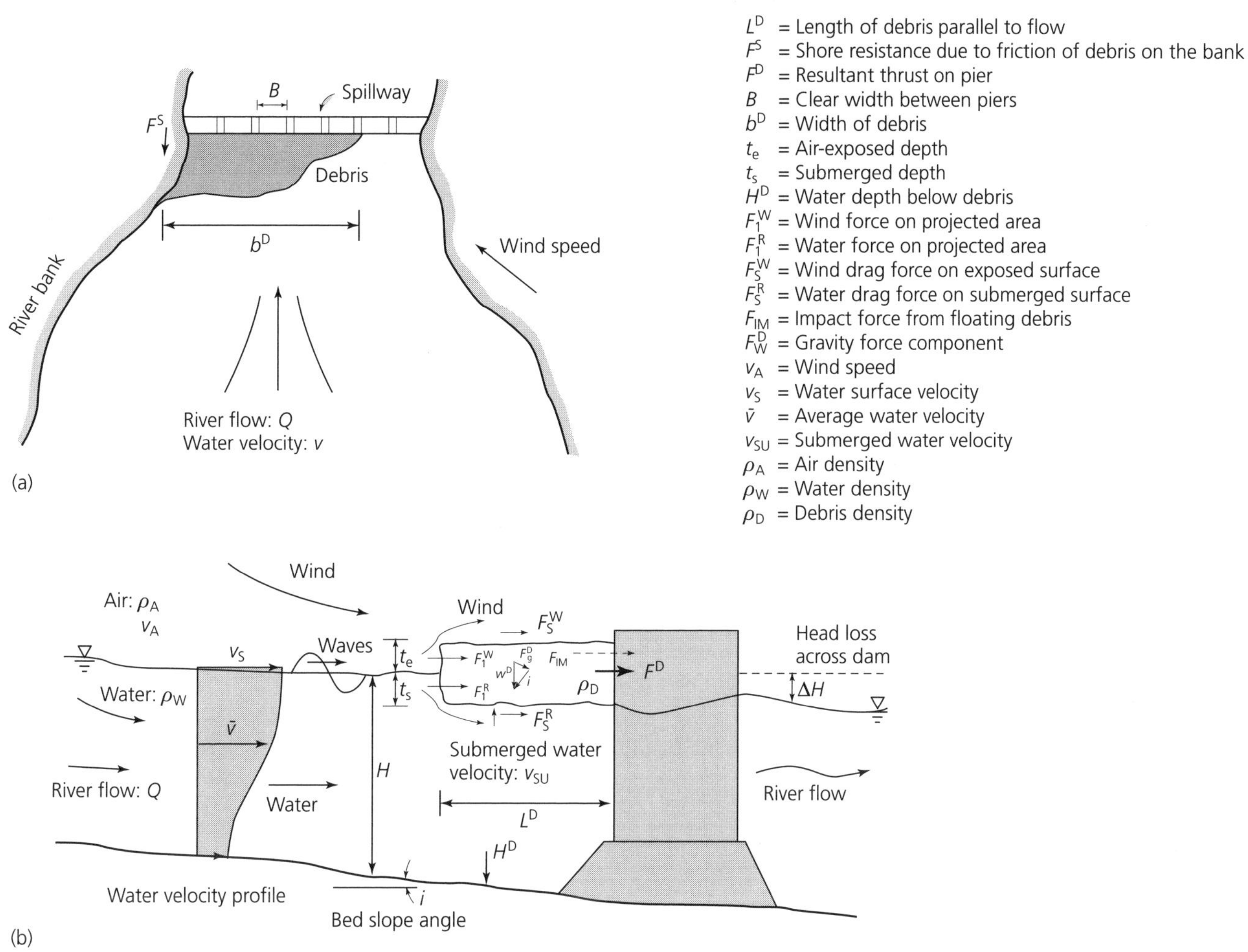


riprap. Ice also affects navigation. Many of these non-flow-control impacts of ice are discussed in the US Army Corps of Engineers' (USACE) *Ice Engineering Manual* (USACE, 2002).

### 7.5.1 Ice incidents at dams

Ice incidents at dams seldom lead to catastrophic structural failure of the dam itself, but can cause operational issues, hinder or block waterways and gates, and cause changes to river hydraulics through bank erosion and blockages (Table 7.7). Spillway gates are susceptible to freezing in place due to leakages or ice interfering with the movement of gates. Ice loads on hoisting chains, trunnion arms or other mechanical equipment may be too great for hoisting equipment to overcome. Frazil ice may form on trash racks and other structures and block intakes to gates or valves or increase river levels downstream in an unforeseen manner. Furthermore, downstream releases from the reservoir may induce flooding if downstream reaches are blocked by ice.

Table 7.6 Hydraulic drag coefficient for various ratios of surface ($v_s$) and submerged ($v_{su}$) water velocities (Godtland and Tesaker, 1994)

| $C_d$ | Water velocity |
|---|---|
| 0.06 | $v_s < v_{su}$ |
| 0.08 | $v_s < v_s < 1.1v_{su}$ |
| 0.10 | $v_s > 1.1v_{su}$ |

Figure 7.18 Overflow spillway at Trollhättan partially blocked by ice

Table 7.7 Ice-affected dam components and related issues (Tuthill, 2002)

| Dam component | Ice issue |
|---|---|
| *Tainter gates* | |
| Trunnion arms | Arms freezing to pier walls |
| Trusswork and enclosures | Spray ice build-up on trusswork |
| Lifting mechanisms, chains, and cables | Icing and freezing-in of lifting mechanism |
| Side and bottom seals | Leakage and freezing-in of seals |
| | Large openings needed to pass ice |
| *Submergible Tainter gates* | |
| Direct lift gates | Allow ice skimming, vibration possible |
| Roller gates | Freezing of lifting chains |
| Pier walls | Submergible, prone to freezing-in |
| Gate sills | Ice build-up problems |
| Downstream bed protection | Freezing-in |
| | Downstream scour possible when passing ice with low tail water |
| | Frazil or brash ice blockages |
| *Hydroelectric intakes* | |
| Electric heaters: | |
| Side and bottom seals | Energy costs, corrosion of connections and elements |
| Side seal plate | Frequent maintenance and replacement |
| Skin plate | |
| Steam heat systems: | |
| Boilers | Personnel requirements, safety |
| Portable hoses | Piping and easy hose connections |
| Insulated piping systems | Freezing of condensate |
| Perforated lances and tees | |
| Seal leak prevention: | |
| Ash hoppers | Personnel requirements and safety |

### 7.5.2 Mechanical properties of ice

Most current practice comes from northern countries. Ekström (2006) summarises current practice in Sweden, Norway and Canada. In these cases ice loads are specified as force per length of dam surface against the upstream pool. Recommended ice loads for design are a function of ambient temperature and the deformability of the structure against which the loads apply, but are reasonably consistent across jurisdictions (Table 7.8).

Static ice forms on water for which flow velocity plays no part. This includes reservoirs and lakes, and streams with low flow velocities (approx. <0.3 m/s). The static ice layer forms as heat is transferred upward from the water, through the ice layer, to the colder atmosphere above. In a still environment the basal plane of ice crystals tends to align horizontally, leading to columnar crystals called 'columnar ice'. Impurities in the water are forced to the crystal boundaries, yielding weaknesses and planes of early thawing. If the weight of snow cover on the ice layer is sufficiently large, the ice will sink somewhat, allowing water to invade the snow cover, creating granular (snow) ice.

Frazil ice forms in turbulent supercooled open water ($T < 0°C$). It appears as small crystals that are distributed more or less uniformly throughout the turbulent layer of water. These crystals slowly mass into larger bodies in the form of floes or flocs, which flow for long distances with the surface currents. An issue with frazil ice is that if streams remain free of packed ice for long durations, the frazil ice may be carried long distances downstream and deposited in large masses against structures or as blockages in streams. Frazil ice crystals in the supercooled water will stick to any object they come into contact with if those objects are at the temperature of the water, including spillway gates and valves, trash racks or other intake structures, quickly reducing the discharge capacity or blocking the waterways entirely. An unheated structure will quickly accumulate frazil ice.

The strength of the ice sheet itself is usually the limiting condition in generating loads on structures. An ice sheet fails by crushing, splitting, bending or bucking, or by some combination of these (Table 7.8). Typically, each of these modes of failure is analysed using analytical models and presumed or measured mechanical properties of the ice. However, it is often necessary to perform model tests to understand the mechanisms of ice-structure interaction creating loads. In some cases loads are created by the collision of drifting ice floes with the structure, in which case the kinetic energy of the flow controls the loads up to the point that the loads overcome the structural strength of the ice (Table 7.9).

**Table 7.8** Factors affecting ice loads on dams

| Factor | Dependence |
|---|---|
| Ice pressure | Velocity and thickness of the ice<br>Aspect ratio (width of the structure to thickness of the ice)<br>Shape of the structure<br>Sloping face of the structure<br>Stiffness of the structure<br>Difference of stiffness between different expose structures<br>Rheology of the ice<br>Extent of cracks and refreezing<br>Restriction from the shore<br>Model of local failure<br>Strength of the ice |
| Model of failure | Velocity of the ice<br>Aspect ration<br>Direction of ice movement<br>Ice thickness<br>Temperature |
| Strength of ice | Types of ice<br>Structure of the ice<br>Size of crystals in the ice<br>Direction of loading<br>Temperature<br>Velocity of the loads |

Table 7.9 Recommended ice loads on dams for design

| Jurisdiction | Recommended design load | Reference |
|---|---|---|
| Sweden | 50 kN/m in south of Sweden<br>100 kN/m in middle Sweden<br>200 kN/m in north of Sweden | The Swedish guidelines for dam safety, *Kraftföretagens Riktlinjer för Dammsäkerhet* (Ekström, 2006) |
| Norway | 100 kN/m | Norges Vassdrag och Energiverk (Ekström, 2006) |
| Canada | 146 kN/m for concrete dams<br>73 kN/m for steel gates<br>29 kN/m for wood gates | Comfort *et al.* (1995) |
| Canada | 60 kN/m 0 to −20°C<br>80 kN/m −21 to −30°C<br>100 kN/m below −30°C | Ko (1999) |
| Canada | 100–200 kN/m for wide structures<br>200–600 kN/m for narrow structures<br>40–90 kN/m for weak structures between stiff structures (e.g. gates) | Comfort (2000) , Comfort *et al.* (2003) |

The *uniaxial compressive strength* of freshwater ice varies from about 0.5 to 20 MPa, depending on temperature, strain rate, grain size and structure, and porosity. The peak strength occurs at a strain rate of about 0.001/s above which brittle fracture occurs. For strain rates lower than 0.001/s, Sinha *et al.* (1987) have proposed the following relationship

$$\sigma_c = 212\dot{\varepsilon}^{0.34} \tag{7.21}$$

for freshwater ice at −10°C, where $\sigma_c$ is the compressive strength (in MPa), and $\dot{\varepsilon}$ is the strain rate (in $s^{-1}$). For ice at other temperatures the following correction factor is suggested:

$$\sigma_c(T) = \sigma_c(-10°C)\left\{\exp\left[7.82\frac{236 - T}{263T}\right]\right\}^{1/3} \tag{7.22}$$

The *flexural strength* of freshwater ice is lower than the compressive strength, varying from 0.5 to 3 MPa, with an average of about 1.73 MPa.

The *elastic modulus* of freshwater ice varies greatly, depending on the loading rate, temperature, and grain size and configuration. The elastic modulus $E$ varies from about 2 GPa at low loading rates to about 9 GPa at high rates (Cole *et al.*, 1995).

*Broken ice* is often assumed to obey a Mohr–Coulomb strength criterion, $\tau = \sigma_n \tan \phi$, in which $\tau$ is the shear strength, $\sigma_n$ is the normal or confining stress and $\phi$ is the friction angle, which varies between $20 \leq \phi \leq 45$ depending on the strain path and strain rate.

Wind and water flow over large fetch areas create drag forces on an ice cover. These forces are often difficult to estimate because relevant fetch distances and widths are difficult to assign. Calculated drag forces are also commonly greater that the crushing strength of ice, and thus the latter determines the load on structures.

The *drag force* on ice due to wind or water is estimated as

$$F_d = C_d \rho A V^2 \tag{7.23}$$

where $C_d$ is the drag coefficient (varying from 0.002 for smooth ice cover to 0.005 for rough cover), $\rho$ is the density of air (1.3 kg/m$^3$) or water (1000 kg/m$^3$), $A$ is the fetch area and $V$ is the air or water velocity measured a standard reference distance above or below the ice.

### 7.5.3 Predicting ice thickness

The growth of static ice cover on a reservoir follows a differential process in which the change in thickness over time $\partial h/\partial t$ depends on the existing thickness, the air temperature and various physical properties of the ice (Ashton, 1986). The standard solution to this differential process is (USACE, 2002)

$$h_j = \sqrt{(B + h_k)^2 + 2A(U_j - U_k)} - B \tag{7.24}$$

Table 7.10 Typical values of $\alpha$. (Michel, 1971)

| Ice cover condition | $\alpha$ (°C) (temperature in °C, ice thickness in centimetres) | $\alpha$ (°F) (temperature in °F, ice thickness in inches) |
|---|---|---|
| Windy lake, no snow | 2.7 | 0.80 |
| Average lake with snow | 1.7–2.4 | 0.5–0.7 |
| Average river with snow | 1.4–1.7 | 0.4–0.5 |
| Sheltered small river | 0.7–1.4 | 0.2–0.4 |

where: $h_j$ is the ice thickness at time $j$; $h_k$ is the ice thickness at time $k < j$; $A = k_i/\rho\lambda$, where $\rho$ is the ice density and $\lambda$ is the latent heat of the ice; $B = k_i/H_{ai}$, where $k_i$ is the thermal conductivity of the ice and $H_{ai}$ is the heat transfer coefficient from the ice surface to the atmosphere; $U_j = \sum_{i=1}^{j}(T_m - T_{ai})$ is the accumulated freezing degree-days between the onset of freezing (day 1) and day $j$; and $U_k = \sum_{i=1}^{k}(T_m - T_{ai})$, where $U_k \leq U_j$.

Presuming that heat conduction through the ice controls the rate of growth, and the initial thickness is zero, the simplified result is

$$h_j = \alpha\sqrt{U_j} \tag{7.25}$$

where $\alpha = \sqrt{2k_i/\rho\lambda}$ (Table 7.10). This model assumes a number of things that may be only approximately reflected in field conditions. These include the effect of snow cover, possible flooding of the ice sheet and variations in the heat transfer rate. Yet this remains the state-of-practice for forecasting ice growth.

Predictions of frazil ice accumulation are more complex and usually rely on numerical computational models of stream reaches and flows that combine hydraulic analysis with thermal models. These are discussed by Shen *et al.* (1991).

### 7.5.4 Predicting ice loads

Comfort *et al.* at Feet Technology Ltd., Ontario (now BMT-Fleet), conducted a long-term empirical study of ice loads on Canadian dams in Quebec, Ontario, Manitoba, Labrador and Newfoundland, and reported their findings in a series of reports sponsored by CEA Technologies (now CEATI). Much of this is summarised in Comfort *et al.* (2003). These studies reinforce the notion that much of the concern with static ice loads on dams relates to loads on gates, pillars and stoplogs.

The conclusion of the studies is that the principal cause of static ice loadings on dams derives from a combination of thermal effects (as ice warms it expands and presses against the dam) and modest pool fluctuations (as the pool rises and falls it cracks the ice sheet and either wedges into the dam or allows infill freezing and pressures against the dam). The second mechanism was newly identified in this series of studies. The greatest increases in load tend to occur with the combination of these two conditions, which also is the source of most spikes in ice loading.

The ice loads depend on – in addition to temperature and pool-level changes – whether the ice sheet is restricted in its movement, the expansion or contraction of the ice sheet, drag forces due to wind or current and fluctuations in pool level. Other factors affecting ice loads include forebay geometry, weather patterns and individual plant operations.

#### 7.5.4.1 Loads caused by thermal effects

Ice expands as it warms. Early in the winter season when there is little or no snow insulation on the ice sheet the thermally induced ice loads on a dam tend to be low. Later in the winter, as warming occurs and there is often an insulating blanket of snow on the ice sheet, the ice loads increase.

Thermal loads are found to depend on ice temperature changes throughout the full thickness of the ice sheet, and are closely dependent on changes in what Comfort *et al.* call $\Delta A$, the ice temperature area (measured in °C-cm) (Figure 7.19).

#### 7.5.4.2 Loads caused by water-level changes

A one-time lowering or raising of the pool by approximately one ice thickness or more tends to reduce the load on a dam by a factor of 3–5. Larger changes in pool level may reduce the load to nearly zero, due to separation of the ice sheet from the dam face. This is caused by changes to the boundary condition (i.e. cracking) between the ice sheet and the dam.

More modest changes in pool level, however, can induce ice wedging and significant consequent loads. Lowering of pool level sets up a wedging action within the ice sheet. The rising pool exposes new cracking on the upper surface, which may fill with

Figure 7.19 Thermal profile and thermal area change. (Adapted from Comfort *et al.*, 2003)

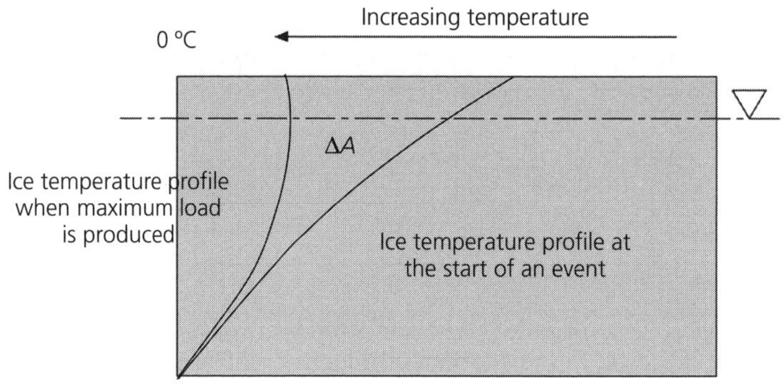


water and refreeze, so that a subsequent lowering of the pool causes further compression in the newly formed ice and increased lateral load. The strength of the wedging mechanism depends on the amplitude and rate of pool-elevation change. As noted by Comfort *et al.* (2003).

> 'Small and slow' water-level changes (i.e. less than about 0.1 m, and occurring longer than about 2 days) have little effect; the dominant ice loads in this situation are primarily thermal. 'Large and frequent' water-level changes (i.e. about 0.3 to 0.6 m/day) lead to hinge-shaped cracks and low ice loads. 'Intermediate' water-level changes (i.e. about 0.05 to 0.10 m/day) lead to near-vertical ice cracks and high ice loads.

#### 7.5.4.3 Quantitative predictions of ice forces

Comfort *et al.* (2003) suggest predicting static ice sheet line loads on the dam (in kN/m) as the simple superposition of the empirically based component line loads:

$$LL_{\text{Total}} = LL_{\text{Residual}} + LL_{\text{Thermal}} + LL_{\text{Water Level}} + LL_{\text{Contingency}} \tag{7.26}$$

in which $LL_{\text{Total}}$ is the total load, $LL_{\text{Residual}}$ is the load prior to warming changes in the pool level, $LL_{\text{Thermal}}$ is the load due to thermal change, $LL_{\text{Water Level}}$ is the load due to changes in the pool level, $LL_{\text{Contingency}}$ is the load added to ensure an upper bound and

$$LL_{\text{Residual}} = 0.37(h - 25)\ \textit{for}\ h > 25\,\text{cm} \quad (\text{in kN/m}) \tag{7.27}$$

$$LL_{\text{Thermal}} = 0.064\Delta A^{0.6} h^{0.88} Dur^{f(\Delta A)} \quad (\text{in kN/m}) \tag{7.28}$$

where $A$ is the initial thermal area of the ice sheet (°C-cm), $h$ is the ice sheet thickness (cm) and $Dur$ is the duration of the event (days) and

$$f(\Delta A) = \min\begin{cases} \log\dfrac{160}{\Delta A} \\ 0 \end{cases} \tag{7.29}$$

In the studies, the term $LL_{\text{Water Level}}$ was back-calculated from the total measured static load by subtracting the calculated values of $LL_{\text{Residual}}$ and $LL_{\text{Thermal}}$. This led to the approximation

$$LL_{\text{Water Level}} = F_1 + \left(7.510^{-8}\Delta A^2 \frac{A^{1.2}}{Dur^{0.4}}\right) \quad (\text{in kN/m}) \tag{7.30}$$

$$F_1 = f(\Delta A) f(drop) f\left(\frac{a}{h}\right) \tag{7.31}$$

$$f(\Delta A) = \max\begin{cases} 0 \\ 1 - 4\left(\dfrac{\Delta A_i}{A_m}\right)^{1.4} \end{cases} \tag{7.32}$$

$$f(drop) = \max\begin{cases} 0 \\ 1 - 20\left(\dfrac{drop}{2h}\right)^2 \end{cases} \tag{7.33}$$

$$f\left(\frac{a}{h}\right) = \left(\frac{1}{\text{max of } 0.22 \text{ or } \left(\dfrac{a}{h}\right)}\right)^2 \tag{7.34}$$

in which *drop* is the absolute drop or rise of the pool during the event (cm), $h$ is the ice thickness (cm), $a$ is the average water-level change in the 2–3 days preceding the event (cm), $\Delta A_i$ is the initial change in the profile area prior to the start of the event (°C-cm) and $A_m$ is the maximum area of the ice temperature profile area, which is defined as $20h$ (cm).

#### 7.5.4.4 Loads between spillway gates and piers

Comfort *et al.* (2003) reported that a large fraction of the far-field ice load concentrates on spillway piers rather than gates, because the ice sheet arches across the piers, protecting the gate structures. They cite the following factors as influencing the arching and thus the proportion of the load taken up by the piers:

- Pier length: the load carried by the piers increases with increasing pier length.
- Gate or stoplog span: more arching accompanies shorter spans.
- Duration of loading: longer durations reduce the amount of arching as the ice creeps.

Empirically, the fractions of far-field load taken up by the piers varies considerably from one dam to another, and from winter to winter, depending on design geometries and the stiffnesses of the piers relative to the stiffnesses of the gates or stoplogs. In the lowest case the ratio was less than 3, while in other cases the ratio approached 27.

#### 7.5.4.5 Forces on steel and wood stoplogs

Surprisingly, the load arching around wooden stoplogs was consistently observed to be the same or less than around steel stoplogs, despite the latter being stiffer. Also, considerably more ice build-up (about two times as much) occurred on the steel stoplogs than on the wood stoplogs, presumably due to the different thermal properties of the two materials. The finite-element analysis done by Comfort *et al.* (2003) suggests a decrease in load arching with increasing stoplog stiffness, and an increase with increasing pier length.

## 7.6. Interference with discharge function

Interference refers to any reduction in the available discharge capacity due to factors that are external to the spillway.

### 7.6.1 Floating debris

Rivers will carry floating debris, especially during floods. Depending on the natural conditions of the catchment the debris may be in the form of uprooted large trees, cut timber, branches, bushes, grass, floating mires, dead animals and various man-made objects such as bridges, boats and houses.

Bogs or mires may, in some catchments, come afloat in chunks the size of football fields and with thicknesses of metres. While the actual floating may be due to gas development, the mires may be carried downstream during larger floods.

Available field experience and laboratory testing show that floating debris may significantly influence the discharge capacity of spillways by sticking to structures such as gates, spillway piers and bridges spanning spillways, and obstructing spillway openings and prevent gate operation. Larger trees may bridge the gap between piers, initially near the water surface due to buoyancy. Further trees may thereafter get sucked down below the water surface and increase the clogging effect. Smaller trees, bushes and grass may thereafter lead to almost complete clogging and loss of capacity.

Measures taken or considered to counteract loss of discharge capacity by floating debris include catchment management, check dams, diversions, floating booms on rivers and reservoirs, and visor structures constructed just upstream of spillways and intakes.

Hydraulic model tests have been performed in northern Europe on several occasions during the last 20 years or so to try to assess the possibility of debris clogging spillway openings. The latest tests are summarised in the CEATI report *Floating Debris and Dam Safety: Hydraulic Model Tests* (CEATI, 2010c).

The model tests, usually using young plants of spruce and pine (European redwood) with only short remaining stumps of branches, generally indicate that openings with a free width up to 80–100% of the relevant tree length are likely to become blocked by the trees if a fairly large number of them reach the openings at about the same time.

All these model tests were performed with a limited number of trees, but generally point to an increased risk of trees getting stuck once one has been caught. As the potential maximum number of trees may be many orders larger than the tested number, the reality can be expected to be (on average) worse than the direct test results indicate in terms of the percentage of trees stuck.

A single spillway opening often provides significant flow acceleration upstream, and generally fares somewhat better than multiple openings because the acceleration tends to line up many, or even most, trees in parallel with the stream lines so that they pass through more easily.

The physical tree strength may provide a cap on the possibility of a single tree bridging a wide opening. Tests on freshly cut trunks of Scandinavian trees (three tests each on birch, spruce and pine) of around 0.1 m diameter (Rosén, 1995) indicated a mean tensile strength of 46 $N/mm^2$, with individual values deviating by up to $\pm 15$ $N/mm^2$.

If the above tensile strength values are also representative of larger tree diameters, a single tree with a trunk diameter of 0.3–0.4 m would have the strength required to bridge and completely block a normal free spillway opening having a width of up to around 20 m. However, no exact 'safe width' can be determined, as the build-up of debris may encompass several trees arriving successively or tangled together and sharing the load.

Hydraulic model tests demonstrate that trees arriving at a spillway in jams of a few or many entangled trees indeed require larger free openings for passage than do single trees. A quite significant loading condition for spillway systems may, therefore, be the collapse of an upstream structure collecting floating debris, such as a low bridge, a visor or a floating boom. This is a significant drawback with booms, with the possible exception of shear booms specifically designed to continuously remove the collected debris out of the main current.

The present understanding is that both the flow velocities and the density of the floating debris significantly influence the behaviour of the debris at spillways, floating booms and visors. If the flow velocities are low and the density of the debris is also low, debris stopped at a spillway, boom or visor will tend to form a single surface layer, growing successively upstream as more and more debris is collected. The effect on the discharge capacity given the release of the debris could then be rather limited. With higher flow velocities or higher debris densities this will no longer be the case, as debris will also pass under the surface layer and block the lower parts of a spillway opening, with much larger effects on discharge capacity.

With fresh tree densities in the range 800–975 $kg/m^3$ common in the northern hemisphere, flow velocities higher than 1.5–1.6 m/s would be sufficient to drive trees under booms and previously collected surface layers of debris. The density of uprooted trees in northern Europe may in fact be expected to vary in the range 800–1050 $kg/m^3$ depending on age, season, time in the water and travel history (soaking and physical grinding away of longer branches and bark), tree species and root configuration. Most spruces could, for instance, be expected to arrive at spillways more or less free of longer branches due to the grinding of trees against each other after any reasonably long river travel. This stripping effect may be largely absent for debris entering wide reservoirs.

Insufficient buoyancy of floating booms or oscillations due to unstable hydraulic conditions may also cause debris to pass over the booms.

Trees may also stick to obstructions located above the water level, such as insufficiently opened gates or bridges placed just upstream of spillway gates. The clearance required between the water-surface elevation and potential obstacles would depend on tree type (species, length and strength of remaining branch stumps), flow velocity and the shape of the structure, gate or bridge. A bell-mouth-shaped upstream edge may reduce the problem. Critical water levels are likely to be those prevailing from about one-third of a tree length upstream of the obstacle down to the actual obstacle, or even somewhat beyond that point.

It follows from the above that *partially* open radial or vertical lift gates may prevent discharge of floating debris. Accordingly, it might sometimes be a better strategy with regard to debris discharge to open one gate fully rather than to open several gates partially. This may, however, be at variance with design assumptions for numerous spillways with more than one spillway gate discharging (over a short chute) into a single energy-dissipation basin.

The possibility to mechanically remove jams of more than a few trees stuck at a spillway opening in real time during flood discharge may only be theoretical, especially given the dangers that may be involved.

Characterisation of debris production is not well developed. It may be assumed that the size of the floating debris yield will depend on the catchment characteristics, such as the type of vegetation and the occurrence of potentially unstable slopes and river banks, and be fairly strongly correlated in time with intense rainfall and large river flows.

### 7.6.2 Ice

Ice can reduce spillway-discharge capacities in different ways. At break up of lake ice or river ice the ensuing ice floes can build up and jam spillway openings, particularly those of limited free dimensions, in the same way as trees or other types of floating debris.

Ice or frazil developed in supercooled water can also adhere to gates and spillway piers, and prevent gate opening or reduce the size of the free opening, thereby limiting the discharge capacity. Leaky seals are a not uncommon reason for ice build-up between gates and piers.

Ice jams, many metres in height, can also develop where frazil adheres to the boundaries of rivers and channels, and completely change the hydraulic capacities and rating curves. Fresh water only requires supercooling of the water to fractions of a degree under 0°C to make frazil production possible throughout a water body. Weather conditions promoting supercooling and frazil growth in shallow waters involve those with very high heat losses from the water due to, among other things, radiation, low air temperature, wind and heavy snowfall.

Where the effects of ice need to be considered specifically, modelling of the ice conditions in the reservoir and river, both upstream and downstream, may be required. Ideally, this should be done in coupled models, taking into account the effects of the flow and weather on the ice processes, as well as the effects of the ice on the flow.

### 7.6.3 Sediments

Many rivers carry significant amounts of solids, especially those flowing through more easily erodible materials. Conversely, in rivers running through old and stable igneous rocks, sediments may often not be present in appreciable amounts. Large volumes of sediments may also enter into reservoirs in close proximity to dams in regions of reservoir slope instability. Very large volumes of sediments arriving into reservoirs due to landslides can also carry a debris load that exacerbates any pre-existing debris on the reservoir. In some cases, the volume of sediment transported by a landslide may be sufficient to create a landslide dam, thereby creating a very significant sediment hazard to any dam downstream.

Coarse sediments, carried as bed load by a river, will initially settle in the upstream end of reservoirs, where flow velocities start to reduce and the carrying capacity of the flow diminishes. Finer sediments carried as suspended load will continue further into a reservoir, and usually tend to create bars sloping by around 1%. Unless regular (annual) flushing of the reservoir is performed, these sediment bars may, in time, reach and start to clog intake areas for water conduits, such as bottom outlets and power intakes, and sometimes even spillway intakes. Sediments have a similar potential to disturb water conduit operation as floating debris and ice, but, in addition, the discharge of coarse sediments may also lead to heavy erosion in the fast-flowing water of low-level outlets.

### 7.6.4 Counter measures

A number of types of structure are presently in use or considered for preventing floating debris from reaching spillways. These are check dams, diversions, floating booms and visors. Check dams can also assist in controlling sediment inflow, while floating booms have long been used to control ice for many years.

Check dams are usually built in smaller streams and tributaries to physically stop large-sized debris while the water flow passes through the structure. The check dams will create a backwater effect and cause inundation where placed. Check dams may be made of steel or concrete, and usually comprise coarse-mesh grids of beams anchored to the foundation. Of course, other types of load transfer are also conceivable. The possibility of bank erosion will have to be considered where check dams are used.

Floating booms have a long history of use in log flotation in the forestry industry. However, much of the experience from that era was not systematically collated and has been lost. On the other hand, the old tactics and methods involved relatively high-risk working conditions, and new methods are required to comply with occupational safety requirements.

There are, in principle, two conceivable types of booms, check booms physically intercepting and stopping floating debris, and shear booms directing or diverting the flow of debris away from the track leading to spillways. Both types of boom may need to be designed and properly maintained for ice conditions, or be removed during the winter season, where lake or river ice can assert significant loads on the booms. Annual removal of booms leads to high costs of operation and may also limit the choice of materials for the structures.

The maintenance of booms for reliability, both in terms of structural condition and proper function, is a significant difficulty, with very significant limits on what can actually be achieved.

The design of check booms is especially difficult, due to the need to predict the maximum conceivable load on the boom, as an overloaded and collapsed check boom may release a debris jam with the potential to clog a downstream spillway. A failure will, accordingly, not only be sudden, it may also be catastrophic.

The design of shear booms is difficult for hydraulic reasons. A position needs to be found where the boom can be placed at an acute angle to the flow, so that the flow velocities along the boom are sufficiently high to carry the debris out of the principal flow path to the essential water conduits and towards an area where it can be effectively managed without being so high that it is carried under the boom. The velocities will, by definition, be successively reduced along the boom to the point where the debris comes to rest in quantities. The different requirements for debris control during all significant (major) river flows may not be achievable either actually or theoretically.

The CEATI reports *Failure Modes for Debris Booms* (CEATI, 2010a) and *Dynamic Loads on Debris Booms* (CEATI, 2010b) provide an overview of the loadings and failure modes of debris booms. Boom failures may be local or global, and involve total or partial loss of function. Failures may be caused by: static loading on the structure through form and friction drag forces created by the water flowing past a boom; dynamic loadings due to oscillations in unstable flow conditions; impact when larger gatherings of debris collide with a boom; or expanding ice. Failures may be influenced by a combination of wear and inadequate maintenance.

Large ice sheets held back by floating booms will increase loads on the booms by the friction losses in the flowing water and the wind. Expanding ice can also influence the loads on booms.

Design requires the definition of a design load; that is, a choice has to be made about how large a load a boom should be able to resist with the debris-control function retained wholly or partially. The determination of a design loading presupposes knowledge of the configuration, extent, type and properties of debris on the boom, as well as the ensuing flow conditions and the current state of maintenance of the boom arrangement. Even assuming that this knowledge is available, the loading is not easy to determine and, as mentioned above, extreme debris production is problematic. A safe choice of design load may be very expensive, as there will often be a large difference between probable and physically possible extreme loadings.

Over time, the combination of wear and inadequate maintenance may be one of the most difficult factors to assess, due to its dependence on current and future of generations of management and personnel.

It follows from the above that floating retention booms provided mainly to facilitate normal year-to-year operation may cause serious threats to spillways under more severe conditions. Shear booms designed to divert the floating debris out of the main current are, from that point of view, to be preferred, but on the other hand they are very difficult to design for proper hydraulic function.

Visors constitute an attempt to avoid some of the potential problems with debris booms by the use of fixed structures located immediately upstream of conduit intakes, so that the horizontal forces can be transferred directly to the dam body or intake structure proper, both of which would on a day-to-day basis be exposed to these forces anyway. Visors comprise some kind of grating of steel or concrete beams with openings smaller than the spillway so that whatever passes the visor will also pass through the spillway.

Visors require a reasonably large water depth and sufficiently low flow velocities to facilitate the collection of debris in a single surface layer. The front of visors has been conceived to be sloping at a low angle, so that the collected debris will tend to slide up the front where it will not obstruct deeper areas and could possibly be removed mechanically, even during floods.

Electric heating systems may be provided for gates and intake piers with the aim of preventing build-up of ice. However, many such systems are designed for still water conditions so that closed gates will not freeze up and will be able to be opened in winter, but this does not mean that the water conduit can remain in operation for any duration. During spilling the heat losses on both gate leafs and water conduit walls will be many times larger, sometimes even reaching values approaching the installed capacity of low-head power stations. Where the structures can be isolated from the flowing water temporarily, de-icing by hot steam may be effective for a limited period.

Control of sediments can be done in different ways. Apart from measures in the catchment aimed at preventing erosion or stopping sediments, there are two different philosophies for handling sediments that have reached a reservoir. Either sediments are allowed to settle in the reservoir as far upstream as possible, or the aim is to pass the sediments along to the river downstream and beyond.

The first of these strategies will inevitably lead to the loss of reservoir volume and area until a balance is reached, when flow velocities have become high enough to carry the sediment along. One disadvantage of this strategy is that nutrient-rich sediments that historically were transported to the river mouth are now held upstream, sometimes with associated negative environmental impacts. After that only the second strategy is feasible.

In order to let as much sediment as possible pass a dam, regular reservoir flushing may be undertaken by reducing the water level temporarily. This will increase the flow velocities through the reservoir and promote erosion of already deposited sediments. This practice is most effective if the drawdown is done in anticipation of the (annual) flood, so that sediments are not allowed to settle but are passed straight down the river. Techniques have been developed for many years, especially in countries where the rivers have a high sediment load such as in India and in China (Ruijin and Xie, 1993).

## 7.7. Landslides

Landslide risk has been widely studied in the context of regional infrastructure development (Glade *et al.*, 2005; Hungr *et al.*, 2005; Lee and Jones, 2004). While the disciplines of soil and rock mechanics have generated many detailed engineering mechanics models of slope instability, most landslide risk modelling is undertaken with more approximate, statistical correlation methodologies drawing on geospatial (i.e. GIS) databases. This is due to the fact that most engineering mechanics models require many more field data, laboratory results and computations that are feasible for a regional analysis.

From an engineering systems view, landslides are more complexly related to spillway and dam risk than are many other disturbances, as they are both a causal result of other disturbances such as earthquakes, high rainfall and forest fires, and they are themselves a cause of disturbances such as debris loading in the reservoir (Figure 7.20). Landslides and rock falls directly into reservoirs have created failures of large dams as a consequence of impact waves with wave run-ups more than 100 m high.

Landslides are of many types: falls, topples, slides, lateral spreadings and flows (Van Asch, 2009). Any of these may be triggered by other disturbances such as excessive rainfall, earthquakes, forest fires or land-use changes. Common factors in slope instabilities are summarised in Table 7.11.

The common approach to mapping and simulating landslide risk over large regional areas is by statistical correlations between observable land features and landslide occurrence. Typical land features used for assessing landslide risks are slope angles, slope aspect (orientation), elevation, geology, bedrock type, soil type, moisture conditions, groundwater elevation and pre-existing landslides.

Figure 7.20 Kootenay Lake, British Columbia. (Courtesy of BC Hydro)

Table 7.11 Common factors of slope movements. (Van Asch *et al.*, 2007)

| Category of factor | Specific causes |
|---|---|
| Increase in shear stress | Erosion and excavation at the toe of the slope<br>Subterranean erosion (piping)<br>Surcharging and loading at the crest (by deposition or sedimentation)<br>Rapid drawdown (man-made reservoir, flood high tide, breaching of natural dams)<br>Earthquake<br>Volcanic eruption<br>Modification of slope geometry<br>Fall of material (rock and debris) |
| Decrease in shearing resistance | Water infiltration (rainfall, snowmelt, irrigation, leakage of drainage systems)<br>Weathering (freeze and thaw weathering, shrink and swell weathering of expansive soils)<br>Physico-chemical changes<br>Fatigue due to static/cyclic loading and creep<br>Vegetation removal (by erosion, forest fire, drought or deforestation)<br>Thawing of frozen soils |
| Possible increase in shear stress and decrease in shearing resistance | Earthquake shaking<br>Artificial vibration (including traffic, pile driving, heavy machinery)<br>Mining and quarrying (open pits, underground galleries)<br>Swinging of trees |
| Development-related disturbances (ignored for most cases) | Changes in vegetation cover<br>Earthen dams<br>Excavation and mining<br>Irrigation<br>Infrastructure and structure construction<br>Liquid disposal (sanitary, sewers, latrines, etc.)<br>Pilings<br>Soils deposits<br>Bedrock<br>Slope steepness<br>Hydrological factors |

### 7.7.1 Quantifying landslide risk

A number of approaches to quantitatively predicting the probability of land sliding based on regional geospatial parameters are in wide use. The most common of these are logistical regression models, Bayes factors models and GIS-based indicator overlays.

#### 7.7.1.1 Logistic regression approach

A common approach used to assess landslide probabilities is *logistic regression* (Chen and Wang, 2007; Gorsevski *et al.*, 2000; Ohlmacher and Davis, 2003). Logistic regression deals with binary variables: one versus zero, or landslide versus no landslide. The $y$-axis of a logistic regression is related to the proportion of ones in a population (e.g. the probability of a landslide) against an $x$-axis of indicator variables.

The logarithm of the odds $\ln[P/(1-P)]$ has the convenient property that it is symmetrical about $P = \frac{1}{2}$. Thus, a linear regression is constructed of the form

$$\ln\left[\frac{P}{(1-P)}\right] = \beta_0 + \sum_i \beta_i x_i \tag{7.35}$$

where $\beta_i$ are regression coefficients and the $x_i$ are independent variables upon which the probability of landslides depends. This yields an S-curve. An example of a logistic regression applied to landslide data for Puerto Rico (Leon, 2012) is given in Table 7.12.

**Table 7.12** Logistic regression of landslide predictive variables, Puerto Rico. (Leon, 2012)

| Term | Logistic regression parameter estimate | |
|---|---|---|
| | Beta | Standard error |
| Intercept | 8.921 | 0.220 |
| Slope degree | −0.0224 | 0.0018 |
| Annual precipitation | −0.0023 | 8.1E-5 |
| Slope aspect | 0.0018 | 0.0001 |
| Annual geopotential of rain | 8.8E-15 | 3.8E-16 |
| Roads within 85 m | 0.054 | 0.016 |

#### 7.7.1.2 Bayes factors approach

Inferences of probability of some event based on new information need to reflect two things: how probable the event was thought to be before the new information was available, and how knowing the new information changes that prior probability. From the Bayes rule (Kass and Raftery, 1995),

$$P(x \mid data) \propto P(x)P(data \mid x) \tag{7.36}$$

where $P(x \mid data)$ is the updated probability of some event, condition, parameter, etc., $x$ after some new 'data' are taken into account, $P(x)$ is the probability of $x$ before the new data were seen, and $P(data \mid x)$ is the conditional probability of observing the new data if $x$ were true. This last conditional probability has a special name in statistical theory, it is called the *likelihood*.

So, the updated probability of $x$ is logically proportional to the prior probability of $x$ multiplied by how likely the data would be if that value of $x$ were obtained. The more likely the data that were observed, the greater the weight of evidence the data provide for $x$ being true. The normalising constant for this equation is the sum over all possible values that $x$ might take – that is, how probable the observed data are irrespective of the true value of $x$ – so that the sum of the $P(x \mid data)$ over all $x$ is 1.0:

$$P(x \mid data) = \frac{P(x)P(data \mid x)}{\sum_i P(x_i)P(data \mid x_i)} \tag{7.37}$$

Several workers have used this approach to assess landslide probabilities (Dahal *et al.*, 2008; Quinn *et al.*, 2010; van Westen *et al.*, 2003). It has also been applied to levee safety (Baecher and Christian, 2013).

The weight of evidence of a set of data can be seen by dividing the updated probability in favour of $x$ by the updated probability in favour of not-$x$ ($\bar{x}$). The normalising constant is the same in the numerator and denominator, so it cancels:

$$\frac{P(x \mid data)}{P(\bar{x} \mid data)} = \frac{P(x)}{P(\bar{x})}\frac{P(data \mid x)}{P(data \mid \bar{x})} \tag{7.38}$$

The term $P(x)/P(\bar{x})$ is the odds in favour of $x$ before the data are taken into account. The second term, $LR = P(data \mid x)/P(data \mid \bar{x})$, is the likelihood ratio, which is the weight of evidence contained in the observations.

Note that the weight of information contained in the likelihood ratio depends on how different $LR$ is from 1.0. If $LR = 1.0$, then $P(data \mid x)/P(data \mid \bar{x})$, and the association of the data observed with $x$ is exactly the same as their association with $\bar{x}$, so the observation does not tell us anything about how probable $x$ is. If $LR > 1.0$, then the data increase the probability of $x$, and if $LR < 1.0$ the data decrease the probability of $x$.

When there is more than one observation, $(z_1, z_2, \ldots, z_k)$, the likelihood becomes the *joint likelihood* of all the observations. This is the joint probability of observing the set of things that were observed, were $x$ true. It is a joint conditional probability, so if the observations are correlated, this must be accounted for.

$$LR = \frac{P(z_k, \ldots, z_2, z_1 \mid x)}{P(z_k, \ldots, z_2, z_1 \mid \bar{x})} = \frac{P(z_1 \mid x)}{P(z_1 \mid \bar{x})}\frac{P(z_2 \mid z_1, x)}{P(z_2 \mid z_1, \bar{x})}\cdots\frac{P(z_k \mid z_2, z_1, \bar{x})}{P(z_k \mid z_2, z_1, \bar{x})} \tag{7.39}$$

In the desirable case that the observations are mutually independent, the joint likelihood ratio reduces to the simple product

$$LR = \prod_i \frac{P(z_i \mid x)}{P(z_i \mid \bar{x})} \tag{7.40}$$

This is sometimes called the *Naive Bayes estimator*.

Figure 7.21 Geospatial approach to predicting landslide probabilities on the basis of regional mapping information. Prior and posterior probabilities of landsliding are shown on the input and output maps; conditional probabilities of the hazard factors are shown on the data maps. (Adapted from Mahler *et al.*, 2012)

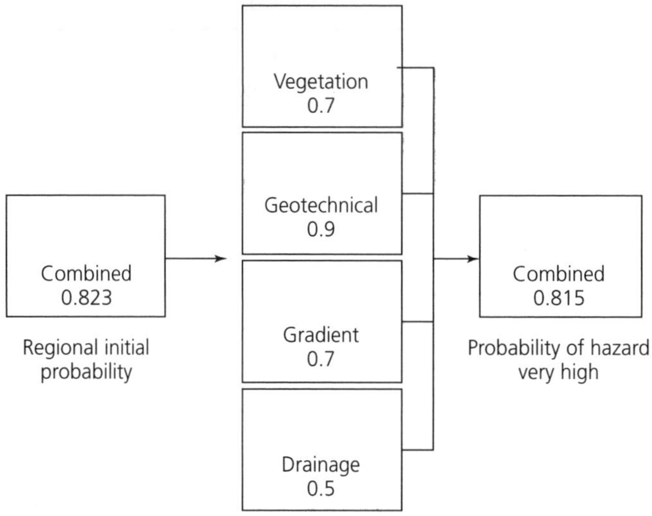


#### 7.7.1.3 Geospatial data processing for landslide risk

Either the logistic regression or the Bayes factors approach lends itself to interpretation through geospatial data processing using GIS. For example, Mahler *et al.* (2012) used a Bayes factors approach to regional landslide prediction implemented using GIS (Figure 7.21).

### 7.7.2 Landslide volume

As for many geophysical phenomena, the size of landslides, as measured by the volume of material in the slide, has a frequency distribution roughly approximated by a power law or Pareto curve (Moon *et al.*, 2005). Fujii (1969) was among the first to study the size distribution of landslides. Based on approximately 800 examples in Japan, he proposed a power-law model for volume. More recent meta-studies of landslide volume indicate a scaling parameter between 1 and 2, the lower slopes applying to soil slopes and the larger to rock slopes (Brunetti *et al.*, 2009; Stark and Guzzetti, 2009). Klar *et al.* (2011) have suggested that the slope of the distribution of landslide volume to surface area in log–log units is about 1.4 (obtained from many different regions around the world).

## 7.8. Forest fires

Each year, forest fires consume thousands of hectares of timberlands across the western regions of North America, in Australia and in southern Europe. Forest fires may taint water-supply reservoirs, as illustrated by the following quote (CBS-KPIX, 2012):

> TUOLUMNE CITY (CBS/AP/BCN) August 26, 2013 – A raging wildfire in Yosemite National Park rained ash on the reservoir that is the chief source of San Francisco's famously pure drinking water, and utility officials Monday scrambled to send more water toward the metropolitan area before it becomes tainted.

The problem has become worse over the past two decades. In an effort to understand better the processes of fire ignition and spread, the forest services of Canada and the USA have invested heavily in efforts to model forest-fire risk (Calkin *et al.*, 2011).

Wildfire risk models typically comprise three components: the probability of fire and its subsequent intensity, identification of the forest inventory that could be damaged, and assessment of the economic and ecological consequences of the fire. In this way, the wildfire risk models follow a protocol similar to that used for other natural hazards (see Figure 7.3). These components are combined in a standard risk relation

$$E(NVC) = \sum_i p(f_i)RF(f_i) \tag{7.41}$$

where $E(NVC)$ is the expected net value change (which might be subdivided by resource type), $p(f_i)$ is the probability of fire of intensity $i$, and $RF(f_i)$ is the response function (consequence) as a function of the fire intensity. Various current approaches to developing the component models of this relationship are surveyed in Thompson and Calkin (2011).

Establishing fire risk requires geospatial assessments of the probability and intensity of a fire, as these parameters interact with the forest resource. For example, Wei *et al.* (2008) present a fire ignition and spread model based on raster cells of forest properties. Finney (2005) simulated fires to calculate the relative burn probability of fire across a landscape and to reflect the topological heterogeneities. A number of authors have applied logistic regression models to empirical fire occurrence data based on ignition type (e.g. lightning versus human caused) and landscape characteristics (Chou *et al.*, 1993; Dickson *et al.*, 2006). In a series of papers, Brillinger, Preisler and their associates have applied random point process models of various types to wildfire ignition (Brillinger *et al.*, 2003, 2004, 2006, 2009; Preisler *et al.*, 2004).

The areal extent of a fire depends on both the landscape characteristics and the wind and weather conditions. The typical approach to estimating fire extent is by means of fire simulation tools that model fire spread under various weather scenarios (Finney *et al.*, 2011a,b). The input to these simulations is partly based on historical data, but inevitably also involves quantified expert judgement. The US Forest Service, in conjunction with the software development company Esri, has developed a special version of ArcMap for geospatial modelling of wildfire behaviour based on spatial data sets (Stephens *et al.*, 2009). There has also been considerable interest in recent years in whether wildfire size is well modelled as a fractal process, leading to Power-law probability density functions (Andrea *et al.* 2010; Malamud *et al.*, 1998; Ricotta *et al.*, 1999; Song *et al.*, 2001). Analyses by and Reed and McKelvey (2002) of six regions in the western USA, however, suggest that power-laws are too simple adequately to describe wildfire extent and behaviour, but better models have yet to appear.

The principal impacts of forest fires on dams are disruptions of the electric grid, interference with communications and potential increases in debris entering the reservoir. In cases where the reservoir provides drinking water, fires may cause chemical contamination.

## 7.9. Tornadoes

In a typical year, the USA experiences some 1200 tornadoes, causing USD 1 billion in damages. Of course, these meteorological events are not uniformly distributed in geographical space but concentrate in belts of high frequency. A tornado is a violent swirling column of air, extending from thunderstorm clouds to the ground, and can cause tremendous damage and loss of life. Tornado intensity is measured using the Fujita scale (Table 7.13), which measures tornado intensity as the maximum wind speeds within the column of air (Fujita, 1971). A tornado is rated according to the maximum damage that occurs in its path, even though the intensity may vary as the tornado moves over the ground. In 2007, the National Weather Service in the USA switched from the original Fujita scale to an *enhanced Fujita scale*, which is somewhat

Table 7.13 Fujita scale for tornado wind speeds

| Fujita scale | 3-s gust speed: mph | Enhanced Fujita scale | 3-s gust speed: mph | Typical damage |
|---|---|---|---|---|
| F0 | 45–78 | EF0 | 65–85 | Light damage: some damage to chimneys; 65–85 branches broken off trees; shallow-rooted trees pushed over; signboards damaged |
| F1 | 79–117 | EF1 | 86–109 | Moderate damage: peels surface off roofs; 86–109 mobile homes pushed off foundations or overturned; moving vehicles blown off roads |
| F2 | 118–161 | EF2 | 110–137 | Considerable damage: roofs torn off frame houses; mobile homes demolished; 110–137 boxcars overturned; large trees snapped or uprooted; light-object missiles generated; cars lifted off ground |
| F3 | 162–209 | EF3 | 138–167 | Severe damage: roofs and some walls torn off well-constructed houses; trains overturned; most trees in forest uprooted; heavy cars lifted off the ground and thrown |
| F4 | 210–261 | EF4 | 168–199 | Devastating damage: well-constructed houses levelled; structures with weak foundations blown away some distance; cars thrown and large missiles generated |
| F5 | 262–317 | EF5 | 200–234 | Incredible damage: strong frame houses levelled off foundations and swept away; 200–234 automobile-sized missiles flew through the air in excess of 100 m (109 yd); trees debarked; incredible phenomena occur |

Figure 7.22 Probabilities of tornado wind speed striking a structure, Calculated for the contiguous USA and for the Western, Central and Eastern regions. (Ramsdell and Rishel, 2007)

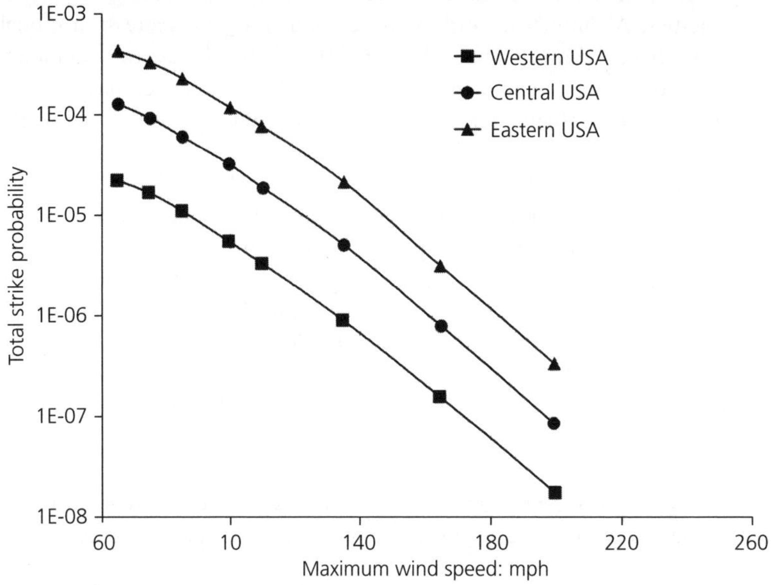


better at discriminating between storms. The frequency of tornados varies substantially from region to region and from year to year.

Ramsdell and Rishel (2007) at the Pacific Northwest Laboratory developed a probabilistic tornado loading model in support of the US Nuclear Regulatory Commission's efforts on probabilistic risk analysis of nuclear power plants. This built on an earlier study by Lawrence Livermore National Laboratory, which developed a probabilistic wind-hazard model (Hossain *et al.*, 1999). The wind-hazard model treats tornados as aleatory Poisson point process events with a spatio-temporal arrival rate $\lambda$. Presuming the probability of more than one tornado striking the same facility in a single year to be negligible (an assumption that is easily relaxed), the annual exceedance probability of wind-speed events of velocity $v$ is then

$$P(V \geq v) = \lambda \sum_i P(V \geq v \mid F_i)P(F_i) \tag{7.42}$$

where $P(V \geq v \mid F_i)$ is the ccdf of wind velocity for a given Fujita-scale value, and $P(F_i)$ is the probability mass function of the scale value. The annual exceedance probabilities of tornado wind speeds modelled using a Weibull distribution, the regional parameters for which are given in Ramsdell and Rishel (2007), are shown in Figure 7.22.

Tornados actually touch ground in 'paths' of varying widths. Figure 7.23 shows tornado paths across the central states of the USA during a 2-day period in April 1974 (Bloomer *et al.*, 2003). The more intense tornadoes typically have both longer average lengths and wider average widths than do less intense tornadoes. Thus, the assumptions of a pure Poisson point process of tornado occurrence is sometimes modified for the finite length and width of the strike zone (Milford and Golinger 1997). Twisdale and Dunn (1983) used a similar approach, although allowed for Bayesian uncertainty in the parameters of both the Poisson point process and the Weibull wind-velocity distribution, as well as for variations in the path geometry and the velocity field around the tornado.

## 7.10. Miscellaneous disturbances

Examples of internal disturbances are shown in Table 7.14. These are things that occur inside the boundary of the system being modelled (e.g. operational exceptions, planned or unplanned maintenance activities, or accidents). There is a large literature on the reliability of power plant operations, and much of this can inform modelling and risk analysis of internal disturbances (Figure 7.24) (Billinton and Allan, 1996).

In an earlier age when most hydropower plants were staffed by onsite personnel, the precursors to many internal disturbances might be detected before a severe disruption occurred. With the increasing automation of facilities and the increasing use of remote operation this is no longer the general case. As a result, it may be expected that the frequency of disruption due to internal disturbances, other things being equal, may increase in the future.

Figure 7.23 Tornado paths 3–4 April 1974 across the central USA. (Courtesy of the US National Weather Service)

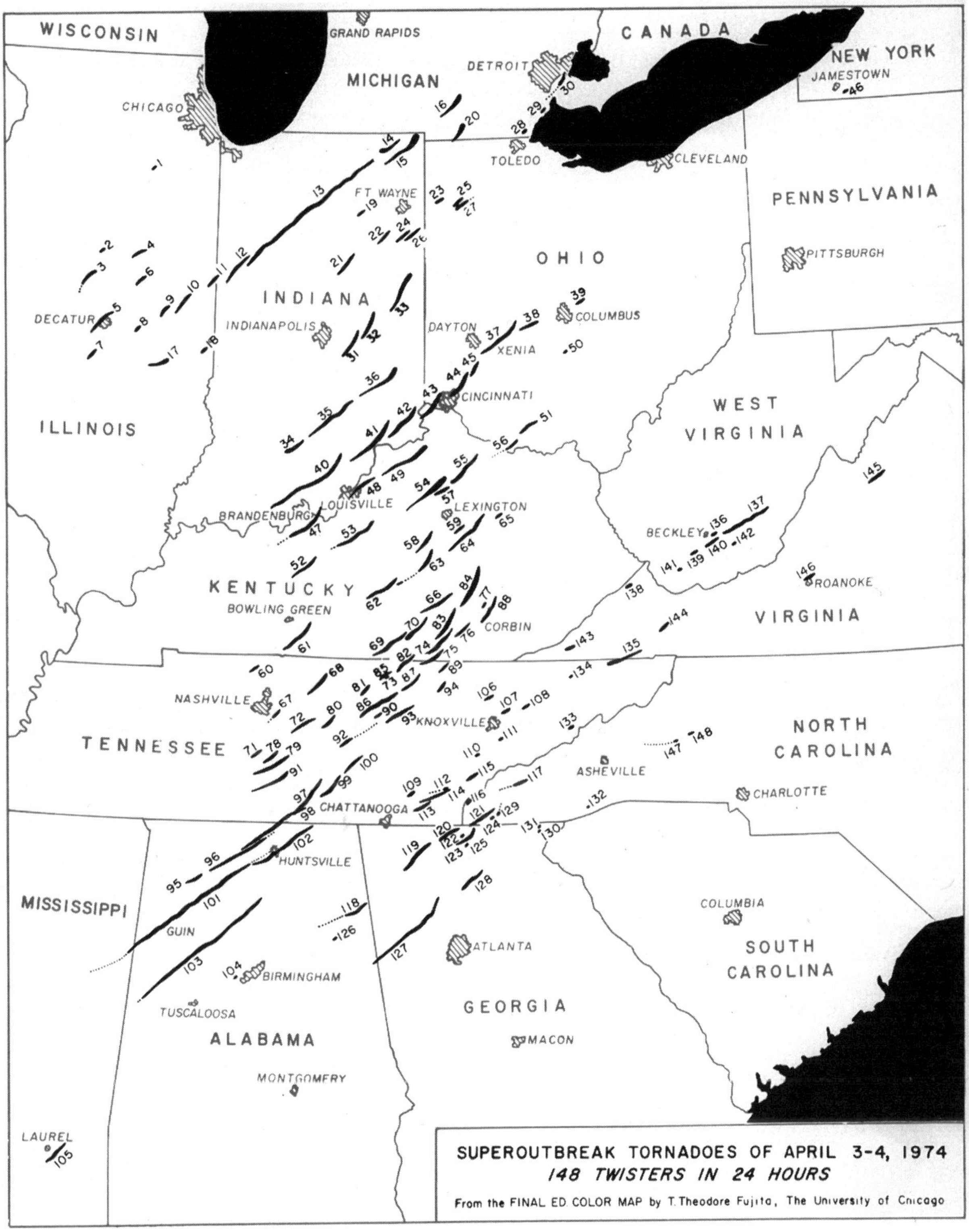


### 7.10.1 Interruption of communications

Interruption of communications is possibly the most critical internal disturbance. The current trend toward remote operation of hydropower facilities means that the ability to maintain open communication between the site and remote operators is extremely important. Loss of communication contact can be caused by many things (e.g. weather, forest fire, mechanical and electrical failures and human error).

Table 7.14 Possible internal disturbances to the flow-control systems

| Disturbance | Possible effect |
|---|---|
| Interruption of communications | Loss of management control<br>Loss of information from off site |
| Fires in the facility | Inhibits operation<br>Damages instrumentation<br>Inhibits power and controls<br>Puts powerhouse out and thus discharge capacity |
| Forced outage | Loss of generation and power plant waterway |
| Planned outages | Loss of generation and power plant waterway |
| Plant upgrades | Loss of generation and power plant waterway |
| Physical attack or vandalism | Loss of management control |
| Deterioration and ageing | Accelerated fragility |
| Disruption of communications or control | Loss of management control<br>Loss of operational control |
| Inadequate inspections | Accelerated fragility |
| Rapid drawdown | Geotechnical load on embankment and spillway structures |
| Repair delays | Accelerated fragility |
| Labour unrest | Facility shutdowns |
| Public safety incident | May force shutdown of generation |

### 7.10.2 Facility fires

Hydropower plants are industrial facilities, and like any other industrial facility they are subject to operational fires (Figure 7.25). The *Union-Bulletin* of Walla Walla, WA, USA, reported on 18 January 2013 that (*Union-Bulletin*, 2013):

> Two small fires caused a shutdown of the powerhouse at Ice Harbor Dam on Thursday. Corps officials said the fires occurred about 9 a.m. inside two electrical cabinets. There were no injuries. Three generators that were producing electricity at the time were stopped. As of today, two generators have been brought back on line and work is continuing on two other units, said a U.S. Army Corps of Engineers public affairs specialist. At the time of the incident, Ice Harbor Dam staff took action to suppress the fires and fire fighters with Walla Walla County Fire District No. 5 responded to the alarm as well. Backup power for operations restored basic facility power within 30 minutes and the dam's navigation lock was returned to service at 12:40 p.m. on Thursday.

Figure 7.24 Forced outage rates averaged across USACE hydropower units. (Corps of Engineers – Civil Works, 2008)

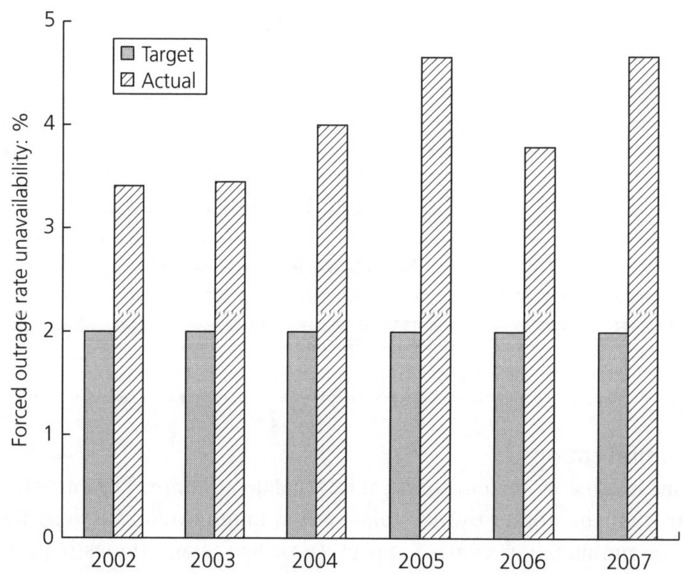

Figure 7.25 Ice Harbor Dam, Snake River, USA. (Courtesy of USACE)

Table 7.15 Typical large-loss fires at hydropower stations. (Marsh Risk Consulting, 1988)

| Date | Location | Description | Loss: USD million |
|---|---|---|---|
| 8 February 1996 | Switzerland | Fire initiated by a malfunctioning overload switch spread to all four units. Entire plant shut down | 55.9 |
| 20 July 1981 | USA (Washington) | Fire involving oil-filled cables in a cable tunnel. Amplified by difficult access and lack of fire-fighting water supply | 48.5 |
| 27 September 2002 | USA (Tennessee) | Rapid-growth fire initiated in vertical cable shaft, spread to control building and eventually self-extinguished. Extensive damage to entire facility | 36.3 |
| 21 March 1997 | Portugal | Electrical fire at unattended station. Extensively damaged cabling, monitoring equipment and control room due to corrosive smoke | 13.7 |

The insurance industry maintains reasonably complete actuarial records on industrial fires, and as a result the probabilities of such fires may be better known than the frequencies of many other disturbances (Table 7.15). In addition to the direct cost of the damages due to the fire, a serious fire commonly forces closure of power generation, with resulting consequences for discharge capacity.

## REFERENCES


Andrea MD, Fiorucci P and Holmes TP (2010) A stochastic forest fire model for future land cover scenarios assessment. *Natural Hazards and Earth System Sciences* **10(10)**: 2161–2167.

ASDSO (Association of State Dam Safety Officials) (2012) *Dam Failures, Dam Incidents (Near Failures)*. See http://www.damsafety.org/media/Documents/PDF/US_FailuresIncidents.pdf (accessed 01/06/2016).

Ashton GD (1986) *River and Lake Ice Engineering*. Water Resources Publications, Littleton, CO, USA.

Baecher G and Christian J (2013) Screening geotechnical risks. In *Foundation Engineering in the Face of Uncertainty* (Withiam JL, Phoon KK and Hussein MH (eds)). American Society of Civil Engineers, Reston, VA, USA, pp. 215–224.

BC Hydro (1992) *Estimating Organic Debris Entering BC Hydro Reservoirs. Guidelines for Management of Reservoir Debris*. BC Hydro, Vancouver, Canada.

Billinton R and Allan RN (1996) *Reliability Evaluation of Power Systems*. Plenum Press, New York, NY, USA.

Bloomer L, Church C, Henry J *et al.* (2003) *Tornado Hazard Risk Analysis: A Report for Rutherford County Emergency Management Agency*. Middle Tennessee State University. See http://mtweb.mtsu.edu/tornado/Tornado_Risk_Analysis_v7part1.pdf (accessed 01/06/2016).

BMT Isis (2010) *Campbell River Engineering Assessment State 3 Report*. BC Hydro, Vancouver, Canada.

Brillinger DR, Preisler HK and Benoit JW (2003) Risk assessment: a forest fire example. *Science and Statistics* **IMS(4)**: 177–196.

Brillinger DR, Preisler HK and Naderi HM (2004) Wildfire chances and probabilistic risk assessment. *Proceedings of Accuracy 2004 and Ties 2004*. Portland, ME, USA.

Brillinger DR, Preisler, HK and Benoit JW (2006) Probabilistic risk assessment for wildfires. *Environmetrics* **17(6)**: 623–633.

Brillinger DR, Autrey BS and Cattaneo MD (2009) Probabilistic risk modeling at the wildland urban interface: the 2003 Cedar Fire. *Environmetrics* **20(6)**: 607–620.

Brown PT and Nielson NM (1992) Dams + Debris = Danger! *Proceedings of the Canadian Dam Safety Conference*, CDA-ACD, Quebec City, Quebec, Canada.

Brunetti MT, Guzzetti F and Rossi M (2009) Probability distributions of landslide volumes. *Nonlinear Processes in Geophysics* **16(2)**: 179–188.

Bruschin J, Bauer S, Delley P and Trucco G (1982) The overtopping of the Palagnedra dam. *Water Power and Dam Construction* **34**: 13–19.

Calkin DC, Finney MA, Ager AA, Thompson MP and Gebert KM (2011) Progress towards and barriers to implementation of a risk framework for US federal wildland fire policy and decision making. *Forest Policy and Economics* **13(5)**: 378–389.

CBS-KPIX (2012) Rim fire closing in on San Francisco's Hetch Hetchy Reservoir in Yosemite. See http://sanfrancisco.cbslocal.com/2013/08/26/officials-filling-area-reservoirs-as-rim-fire-approaches-hetch-hetchy/ (accessed 01/06/2016).

CEA Technologies (1982) *Analysis of Procedures Used in the Design of Log and Trash Booms to Protect Hydro Electric Intakes*. Report 028 G 227. Canadian Electrical Association, Montreal, Canada.

CEATI (Centre for Energy Advancement through Technological Innovation) (2010a) *Failure Modes for Debris Booms*. Report T052700-0209A/2. CEATI International.

CEATI (2010b) *Dynamic Loads on Debris Booms*. Report T052700-0209A/3. CEATI International.

CEATI (2010c) *Floating Debris and Dam Safety: Hydraulic Model Tests*. ReportT-052700-0209A/4. CEATI International.

Chen Z and Wang J (2007) Landslide hazard mapping using logistic regression model in Mackenzie Valley, Canada. *Natural Hazards* **42(1)**: 75–89.

Chou YH, Minnich RA and Chase RA (1993) Mapping probability of fire occurrence in San Jacinto Mountains, California, USA. *Environmental Management* **17(1)**: 129–140.

Cliff AD and Ord JK (1981) *Spatial Processes: Models and Applications*. Pion, London, UK.

Cole DM, Shapiro LH, Weeks WF *et al.* (1995) Overview of recent program on mechanical properties of sea ice. *Journal of Cold Regions Engineering* **9(4)**: 219–234.

Comfort G (2000) Ice loads against dam faces. *Journal of Hydraulic Engineering* **126(12)**: 880.

Comfort G and Abdelnour R (2009) *Debris at Hydro-Electric Structures: Three-part Investigation*. BMT Fleet Technologies, Dam Safety Interest Group of Centre for Energy Advancement through Technological Innovation (CEATI) Inc., Montreal, Canada.

Comfort G and Abdelnour R (2010) *Debris Management in Spillways and Waterways during Floods, Debris Yield Reduction by Management*. BMT Fleet Technologies, Dam Safety Interest Group of Centre for Energy Advancement through Technological Innovation (CEATI) Inc., Montreal, Canada.

Comfort G, Abdelnour R, Gong Y and Dinovitzer A (1995) Static ice loads on hydro-electric structures: field measurements and analysis. *Proceedings CDSA Conference*, Canadian Dam Safety Association, Winnipeg, Canada.

Comfort G, Gong Y, Singh S and Abdelnour R (1997) Static ice loads on dams: Loads along a long dam face; and a comparison of the loads exerted on wooden and steel stoplogs. *Proceedings of the Canadian Congress on Large Dams*, CDA, Edmonton.

Comfort G, Gong Y, Singh S and Abdelnour R (2003) Static ice loads on dams. *Canadian Journal of Civil Engineering* **30(1)**: 42–68.

Consecru (1966) *Projekt Consecru – Teilprojekt Schemmholz*. Schlussbericht Teilprojekt Schwemmholz, Projektgemeinschaft Glenz & Walther AB, Zurich, Switzerland.

Corps of Engineers – Civil Works (2008) Detailed Information on the Corps of Engineers: Hydropower Assessment. See http://www.whitehouse.gov/sites/default/files/omb/assets/OMB/expectmore/detail/10000002.2004.html (accessed 01/06/2016).

Cornell CA (1968) Engineering seismic risk assessment. *Bulletin of the Seismological Society of America* **58**: 1583–1606.

Cox DR (1955) Some statistical methods connected with series of events. *Journal of the Royal Statistical Society, Series B – Statistical Methodology* **17(2)**: 129–164.

Cox DR (1962) *Renewal Theory*. Methuen, London, UK.

Cox DR and Isham V (1980) *Point Processes*. Chapman and Hall/CRC Press, Virginia Beach, VA, USA.

Cressie N (1993) *Statistics for Spatial Data*. Wiley-Interscience, New York, NY, USA.

Cruden D (1977) Length distribution of strike–slip faults and the process of breakage in the continental crust: discussion. *Canadian Journal of Earth Sciences* **14(3)**: 508–509.

Dahal RK, Hasegawa S, Nonomura A, Yamanaka M, Masuda T and Nishino K (2008) GIS-based weights-of-evidence modelling of rainfall-induced landslides in small catchments for landslide susceptibility mapping. *Environmental Geology* **54(2)**: 311–324.

Daykin CD, Pentikainen T and Pesonen M (1993) *Practical Risk Theory for Actuaries*. Chapman and Hall/CRC Press, Virginia Beach, VA, USA.

Dickson BG, Prather JW, Xu Y, Hampton HM, Aumack EN and Sisk TD (2006) Mapping the probability of large fire occurrence in northern Arizona, USA. *Landscape Ecology* **21(5)**: 747–761.

Eden UT (2005) *Point Process Filters in the Analysis of Neural Spiking Models*. PhD thesis, Harvard University, Cambridge, MA, USA.

Ekström T (2006) A model of ice loads on dam structures. *Dams and Reservoirs, Societies and Environment in the 21st Century*. Taylor & Francis, London, UK.

Eschenbach TG, Harper WV and Anderson CM (2010) Estimating oil spill occurrence rates: a case study for outer Continental Shelf areas of Gulf of Mexico. *Journal of Environmental Statistics* **1(1)**: 1–19.

FHWA (Federal Highway Administration) (2005) *Debris Control Structures Evaluation and Countermeasures*, 3rd edn. FHWA, Washington, DC, USA.

Field EH (2001) *Probabilistic Seismic Hazard Analysis (PSHA) – A Primer*. See http://www.world-earthquakes.com/index.php?option = view&cat = 1&cid = 227&ext = pdf (accessed 01/06/2016).

Finney MA (2005) The challenge of quantitative risk analysis for wildland fire. *Forest Ecology and Management* **211(1–2)**: 97–108.

Finney MA, Grenfell IC, McHugh CW *et al.* (2011a) A method for ensemble wildland fire simulation. *Environmental Modeling and Assessment* **16(2)**: 153–167.

Finney MA, McHugh CW, Grenfell IC, Riley KL and Short KC (2011b) A simulation of probabilistic wildfire risk components for the continental United States. *Stochastic Environmental Research and Risk Assessment* **25(7)**: 973–1000.

Friedman DG (1984) Natural hazard risk assessment for and insurance program. *Geneva Papers on Risk and Insurance* **9(30)**: 57–128.

Fujii Y (1969) Frequency distribution of landslides caused by heavy rainfall. *Journal Seismological Society Japan* **22**: 244–247.

Fujita TT (1971) *A Proposed Characterization of Tornadoes and Hurricanes by Area and Intensity*. Satellite and Mesometeorology Research Project, University of Chicago, Chicago, IL, USA.

Gaetan C and Guyon X (2009) *Spatial Statistics and Modeling*. Springer, Berlin, Germany.

Galeati GA (2009) Hydraulic safety of ENEL dams. See http://www.barrages-cfbr.eu/BackUp/Info/documentation/texte/col2009/session2/10)%20G.%20Galeati%20full%20paper%20.pdf (accessed 01/06/2016).

Gatwood E, Pederson J and Casey K (2000) *Debris Method: Los Angeles District Method for Prediction of Debris Yield*. US Army Corps of Engineers, Los Angeles, CA, USA.

Glade T, Anderson M and Crozier MJ (2005) *Landslide Hazard and Risk*. John Wiley, New York, NY, USA.

Godtland K and Tesaker E (1994) Clogging of spillways by trash. *Transactions of the 18th International Congress on Large Dams*, Durban, South Africa, vol. 1, pp. 543–557, Q68 R36.

Gorsevski PV, Gessler P and Foltz RB (2000) Spatial prediction of landslide hazard using logistic regression and GIS. *GIS in the Rockies 2000 Conference and Workshop: Applications for the 21st Century*, Denver, CO, USA.

Government of China News Agency (2006) China struggles to treat floating debris at Three Gorges. See http://www.chinadaily.com.cn/china/2006-10/22/content_713882.htm (accessed 01/06/2016).

Gumbel EJ (1958) *Statistics of Extremes*. Columbia University Press, New York, NY, USA.

Hartford DND and Baecher GB (2004) *Risk and Uncertainty in Dam Safety*. Thomas Telford, London, UK.

Hogg RV and Klugman SA (1984) *Loss Distributions*. Wiley, Chichester, UK.

Hossain Q, Mensing R, Savy J and Kimball J (1999) *A Probabilistic Tornado Wind Hazard Model for the Continental United States*. Lawrence Livermore National Laboratory, Livermore, CA, USA.

Hungr O, Fell R and Couture R (2005) *Landslide Risk Management: Proceedings of the International Conference on Landslide Risk Management, Vancouver, Canada, 31 May–3 June 2005*. Taylor & Francis, London, UK.

ISAB Dammar & Kraft. (2005) *Debris in Reservoirs and Rivers – Dam Safety Aspects*. CEA Technologies, Montreal, Canada.

Ishikawa Y (1990) *Studies on Disasters Caused by Debris Flows Carrying Floating Logs down Mountain Streams*. PhD thesis, Kyoto University, Kyoto, Japan.

Jaynes ET (1996) *Probability Theory: The Logic of Science*. See http://omega.albany.edu:8008/JaynesBook.html (accessed 01/06/2016).

Kass RE and Raftery AE (1995) Bayes factors. *Journal of the American Statistical Association* **90(430)**: 773–795.

Kaufman G and Barouch E (1976) Estimating undiscovered oil and gas. *Siam Review* **18(4)**: 812–812.

Keller KD (2013) Density of ice and water. *Wikimedia Commons*. See https://commons.wikimedia.org/wiki/File:Density_of_ice_and_water_(en).svg (accessed 01/06/2016).

Kennedy RJ (1962) *Manual of Force Calculation for Pulpwood Holding Grounds*. Technical Report 293. Pulp and Paper Research Institute of Canada, Vancouver, Canada.

Klar A, Aharonov E, Kalderon-Asael B and Katz O (2011) Analytical and observational relations between landslide volume and surface area. *Journal of Geophysical Research-Earth Surface* **116**: F02001.

Kleiber C and Kotz S (2003) *Statistics Size Distributions in Economics and Actuarial Sciences*, 1st edn. Wiley-Interscience, New York, NY, USA.

Ko SR (1999) *Cracking During the Indentation Failure of S2 Columnar Ice* (Shen HT (ed.)). Balkema, Leiden, Germany.

Kotz S, Kozubowski TJ and Podgórski K (2001) *The Laplace Distribution and Generalizations: A Revisit with Applications to Communications, Economics, Engineering, and Finance*. Birkhäuser, Berlin, Germany.

Kozubowski TJ and Podgorski K (2002) Log-Laplace distributions. *International Mathematical Journal* **3**: 467–495.

Larivière R, Léger P, Tinawi R and Roussel M (1997) Safety of gravity dams and spillways against floods. *9th Canadian Dam Safety Conference*, Montreal, Canada, 1.17–1.30.

Lee EM and Jones DKC (2004) *Landslide Risk Assessment*. Thomas Telford, London, UK.

Léger P, Tinawi R, Rheault S and Leclerc M (1996) A progressive methodology for strtuctural safety evaluation of gravity dams subjected to floods. *Proceedings of Canadian Dam Safety Conference*, Niagara Falls, pp. 1–15.

Leon M (2012) Real-time Rainfall-induced Landslide Risk Assessment System for Northeastern Puerto Rico. See http://his.cuahsi.org/conference2011/Presentations/P02.Leon.pdf (accessed 26/06/2016).

Lewis PAW (1972) *Stochastic Point Processes: Statistical Analysis, Theory, and Applications*. Wiley-Interscience, New York, NY, USA.

Mahler C, Varanda E and Oliveira L (2012) Analytical model of landslide risk using GIS. *Open Journal of Geology* **2(3)**: 182–188.

Malamud BD, Morein G and Turcotte DL (1998) Forest fires: an example of self-organized critical behavior. *Science* **281(5384)**: 1840–1842.

Marsh Risk Consulting (1988) *A 30 Year Review of Large Losses in the Power Industry (1969–1998)*, 7th edn. Marsh & McLennan Companies, New York, NY, USA.

McFadden T and Stallion M (1976) *Debris of the Chena River*. US Army Corps of Engineers, Hydrologic Engineering Center, Hanover, NH, USA.

Michel B (1971) *Winter Regime of Rivers and Lakes*. U.S. Army Corps of Engineers, Hanover NH, USA.

Milford R and Goliger A (1997) Tornado risk model for transmission line design. *Journal of Wind Engineering and Industrial Aerodynamics* **72(1–3)**: 469–478.

Moon A, Wilson R and Flentje P (2005) *Developing and Using Landslide Size Frequency Models*. Faculty of Engineering Papers, No. 384. See http://ro.uow.edu.au/engpapers/384/ (accessed 01/06/2016).

NPDP (National Performance of Dams Program) (2015) http://npdp.stanford.edu/ (accessed 01/06/2016).

NRC (US National Research Council) (1983) *Safety of Existing Dams: Evaluation and Improvement*. National Academies Press, Washington, DC, USA.

Ohlmacher G and Davis J (2003) Using multiple logistic regression and GIS technology to predict landslide hazard in northeast Kansas, USA. *Engineering Geology* **69(3–4)**: 331–343.

Ott RA, Lee MA, Putman WE, Mason OK, Worum GT and Burns DN (2001) *Bank Erosion and Large Woody Debris Recruitment along the Tanana River, Interior Alaska*. Alaska Department of Natural Resources, Division of Forestry, Juneau, Alaska.

Perham RE (1988) *Elements of Floating-debris Control Systems*. US Army Corps of Engineers, Washington, DC, USA.

Preisler HK, Brillinger D, Burgan R and Benoit J (2004) Probability based models for estimation of wildfire risk. *International Journal of Wildland Fire* **13(2)**: 133–142.

Preisler HK, Westerling AL, Gebert KM, Munoz-Arriola F and Holmes TP (2011) Spatially explicit forecasts of large wildland fire probability and suppression costs for California. *International Journal of Wildland Fire* **20(4)**: 508–517.

Quinn PE, Hutchinson DJ, Diederichs MS and Rowe RK (2010) Regional-scale landslide susceptibility mapping using the weights of evidence method: an example applied to linear infrastructure. *Canadian Geotechnical Journal* **47(8)**: 905–927.

R2 Resource Consultants (2003) *Baker River Project Relicensing: Large Woody Debris Budget*. Puget Sound Energy, Inc., Belevue, WA, USA.

Ramsdell JV and Rishel JP (2007) *Tornado Climatology of the Contiguous United States*. PNNL-15112, Rev. 2. US Nuclear Regulatory Commission, Washington, DC, USA.

Reed W and McKelvey K (2002) Power-law behaviour and parametric models for the size-distribution of forest fires. *Ecological Modelling* **150(3)**: 239–254.

Rickenmann D (1997) Schwemmholz und hochwasser. *Wasser, Energie, Luft* **89**: 115–119.

Ricotta C, Avena G and Marchetti M (1999) The flaming sandpile: self-organized criticality and wildfires. *Ecological Modelling* **119(1)**: 73–77.

Rimböck A (2002) *Schwemmholzrückhalt in Wildbaechen*. Grundlagen zu Planung und Berechnung von Seilnetzsperren, Berichte des Lehrstuhls und der Versuchsanstalt für Wasserbau und Wasserwirtschaft, Techniche Universitat Muenchen, Munich, Germany.

Ripley BD (2004) *Spatial Statistics*. Wiley-Interscience, New York, NY, USA.

Rogers A (1974) *Statistical Analysis of Spatial Dispersion: The Quadrat Method*. Pion, London, UK.

Rosén A (1995) Hållfasthetaprovning av träd, Vattenfall Utreckling AB (in Sweden).

Ruijin Z and Xie J (1993) *Sedimentation Research in China. Systematic Selections*. Wuhan University, Beijing, China.

Rundqvist J (2005) *Debris in Reservoirs and Rivers – Dam Safety Aspects*. CEATI International, Montreal, Canada.

Shen HT, Bjedov G, Daly SF and Wasantha Lal AM (1991) *Numerical Model for Forecasting Ice Conditions on the Ohio River*. US Army Corps of Engineers, Hanover, NH, USA.

Simons DB, Andrew JW, Li RM and Alawady MA (1979) *Connecticut River Streambank Erosion Study, Massachusetts, New Hampshire and Vermont*. US Army Corps of Engineers, Waltham, MA, USA.

Sinha NK, Timco GW and Frederking R (1987) *Recent Advances in Ice Mechanics in Canada*. NRCC 28452. National Research Council Canada, Ottawa, Canada.

Song W, Fan W, Wang B and Zhou J (2001) Self-organized criticality of forest fire in China. *Ecological Modelling* **145(1)**: 61–68.

Stark CP and Guzzetti F (2009) Landslide rupture and the probability distribution of mobilized debris volumes. *Journal of Geophysical Research – Earth Surface* **114**: F00A2.

Stephens SL, Moghaddas JJ, Edminster C *et al.* (2009) Fire treatment effects on vegetation structure, fuels, and potential fire severity in western US forests RID A-8762-2009. *Ecological Applications* **19(2)**: 305–320.

Thompson MP and Calkin DE (2011) Uncertainty and risk in wildland fire management: a review. *Journal of Environmental Management* **92(8)**: 1895–1909.

Tuthill AM (2002) *Ice-affected Components of Locks and Dams*. US Army Corps of Engineers, Hanover, NH, USA.

Twisdale L and Dunn W (1983) Probabilistic analysis of tornado wind risks. *Journal of Structural Engineering – ASCE* **109(2)**: 468–488.

Uchiogi T, Shima J, Tajima H and Ishikawa Y (1996) Design methods for wood-debris entrapment. Garmisch-Partenkirchen **5**: 279–288.

*Union-Bulletin* (2013) Fires cause shutdown of Ice Harbor Dam power house. *Union-Bulletin*, Walla Walla, WA, USA.

USACE (US Army Corps of Engineers) (2002) *Ice Engineering Manual*. USACE, Hanover NH, USA.

USCOLD (United States Committee on Large Dams) (1975) *Lessons from Dam Incidents, USA*. American Society of Civil Engineers, New York, NY, USA.

USNRC (2012) *Practical Implementation Guidelines for SSHAC Level 3 and 4 Hazard Studies (NUREG-2117, Revision 1)*. US Nuclear Regulatory Commission, Washington DC.

Van Asch T, Malet JP, Van Beek L and Amitrano D (2007) Techniques, advances, problems and issues in numerical modelling of landslide hazard. *Bulletin de la Société Géologique de France* **178(2)**: 65–88.

Van Asch T (2009) The landslide problem, presentation to the Royal Dutch Geographical Society Education Day, 6 November 2009, Nijmegen, the Netherlands.

Vansickle J and Gregory SV (1990) Modeling inputs of large woody debris to streams from falling trees. *Canadian Journal of Forest Research (Revue Canadienne De Recherche Forestiere)* **20(10)**: 1593–1601.

van Westen CJ, Rengers N and Soeters R (2003) Use of geomorphological information in indirect landslide susceptibility assessment RID A-4043-2010. *Natural Hazards* **30(3)**: 399–419.

Wallerstein NP and Thorne CR (2004) Influence of large woody debris on morphological evolution of incised, sand-bed channels. *Geomorphology* **57(1–2)**: 53–73.

Wallerstein NP, Thorne CR and Abt SR (1997) *Debris Control at Hydraulic Structures in Selected Areas of the United States and Europe*. US Army Corps of Engineers Development & Standardization Group, London, UK, 109pp.

Wei Y, Rideout D and Kirsch A (2008) An optimization model for locating fuel treatments across a landscape to reduce expected fire losses. *Canadian Journal of Forest Research* **38(4)**: 868–877.

Wold HOA (1948) On stationary point processes and Markov chains. *Skandinavisk Aktuarietidskrift* **31**: 229–240.

WPDC (International Water Power and Dam Construction) (2012) Dam safety – dam safe. *Water Power and Dam Construction*, 30 June, p. 44.

Zollinger F (1983) *Die Vorgänge in einem Geschiebeablagerungsplatz* (Processes in a sediment retention basin). PhD thesis, ETH Zürich, Zurich, Switzerland.

Part III

# Engineered systems of flow control

**Operational Safety of Dams and Reservoirs**
ISBN 978-0-7277-6121-7


http://dx.doi.org/10.1680/osdr.61217.223

# Chapter 8
# Release function and human factors

The discharge in free overflow (ungated) spillways in the operating phase is normally only a function of water level(s), while the influence of physical dimensions is determined by design and construction (but these may, of course, be influenced by erosion, sedimentation, ice or floating debris). This *passive control* is generally a significant advantage for reliability.

When greater flexibility of operation is required, *movable water barriers*, such as gates, valves or turbines, are used to provide *active control* of the discharge rate. Provision of movable water barriers is (almost) compulsory for low-level or bottom outlets, and is also very common for surface spillways.

The greater flexibility of operation provided by movable water barriers makes it possible to regulate either the upstream water level or the water conduit discharge in a narrow band. At the same time the variations downstream will be more accentuated. The price to pay for the introduction of a movable water barrier is a significant reduction in spillway function reliability.

## 8.1. Gate systems

Gates are a critical part of dam, spillway and hydropower plant operations as they are the only control outlet that can prevent overtopping and failure of the dam or uncontrollable release of the water causing problems downstream. These systems are composed of an integral number of structural, mechanical and electrical components that have to function on demand. Unfortunately, most of these demands occur when it is critical to open the gates to release water before or during a flood event, leaving the potential for catastrophic failure of the spillway system. However, there are also a number of events for gated spillway where the demands made during 'sunny day' operations resulted in an unexpected release of water downstream, creating a potential effect on public safety downstream of the spillway.

These gated systems are generally designed to a set of defined engineering standards. However, the effects of ageing, exposure, inadequate preventive maintenance, changing operating regimes, resulting in a significant increase in gate operation, and lack of frequent operations, in combination with human error, seem to make these systems more vulnerable than one would think. Lewin *et al.* (2003) discuss that most new spillway gate systems are designed to reach a target probability failure on demand of anywhere from $10^{-3}$ to $10^{-7}$. These probabilities of on-demand failures are also supported by the design of spillway gates, as discussed by Klohn (1994), van Leperen (1984) and Lewin (2001a,b). From a number of gate studies, Lewin *et al.* (2003) concluded that the design of new gate systems should be aiming toward the target probability of one failure in every 10 000 demands ($10^{-4}$). However, Lewin *et al.* also noted that many existing gate systems currently exhibit a probability of failure of one in every 10 ($10^{-1}$) to one in every 100 ($10^{-2}$) demands. These on-demand failures are complex, and may be caused by a gate component that can be repaired in minutes to hours, or a component that may cause complete failure of the gate system and unexpected release of the reservoir containment.

The poor performance of well-designed gated spillways has always been a concern for owners but has come more into the dam-safety limelight since the failure of gate No. 3 at Folsom Dam in California in July 1995 (Todd, 1999). Structural gate failure of the gate occurred during a 'sunny day' raising of the gates and caused an uncontrollable release of 1133 $m^3/s$ downstream into the American River. While no one was injured during this event, this is one of the keystone events that led dam owners to examine more closely the design, maintenance and operation of their gated spillway systems. Other recent events, such as the Lake Delhi failure (Fiedler *et al.*, 2013) in Iowa, have again increased the concern about performance problems of dams caused by gate systems.

There are a number of readily available excellent publications on the reliability of spillway gates, such as Hobbs and Azavedo (2000), Patev and Putcha (2005), Barker *et al.* (2003) and New South Wales Dam Safety Committee (2010). This chapter presents information on the historical failures of dams and summarises the different methodologies available for estimating the probability of on-demand gated spillway systems function.

## 8.2. Spillway accidents

Most work in dam safety risk analysis addresses the reliability of system components and their interactions under random and usually extreme loadings. In practice, many or most accidents involving flow-control systems do not originate in component (un-)reliabilities or component failures under random loadings. Rather, they are caused by chains of events involving not only technical system components but also control functions and communications. In gated spillway operation, the release function, unlike the conveyance function, involves active control by a human being, and therefore adds an important dimension to the associated risks.

### 8.2.1 Folsom Dam, 1995

Folsom Dam is a concrete gravity structure designed and constructed by the US Army Corps of Engineers (USACE). Construction was begun in 1948 and the dam was put into service under US Bureau of Reclamation (USBR) management in 1956. The dam, together with its reservoir, Folsom Lake, is a component of the Central Valley Project in California, which provides flood control, hydropower and irrigation water to California's Sacramento and San Joaquin Valleys.

The original spillway works consist of five Tainter-type service gates, each 50 ft (15 m) high by 42 ft (13 m) wide, and three emergency gates of the same width but 53 ft (16 m) high. On 17 July 1995, the gate No. 3 structure failed due to overstressing, releasing nearly 40% of the water in Folsom Lake before the gate could be repaired (Figure 8.1). Temporary closure of the stoplogs at the time of failure was not possible because the bracketing for such an emergency closure had not be designed into the spillway system. About 1132 $m^3/s$ (40 000 $ft^3/s$) flowed through the broken gate. The full capacity of the channel downstream of the dam is approximately 3500 $m^3/s$ (125 000 $ft^3/s$), so this was a significant discharge. The repair cost about USD 20 million (1995 price).

The after-incident report conducted by the USBR attributed the failure to a design flaw in the gate structure that allowed stresses to concentrate in one of the strut arms of the gate (Figure 8.2). While that design situation certainty contributed to the failure occurrence, the chain of events leading to the failure was more complex (Todd, 2009) (Box 8.1).

Figure 8.1 Failure of gate No. 3, Folsom Dam, CA, USA, July 1995. (Courtesy of USBR)

Figure 8.2 Tainter gate section of Folsom Dam, CA, USA, showing the location of the strut failure due to overloading. (Courtesy of USBR)

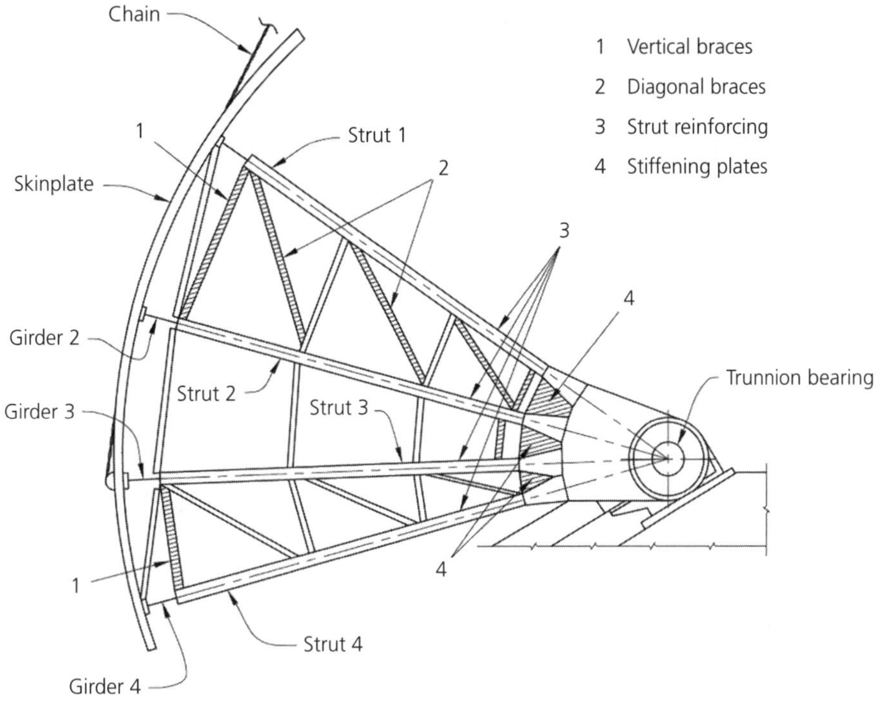


Box 8.1 Chain of events preceding the Folsom Dam gate failure in 1995

1. About 8:00 a.m. on 17 July 1995 gate No. 3 was being opened to maintain flow in the river during a power-plant shutdown. The operator attempted to open the gates to effect a routine release. This involved starting the electrical hoist motors, which in turn engaged the lift chains and cables.
2. Normally, the gate is allowed to stop at 6 in. and 1 ft automatically. The auto-stop function was overridden at the 2 ft level, as was the normal procedure.
3. The trunnions (at least at gate No. 3) were inadequately lubricated and inadequately protected from water and the weather, and thus were corroded (inadequate maintenance).
4. When the attempt to open the radial gate was made, friction at the trunnion pin was transferred as bending moment into the gate arms. 'As the gate opening approached 2.4 ft, the gate operator felt an "unusual vibration" and he stopped the gate hoist motor. As the operator turned to check the gate, he saw the right side of the gate swing open slowly, like a door hinged on the left side, and saw water pouring around both sides of the gate leaf. The time from the operator's initial awareness of the vibration to observing gate displacement and uncontrolled flow of water was estimated to be no more than 5 s.'
5. Design flaws made the gate increasingly susceptible to failure (a large-diameter pin, the size of the radial gates and the mechanics of the pin friction).
6. The gates had not been inspected and exercised on a regular schedule.
7. The gate sensors did not indicate a problem (there was no sensor for strut bending).
8. The operator was not able to stop the gate once trouble was suspected.
9. Stoplogs could not be installed because provisions had not been made for them in the original design and construction, which was completed in the early 1950s. Stoplog slots were incorporated in hastily fabricated steel frames for installation on the upstream side of the spillway piers. Ultimately, frames were fabricated and installed on all eight spillway bays. Although the frame then provided a means for closing off the gate bay and isolating the failed gate, it now disrupts flow going to the gates and has increased the size of the vortices that form in front of the gates when they are operating at partial openings.

Figure 8.3 Folsom Dam, CA, USA – failure of right side of gate. (Todd, 2009)

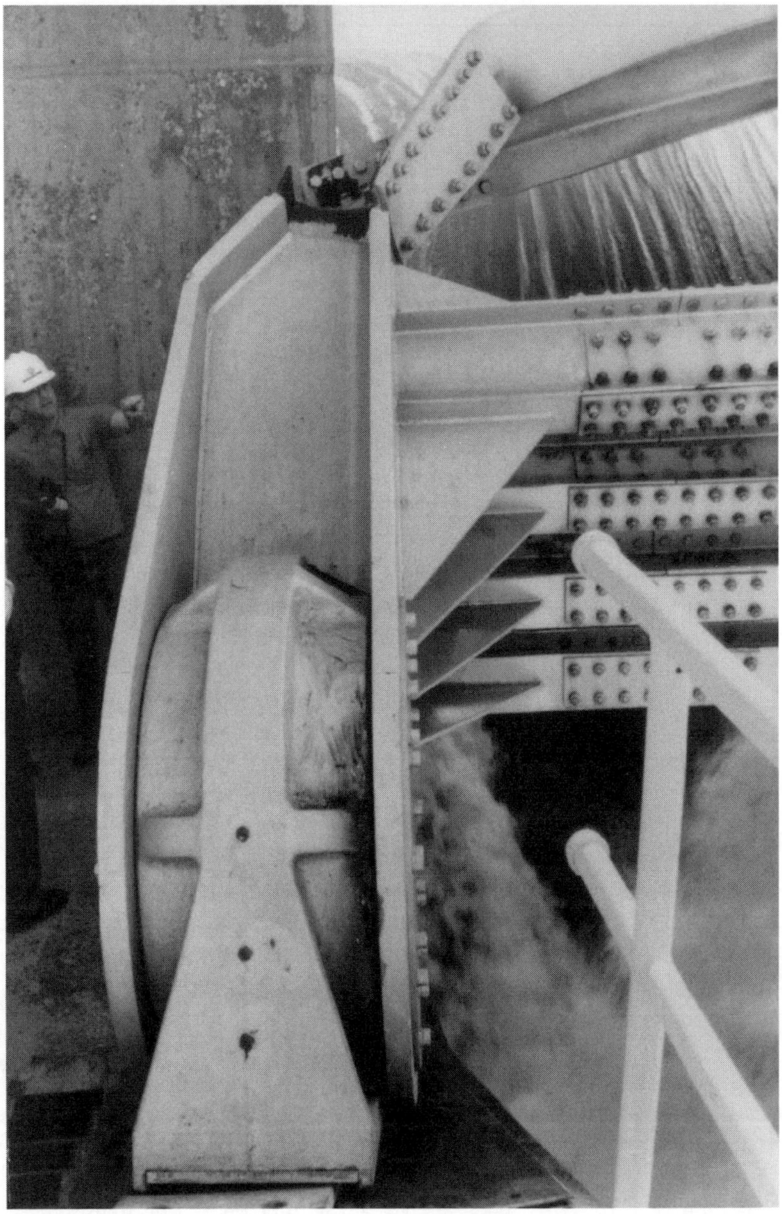

The gates rotate on 32 in. (0.82 m) brass and steel trunnion pins. These trunnions were inadequately lubricated and began to corrode. This increased friction on the bearings and transferred load to the struts. The gates are raised by a high-torque electric motor, which engages hoisting chains attached to each side of the gate structure. The capacity of the motor allowed for the continued lifting of the gate despite the resistance of the bearings. This caused bending stresses in the gate structure, which was designed principally to resist axial stress. At some point the bending stresses exceeded the capacity of the struts and the structure buckled.

The investigation team determined that the failure loading was due to trunnion friction moment, which increased over time due to corrosion of the pins. This had been overlooked in the initial design, and consequently the strut bracing and struts had not been sized adequately. With the failure of the joint connecting the first diagonal brace on the right side of the gate, the two lower struts failed around their weak axes (Figure 8.3). A more detailed explanation of the failure and information on the various reports and tests appear in the forensic report (Todd, 2009).

Like many accidents or failures, this one involved what Reason (Reason, 1997; USBR, 2010) calls latent conditions and active errors. Latent conditions are those aspects of the system such as design errors or inadequate maintenance or flawed operating rules that develop over time and may be present long before the accident or failure occurs. Active errors are those aspects of the accident or failure that involve human actions immediately preceding or during the failure.

Figure 8.4 Delhi Dam, OH, USA: (a) failed vertical lift gate No. 3; (b) dam failure. (Courtesy of USACE)

(a)

(b)

### 8.2.2 Delhi Dam, 2010

Delhi Dam in Delaware County, Ohio, USA, was a low (18 m) earth embankment dam with a concrete spillway and ogee gate section. The dam was built between 1922 and 1929 by the Interstate Power Company as a hydropower dam. Ownership was transferred in 1973 to the local riparian home-owners' association, and the dam ceased to produce power. In 2008, an agreement was entered into between the home-owners' association and the hydropower producer, Modern Hydro Inc., to refurbish the turbines and decommission the dam as a power producer under US Federal Energy Regulatory Commission (FERC) regulation. The new power plant was to have two turbines operating at 1.5 MW capacity.

The southern embankment of the Delhi Dam failed on 24 July 2010, the failure being associated with about 250 mm (10 in.) of rainfall in 12 hours (Figure 8.4). Before the breach, river levels upstream of the dam had reached 7.38 m (24.22 ft), 3 m (10 ft) above flood stage and breaking the May 2004 record of 6.60 m (21.66 ft). However, the stream flow at the time was 40 $m^3/s$ (1400 $ft^3/s$) less than the 2004 record peak stream flow of 740 $m^3/s$ (26 000 $ft^3/s$) (Box 8.2).

The dam had three vertical lift gates, but only two of these could be opened during the high pool elevation, and the embankment subsequently overtopped, leading to a 60 m long section of the embankment washing away. The downstream towns of

Box 8.2 Chain of events preceding the Delhi Dam gate failure in 2010

**Thursday 22 July 2010**

04:45 Delaware County Emergency is aware of heavy rain since midnight.

**Friday 23 July 2010**

00:15 David Fink and Mike Russell (dam operators for the Lake Delhi Recreation Association) discuss lowering Lake Delhi.

01:00 David Fink notifies Brenda Leonard (Jones County Emergency Manager) of the first gate opening at 7% (meaning 7% of the total opening of all three spillway gates).

07:00 David Fink notifies Brenda Leonard that gate opening has been changed to 15%.

08:30 David Fink notifies Brenda Leonard that gate opening has been changed to 20%.

10:00 David Fink notifies Brenda Leonard that gate opening has been changed to 33% (gate No. 2 fully open).

11:40 David Fink reports reservoir to be at 121.0 ft (local datum); successful lowering of the reservoir by 0.8 ft.

12:45 David Fink reports reservoir to be at 120.7 ft.

14:30 David Fink notifies Brenda Leonard that gate opening has been changed to 40%.

15:05 David Fink notifies Brenda Leonard that gate opening has been changed to 50%.

17:30 David Fink notifies Brenda Leonard that gate opening has been changed to 66% (gate Nos 1 and 2 now fully open); reservoir at 122.0 ft (local datum).

20:00 Mike Ryan (Delaware County Emergency Management Coordinator) receives notice from the weather service that 11–14 in. of rain is predicted over the next 48 hours.

20:30 David Fink reports reservoir to be at 122.17 ft and that gate No. 3 is wedged at 6 ft open.

22:30 Mike Ryan reports heavy rain in Manchester. David Fink reports 3–3.25 in. of rain at Delhi.

23:00 David Fink reports reservoir to be at 122.9 ft (local datum).

**Saturday 24 July 2010**

01:00 Mike Ryan orders sandbags to be sent to Delhi.

02:15 Weather service notifies Mike Ryan of a forecast for three more hours of heavy rain.

03:30 David Fink reports on his inspection of Delhi Dam: sagging in the upstream crest chain-link fence; 4 ft diameter whirlpool and 5–6 in. vortex located 40–50 ft south of the spillway section; no water on the downstream side as far as he can see; reservoir estimated to be around 124–125 ft (local datum).

03:40 David Fink reports to Mike Ryan: 3 ft of freeboard at the dam.

03:45 David Fink reports to Brenda Leonard: dam issues developing and could be big. Mark Stoneking (Monticello Fire Chief) notified.

04:00 T. Gorman, a resident on the river less than a mile downstream of the dam, called by a neighbour and told to evacuate immediately.

04:30 Brenda Leonard picks up Mark Stoneking and they drive to the dam, arriving at 05:00 h.

05:30 David Fink asks for riprap, and Anthony Bardgett (County Engineer) calls for rock and a loader. Brenda Leonard notes: a 6 ft long, 3 ft wide hole above the waterline at the upstream shoulder in line with the steep section of the embankment, fairly close to the dam structure; soil is falling into the hole above the waterline, water appears to be coming up from the bottom and all sand pieces are sliding in; the bottom of the hole is filled with water and sandy material; she didn't get a look at the downstream face of dam; there is no evidence of a bump, dip or hole in the road.

05:45 David Fink calls Mike Ryan: there is a hole in the north embankment.

06:00 David Fink issues a potential dam-failure alert to Mike Ryan. Road settlement is occurring, but not in line with the whirlpool; water is discharging from the riprap 40–50 ft south of the wall on the downstream side, above the toe and near some trees; the flow is clear, exiting 5–10 ft (120–125 ft local datum) down from the crest; the reservoir is at 125–126 ft; water is running in the new riprap above the berm area; no water is passing over the dam. Anthony Bardgett arrives at the dam site on the south side and notes dipping of the road, first in the northbound lane (downstream side, nearer to the white work tent, initially 6–8 in. dip) and then in the southbound lane (upstream side, further south in the roadway from the whirlpool); he does not want to send equipment across the road for safety reasons. T. Gorman, the resident, is at the dam at daylight, and reports a 6–8 ft diameter sinkhole in the upstream slope (about 100 ft from the concrete spillway, 15 ft up from the lake and 6–8 ft from the road) and a dip in the road in the area left (north) of the whirlpool

06:10 Anthony Bardgett calls Mike Ryan: delivery of the riprap and the loader has been confirmed and the equipment will be in place in 45 minutes. A 2 ft by 4 ft hole with caving has been spotted above the water line. David Fink calls Mike Ryan to tell him to contact/alert the public about potential dam failure; firemen are asked to go door to door to inform residents; National Weather Service is asked to put into action a dam watch; the Public Information Office is asked to alert the media; the riprap will arrive soon; the situation is dangerous. The city administrations at Delhi and Hopkinton are contacted.

Box 8.2 Continued

| | |
|---|---|
| 06:15 | Anthony Bardgett: there is a whirlpool between the embankment and the lake association work barge, a dip in the pavement in line with the whirlpool and the work tent, and surface drainage on the berm surface. Tension cracks have appeared in the parking lot area between the pavement and the work tent; the cracks are parallel to the road and 2 in. at their widest. A general swirl (not a deep vortex, circular motion) is observed in the water on the upstream side of the embankment; the water level is 1–2 ft below the road surface at this point; the berm on the upstream side is still above the water level; water is seeping through the area near the waterline and the toe, around trees/bushes; there is a small seepage of water through part of the slope. |
| 06:20 | The National Weather Service confirms the dam watch. |
| 06:25 | David Fink calls Mike Ryan: roadbed subsidence has started (1–2 in.); does not want to risk putting an end loader on the road; it is too dangerous to dump riprap in the hole. Mike Ryan moves to the Emergency Operations Centre. |
| 06:30 | National Weather Service dam watch is confirmed to Mike Ryan. Anthony Bardgett: nothing of note is exiting downstream of the trees, a lot of water is coming through; minor seepage at the downstream slope. Morris Wruck (Delaware County Supervisor): water is running out of the riprap 40–50 ft south of the wall. |
| 07:30 | Anthony Bardgett moves to the north side. T. Gorman notes road subsidence in line with the sinkhole in the downstream direction; subsidence is noticed when the county equipment is arriving on site. |
| 08:00 | David Fink: the reservoir is approximately 3 ft below the crest of the dam (126.8–127 ft on visual inspection); there is a dip in the road at the same location as the fence sag and whirlpool and a hole in the north embankment at the culvert; immediately along the fish ladder, water has started going over the road and toppling trees. Scott Kinseth (Iowa State Department of Natural Resources (DNR) Conservation Officer) arrives at the dam; he notes that the water is high but not over the dam. Morris Wruck reports 3 ft of freeboard at the embankment section. |
| 08:15 | Jon Garton at the State DNR is notified. Report of 1.5 ft of freeboard at the dam; the road has settled 6–12 in. |
| 08:30 | Scott Kinseth: water is percolating through the grass at the downstream edge of the northern concrete non-overflow section; scour has started on the north side, left of the powerhouse. |
| 09:00 | David Fink: dirty flow at the toe of the downstream slope, 5–10 ft (horizontal) from the bottom of the fish ladder, orange in colour, exiting over a rock outcrop at the toe of the training wall; flow in the riprap has diminished. David Fink sends out a full warning to the Delhi, Ryan and Hopkinton Fire Departments and Jones County Emergency management Agency; a whirlpool 50–75 ft from the spillway training wall, 3–4 ft downslope from the guardrail, 8–9 ft from the edge of the pavement; cloudy/dirty water coming/blowing out of the bank/riprap on the downstream side near the bushes/trees; seepage to the left and up higher, and running water at other locations; issues mandatory evacuation downstream; water not over roadway; running water exiting above the tail water; a hole has appeared on the north side of the upstream face, 3 ft below the dam. Jon Garton calls Chip Hughes (Lake Delhi Recreation Association): reservoir 6 in. below the crest and rising. |
| 09:30 | Dirty water at the toe of the short section, exiting over the rock outcrop at the toe; road dip is well defined; water exiting above the tail water; a hole has appeared on the north embankment upstream face; water is 3–3.5 ft higher than normal (around 127 ft local datum). |
| 09:40 | Call to Chip Hughes (LDRA): 1–2 in. of water over the road; reservoir is rising 1 ft/h; 1.5 miles upstream Hartwick bridge deck is under 4–5 ft of water. David Fink calls Brenda Leonard: water is over the road at the narrow part of the levee. David Fink orders everyone to evacuate. |
| 10:00 | The downstream shoulder is starting to develop a hole, widening within 10 ft of the concrete wall/structure, not far from the wall; water is going over the roadway – a minor sheet of water/flow 10 ft wide (not where the road has dipped), stopped and started; an area right next to the concrete section is being overtopped, 10 ft; at the overtopped area there is minor sheet flow, south of the settled area; the shoulder area is eroded at the downstream end of the sheet flow; water is coming through the north side of the downstream wall and bubbling up through the fill; water is coming through cracks in the north downstream wall, water is bubbling up through the ground near the wall at the dam location near the curve on the north downstream side. |
| 10:37 | Chip Hughes calls Scott Kinset: there is water over the road. |
| 11:30 | Anthony Bardgett calls Mike Ryan: water is leaving the south side of dam and flowing through the roadbed; the roadbed is going to washout. |
| 12:00 | Antony Bardgett observes a lot of water going around the edge of the parking lot area, over road and maybe under road as well (near/around dipped area); there is heavy flow over the top of the dam and underneath the road. |
| 12:10 | Tom McCarthy (an environmental specialist with the Iowa DNR): water has been flowing over the road since 10:30 h and has trenched through the road; water is no more than 1.5 ft over road; has arrived at Hartwick Marina. |
| 12:15 | Mike Ryan and Anthony Bardgett: roadbed has collapsed. |
| 12:22 | Report to Jon Garton from Chip Hughes: erosion has stalled around 50% road washout, 100–150 ft south of the spillway. Tom McCarthey reports 50% road washout. |

Box 8.2 Continued

| | |
|---|---|
| 12:30 | David Fink: water has started going over the roadway, has washed into the trees and is eroding a large area of trees downstream; 15–30 ft trees are toppling; 25% breach; roadway has caved in, beginning at the dam and moving wider. Tom McCarthy at Delhi Dam: water has trenched through the road; the asphalt surface downstream is chipping off, midway within the tree section. |
| 12:45 | Antony Bardgett: the crest road is 75% breached. |
| 12:55 | The entire width of road over a 8–12 ft section has dropped gradually, and the drop is progressing to the south; an asphalt/pavement section of the road/embankment has dropped 6–7 in., and has started to crumble. It seems the first section to go was close to the dam/concrete wall, but it is not certain exactly where this started. The drop is propagating and continuing to the other side along the embankment section by section. |
| 13:00 | Tom McCarthy: the remainder of the embankment and the upstream chain-link fence collapses (*Authors' note – This is the first indication of a complete upstream to downstream breach.*) |
| 13:02 | Tom McCarthy reports a 60 ft wide breach through the roadway. |

Hopkinton and Monticello were evacuated, affecting some 8000 residents. In Monticello, 50 homes and 20 businesses were destroyed, leading to several tens of millions of dollars' worth of damage. The city's sewage treatment plant was flooded, leaving residents without services.

In December of the same year, a consulting panel of engineers released a report concluding that the cause of the failure 'was internal erosion in the embankment coupled with overtopping flow'. The study identified malfunctioning spillway gates and increase reservoir pool level as exacerbating an existing problem with the dam's design. In April 2014, ground was broken on a \$16 million project to replace the failed dam.

No adverse performance of the dam was reported before the event, and no significant seepage had been reported at the downstream toe or on the downstream face of the dam. Although the embankment performed well up to the recent event, it is very possible that prior loadings did not achieve a water-surface elevation that exceeded the top of the core wall (elevation 274 m, (898.8 ft)) or were not of sufficient duration to develop internal erosion.

The spillway gates had been difficult to operate in the past. The gate guides are tapered at the bottom, and sometimes the gates would stick in the closed position or at small gate openings. A crane had been used in previous floods to operate the spillway gates. The lack of maintenance of the embankment section immediately south of the spillway and the 2H : 1V downstream slope made inspection of the dam for seepage flows difficult.

During the 22–24 July flood, gate No. 3 could only be opened 6 ft, which meant there was a significant reduction in the spillway capacity. Gate No. 3 was never opened any more than 1.30 m (4.25 ft) during the entire 24 July 2010 flood.

Nothing out of the ordinary with regard to the dam performance was observed during the 22–24 July flood, until the reservoir water surface exceeded the top of the core wall, at an elevation of 898.8 ft. Within about 8 hours of this occurring, a vortex in the reservoir and sinkholes in the upper portion of the upstream face of the dam were observed. The first vortex was noticed about 40–50 ft south of the concrete structure; the second was noticed later and was estimated to be about 100 ft south of the concrete structure. Seepage from the downstream slope was first observed around 06:00 h on 24 July, 40–50 ft south of the spillway training wall. At 06:00 h settlement of the dam crest was observed in the areas where the vortices and sinkholes were first observed. All this evidence is consistent with internal erosion occurring in the portion of the embankment above the top of and downstream of the concrete core wall (see Box 8.2).

The dam breach began to accelerate around 12:30 h on Saturday 24 July. The breach was caused by internal erosion of the embankment, flows over the embankment and structural failure of the thin concrete core wall. The breach of the embankment dam occurred at about 13:00 h on 24 July. The concrete core wall appeared to have failed due to differential loading caused by the flood and erosion of downstream embankment soils. As erosion of embankment soils continued, sections of the core wall also toppled, ceasing at a maximum breach width of 235 ft. It is likely that the concrete core wall slowed down the rate at which the embankment dam was breached.

The breach of Delhi Dam did not cause any loss of life. This is attributed to several factors: the concrete core wall probably slowed down the rate of the dam breach; the warning of dam failure was sent several hours before the breach; the flood wave was dissipated in farm fields, which reduced the level of flooding in the downstream communities of Hopkinton and Monticello; door-to-door warnings were issued in Hopkinton and Monticello, and residents whose homes would have been inundated were evacuated.

Table 8.1 Survey of gate failure incidents. (Hobbs and Azavedo, 2000)

| Category | Recorded incidents (60 total) | Contribution: % |
|---|---|---|
| Human factors | Maintenance related (10)<br>Human error during gate operation (3) | 22 |
| Loss of motive power | Loss of mains (5)<br>Motor failure (2)<br>Failure of back-up power (2) | 14 |
| Control and communication | Failure of limit switches (7)<br>Position indicators (3)<br>Controls failure (3) | 22 |
| Gates | Rollers or bearing failure (4)<br>Rope failure (3)<br>Miscellaneous (3) | 17 |
| Hoist units | Screw thread nut failure/problems (2)<br>Mechanical failure of hoist equipment (3) | 8 |
| Miscellaneous | De-icing/heater failure (4)<br>Operation manuals (1)<br>Concrete movement/deterioration (2)<br>Trunnion anchor inspection problems (1)<br>Float-well design/control issue (1)<br>Truck loses load, damaging hoist units (1) | 17 |

### 8.2.3 Statistical history of gate failure incidents

An international survey of dam owners undertaken by Hobbs and Azavedo (2000) reported the 'causes' of gate accidents and failures shown in Table 8.1.

Flow-control in hydropower and related dam systems is a complex system involving civil, mechanical, electrical, communications and control components. These systems components interact in complex ways, and failures of flow-control systems involve not only reliability failings of physical components but also the interactions of these components with communications and control, and with human actions and errors. An analysis of the performance of the system needs to incorporate all these aspects.

## 8.3. Gated spillways

Gated spillways contain a complex integration of structural, mechanical and electrical (SME) components that must operate on demand. There is a wide variety of such systems with differing SME components that have been designed all over the world. USACE (1992) states that the inclusion of crest gates allows the spillway crest to be placed significantly below the maximum operating reservoir level, in turn permitting the entire reservoir to be used for normal operating purposes. This results in a much narrower spillway facility than an uncontrolled spillway avoiding the problems associated with high unit discharge/high-velocity flow and increased operation and maintenance costs. USACE states that a gated spillway must include, as a minimum, two or preferably three spillway gates in order to satisfy safety concerns.

The American Society of Civil Engineers (ASCE) defines the function of a water control gate as being to control the flow of water through a conduit or opening. ASCE (2012) states that these gates may be used to perform one or more of the following functions: regulating, flood control, emergency closure or maintenance closure. In his survey of USACE flood-control projects, Schultz (2013) differentiated gates as high flow (service, spillway and maintenance gates), low flow (gates, valves and bulkheads) and powerhouse (gates and valves). The reader is referred to a number of publications that define the different types of gate and hoist systems used for spillway gate systems (ASCE, 2012; USACE 1992, 1997, 2000).

### 8.3.1 Types of spillway gate

As stated earlier, a number of different types of gates are used in spillway projects to control the flow of water. Table 8.2 lists typical uses of the most common gate types. However, the primary control gates used in spillway crests around the world are the vertical lift gate (wheeled or slotted) and Tainter or radial gates. This is due to their reliability and simplicity of design. Smaller and older structures often use simple stoplog gates (Figure 8.5).

Table 8.2 Gate types and common uses (after ASCE 2012)

| Type of gate | Regulating | Flood control | Emergency closure | Maintenance/dewatering |
|---|---|---|---|---|
| Bulkhead/stoplog | | | ✓ | ✓ |
| Drum | ✓ | ✓ | | |
| Flap (hinged crest) | ✓ | ✓ | | |
| Flashboards | | ✓ | | |
| Wicket (navigation dam) | ✓ | | | |
| Floating bulkhead (caisson) | | | | ✓ |
| Inflatable | ✓ | ✓ | | |
| Radial (Tainter) | ✓ | ✓ | | |
| Roller dam | ✓ | ✓ | | |
| Wheel mounted: | | | | |
| ■ Fixed wheel | ✓ | ✓ | ✓ | ✓ |
| ■ Coaster/caterpillar | | | ✓ | ✓ |
| ■ Ring seal/paradox (bonneted) | ✓ | | | |
| ■ Stoney | | | | ✓ |
| Slide: | | | | |
| ■ Unbonneted | ✓ | | ✓ | ✓ |
| ■ Bonneted | ✓ | | ✓ | |
| ■ Jet flow (bonneted) | ✓ | | | |
| ■ Ring follower (bonneted) | | | ✓ | ✓ |
| ■ Sluice | ✓ | ✓ | ✓ | ✓ |
| ■ Shear | ✓ | | | |
| Other types: | | | | |
| ■ Bear trap | | ✓ | | |
| ■ Clamshell | ✓ | | | |
| ■ Cylinder | | ✓ | ✓ | |
| ■ Ring gate (morning glory outlet) | ✓ | ✓ | | |
| ■ Fuse | | ✓ | | |
| ■ Flap gate | | ✓ | | |

Figure 8.5 Stoplog gates at Whatshan Dam. Whatshan Dam was built by the British Columbia Power Commission in 1952; it is a concrete dam on the Whatshan River. (Courtesy of BC Hydro)

Figure 8.6 Vertical lift gates, Keswick Dam, Sacramento River, CA, USA. (Courtesy of USBR)

Vertical lift gates have a roller-type support system that transfers the hydraulic load from the gate to the sides of the water passage (Figure 8.6). The rollers minimise the friction forces when moving the gate, thereby keeping the operating-mechanism capacity as low as possible. These gates can be used in any range of head for spillway or intake gates. Wheel-mounted gates can generally be operated under full or partial differential head, and are used for emergency closure, flood control and, sometimes, flow regulation.

USACE (1992) considers Tainter gates to be the most economical, and usually the most suitable, type of gate for controlled spillways due to their simplicity, light weight and low hoist-capacity requirements. A Tainter gate is a segment of a cylinder mounted on radial arms that rotate on trunnions anchored to the piers. Spillway flow is regulated by raising or lowering the gate to adjust the discharge under the gate. Figure 8.7 shows the configuration of Tainter gates with wire rope and hydraulic hoist systems.

USACE (2000) states that the advantages and disadvantages of Tainter gates are as follows:

> Tainter gates (Figure 8.8) have several unique advantages compared to other spillway gate types (lift gates, roller gates, hinged or flap gates).
>
> 1. The radial shape provides efficient transfer of hydrostatic loads through the trunnion.
> 2. A lower hoist capacity is required.
> 3. Tainter gates have a relatively fast operating speed and can be operated efficiently.
> 4. Side seals are used, so gate slots are not required. This reduces problems associated with cavitation, debris collection, and build up of ice.
> 5. Tainter gate geometry provides favourable hydraulic discharge characteristics.
>
> Disadvantages include the following:
>
> 1. To accommodate location of the trunnion, the pier and foundation will likely be longer in the downstream direction than would be necessary for vertical gates. The hoist arrangement may result in taller piers especially when a wire rope hoist system is used. (Gates with hydraulic cylinder hoists generally require shorter piers than gates with wire rope hoists.) Larger piers increase cost due to more required concrete and will usually result in a less favourable seismic resistance due to greater height and mass.
> 2. End frame members may encroach on water passage. This is more critical with inclined end frames.

**Figure 8.7** Tainter Gates with (a) a wire-rope configuration and (b) a hydraulic cylinder configuration

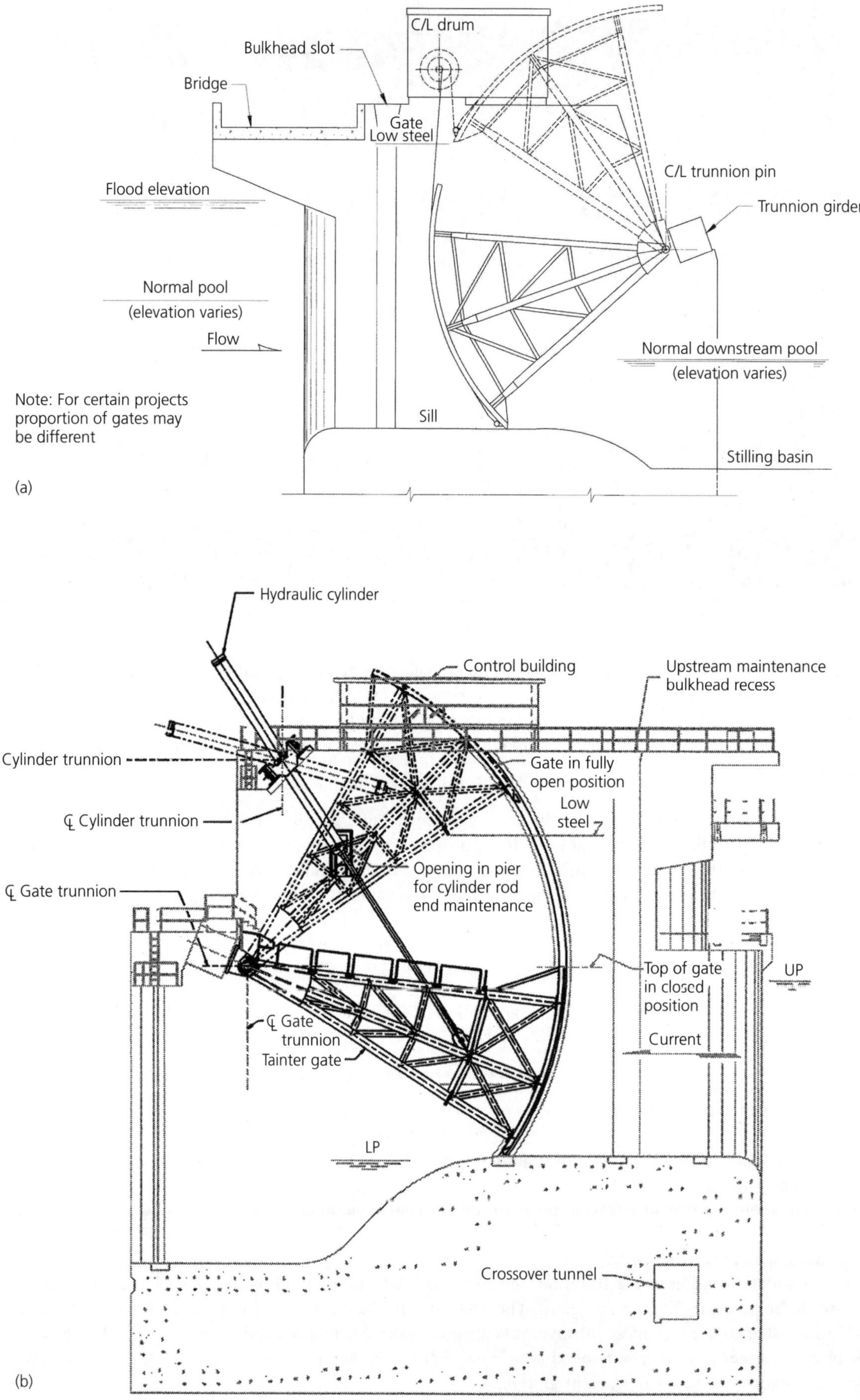

Figure 8.8 Spillway gates at Thurmond Dam, Savannah River, GA, USA. (Courtesy of USACE)

### 8.3.2 Inventory of gate systems

This section highlights the different types of gate and hoist system that exist in the portfolio of the inventory of one of this book's sponsors.

Schultz (2013) undertook an inventory of 295 of the 328 USACE flood-control dams with the aim of determining the types of gates, valves and bulkheads in current use at USACE flood-control structures. Schultz also collected data on the time to repair and failure data for a wide variety of components in the fault trees used for these projects. The gates types and their associated hoist systems are summarised in Table 8.3.

### 8.3.3 Historical performance of gated spillways

Recent failures of gated systems such as Folsom Dam in California (see Figure 8.3) and Delhi Dam in Iowa (see Figure 8.4) show that these complex SME systems have numerous flaws due to limited or poor maintenance, inadequate design or lack of operation, which result, in addition to human error, in the potential for catastrophic failure of the systems.

There are a number of surveys reported in the literature on the historical failures of gate systems and their resulting consequences (Erbisti, 2004; Lewin, 2001a; Lewin *et al.*, 2003). New South Wales Dam Safety Committee (2010) gives the details of six dam projects where gate failures were the primary cause of significant events. These data (adapted from Hobbs and Azavedo, 2000) are shown in Table 8.4.

Hobbs (2003) also inventoried dam owners from all over the world to get a perspective on incidents involving dam gates. He recorded over 60 incidents of gate problems; a summary of these data is given in Table 8.5. It is interesting to note that the percentage of human factors causing dam-gate incidents is nearly 1 in 4.

National Performance of Dams Program (NPDP, 2015) has reported data from the USA on gate incidents prior to the Folsom Dam failure in 1995 (Table 8.6). The incident data are broken down into failure modes and by the consequences of the incidents (Figure 8.9).

## 8.4. Fault-tree analysis

The fault-tree approach is a deductive process whereby an undesired state (usually the state that is critical from the reliability and safety point of view) of the system is postulated, and the possible ways in which the undesired event (top event) can occur are systematically deduced in the context of the environment and operation of the system. Indeed, unlike inductive approaches, in which one starts with causes and traces up to the top event, in the fault-tree method one starts with the top event and traces backwards to the primary causes.

The fault tree is a graphical representation of various parallel and sequential combinations of failures that result in the occurrence of the top event. The faults can include events related to hardware components, software errors, human errors or any other relevant events that can lead to the undesired event.

**Table 8.3** Inventory of USACE flood-control projects (Schultz, 2013)

| | Hoist system | No. |
|---|---|---|
| **Service gates (total gates = 936)** | | |
| Fixed-wheel roller gates | Chain | 0 |
| | Hydraulic | 37 |
| | Wire ropes | 59 |
| | Electric screw | 9 |
| Caterpillar/tractor/roller train | Chain | 0 |
| | Hydraulic | 2 |
| | Wire ropes | 112 |
| | Electric screw | 0 |
| Sluice gates | Hydraulic | 18 |
| | Electric screw | 20 |
| | Manual screw | 1 |
| Radial/Tainter | Hydraulic | 6 |
| **Spillway gates (total gates = 784)** | | |
| Radial/Tainter gates | Chain | 190 |
| | Hydraulic | 16 |
| | Wire Rope | 508 |
| Fixed-wheel roller gate | Chain | 16 |
| | Wire rope | 44 |
| Slide gate | Electric screw | 0 |
| Bladder gate | Air operated | 10 |
| **Emergency gates (total gates = 378)** | | |
| Fixed-wheel roller gates | Chain | 5 |
| | Hydraulic | 19 |
| | Wire ropes | 73 |
| | Electric screw | 0 |
| Caterpillar/tractor/roller train | Chain | 1 |
| | Hydraulic | 3 |
| | Wire ropes | 26 |
| | Electric screw | 0 |
| Slide gates | Hydraulic | 527 |
| | Electric screw | 145 |
| Sluice gates | Hydraulic | 237 |
| | Electric screw | 14 |
| **Maintenance bulkhead/gates (total gates = 149)** | | |
| Fixed-wheel roller gates | Chain | 29 |
| | Hydraulic | 0 |
| | Pneumatic | 4 |
| | Wire ropes | 73 |
| Slide gate | Hydraulic | 237 |
| | Wire rope | 39 |

It is important to point out that a fault tree does not necessarily involve either all possible models of the system failure or all possible causes of the system failure. Therefore, only those failures contributing to the occurrence of the top event are modelled. One of the important applications of fault-tree analysis is the identification of vulnerable areas of a system, which helps to prevent the occurrence of the top event.

A fault tree is a logical representation of basic events; it represents qualitative aspects of the model and can also be evaluated quantitatively. Qualitative evaluation of a fault tree is performed by the transformation of the fault-tree logic into an equivalent form that is obtained based on minimal cut sets. Each cut set includes basic events leading to the top event. Quantitative

Table 8.4 Examples of dam gate failures (Hobbs and Azavedo, 2000)

| Dam name | 1 | 2 | 3 | 4 | 5 | Result |
|---|---|---|---|---|---|---|
| Belci, Romania | Heavy silting of dam | Poor operating instructions for extreme events | Main power failure | Gate motors flooded, due to poor location, and failed to operate | Manual opening failed due to gate blockage | Dam overtopped and failed; 17 deaths |
| Mavici, Slovenia | Condensed moisture on contact terminals | Redundant relay activated | | | | Gates opened unexpectedly; flood wave downstream |
| Ord River Diversion, Australia | Gate transmitter failure | Control contractor failure | | | | Gates failed to open; dam overtopped |
| Machhu II India | Under capacity spillway | Main power failure | | | | Delayed opening of gates; dam failed; hundreds of deaths |
| Folsom, USA | Reduced bearing maintenance | Reduced structural maintenance | Poor gate design | | | Gate failed; uncontrolled discharge |
| Tarbela, Pakistan | High ambient temperature caused gate expansion | Dramatically reduced seal clamp to side plate clearance | No slack rope switch provided on gates | | | Gate jammed and then dropped; snagged ropes and damaged gates |

evaluation of a fault tree consists of specifying the probability of the top event based on probability of the basic failure events. There are many quantitative algorithms that can be used to evaluate fault trees. Typically, fault trees are quantified by calculating the probability of each minimal cut set and summing the probabilities of all the cut sets. In addition to the probability of the top event, the importance of events can also be extracted from fault-tree quantification. This allows the required actions and resources to be prioritised. For example, the events that have negligible influence on the top event can be ignored.

The outcome of a fault tree is a success/failure (binary) event, and as success and failure are related, each can be transformed into the other. In fact, the success tree is the complement of the fault tree. Thus, the top event of the success tree is the complement of the fault tree.

Table 8.5 Examples of dam gate failures in the USA (Hobbs, 2003)

| Dam name | State | Year | Event description |
|---|---|---|---|
| Bonneville | Oregon | 1945 | Vibration damage to spillway gates |
| Wanapum | Washington | 1964 | Spillway gate hoist failed to stop |
| Belden | California | 1966 | Gate frame collapsed. The gate controls were inadvertently interchanged; thus, when throttled releases went through the unvented gate, the steel liner collapsed and the concrete around the gate frame collapsed |
| Amisted | Texas | 1976 | During a gate operation, the upstream roller intake gate rose, ultimately breaking a suspension cable |
| Milks Camp | South Dakota | 1976 | A hydraulic gate jammed in the open position, allowing water to drain continuously |
| Coolidge | Arizona | 1978 | Drum gates failed to operate as a result of not being used in 45 years |
| Jackson Lake | Colorado | 1984 | A spillway gate broke; one-half of gate leaf missing |
| Hatfield | Wisconsin | 1993 | Tainter gate hoist chain failure |
| Warm Springs | California | 1995 | Failure of a bulkhead gate hoist system during normal operations, resulting in dropping of gate |

Table 8.6 Summary of spillway gate failure modes and incidents

| Failure mode | Incident |
|---|---|
| Spillway gates fail to open/close when directed. Loss of electrical power, undersized motors, failure of automated control systems, corrosion of wire ropes, rope connections, etc. | ■ San Teresa Dam, Spain – automatic controls<br>■ Picote Dam, Portugal – hoist chain<br>■ Tous Dam, Spain – power outage, no backup<br>■ Upriver Dam, USA – lightning affected generators |
| Spillway gates open accidentally through failure of automation equipment or unexpected occurrence | ■ Maveice Dam, Slovenia – hydraulic valve<br>■ Seton Dam, Canada – freezing water in electrical conduit |
| Spillway gates opened intentionally during a major flood to prevent overtopping | ■ Sabaneta Dam, Dominican Republic<br>■ Jebba and Shiriro Dams, Nigeria<br>■ Penitas Dam, Mexico |
| Spillway gates fail structurally due to deficiency in design or lack of maintenance | ■ Guernesy Dam, USA – plugged drain lines; loss of buoyancy<br>■ Singar Dam, India – trunnion girder<br>■ Folsom Dam, USA – trunnion friction, maintenance, design deficiency<br>■ Cresta Dam, USA – floatation |
| Debris blockage of spillway gates impedes outflow, leading to damage to gates or guides | ■ Chute Garneau Dam, Canada<br>■ Mud Mountain, USA |
| Spillway gates operated incorrectly | ■ Human error and judgement – too many to list! |

The success tree introduces the path sets. Path sets are the minimal sets of basic events that need to be inhibited to prevent the occurrence of the top event. An interesting analytical point is that there are several advantages of failure space overriding success space. First, finding failure factors is generally easier than finding contributors to success. In other words, success associated with factors such as the degree of efficiency, the amount of production and marketing characteristics, which are typically described by continuous parameters, cannot be easily defined by discrete events such as 'valve does not work'. Therefore, it is easier to work in the failure space than the success space. Second, the system failure is less probable than the success, which makes the failure-based calculations more efficient. Finally, the small value of the failure probabilities (usually less than 0.1) allows for approximation, resulting in a reduction in the calculation costs.

Figure 8.9 Gate incident data by failure mode and consequence. (NPDP, 2015)

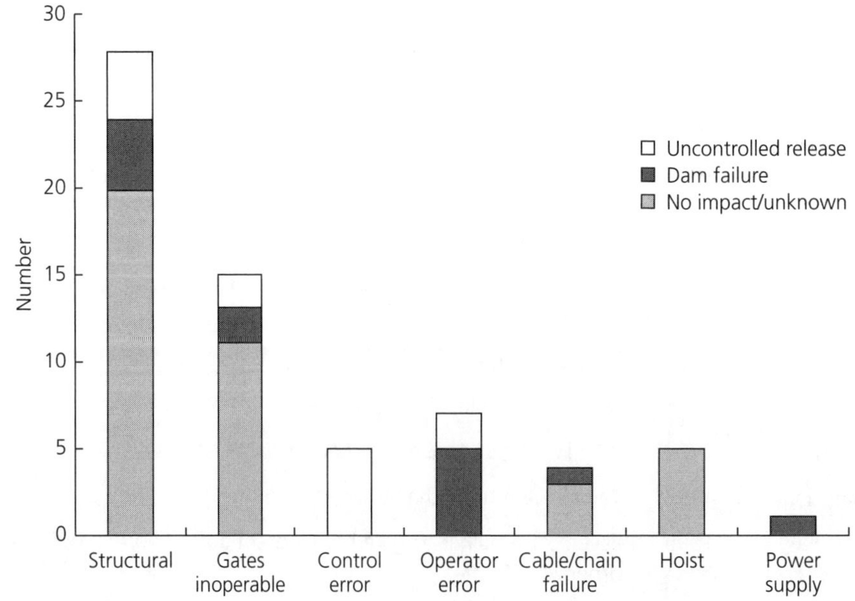

### 8.4.1 Fault-tree analytics

Many different methods can be used to model a fault tree, but the most popular way leading to successful fault-tree analysis (FTA) can be summarised in a few steps:

1. Identify the purpose of the FTA.
2. Obtain an understanding of the FTA.
3. Construct the FTA.
4. Evaluate the fault tree.
5. Interpret and present the results.

The first step in a successful FTA is to define an objective for the FTA. Although this may seem obvious, there are many cases under which FTA is not performed well due to dissatisfaction with the objective of the decision-maker. To undertake a successful FTA, the purpose of the system should be defined in terms of the failure of the system.

Once the objective of the system has been identified, the decision-maker should obtain an understanding of the whole system in terms of determining the top event, defining the scope of the analysis, defining the resolution of the FTA and defining some ground rules for the FTA. The scope of the system suggests which of the contributors will be included in the FTA. In addition, it includes the special design and boundary conditions for the analysis. The resolution of the system indicates the level of detail to which the failure causes will be developed. This development is often below the level where data are available. The ground rules are procedures whereby events and gates are named that are very important in understanding the fault tree. For example, to be consistent, modelling different individual factors in the fault tree – such as failure of a particular component, human errors and common cause failures – can be done using ground rules.

Step 3 involves the construction of the fault tree. The logic of constructing the fault tree along with system schematics is described in the subsequent sections of this chapter.

Step 4 is the evaluation of the fault tree, which includes both the quantitative and qualitative evaluation.

The last step emphasises the interpretation of the fault tree in terms of providing potential impacts based on the objectives.

### 8.4.2 Fault-tree building blocks

Three main symbols are employed in an event tree: events, logical gates and transfers. An event symbol is used for primary events that are not developed further on the fault tree, and for intermediate events found at the output of a logical gate. The function of a logical gate is to either permit or prevent the passage of the fault logic up to the top, and the logical gate symbol reveals the type of relationship that input events require for the output event. The function of a logical gate is derived from Boolean logic symbols. Transfer symbols are used to connect the inputs and outputs of relevant trees, connecting the fault tree of a subsystem to its system. Figure 8.10 shows a typical fault tree composed of a number of symbols. Standard fault tree symbols can be found in the US Nuclear Regulatory Commission (USNRC) guidance document, NUREG-0492 (Vesely *et al.*, 1981). These failure events are defined in Table 8.7.

### 8.4.3 Component fault categories: primary, secondary and command

The classification of faults into primary, secondary and command faults can be helpful for better analysis of the system. A primary fault occurs as a result of the system's malfunction while the environmental factors are fully controlled. For example, a defective weld could cause a pressure tank, designed to tolerate the pressure up to point P0, to rupture at some pressure less than P0.

A secondary fault is not caused by the system's malfunction, but results from environmental factors that negatively affect the system. That is, the component fails under unplanned conditions, such as excessive vibration, higher than expected temperature, etc.

In contrast to primary and secondary faults, which are cause by component failure, a command fault is attributed to the proper operation of the component but at the wrong time or in the wrong place (e.g. loss of signal or control power causes a component failure).

These categories often serve as a checklist to ensure the completeness of a fault tree in the sense of covering different types of fault.

### 8.4.4 Passive versus active components

In most cases it is useful to divide categories of components into passive and active. A passive component contributes to the functioning of the system.

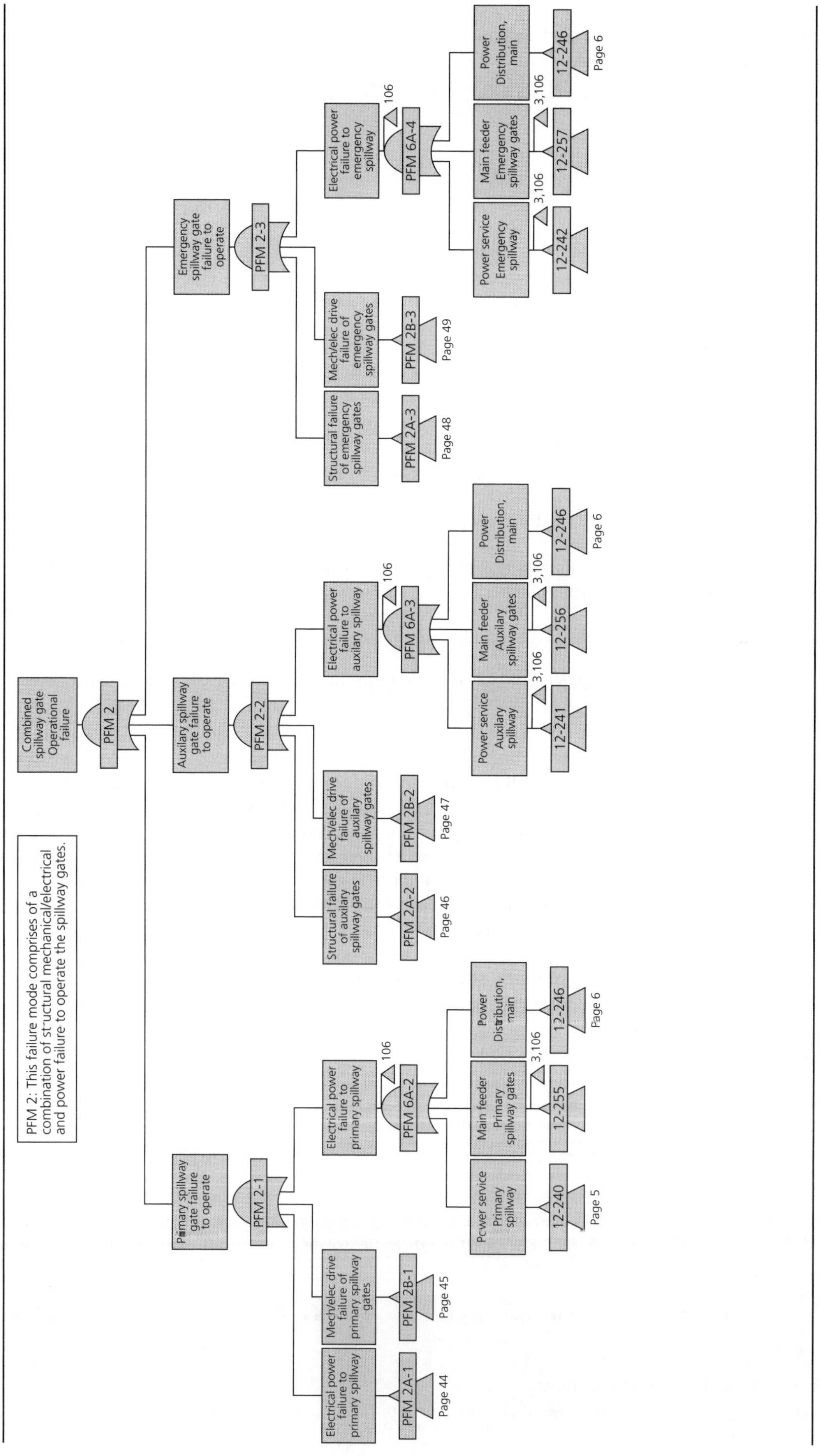


**Figure 8.10** Fault tree for a pumping system

This component may operate as a transmitter of energy or load from place to place (e.g. current or force; or a mechanism, such as a pipe, in which the output of one active component becomes the input for a second active component). To evaluate the operation of a passive component, tests such as stress analysis and heat-transfer research may be conducted. The failure of the passive component will cause the loss of signal transmission.

An active component contributes to the operation of the parent system and can modify the performance of the system. Generally, the function of an active component depends on its input signal (e.g. a valve or a switch the opening or closing operations of which can improve the system's behaviour). The failure of an active component might result in either an incorrect signal or no signal. To assess an active component, a study of the operating parameters is required.

From the reliability perspective, the most important difference between failure of an active component and failure of a passive component is the difference in their failure rates, which is often about two or three orders of magnitude.

## 8.5. Human reliability analysis for spillway systems

Spillway systems are composed of a variety of mechanical, electrical and structural components that have to function and interact together as a system to control the inflows and outflows of water into a reservoir. All the components in a spillway system are required to function successfully under a wide variety of natural hazards such as floods, icing, debris and earthquakes. Figure 8.11 shows a typical USACE spillway system at Gavin Point in Yankton, South Dakota, USA.

Dam projects have a number of beneficial purposes, including flood control, hydropower, navigation, recreation and water supply (public water or irrigation). Many of these projects are designed to be multi-purpose, such that they provide many benefits to society. However, spillway systems are inherently complex, and human error plays a critical role in the success or failure of spillway systems. Human reliability analysis (HRA) will be an important part of the overall risk-analysis process.

In this section we discuss various HRA methods that are currently available and could be adapted for use in the risk analysis of spillway systems. We outline human error and performance-influencing factors or performance-shaping factors, HRA methodologies and methods for estimating the probability of human errors. Recommendations are made with regard to methodologies that are applicable when undertaking a HRA for a spillway system.

### 8.5.1 Overview of spillway systems

A typical spillway system can be composed of the following subsystems:

- spillway gates, weir (controlling the overflow), discharge chute and energy dissipator
- hydropower facilities (penstock, turbines and outlet works)
- low-level outlet works.

Figure 8.11 Spillway system of John Day Dam on the Columbia River, WA, USA

Figure 8.12 Example of a spillway system

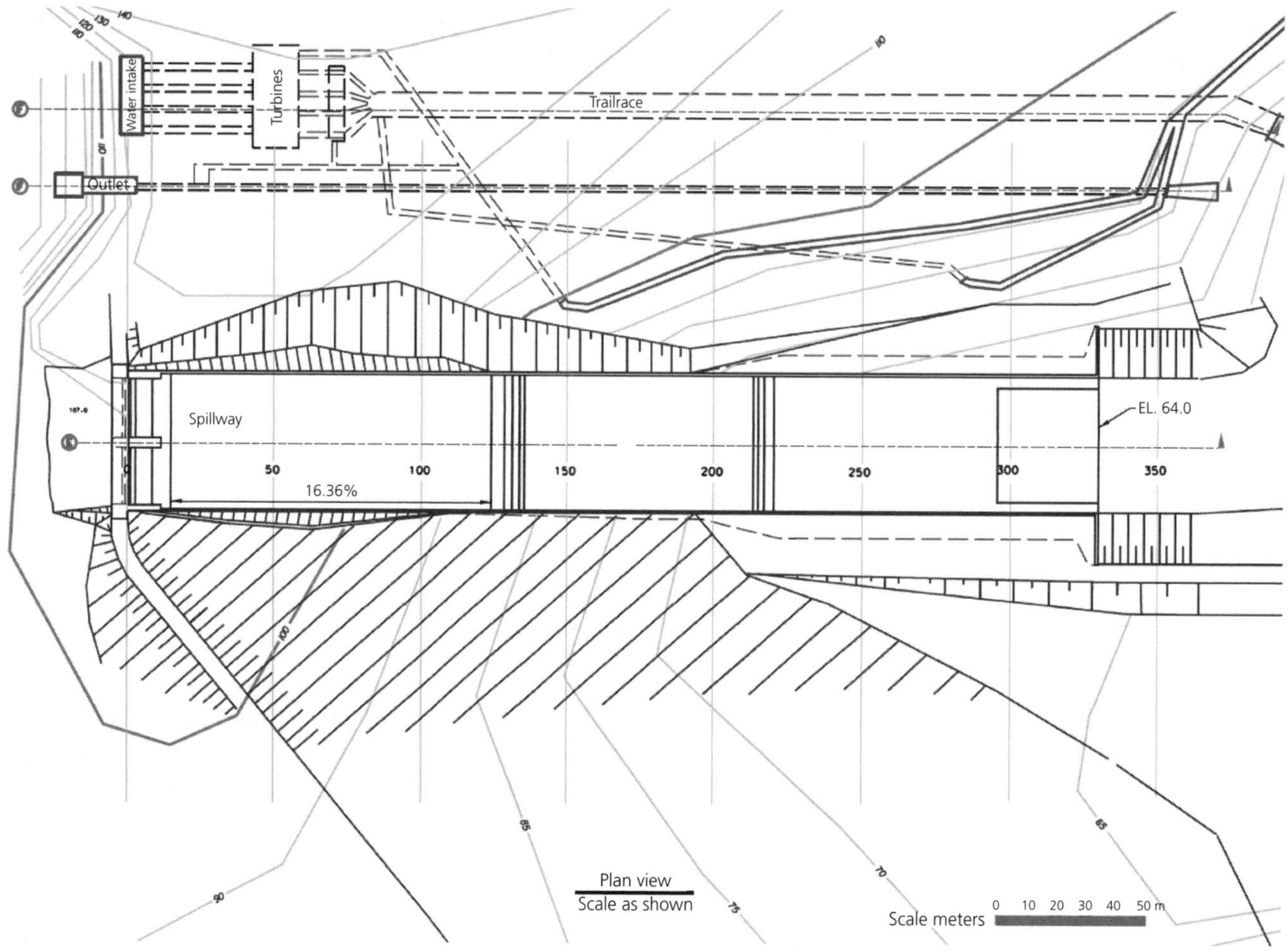


Each of these subsystems has SME components that function separately from each other. However, the subsystems are highly dependent on each other for the overall successful operation of the spillway project, as they have to be operated in concert with one other. For example, if the inflows to the spillway are greater than the capacity of the hydropower facility, the spillway gates will have to be opened to allow the additional inflow to be discharged down the spillway. Another example is the use of the spillway as a safety valve in cases when the power plant unexpectedly shuts down. A certain reliability level is required for the gate operation in a scheme (typically run-of-the-river schemes) where the reservoir is small and the water is rising rapidly. A plan view of an example system is shown in Figure 8.12.

The primary human interaction with the spillway comes into play throughout the daily operations process. The system is modelled using a watershed–reservoir–spillway model (International Commission on Large Dams (ICOLD)) and is shown in Figure 8.13. Inflows are read using a computer system that is composed of river gauges upstream of the reservoir, gauges within the reservoir and gauges at the spillway. Operating rules are written for the spillway to programmatically decide what the sequence of the subsystems will be in order to permit the safe passage of the outflows downstream of the spillway. The primary goal of the system is to continue the generation of power at the maximum output possible without having to exercise the spillways gates. However, if inflows exceed the capacity, the spillway gates are opened, releasing water down the discharge chute. The outflows from the spillway gates have to be monitored carefully to ensure that the hydraulic jumps are controlled and cavitation on the spillway slope does not cause a failure of the spillway chute.

Humans are tasked with making the decisions and implementing actions to operate the spillway system in a safe and efficient manner. This involves both cognitive and physical responses that could lead directly to a human failure event. Depending on the event sequence, the human error could either create consequences ranging from minor spillway erosion, complete shutdown of the hydropower facility or even catastrophic failure of the spillway system. It is important to incorporate in the system model a full understanding of human errors in spillway systems and to develop a proper methodology for the HRA, as discussed later in this chapter.

Figure 8.13 Watershed–reservoir–spillway model. (Hartford and Baecher, 2004)

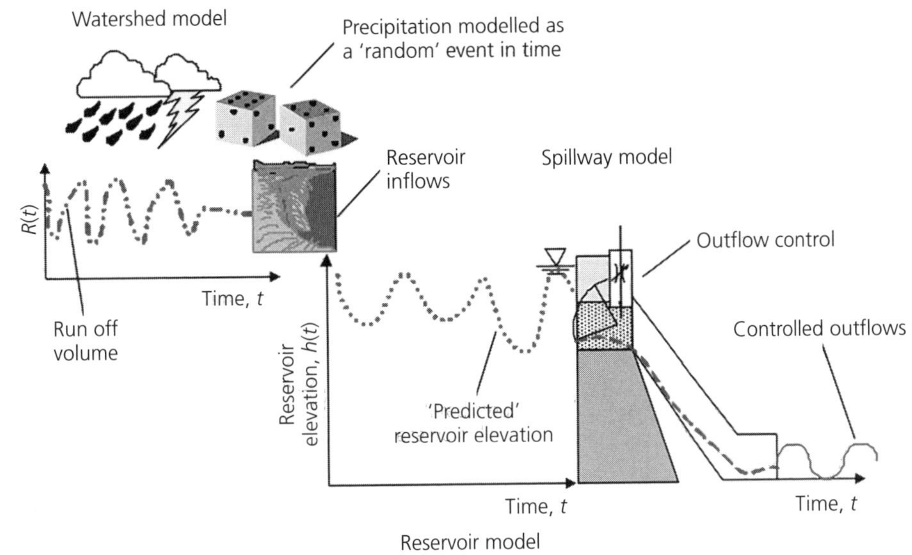


### 8.5.2 Human reliability analysis

Human error has been classified by Reason (1990) into three levels: behavioural, contextual and conceptual. Reason concludes that the first two types include the different errors that can result from the same process, but the conceptual level leads to understanding the causal mechanisms using more theory than observation. Therefore, the conceptual level is more robust in terms of understanding human error in spillway systems.

Some examples of human error are skill-, rule- and knowledge-based errors, intentional and unintentional errors, errors of omission (failure to take actions) and errors of commission (actions that should not be taken). For spillway systems, many of the human errors occur during the operations phase, but they also occur in design deficiencies, maintenance practices or strategies, lack of updated safety manuals and upper management decisions regarding such systems.

A method for identifying human error in spillway processes has been developed by Boeing for the US National Aeronautics and Space Administration (NASA) (Broughton *et al.*, 1999). The methodology is called the Human Factor Process Failure Modes and Effects Analysis (HF PFMEA). This process is qualitative in nature but identifies the behavioural functions, potential human errors and performance-shaping factors for the task that is being investigated. This process is similar to a typical failure modes and effects analysis (FMEA) but also accounts for the frequency of human error and the consequences of such errors. HF PFMEA would be a good screening process to develop a risk matrix to assist with identifying which human errors need to be defined better in the systems analysis.

A more robust procedure for examining human errors associated with spillway gates is the Information, Decision and Action in Crew (IDAC) model discussed in Chang and Mosleh (2007). IDAC is a dynamic simulation environment that models an operator's natural responses to a potential failure situation. The responses are influenced by the use of performance-influencing factors (PIFs) that are internal (cognitive, psychological, physical, etc.) or external (organisational, environmental, etc.) These higher level PIFs are shown in Figure 8.14.

Even though the application of the IDAC model is directed toward nuclear plant operations, the methodology is useful in identifying the critical human errors that could occur during the operation of a spillway during critical natural and man-made events. The operations of the two types of system are very similar and the IDAC process should be investigated and developed further for spillway system operations.

NASA (Chandler *et al.*, 2006) defines human reliability as: (1) the probability that the human elements will function as intended over a specific period of time under specified environmental conditions, and (2) the probability that no extraneous human actions detrimental to the system reliability or availability will be performed. Key to these definitions of human reliability are time and specified environmental conditions. These are important because past human failure events in spillway systems have been directly correlated with both of these performance factors (Dandan and Lei, 2011). Therefore, HRA is an important methodology that should be applied to spillway systems to account for the human errors that occur. The HRA process will need to be tied back into the system model that is developed as part of this process.

Figure 8.14 High-level representation of the IDAC performance-influencing factors. (Chang and Mosleh, 2007)

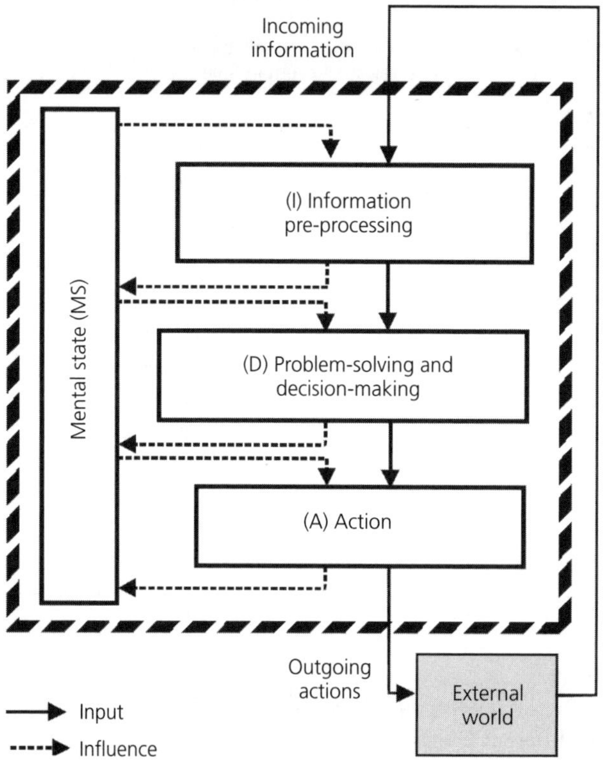


The HRA process established by NASA for use in their Exploration Systems Mission Directorate programme is composed of the following steps (Chandler *et al.*, 2006):

1 Problem definition – determine what type of analysis, tasks and human actions are considered.
2 Task analysis – a systematic process to identify, list and break down each task.
3 Error identification – what and where are the human errors and their consequences?
4 Error modelling – visualisation of the data and information obtained in the previous tasks using master logic diagrams, event trees or fault trees.
5 Quantification and integration in the probabilistic risk assessment (PRA) process (NASA, 2002) – assignment of human-error probabilities (HEPs) in the risk calculations and tying these to the PRA, event-tree analysis or fault-tree analysis.
6 Error management – minimisation of the errors of the potential human errors, and mitigation to increase overall system reliability.

This process can also account for the dependencies between human errors. NASA also recommends including in the HRA the assessment of errors of both omission and commission. The process takes into account all initiating actions, and considers both cognitive responses and physical actions. The process can also incorporate the HP PFEMA processed described previously. The HRA process is illustrated schematically in Figure 8.15.

### 8.5.3 Estimation of human-error probabilities

A significant number of HRA methods that have been developed over the past 25 years can be found in the literature. Many of these applications have been focused on nuclear power plant or the aerospace field, but their adaptation to other lines of engineering is very feasible. These HRA methods were developed either on the basis of predefined task analysis or by using expert elicitation methods. The methods also include the ability to include both error recovery as well as the task dependencies of subsequent tasks in the same event sequence.

A NASA report (Chandler *et al.*, 2006) documents many of the HRA methodologies that have been developed to estimate HEP. The report also gives an excellent comparison of these methods, and provides a selection of four methods that could be used for the NASA risk analysis programme: the Technique for Human Error Rate Prediction (THERP) (Swain and Guttmann 1983), the Cognitive Reliability and Error Analysis Method (CREAM) (Hollnagel, 1998), the Nuclear Action Reliability Assessment (NARA) (Kirwan 1994) and the Simplified Plant Analysis Risk Human Reliability Assessment (SPAR-H) (Gertman *et al.*, 2005). The report recommends that all four methods be deployed and their results documented to be able to select the most suitable programme for HEP estimation for future NASA endeavours.

Figure 8.15 The NASA HRA process. (Chandler *et al.*, 2006)

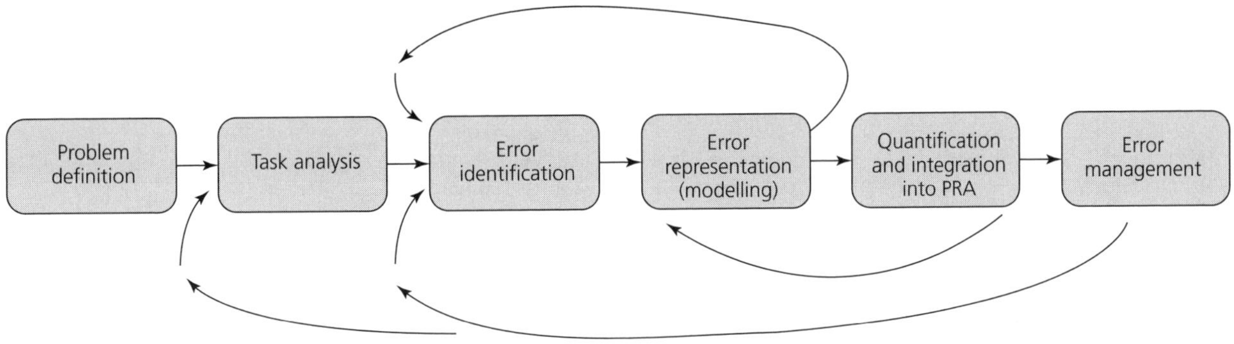


With regard to spillway systems, these HRA methods are very focused on nuclear power plant and would need significant adaptation before they could be applied to the operation of spillway subsystems.

Based on this NASA HRA study, there are two methods, THERP and ATHEANA (A Technique for Human Event Analysis), that are recommended for use in estimating HEPs for spillway HRAs. Both have their benefits and drawbacks and are suitable for application to certain parts of spillway systems. However, these methods could be adapted to provide reasonable estimates of HEPs.

The methodology of THERP (Swain and Guttmann, 1983) permits tasks to be studied during normal operating conditions as well as abnormal operating conditions that are important for spillway systems. The method also has the element of time and incorporates a time–reliability curve to modify the HEP based on the time it takes to perform a task or action. This HRA is task-analysis oriented and contains 27 tables of HEPs for a wide variety of internal and external performance-shaping factors (Box 8.3). The task actions defined in THERP would be more applicable to hydropower facilities than other substructures in a spillway project because the operation and staffing at such facilities are similar to nuclear power plant.

However, there are a few pitfalls to this methodology. THERP is a very time-consuming and costly process. It requires the development of HRA event trees that can sometimes be very complex, especially if there are a large number of events and possible recovery paths in the critical sequence. An example HRA event tree from THERP is shown in Figure 8.16. Also, the HEPs and error factors in the table have not been developed for new types of equipment that might be found at spillway projects, and extensive calibration of the HEPs would be required to check if they are within an order of magnitude. However, the methodology should be adaptable for use in the estimation of the HEPs for the hydropower facilities, as these are very similar in design and operation to nuclear power plant.

The other method recommended for use with the spillway gates and outlet works is ATHEANA (USNRC, 2000). This HRA technique is based on error-forcing contexts (EFCs), which are characterised into two groups: those errors that are directly characteristic of the initiation of the event sequence, and those errors that are characteristic of the system. The ATHEANA methodology uses expert elicitation as the cornerstone to developing HEPs. Expert judgement is also used to estimate the weighting factor for each of the EFCs and HEPs.

In particular, ATHEANA covers the following steps in its HRA process (USNRC, 2000): identify the human actions to be assessed; define human failure vents (HFEs) pertinent to performing these human actions incorrectly; and determine the HEPs for the defined HFEs, including consideration of likely recovery actions. The ATHEANA process is developed primarily for nuclear power plant, but is a framework that could be developed for other applications such as spillway systems.

ATHEANA follows a more formalised and structured approach and has a set documented process. The ATHEANA HRA process is shown in Figure 8.17. USNRC (2000) also states that this process helps to identify the following critical HRA and HEP information:

Identify operational vulnerabilities that could set up potentially unsafe actions (UAs) (e.g. procedure weaknesses and operator knowledge limitations and biases).

1. Identify plausible deviations from nominal conditions or plant evolutions that might cause problems or misunderstandings.
2. Identify important performance-shaping factors that are relevant to both nominal and deviation conditions.
3. Identify other aleatory factors that could significantly affect the likelihood of the HFEs and their uncertainties (i.e. investigating a broad range of potential influences).

Box 8.3 External and internal performance-shaping factors (PSFs) for THERP (Chandler *et al.*, 2006)

**External PSFs**

*Situational characteristics*

- Control room architectural feature
- Quality of the working environment
- Work hours and work breaks
- Shift rotation and night work
- Availability/adequacy of special equipment/tools and supplies
- Manning parameters
- Organisational structure and actions by others
- Rewards, recognition and benefits

*Tasks and equipment characteristics*

- Perceptual requirements
- Motor requirements
- Control–display relationships
- Anticipatory requirements
- Interpretation
- Decision-making
- Complexity/information load
- Frequency and repetitiveness
- Task criticality
- Long- and short-term memory
- Calculation requirements
- Feedback
- Dynamic versus step-by-step activities
- Team structure
- Man–machine interface structures

*Job and task instructions*

- Operating procedures
- Oral instructions

**Internal PSFs**

*Psychological stressors*

- Suddenness of onset
- Duration of stress
- Task speed
- Task load
- High jeopardy risk
- Threat of failure/loss of job
- Monotonous, degrading or meaningless work
- Long, uneventful vigilance periods
- Conflicts of motives about job performance
- Reinforcement absent or negative
- Sensory deprivation
- Distraction (noise, glare, movement, flicker, colour)
- Inconsistent

*Physiological stressors*

- Duration of stress
- Fatigue
- Pain or discomfort
- Hunger or thirst
- Temperature extremes
- Radiation
- G-force extremes
- Atmospheric insufficiency
- Vibration
- Movement constriction
- Lack of physical exercise
- Disruption of circadian rhythm

*Organisational factors*

- Previous training/experience
- State of current practice or skill
- Personality and attitudes
- Motivation and attitudes
- Knowledge of required performance standards
- Sex differences
- Physical conditions
- Attitudes based on influence of family and other outside persons or agencies
- Group identifications

Unfortunately, like THERP, there are drawbacks to this methodology, as it was not directly developed for spillway systems. The biggest problem with ATHEANA is that it is heavily reliant on expert opinion. The procedure is both time and labour intensive, and the results can be both hard to replicate and calibrate. However, USACE has had good experience using Expert Opinion Elicitation (EOE) procedures in risk assessments of their civil works infrastructure. USACE is currently assembling formal guidance for EOE procedures and processes (Patev, 2011). The ATHEANA process would, however, be considered a viable HRA methodology for spillway systems because these systems are not newly designed and the ability of experts to estimate probabilities should be realistic.

### 8.5.4 Recommended HRA applications for spillway systems

The incorporation of HRA into a system model for spillways is imperative to show how human error can cause failure of the entire system. The documentation of existing HRA methods is very important in understanding the human performance functions from both the cognitive and physical perspective. These performance functions are critical to developing a realistic framework that can be used to understand the human failure events and to estimate the HEPs for the system.

There are many HRA options available to analyse spillway systems, even though none have a direct application. The development of a new holistic HRA model is a critical task for the future of spillway risk assessment, but there are some robust options that can be used in the HRA. HP CFEMA and IDAC are two methods for developing the causal events and root causes, including the error modes (internal or external). THERP and ATHEANA can, with some modifications, easily be used to estimate the HEP for hydropower facilities and spillway systems.

Figure 8.16 Example of an HRA event tree. (Swain and Guttmann, 1983)

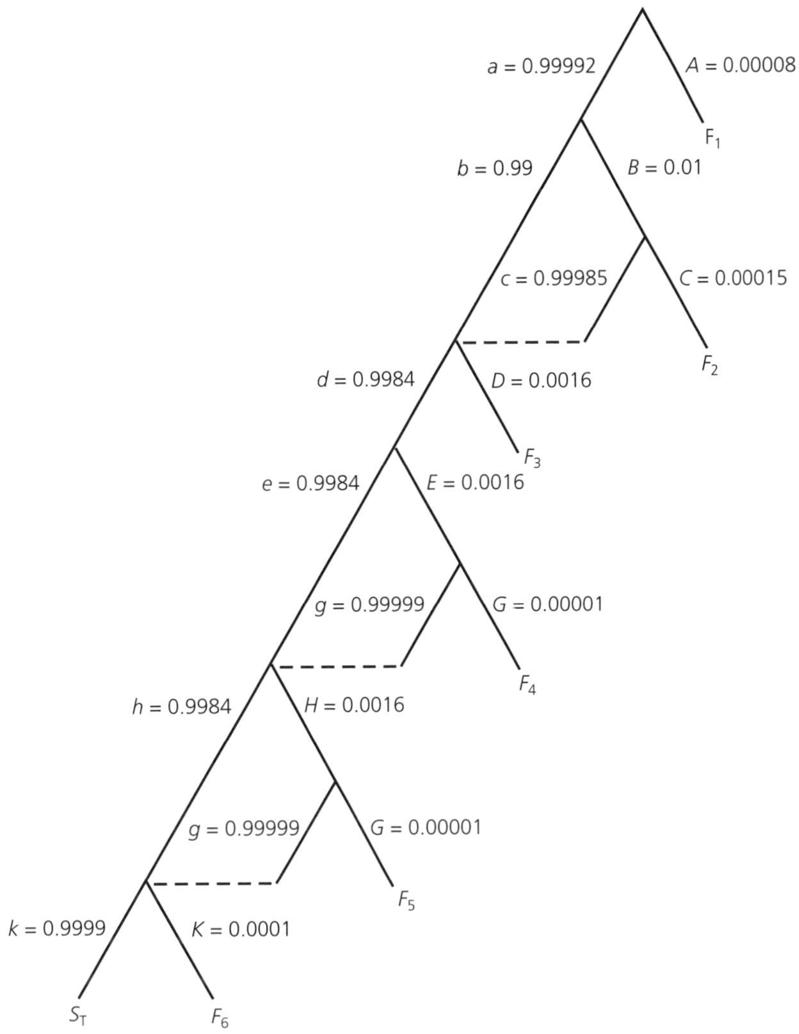


| JHEP | Event | Joint HEPs for 3 operators |
|---|---|---|
| A | Fail to initiate action to ANNs | 0.00008 |
| B | Misdiagnosis | 0.01 |
| C | Fail to initiate action to ANN | 0.00015 |
| D | Omit step 2.4 | 0.0016 |
| E | Omit step 2.5 | 0.0016 |
| G | Fail to initiate action to ANN | 0.00001 |
| H | Omit step 2.6 | 0.0016 |
| K | Fail to initiate HPI | 0.0001 |

ANN, annunciator; HPI, high-pressure injection

Other HRA procedures could be useful, but it is recommended that a study similar to the NASA (Chandler *et al.*, 2006) methods study be completed by USACE for spillway systems.

Another emerging modern technology that seems to be promising when it comes to simulation of the human aspects of the operation of a hydropower system is the agent-based simulation. This methodology treats the entities/departments/operators of an organisation as independent agents, a situation that is realistic in terms of how modern organisations are set up. The advantage of this method is that it allows simulation of decision-making and output control generation in all kinds of situations. Unlike the older methods where it was only possible to evaluate the failure rate of the individuals, agent-based technology allows even potential corrective actions by departments or individuals involved in the different operating situations to be taken into account. And that is how the real situation usually looks – organisations often manage human error by counting on clever individuals solving challenging problems, which must be accounted for in the simulation. Agent-based simulation opens up a new dimension of dynamic simulation where the socio-technical system is addressed in terms of behavioural science and the advances in this technology are used.

**Figure 8.17** ATHEANA methodology (USNRC, 2000). EOP, emergency operating procedures; FSAR, final safety analysis report; HMI, human–machine interface; UA, unsafe action

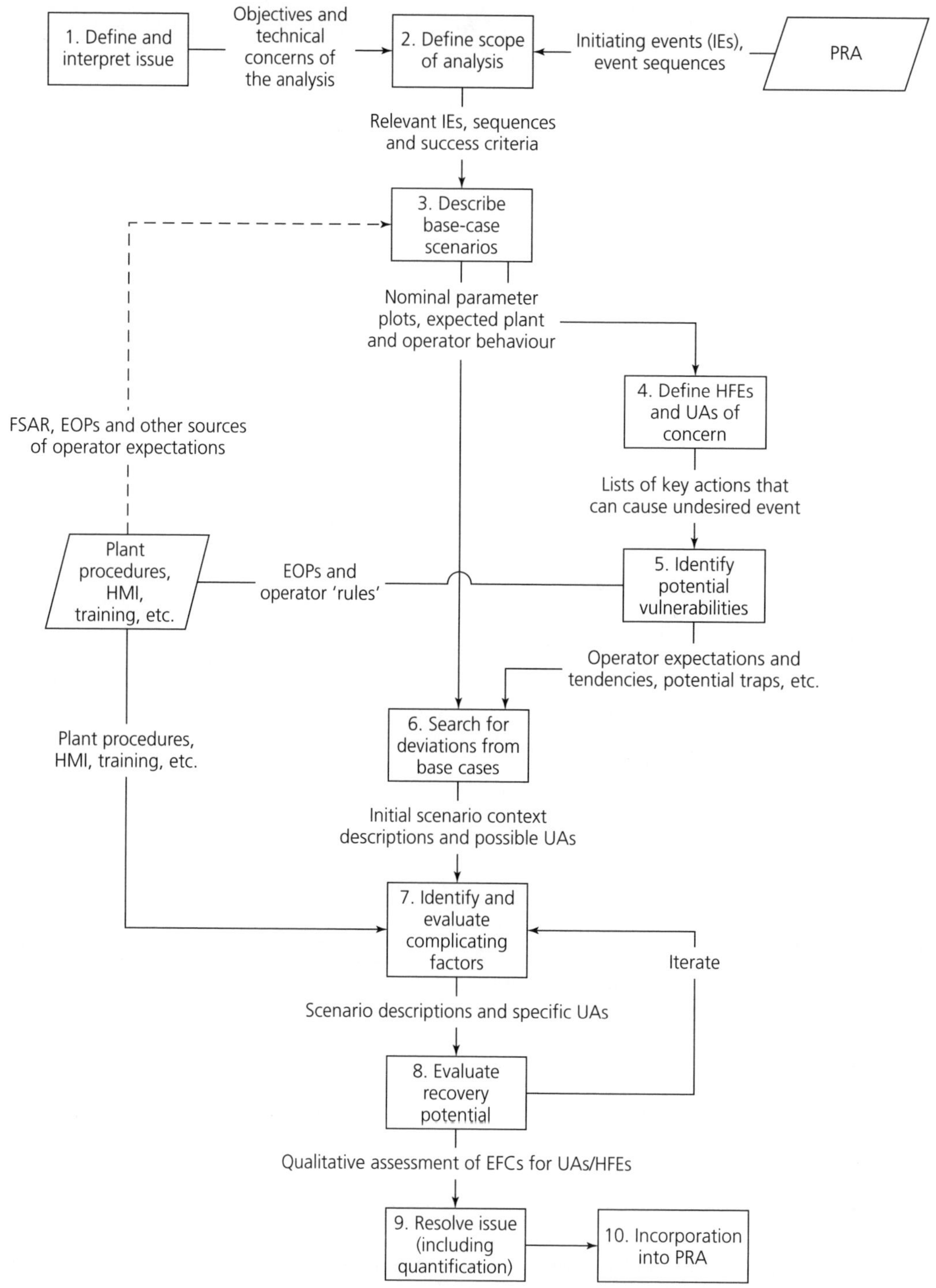


There are other emerging and currently used technologies that offer a very different approach to determining the reliability of human agents. Most of the currently used approaches originate from the chain-of-events thinking, based on the assumptions that:

- uncommon scenarios occur in succession, pass through multiple barriers of defence and result in an accident
- human operators can be treated as just another component in the system
- probabilities of human errors in these causal scenarios can be calculated or estimated.

All these assumptions are now being questioned. More on the different approaches based on a control-theoretic view of human and organisational error can be found in Hollnagel (2004) and Leveson (2004).

## REFERENCES


ASCE (American Society of Civil Engineers) (2012) *Water Control Gates. Guidelines for Inspection and Evaluation*. ASCE, Reston, VA, USA.

Barker M, Vivian B, Matthews J and Oliver P (2003) Spillway gate reliability and handling of risk for radial and drum gates. In *NZSOLD/ANCOLD 2003 Conference on Dams*, Wellington, New Zealand.

Broughton J, Carter R, Chandler F *et al.* (1999) *Human Factors Process Failure Modes & Effects Analysis*. Report Number PGOC91-F050-JMS-99286. The Boeing Corporation, Seattle, WA, USA.

Chandler F, Chang J and Mosleh A (2006) *Human Reliability Analysis Methods Selection Guidance for NASA*. National Aeronautics and Space Administration (NASA), Washington, DC, USA.

Chang YHJ and Mosleh A (2007) Cognitive modeling and dynamic probabilistic simulation of operating crew response to complex system accidents – Part 1. Overview of IDAC model. *Reliability Engineering and System Safety* **92**: 997–1013.

Dandan LI and Lei LI (2011) Suggestions of introducing the human reliability into dam risk analysis. *Advanced Materials Research* **225–226**: 395–398.

Erbisti PCF (2004) *Design of Hydraulic Gates*. Balkema, Lisse, the Netherlands.

Fiedler W, King W, Schwanz N and Holman W (2013) Lake Delhi dam failure. In *Changing Times – The Challenges and Risks of Managing Aging Infrastructure Under a New Financial Reality. 33rd Annual USSD Conference*, Phoenix, AZ, USA, pp. 675–692.

Gertman D, Blackman HS, Marble JL, Byers JC and Smith CL (2005) *The SPAR-H – Human Reliability Analysis Method*. U.S. Nuclear Regulatory Commission, Washington, DC, USA.

Hartford DND and Baecher GB (2004) *Risk and Uncertainty in Dam Safety*. Thomas Telford, London, UK.

Hobbs G (2003) Spillway gates – a survey of operational and reliability issues. *Proceedings of 43rd ANCOLD Conference*, Hobart, Australia.

Hobbs G and Azavedo D (2000) Spillway gate reliability. *Proceedings of ANCOLD Annual Meeting*, Cairns, Australia.

Hollnagel E (1998) *Cognitive Reliability and Error Analysis Method: CREAM*, 1st edn. Elsevier, Oxford, UK.

Hollnagel E (2004) *Barriers and Accident Prevention*. Ashgate, Farnham, UK.

Kirwan B (1994) *A Guide to Practical Human Reliability Assessment*. CRC Press, Boca Raton, FL, USA.

Klohn EJ (1994) Damage and remedial work during operation of several bottom outlets. In *18th ICOLD Congress*, Durban, South Africa, pp. 411–428.

Leveson N (2004) A new accident model for engineering safer systems. *Safety Science* **42(4)**: 237–270.

Lewin J (2001a) Failures and vulnerabilities of reservoir control structures. In *Learning from Construction Failures: Applied Forensic Engineering* (Campbell P (ed.)). John Wiley, London, UK, pp. 168–193.

Lewin J (2001b) *Hydraulic Gates and Valves in Free Surface Flow and Submerged Outlets*, 2nd edn. Thomas Telford, London, UK.

Lewin J, Ballard G and Bowles DS (2003) *Spillway Gate Reliability in the Context of Overall Dam Failure Risk*. USSD Annual Lecture, Charleston, SC, USA.

NASA (National Aeronautics and Space Administration) (2002) *Probabilistic Risk Assessment Procedures Guide for NASA Managers and Practitioners*. National Aeronautics and Space Administration, Washington, DC, USA.

New South Wales Dam Safety Committee (2010) *Reliability of Spillway Flow Control Systems*. See http://www.damsafety.nsw.gov.au/DSC/Download/Info_Sheets_PDF/Dam/DSC3D.pdf (accessed 01/06/2016).

NPDP (National Performance of Dams Program) (2015) http://npdp.stanford.edu/ (accessed 01/06/2016).

Patev R (2011) Vulnerability assessment of complex civil works systems using expert-opinion elicitation. In *NATO International Conference on Security Risk*, Moscow, Russia.

Patev RC and Putcha CS (2005) Development of fault trees for risk assessment of dam gates and associated operating equipment. *International Journal of Modelling and Simulation* **25(3)**: 190–201.

Reason JT (1990) *Human Error*. Cambridge University Press, Cambridge, UK.

Reason JT (1997) *Managing the Risks of Organizational Accidents*, 1st edn. Ashgate, Farnham, UK.

Schultz R (2013) Survey of USACE flood control projects. US Army Corps of Engineers, Pittsburgh, PA, USA (unpublished).

Swain AD and Guttmann HE (1983) *Handbook of Human Reliability Analysis with Emphasis on Nuclear Power Plant Applications*. US Nuclear Regulatory Commission, Washington, DC, USA.

Todd RV (1999) *Spillway Tainter Gate Failure at Folsom Dam, California*. See http://cedb.asce.org/cgi/WWWdisplay.cgi?126726 (accessed 01/06/2016).

Todd RV (2009) *Dam Safety Best Practices: Trunnion Friction Radial Gate Failure*. US Bureau of Reclamation, Denver, CO, USA.

USACE (US Army Corps of Engineers) (1992) *Hydraulic Design of Spillways*. USACE, Washington, DC, USA.

USACE (1997) *Vertical Lift Gates*. USACE, Washington, DC, USA.

USACE (2000) *Engineering and Design – Design of Spillway Tainter Gates*. USACE, Washington, DC, USA.

USBR (US Bureau of Reclamation) (2010) *Appurtenant Structures for Dams (Spillway and Outlet Works) Design Standards*. USBR, Denver, CO, USA.

USNRC (US Nuclear Regulatory Commission) (2000) *Technical Basis and Implementation Guidelines for A Technique for Human Event Analysis (ATHEANA)*. NUREG-1624, Revision 1. Division of Risk Analysis and Applications, Office of Nuclear Regulatory Research, USNRC, Washington DC, USA.

van Leperen A (1984) Design of the new waterway storm surge barrier in the Netherlands. *Hydropower and Dams* **May**: 66–72.

Vesely WE, Goldberg FF, Roberts NH and Haasl DF (1981) *Fault Tree Handbook*. NUREG-0492. US Nuclear Regulatory Commission, Washington, DC, USA.

**Operational Safety of Dams and Reservoirs**
ISBN 978-0-7277-6121-7


http://dx.doi.org/10.1680/osdr.61217.251

# Chapter 9
# Discharge conveyance function

This chapter deals with hydraulic loading conditions on spillways that are normally encountered during the design life and within the design envelope of a dam and spillway structure. It excludes those conditions where the discharge conveyance function has been disturbed by external hazards such as earthquakes (see Chapter 7).

The focus of this chapter is on incorporating the discharge conveyance function – which is principally hydraulic in nature – within the system simulation of dam operations. To this end, the chapter begins with a consideration of conveyance failure processes. It then turns to hydraulic analysis, including open channels, closed conduits and energy-dissipation structures. Finally, it returns to the simulation of hydraulic systems, the use of computational fluid dynamic models and to modern day examples.

The discussion of hydraulic modelling and the inclusion of hydraulic equations are intended to guide the reader back to the fundamental hydraulics rather than to repeat what is readily available in many fine textbooks. The purpose is not to repeat the basics of hydraulics but to demonstrate how hydraulic considerations become part of the system simulation analysis.

The vast majority of the functionality of spillways and conduits is determined in the design of the spillway. The gated-spillway and conveyance away from the spillway for BC Hydro's Bennett Dam is shown in Figure 9.1. The approach to the analysis of the discharge conveyance function and the functional failure of spillways and conduits is intrinsically linked to design assumptions and the design process. The release procedures and the maintenance processes and procedures are also important determinants of the adequacy of the conveyance function. As design is such a dominant factor in the conveyance function, any performance or failure analysis must properly account for all aspects of interactions and interdependencies. While this chapter contains a considerable amount of content relevant to physical design considerations, it is not a chapter on the hydraulic design of spillway and conduits structures. Many excellent texts already exist on that subject. Rather, the purpose here is to highlight aspects of design relevant to functional performance analysis.

Three principal functions for the hydraulic control system were defined in Chapter 3: store, pass and divert. *Pass* is represented by the conveyance and refers to those hydraulic functions that safely pass water from the reservoir to the river channel or lower reservoir, around, over, through or under the body of the dam downstream. In terms of the gate–spillway interface, the discharge conveyance function is dependent on the release function, which is described in Chapter 8.

Figure 9.1 (a) Conveyance structure at Bennett Dam, upper right in (b). (Courtesy of BC Hydro)

Figure 9.2 Strathcona Dam, Campbell River, Victoria, BC, Canada: spillway approach. (BC Hydro, 2015)

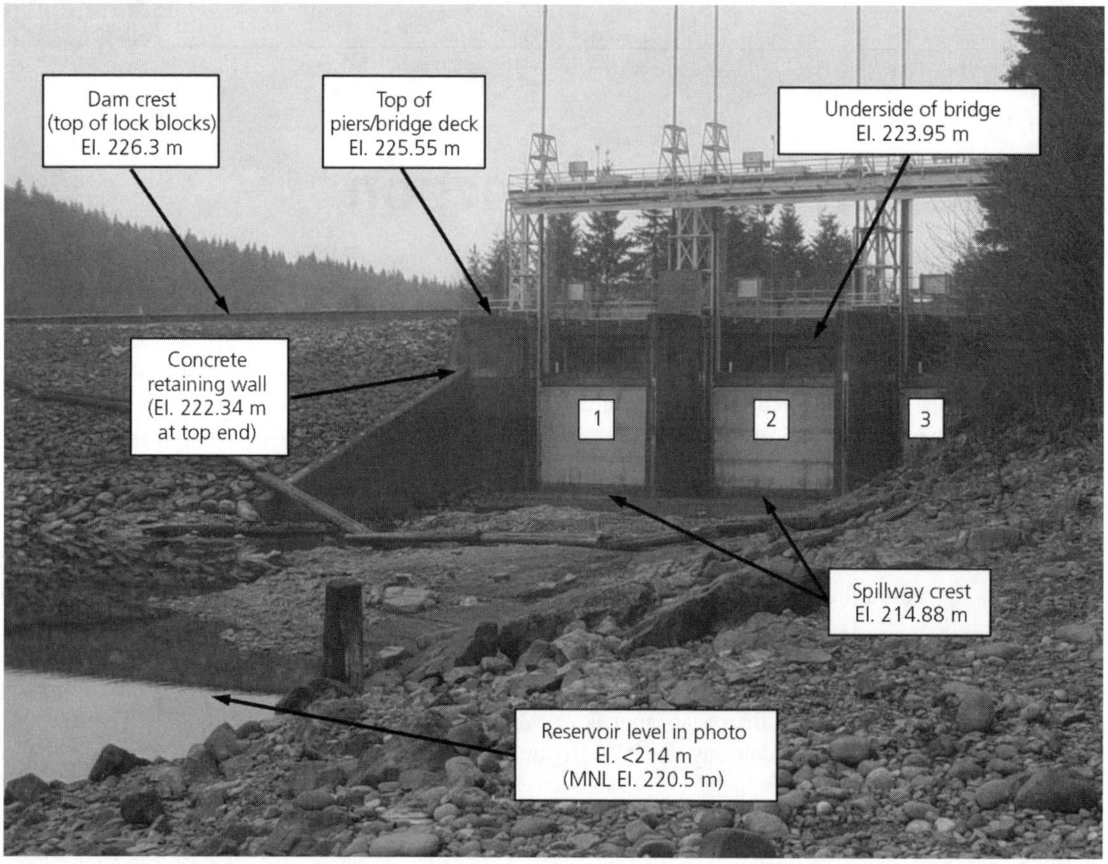


The safe passage of water requires proper control of

- the rate of flow
- the physical conveyance of the flow
- the kinetic and potential energy of the flow.

The last two of these three hydraulic functions are considered in this chapter. The rate of flow is considered here as an input to the system (Chapter 5), which must fit into outflow requirements (Chapter 6), and, to achieve this task, is controlled by the release function (Chapter 8). The release function is commonly, but not always, controlled by a gated spillway typified by that at Strathcona Dam, shown in Figure 9.2. The proper control of the flows requires that the release function and the conveyance function work together to provide the safe passage of the water. However, the components should be designed such that there are no physically possible discharge flow release conditions that could overwhelm the conveyance components. That is, they should be designed for the worst-case release condition (e.g. gate bursts at the maximum permissible reservoir elevation).

While, in most cases, water flow is controlled by actively actuating the release function by opening the gates, there is a growing portfolio of ungated spillway structures, where there is a passive control of flow. This means that the rate of flow to be conveyed is determined as a function of the reservoir level and the sill geometry of the structure. As these structures become ever more common, especially for the passage of extreme floods, the requirements regarding their functionality for conveyance is an important consideration in the design process.

The physical function of conveying water, the pass function, is achieved by means of

- chutes, canals and tunnels, which direct the conveyance of water flow
- energy-dissipation arrangements, of which there are numerous forms.

These components can be designed and combined in many ways to produce water conduits with different layouts and qualities, from simple fixed sill spillways with free surface flow, to gated low-level outlets or power lines with closed conduit flow and strict performance demands with regard to power production and regulation properties.

Figure 9.3 Strathcona Dam, Campbell River, Victoria, BC, Canada: spillway chute. (BC Hydro, 2015)

## 9.1. The conveyance process

Conveyance of the water flow in spillways and conduits may be through natural river channels, artificial canals, culverts and tunnels, and these may be either unlined or have erosion protection of steel, concrete, rock revetment, groins or vegetation cover (Figure 9.3). The suitability of different kinds of erosion protection depends on the flow velocities, ice and floating debris loads, turbulence and wave action, and, of course, the potential consequences of failure.

In spillways intended for use only under special conditions, such as emergency or auxiliary spillways, the conveyance arrangements have not always been pre-constructed to cater for the maximum foreseen discharge. Designers have instead relied on the water flow to erode its own channel through rock or soil, a concept that may involve considerable risk of failure, as both rocks and soils are heterogeneous materials, usually with limited site-specific characterisation available. The rate and direction of the erosion process may, therefore, be difficult to foresee.

### 9.1.1 Rate of flow considerations in the conveyance process

The volume of water stored in a reservoir, and therefore the rate of change of storage, is given by

$$\Delta S = Q_i \Delta t - Q_o \Delta t \qquad (9.1)$$

where $\Delta S$ is the change in storage, $Q_i \Delta t$ is the rate of change of the inflow and $Q_o \Delta t$ is the rate of change of the outflow.

The rate of change of the reservoir elevation can be derived from the characteristics of the reservoir and how the flows are controlled (i.e. the operating rules). The characteristics of these parameters are often represented as 'curves' of various forms, these being graphical representations of the corresponding time-series integrals. The curve of the rate of inflow versus time is represented by the inflow design flood (IDF) hydrograph; the rate of outflow is represented by a curve of the spillway discharge versus reservoir elevation; and storage is represented by the curve of reservoir storage versus reservoir elevation.

The IDF hydrograph is fixed by the selection of the IDF. The maximum volume of storage is set by the height of the dam and the features of the reservoir. The available capacity at any time is predetermined by the manner in which the reservoir is operated. The spillway discharge curve is variable: it depends not only on the size and type of spillway but also on the manner of operating the spillway (and the outlets in some instances) to regulate the outflow. The quantity of water that a spillway can discharge depends on the type of control device. For a simple free overflow crest the flow varies with the head on the crest, and the surcharge storage capacity increases with an increase in spillway discharge. For a gated spillway, however, the outflow can be varied with respect to reservoir head by operation of the gates. For example, in the case of an operating gate-controlled spillway it might be determined that the gates will be regulated so that the inflow and outflow are equal until the gates are wide open. Alternatively, it might be determined that the gates will be opened at a slower rate so that surcharge storage will accumulate before the gates are wide open. Outflows are not necessarily limited to discharges through the spillway; they may be supplemented by releases through the outlets. In all such cases the size, type and method of operation of the spillway and outlets with reference to the storages or to the inflow must be predetermined to establish an outflow-elevation relationship.

As discussed in Chapter 8, the rate of flow in a water conduit such as a spillway (the discharge) is usually controlled at a specific point or cross-section at the intake, where the flow changes from subcritical upstream to supercritical downstream. Supercritical flow is that with a velocity larger than the wave velocity. The energy of the flow and the free dimensions of this control section control the rate of flow. The control section, therefore, needs to be reliably stable. Depending on the flow velocities this usually entails provision of invert and sidewalls (sills and piers) of reinforced concrete or steel. In low-level outlets or bottom outlets a stable soffit (upper boundary) is also required.

(*Note:* In subcritical and supercritical flow water has appreciable internal cohesion. As water accelerates, it moves away from water flowing more slowly behind it. It does not initially separate, but stretches, as the forces of cohesion, including gravity and viscosity, are overcome by the increasing movement of the water. This can be seen as water passes over a weir or approaches a constriction in a channel. The water falling over the crest of a weir is accelerating and may break up, but as it does so its store of kinetic energy increases until it is released in a chaotic explosion of activity. This increase in kinetic energy can be seen by a noticeable dip in the water surface just upstream of the crest of the weir. Prior to this dip the velocity of the water is said to be subcritical; at the dip it is critical, and beyond the dip it is supercritical.)

The shape of the spillway crest is often used to reduce the pressure head on the crest to below atmospheric, thereby increasing the discharge capacity. A more unusual type of control of the rate of flow in spillways is by backwater from downstream. This occasionally occurs in surface spillways at fairly low dams and in some low-level outlets. The flow rate is then controlled by the water level downstream and, in principle, by the dimensions of the entire length of the spillway. This is a case of 'physical feedback' control of flow.

### 9.1.2 Energy dissipation

Effective control of the discharged flows requires that the potential and kinetic energy of the water is controlled within limits appropriate for the water conduit, adjacent structures and terrain, and where the flow re-enters the river downstream. Some kind of dissipation arrangement is usually required to control the energy and avoid erosion of riverbanks, which could otherwise erode the dam toe and cause the dam to fail.

Often the dominating part of the excess energy is dissipated by the intense turbulence created by a hydraulic jump, either in a specifically designed energy dissipator (stilling basin) or in a plunge pool. (Hydraulic jump is the change in the water-surface profile that occurs a higher velocity: supercritical flow upstream meets a subcritical downstream flow with a decreased velocity and sufficient depth.) A well-designed plunge pool may be viewed hydraulically as a drowned jump, where the most intense turbulence is created at some distance from the boundaries. Other types of dissipation arrangements are occasionally used. A significant part of the excess energy of the flow may, in some cases, be dissipated as friction, especially in rocky canals and stepped spillways.

The nature and form of the energy-dissipation arrangement is determined by a number of factors, one of which is the amount of energy to be dissipated, which itself is determined to a significant degree by the head on the spillway. Mason (1982) investigated the historical performance of different energy-dissipation arrangements. His survey indicated that spillways with free-trajectory jets directed to strike the free water surface in a plunge pool dominated successful energy-dissipation arrangements for heads larger than approximately 50 m. Simple, rectangular concrete-lined hydraulic jump stilling basins were found suitable for head sizes in the range 30–50 m. In the head-size range 10–30 m, stilling basins of the mixed impact type, with chute or baffle blocks or other features, of steel or concrete increasing the level of turbulence to permit reduced stilling basin dimensions, were also reported to have performed reasonably well. Simple, unlined rock basins appeared to have a good track record only for still smaller heads, depending on local geological conditions.

The turbulence that accompanies the dissipation of kinetic energy involves energy transformation into heat, vibration and impact forces, all of which are necessarily destructive to some degree. Careful control of these destructive forces is required to limit intense pressure fluctuations and the generally increased potential for cavitation, erosion and other damages. Studies have shown that these destructive forces may overload or wear down structural elements of the stilling basins placed directly in the high-velocity flow zone, thereby completely changing the water–structure interaction and hydraulic loading on the structures. Accordingly, effective inspection, maintenance, repair and restoration procedures are required to ensure that the conveyance function is maintained as intended in the design. Knowledge of the nature of these forces, where they are applied, and how they change over space and time, is essential for the conveyance function to be analysed effectively.

Strict specification requirements, high-quality construction control and the use of prestressed rock anchors have all contributed to raising the upper limits of the performance of concrete-lined stilling basins. Alternatively, the use of artificial aeration in spillway chutes is a way of reducing the vulnerability of high-head spillways to cavitation by reducing the dependence on exceptionally high standards of design, construction and maintenance. Simple concrete-lined stilling basins have now been built with artificial aeration for spillways (Pinto, 1994).

## 9.2. Conveyance failure process

In the context of the conveyance function, failure means that the required hydraulic function cannot be achieved in terms of the as-designed rate of flow, the nature of the flow after release, the conveyance or the energy dissipation. In other words, the water is not conveyed to the location of interest in the condition required. Failure of the conveyance function does not necessarily mean physical collapse of the conveyance structure, although structural collapse will certainly result in the failed conveyance condition.

Conveyance function failure may be considered in terms of the inability to discharge the flows as required by, say, the environmental permit, due to flow disturbances during the conveyance process. While the rate of flow at the point of release is determined by the upstream conditions, and in the case of gated spillways the rate of gate opening, the rate and nature of the flow through the conveyance structure and the stilling basin are determined in part by the geometry of the conveyance structure, the construction and the maintenance and repair.

Failure may be induced by any one of a number of natural hazards or other external or internal disturbances (hazards), but in the end failure always comes down to structural insufficiency – either insufficient dimensions to contain the flow or insufficient integrity to withstand the inherent hydraulic forces, or a combination of these insufficiencies.

As is the case for all these types of analyses, it is necessary to recognise that there are failures at component level and failures at system level. A water conduit such as a spillway may be considered both as a system (a subsystem) and as a component of a larger system.

There may also be different categories of failure-related events, both at component and at system level, so that one type of event may cause the function to be impaired or limited but not necessarily be totally absent or immediately insufficient (e.g. signs of structural damage, which although not immediately decapacitating may rapidly become so unless quickly remedied).

In the analysis of conveyance function performance the analyst is required to utilise knowledge of the design and construction of hydraulic structures in general, as well as the design, construction, operation and maintenance of the particular conveyance subsystem. This is so that proper account is taken of all the physical and functional characteristics of the subsystem as are required to determine how failure and failure categories should be defined for the purpose of the analysis. In this regard, the analysis of the conveyance function requires a broader array of information than might be required for other structures.

For example, failure may be any occasion when the operational demand results in a loading that exceeds the current capacity of the water conduit with regard to discharge, conveyance or energy dissipation, whether due to a high demand, low capacity or a combination thereof. The critical parameter may be excessive flow velocity or excessive or low water level or pressure somewhere as compared with the structure's current capability of resistance. Both the loading and the capacity may be influenced by *external disturbances* or the consequences thereof, such as floating debris, slides and rock falls.

Depending on the purpose of the analysis, the analytical model of the water conduit may include the effect of some changes in the state of the water conduit, so that its changed function can still be described and modelled, while the effect of other changes in the state are not modelled but are considered outside the acceptable range, and therefore constitute a *failure of control*. Typical changes in state to include in modelling are changes in hydraulic roughness of boundaries and limited erosion due to operation, wear, the freeze–thaw cycle, thermal expansion and contraction, and maintenance.

### 9.2.1 Intakes and weirs

As set out in ICOLD Bulletin 154 (ICOLD, 2011), insufficient hydraulic capacity manifests in essentially two ways: inadequate installed capacity and inadequate available capacity. Inadequate installed capacity is principally a matter of spillway sizing at the design and construction stages, whereas inadequate available capacity can be due to a wide range of factors, some of which are within the control of the owner. This straightforward distinction applies in cases where the spillway is sized to pass the IDF. The situation with respect to those cases where passage of the IDF requires preconditioning of the reservoir is less straightforward, as it may be that loss of control of the reservoir elevation could be due to a breakdown in the reservoir conditioning process. For the purposes of this discussion, it is taken as given that the spillway has been sized to pass the IDF without preconditioning of the reservoir, as such a design would recognise the fallibility of the operational conditioning of reservoirs.

#### 9.2.1.1 Insufficient installed hydraulic capacity

Insufficient installed hydraulic capacity of the intake is the classic failure mode to be considered. It will (may) be manifested as a raised reservoir water level, and the final failure may then take on the form of a dam overtopping. Other than in exceptional conditions when the design basis is exceeded, the problem is rooted in the design of the spillway in terms of either estimation of the design flood or errors in the design process. It is not simply a matter of making the spillway opening wide enough, as the geometry of the spillway must be such that the entering water moves rapidly enough away from the opening.

Physical model testing at appropriate scale currently gives the most accurate estimates of potential spillway capacity (i.e. the discharge capacity of the fully open spillway). The inaccuracies in established capacities may then, depending on the equipment and model scales used, be in the range 2–4% (standard deviation) due to model measurement inaccuracy. Other limitations of physical models are well known. Froude models at scales between 1 in 25 and 1 in 75 may be sufficient unless the effects of air entrainment need to be considered, in which case scales of around 1 in 10 to 1 in 15 may sometimes be required.

Computational fluid dynamics (CFD) modelling provides the next most accurate approach to sizing a spillway. Mathematical models currently in use for CFD modelling have difficulty handling the effects of true two-phase flow (aeration) in an efficient and effective manner, and turbulence is modelled by semi-empirical methods.

The use of handbook references provides estimates of even lesser quality. Capacity estimates made by trained experts may be the best approach for layouts with unusual shapes of ogee crests and piers or those significantly influenced by approach flow conditions (e.g. low ogee crests), but it is important to recognise that the expert's opinions may occasionally be as far off as 15–20%.

A more unusual reason for deviations from expected discharge capacity is physical differences between the spillway as shown on drawings and the spillway as actually built, because of layout changes not shown on drawings or because of construction errors.

#### 9.2.1.2 Hydraulic capacity not available

The hydraulic capacity of intakes may be reduced by the influence of floating debris, ice or sediments restricting the free opening size or by gates not opening.

The failure of the spillway gate to open on demand can arise for many reasons, which can be broadly grouped as follows.

- Command failure – the command to actuate the spillway is not issued, either at the control centre or at the control unit at the dam.
- Drive power failure – a loss of power at the spillway gate hoist subsystem due to a failure at any point along the power-supply path.
- Mechanical movement failure – a breakage or seizing of a gate-opening mechanism
- Structural obstruction – in this book, a structural obstruction refers to either structural instability or incompatible geometry between the gate and the supporting structure (e.g. the swelling that occurs due alkali–aggregate reaction, or due to differential movement).

#### 9.2.1.3 Structural instability

Available experience indicates that the erosion resistance of approach channels may be taxed during high spillway discharges, for instance where the upstream slope of an embankment dam constitutes one of the approach-channel banks. The failure mechanism would be external erosion, which may lead to a complete dam failure more quickly than erosion damages elsewhere on the dam, because the dam section is often a bit cramped just in this area. Sometimes approach channels have guide walls of concrete founded on fill below normal reservoir water levels. Their condition may be very difficult to check and maintain, as dewatering may not be realistically possible. It is not unusual for such guide and retention walls to be dependent on rock anchors or bolts, the reliability of which may become endangered in the long term. The failure mechanism in this case would be chemical deterioration.

Intake structures may fail due to local or global loss of stability, such that static or dynamic forces overtax the resistance capacity of the structure itself, of the intersection between the structure and the foundation, or of weak strata in the foundation. Depending on the constitution of the foundation, bearing capacity failures may also be possible. Significant uncertainties may exist with regard to foundation conditions.

Failures may also be caused by mistakes in design, construction or maintenance, and such failures are generally manifested at moments of increased loading or reduced resistance.

The cause of immediate failure may be unbalanced exterior loads from the reservoir or the surrounding terrain. Typical failure conditions include a high upstream (and/or downstream) water level, with spillway openings closed and gates or stoplogs clogged by debris. Also, waves caused by strong winds and impact waves and pressures caused by rock falls, slides or debris flows into the reservoir may overtax structural stability. The reflection of waves against the structure and gates might then lead to significant increases in the horizontal loads.

Failure may also occur due to unbalanced interior loadings caused by static or dynamic water pressure aggravated by unsuitably shaped ogee crests, insufficient drainage, leaky joints, erosion, cavitation or physical/chemical deterioration, such as corrosion or loss of prestressing by relaxation or anchor slippage.

### 9.2.2 Chutes and conduits

Failures due to lack of conveyance capacity will primarily occur at high discharges, but aeration and cross-waves may lead to high water surfaces overtopping the chute or open channel walls at more moderate flows. Figure 9.4 shows low but supercritical flows which nonetheless ride up the retaining walls. Also, changes in boundary roughness or downstream water levels will influence the conveyance capacity.

#### 9.2.2.1 Hydraulic capacity

Flow uncertainties in head losses will have a larger influence on the hydraulic capacity in closed conduits than in free surface water conduits. Also, potential transient effects may limit the useable steady conveyance capacity of tunnels and culverts. Effects to consider include two-phase air/water phenomena, surging and water hammer.

The rock around a spillway or discharge tunnel has positive and negative attributes. Obstructions such as rock falls, slides or debris may reduce the capacity of a water conduit so that open channel sidewalls are overtopped and tunnel capacities reduced.

Figure 9.4 Glenmaggie Dam, Macalister River, southern Victoria, Australia. The weir is topped with 14 radial gates. (Courtesy of the State Library of Victoria)

Rock tunnels are perhaps more vulnerable to obstruction, due to the presence of significant amounts of rock directly above the water conduits. At the same time, the consequences may be worse, as a much larger increase in head will be required to balance partial blockage. Power stations have been inundated and put out of operation for long periods by blocked tailrace tunnels.

#### 9.2.2.2 Structural instability

Structures along canals and other conduits tend to fail by similar processes as intake structures – that is, due to local or global loss of stability, such that static or dynamic forces overtax the resistance capacity of the structure itself; the intersection between the structure and the foundation; or weak strata in the foundation. Depending on the nature of the foundation bearing capacity, failures may also be possible. Failures may be due to mistakes in design, construction or maintenance, and are prone to occur when conditions have worsened due to increased loadings or reduced resistance capacity. The direct cause may be: unbalanced exterior loading due to rock falls or slides; seepage from the reservoir or surrounding terrain; leaky joints; unbalanced interior loading due to static or dynamic water pressure from turbulent pressure fluctuations; or insufficient drainage, erosion, cavitation, physical or chemical deterioration, such as corrosion or loss of pre-stressing by relaxation or anchor slippage.

Failure of low-level outlets and pressure tunnels belonging to power stations have occurred on a number of occasions due to hydraulic fracturing, which essentially takes place when water pressures higher than the minor principal stress are fed into a (jointed) rock mass. This may cause the movement of large (huge) blocks of rock opening up wide joints or destroying tunnels, or causing rockslides, leakage and loss of hydraulic pressure.

High dams result in high flow velocities and, typically, thin boundary layers. These two factors combine to make some parts of water conduits, especially low-level outlets, vulnerable to cavitation damages.

Failures of the water conduits of power stations are sometimes caused by sequences of events involving automated/operator-initiated responses to unforeseen developments and transients. Rapid valve closures and 'rapid load off–load on' sequences have been known to cause such high or low water pressure (water table) that steel penstocks have failed in compression or tension and power stations have been inundated from both upstream and downstream. In rock tunnels, water hammer effects may cause successive loosening of the rock around tunnels that are not protected by watertight linings.

### 9.2.3 Energy dissipation arrangements

Lack of energy-dissipation capacity is usually caused by insufficient tailwater level or dissipation basin length in combination with high hydraulic load in terms of energy level and discharge of flow into the dissipator. The hydraulic capacity is contingent on the downstream conditions assumed in the design and on higher floods than previously recorded, but also on upstream pool level and head losses.

#### 9.2.3.1 Hydraulic jump energy dissipators

Failure to provide the capacity required may lead to relocation of a hydraulic jump compared with the designed position (Figure 9.5). If the tailwater is too low, no hydraulic jump may occur in the basin, but may possibly occur somewhere downstream in an unforeseen, unprepared location. This could then impose supercritical flow on the intermediate stretch of water conduit with high flow velocities and turbulence conditions that are greater than have been designed for. Moreover, potentially unstable back eddies may develop where the hydraulic jump finally occurs unless the riverbanks are very steep and high enough to contain the tailwater. Sloping banks may allow erosive back eddies to develop.

The hydraulic head driving the back eddies is the difference between the prevailing subcritical and supercritical water levels at the jump, which for a normal free jump may be obtained from

$$y_2 - y_1 = \tfrac{1}{2} y_1 \left( \sqrt{1 + 8F_1^2} - 3 \right) \tag{9.2}$$

Figure 9.5 Hydraulic jump in open-channel flow

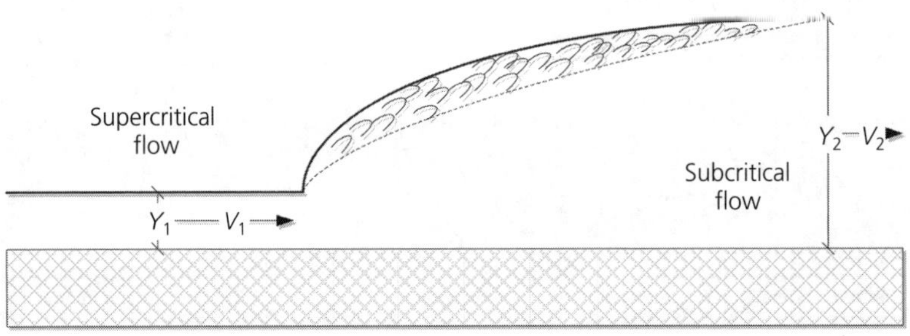

This water-level difference may theoretically produce flow velocities of up to a maximum of

$$u = \frac{\left[\frac{1}{2}y_1\left(\sqrt{1+8F_1^2}-3\right)\right]^2}{2g} \tag{9.3}$$

which for a fairly normal Froude number $F_1 = 6$ and $y_1 = 2.0$ corresponds to $u = 10$ m/s.

Reductions in tailwater may also lead to worse exposure of the fixed boundaries of the flow without essentially relocating the jump. Furthermore, increased tailwater levels may also relocate a hydraulic jump upstream, possibly to an area less prepared to accept the rather harsh conditions, and therefore in effect with reduced capacity. Indeed, high tailwater is one of the limiting parameters for hydraulic capacity of dissipators, as overtopping of sidewalls may occur and the surrounding areas, possibly including the downstream toe of an adjacent dam, may become exposed to erosion.

The water surface in hydraulic jump dissipation basins is often quite unstable and wavy. Significant freeboard may be required to contain instantaneous water levels higher than the mean air–water levels.

Several methods exist for shortening the required length, volume or downstream water depth for hydraulic jump stilling basins to create a forced hydraulic jump. Most of them rely on the artificial increase of turbulence and direct impact by placing fixed objects, such as blocks of concrete or steel, splitters, sills, walls, etc., directly in the path of the incoming high-velocity flow (Figure 9.6). These fixed objects will then be subjected to wear by erosion, and often also to cavitation pitting, and they may therefore not automatically be assumed to be functioning over the duration of large floods. Aggressive pro-active maintenance may or may not be sufficient to guarantee the proper long-term function.

Variations of the original Unwin formula exist to describe necessary conditions for the creation of hydraulic jumps in basins of other shapes, such as basins with sloping inverts or diverging side walls, and basins provided with various turbulence-increasing devices of standardised shapes (Chow, 2009). Mason's studies indicate some definite limitations in their application.

One such basin shape is where an abrupt increase in basin width (which cannot be followed by the high-velocity flow) opens up the flanks of the supercritical flow also to lateral attack by back eddies in the higher downstream water table.

Figure 9.6 Layout of a US Bureau of Reclamation (USBR) type III stilling basin. (Peterka, 1978)

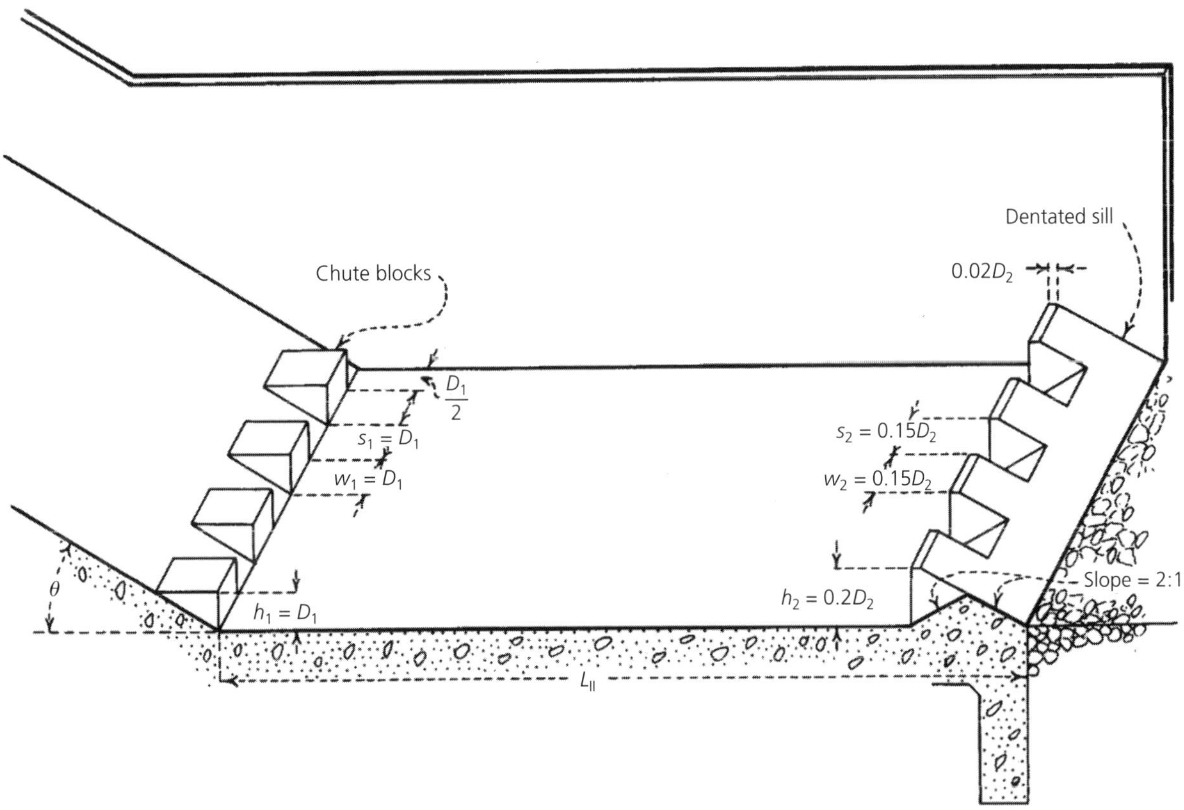

This layout of energy dissipator has been shown to hydraulically permit a reduction in tailwater depth up to a maximum of some 40%, roughly proportional to the square root of the ratio of widths.

For energy dissipators of the impact type, changed hydraulic condition may or may not lead to changed impact exposure for structures. Such dissipators may, therefore, be more independent on tailwater elevations, albeit at the price of constant exposure of the structures during any kind of operation.

So-called supercavitating chute blocks rely on the same principle as plunge pools; cavitation is permitted, but only at a distance from fixed boundaries so that no damage will result. These designs are dependent on the incoming flow being parallel and axial in order to avoid cavitation damage.

Structural failures of energy dissipators are often, but not always, the second phase of a failure sequence starting with hydraulic insufficiency. The geometry and the particular hydraulics conditions may result in quite divergent mean and transient loads on structures. Generalisations of loadings are, therefore, difficult. Wear, degradation and lack of structural maintenance may also lead to failures.

Energy dissipators tend to be designed for symmetrical flow, but are very often operated with clearly unsymmetrical flow – a clear violation of the rather common (but perhaps unrealistic) design basis. If sidewalls in energy dissipators are not vertical or steep, or if the flow is not well distributed over the entire width of the basin, for instance where several spillway openings discharge into a common basin, back eddies will be created. Such eddies may tend to draw solids such as stones or rocks into and around the inside of the stilling basin, and these will erode the invert and disturb the flow pattern.

A similar effect may be created by the surface roller of a traditional energy dissipator, which may keep floating debris being thrown around for an unnecessarily long time inside the basin. This type of dissipator is hydraulically more efficient than the surface jet–bottom roller type, but is probably more taxing for fish in terms of supersaturation. A combination type of arrangement has sometimes been chosen where the high-velocity flow is directed to the surface for minor flows but to the bottom for (very) large flows. This necessitates both flow separation and reattachment, which are both sensitive areas with a potential for large pressure fluctuations and cavitation.

Energy dissipators for higher heads are usually provided with a concrete invert slab to protect the foundation from erosion by high flow velocities. Due to the significant difference in water depth between the two ends of a hydraulic jump, steady-state leakage under the invert slab from the downstream end to the upstream end may destabilise the slab in the upstream part.

Where water is flowing along a straight, fixed boundary, the local water level is a measure of the mean hydraulic pressure against that fixed boundary. If the fixed boundary or the path of the flow deviates from a straight line, the mean pressure may also deviate from the local elevation of the water surface. Thus a vertical curvature, such as is often found in the upstream end of a dissipator, will add a centrifugal force to the pressure on the invert. An offset turned into the flow may significantly increase the pressure.

The pore pressure underneath the invert slab will be dependent on the often limited possibilities of drainage through or around the upstream part of the invert slab. Seepage will occur from areas with higher water tables, such as the downstream river, the downstream end of the dissipator, the reservoir and the surrounding terrain. High pressures under the slab and low pressure (water depth) above it may actually lift the slab. Similar conditions will apply over the entire area if the basin is dewatered for inspection and maintenance. The pore pressure under the slab can only be estimated on the basis of a number of assumptions regarding these different seepage and drainage paths, which may be difficult, especially as they may change during spillway operation due to clogging and/or changed effective stresses in the foundation.

Apart from erosion, the main reason for structural instability is the rapidly varying pressures on the dissipator invert and side walls.

In flows that are not sufficiently aerated these instantaneous low pressures may reach point values below the partial pressure of water vapour and initiate cavitation in energy dissipators, with the incoming velocity head exceeding around 20–25 m. For higher heads the exposure of the invert slabs to cavitating conditions will remain for successively longer periods of time and the potential for damages will increase.

There is also significant spatial correlation in the fluctuations in stilling basins, which may create substantial fluctuating forces of uplift. Where there are joints or cracks in the invert slabs the fluctuations in pressure may reach, and also be propagated to, the lower face of the invert slabs. This propagation appears to take place very rapidly and without appreciable damping.

Figure 9.7 Service spillway with ogee crest and plunge pool, Crystal Dam, Gunnison River, CO, USA. (Courtesy of USBR)

The design of inverts and the resistance to uplift have been investigated by Bellin and Fiorotto (1995), Fiorotto and Salandin (2000), Pei-Qing and Ai-Hua (2007) as well as Mahzari and Schleiss (2010). With regard to the possibility of cavitation in hydraulic jump basins the investigations made by Chanson (2011) are also relevant. The sensitive area is the upstream part of the invert, which is exposed to high flow velocities but is not well protected by the spontaneous aeration of the jump, which may only reach the invert some distance further downstream. Also, the area downstream of the energy dissipators will be exposed to a somewhat raised level of turbulence until decay has normalised the flow (Urban *et al.*, 2005).

#### 9.2.3.2 Plunge pools

The energy-dissipation capacity of plunge pools is a function both of the magnitude and intensity of the plunging jet of air/water and of the pool dimensions (Figure 9.7). The air content and the turbulence of the jet also have considerable impact on the dissipation capacity. Perhaps the most important parameters describing the function of a pool are the water depth at the jet impact point and the inclination of the jet, as the passage of the jet through the water is where most of the dissipation takes place. This will, of course, be governed by the tailwater elevation and the pre-existing elevation of the pool bottom. However, the effect of the plunging jet itself will tend to lower the water surface locally, and this must be taken into account. Also, the areal extent of the plunge pool is relevant, as is the general flow pattern in the pool and its outfall. The depth of water may change significantly during operation due to erosion/deposition.

The meaning of the term ‘failure’ needs a specific definition for a plunge pool, as it is often (usually) expected that there will be some physical damage (erosion) to plunge pools. The definition may be tied to the occurrence of incapacitating damages to the ski jump or trajectory bucket delivering the flow to the plunge pool, or damages to the surrounding structures or terrain. Alternatively, failure may be defined as an occasion when there is a (immediate) need for repair and upgrading works.

### 9.2.4 Structural analysis

Analysis of the static behaviour and overall stability of civil engineering structures, such as different parts of spillways and conduits, retaining walls and embankments, has been performed for a very long time, using methods of varying complexity aiming to describe failure modes such as sliding, overturning and local overstressing. For hydraulic structures the functional requirements usually also include water-tightness. This places increased focus on limiting deformations. Specifically, the following question arises: ‘Will a structure remain water-tight after the deformations needed to produce the required shear resistance in the foundation or anchor tension?’

For hydraulic structures there are two additional failure modes to consider; namely, erosion and cavitation damages, which lead to more gradual failures. The affected structure may also change its capacity to resist other failure modes. In other words, erosion and cavitation may change the state of a structure, as will deterioration due to ageing, weather or other physical, chemical or biological processes.

Classic analysis centres on finding the limits of a structure's ultimate load-bearing capacity or serviceability limit in accordance with norms and standards. These do typically not consider the possibility of new or different loads or changes in loadings on structures caused by other components of a system not operating as originally assumed. In more recent decades analysis has also been performed using stochastic methods to determine the probability of failure or serviceability of structures under different loadings.

Structural models have usually been designed to model a particular failure mode, and sometimes more than one mode. The analysis is often performed in a pre-construction design stage and deals with the structures 'as designed', and not always with much regard to possible construction deficiencies and future deterioration of structures or foundations. Nor are the possible design faults considered.

All analysis is based on certain modelling assumptions regarding the behaviour of the structural materials and the loading. These include, for instance, (linear) elastic and plastic behaviour. One of the more important assumptions is that the superposition principle is valid, so that the analysis can be broken down in parts and the results then added up (superimposed) to get the complete picture. Superposition is, however, strictly valid only for structures that exhibit entirely linear behaviour, where there is no correlation between (previous) loadings or time of operation and the structural behaviour. Rock, soil and reinforced concrete do not exhibit a true linear behaviour, and civil engineering structures used for spillways will certainly wear and deteriorate with use.

Modelling of bearing capacity and erosion of soils has generally been more successful than modelling of rock. As rocks are usually much more inhomogeneous than soils, the available information on rock properties tends to leave considerable room for uncertainty. Models of rock stability have, therefore, often been based on classification of rocks by means of systems such as the Q or RMR systems. Methods for assessing different kinds of erosion in rock have been proposed over the years by, among others, Schoklitsch, Jaeger, Veronese, Doddiah, Ervine, Hartung, Häusler, Martins, Mason, Canepa and Hager (see discussions and references later in this chapter).

Today, both static and dynamic analysis is done using computer-aided numerical methods based on finite elements or finite differences and computer-aided design (CAD), which permits analysis also of non-linear sequential failures.

In system simulations, the correlation between and the changes in structural properties can and should be taken into account by updating the state of the structures and their surroundings. Complex behaviour, however, remains a difficulty in systems modelling given the necessary restraint on computation time for the main loop. Pre-analysis may often be the one viable method of handling non-linear behaviour in components when simulating large systems.

#### 9.2.4.1 Present status

In order to assess the future reliability of a facility such as a water conduit its present status must be determined and its future development described.

The present status may be determined with the help of visual site inspections and/or detailed physical measurements, as required. Parameters to be addressed are those that influence structural behaviour or flow. These include structural damages that may have occurred due to physical, chemical or biological processes, such as cracking, spalling, crushing, sliding, lifting or tilting, as well as corrosion and alkali–aggregate reactions such as alkali–silica reactions.

The historical record of construction, operation and weather conditions may show when the damages occurred and whether they were caused by external disturbances (e.g. frost–thaw cycle, ice, fire, seismic action, slides, excessive rainfall and vandalism (sabotage)) or by internal disturbances related to the operation (or misoperation) of the facility, the use of less suitable materials, or other weaknesses emanating from the construction or design. Operational causes include cavitation, uplift, erosion or abrasion by debris and sediments.

The determination of the historical damage mechanisms is important for the assessment of the facility's future behaviour. It is equally important, of course, to establish the nature, timing, quantity and quality of the maintenance work performed, as the structures' present status and future behaviour are the result of all these forces. Some design and construction defects may have already manifested themselves in physical damages; others may not have done so, as they are coupled to specific events that have not yet materialised.

#### 9.2.4.2 Future status

Future damages and deterioration need different kinds of modelling in time and space.

Some types of damages can be expected to occur at specific locations because of high or variable loading leading to fatigue due to difficult working conditions during construction, recognised weaknesses in the foundation, potential slide zones or known groundwater outflows (Thun, 2006).

Other types of damages may best be modelled as being spatially random in occurrence, such as possible unknown local weaknesses in concrete structures due to insufficient batching and segregation of aggregates, insufficient bonding between individual anchors and rock or unknown weaknesses in the rock itself.

A third type of damage could initially be portrayed as spatially random, whereas new damages might be more likely to occur primarily in areas where there are previous damages of a similar nature.

The occurrence and timing of future wear and damages may be modelled based on certain operation, time, climate and maintenance. Also, the outage time for planned or unplanned future maintenance work needs to be modelled. The correlation between different kinds of damages and operation, weather or time needs to be considered.

## 9.3. Hydraulic analysis

Hydraulic analysis, like structural analysis, geotechnical analysis and hydrology, is one of the major tasks in the engineering effort involved in constructing dams, power plants and spillways. Hydraulic analysis involves different types of flow interacting with the structures and creating phenomena that must be addressed using different analysis techniques. This subsection provides a broad overview of the techniques involved in hydraulic analysis, important considerations and the theoretical basis of the hydraulic phenomena involved. The subsection on the theory may attract readers who are especially interested in the physics of the hydraulic phenomena. However, it is not possible here to cover all aspects of hydraulic theories, and for complete coverage of each of the phenomena touched upon in this chapter the reader is should consult the numerous textbooks and special literature on this topic.

There are two different aims of analysing the flow in water conduits such as spillways: to establish the general flow pattern along the conduit, and to determine the detailed loadings the water flow will exert on the different parts of the spillway and its surroundings. Both these objectives should be met in a computationally efficient way. Depending on the layout of the system, this may call for an advanced, and therefore time-consuming, hydraulic analysis, which may have to be done entirely as a pre-analysis to the main simulation loop or, alternatively, as a more basic analysis inside the loop. In this section the nature of the problem to be analysed is described qualitatively. A flavour of the computational dimensions of these considerations is given in Section 9.7, with more details being available in the textbooks and papers referenced therein (e.g. a more complete treatment of fluid mechanic and hydraulic analysis may be found in Granger (1995) and in the standard text by Chow (2009)).

Irrespective of the method used to analyse a flow field, adequate boundary conditions need to be established. At fixed boundaries, such as the structural walls of a water conduit, the condition is usually zero flow velocity across and along the boundary. A separate question is whether the fixed boundaries can be considered fixed in the same position during the period of simulation. Gate positions are, of course, not usually fixed.

The ambient pressure is normally assumed to be atmospheric pressure (i.e. depending on the elevation above sea level and the weather situation at the free water surface). If airflow is included in the analysis, conditions and mechanisms regarding air velocities, air entrainment and transfer of energy and momentum may also be needed.

At the invert of a straight chute the (mean) pressure may be taken as being equal to the sum of the water depth and the atmospheric pressure. Where the invert has another shape so that flow lines are curved, the centrifugal force will modify the pressure by an amount proportional to the depth, the square of the velocity and the inverse of the flow radius, as integrated over the depth of flow. This mechanism is used to raise the discharge efficiency of appropriately shaped overflow sills by reducing the pressure on the sill.

At the upstream end of a water conduit to be analysed, the water level, pressure level and flow velocities will have to be specified. In the case of a large reservoir this may default to only a water level.

Similarly, at the downstream end of the analysed field a tailwater condition needs to be applied. This is an area where mistakes frequently occur. It is often (erroneously) assumed that tailwater conditions will remain stable even during adverse hydraulic conditions, such as spillway discharges larger than those for which the downstream river can reasonably be expected to be stable.

Moreover, uncertainties are sometime caused by the interface between different specialists. The precision of an estimate that may be considered reasonable in a hydrological analysis is often totally unacceptable to serve as boundary condition in a hydraulic analysis. In a more advanced analysis conditions regarding heat flow may also be required across boundaries.

### 9.3.1 Methods of analysis

The methods available for characterising the flow in spillways are theoretical analysis from first principles and empirical methods, but usually a mixture involving both deduction and induction is used. Empirical data may be based on precedents from individual water conduits in operation or from hydraulic model tests. Systematic laboratory investigations have been published in periodicals, textbooks and manuals.

The following text may be seen as an introduction to the analysis of the flow specifically in water conduits such as spillways.

#### 9.3.1.1 Theoretical fluid mechanics

Theoretical fluid mechanical analysis is founded on a number of assumptions and simplifications of the true behaviour of the flow, the primary one being that the water is treated as a continuum (rather than a discrete fluid) comprising separate parcels of water molecules packed together in an indeterminate manner. Thus fluid properties such as mass, density, pressure, viscosity, velocity and temperature are assumed continuous in the analysis.

No closed-form solution is known to the general equations of flow in spillways. A number of additional assumptions and conditions can, however, often be made to simplify the problem. Often, for the flow analysis the flow in spillways can be treated as single-phase, isotropic, incompressible and steady.

The state of a fluid defines the relationships between pressure, density and temperature. For water, which can be considered incompressible (at pressures well above the partial pressure of vapour), only the dependence on temperature remains, and that may not always be important for water flow in spillways.

Quite often the flow can also be treated as steady state, which generally means independent of time. In fluid mechanics, however, the term 'steady' only means that the time averages of the studied variables are independent of time.

Turbulent flow is characterised by rapid fluctuations in variables such as velocities and pressures. Turbulence is a complex interaction between vorticity and velocity created by friction, for instance when a viscous fluid moves along a surface such as the invert or sidewalls of a water conduit. The fluid layer next to the surface will cling to the surface, meaning that there will be a gradient in velocity, from zero at the fixed boundary (wall) to the free stream velocity at a distance from the wall. This gradient will induce shear in the fluid, which will continue to deform as long as the gradient exists.

#### 9.3.1.2 Analytical and empirical methods

Due to the difficulty of finding closed-form solutions to the available equations, simplifications are usually sought. Among the more useful ones are situations where flow processes can be treated as one-dimensional, isothermal and irrotational. For flow in spillways all these simplifications can often be applied to part of the supercritical flow, but not to boundary layers or developed turbulent flow, and in particular not to hydraulic jumps.

In isothermal processes there is no change in total internal energy, as friction and viscosity will result in temperature changes in the fluid that, by definition, are totally lost to the control volume by heat exchange with the surroundings.

Usually, the conservation of mass and conservation of energy or momentum is the basis for a somewhat simplified analysis. Direct numerical simulation of these equations has only been done for very simple flow situations, and requires the use of supercomputers and extreme computation times.

#### 9.3.1.3 Computational fluid dynamics (CFD)

A variety of computer codes are commercially available for analysis of one-, two- or three-dimensional steady or unsteady flow in spillways using reasonably stable numerical integration methods. However, the computer time and storage needed to treat the differential equations for the smallest eddies in, for instance, turbulent boundary layers are extremely large (although advances in computing speeds are changing this), which implies that empirically based approximations are used, such as different versions of the $k - \varepsilon$ model originally proposed by Launder and Spalding (1974) for modelling turbulence. Layered flows, erosion and sedimentation can be handled using these codes, albeit approximately (empirically). Numerical instability is still problematic in highly turbulent areas such as hydraulic jumps.

The assumptions and shortcuts that are required to solve the flow equations economically using computer codes do not always produce good results for flow in spillways because the computational methods are general in character, having been developed

Figure 9.8 Hydraulic jump at Saint Anthony Falls, Minneapolis, MN, USA

to cover a wide spectrum of flow situations, not specifically for flow in spillways. Movement of the phases in different directions at the same location cannot be handled analytically from first principles by the available codes. The codes generally treat the fluid at a particular location as being a mixture of the different phases present at that point (e.g. air and water). Two-phase air–water flows in spillways, including the flow on stepped chutes, are therefore handled empirically by using the presently available three-dimensional codes. Chaotic turbulent transitions between supercritical and subcritical flow (hydraulic jumps) are quite difficult to handle in a reasonable manner (Figure 9.8).

Especially for gradually varied one- or two-dimensional flow, the analysis is sometimes based on modified equations for conservation of mass and linear momentum, such as the depth-integrated St Venant's equations or the shallow water equations. They cannot treat hydraulic jumps or two-phase air–water flows.

Purpose-built one-dimensional codes supported by empirical data for hydraulic jump zones and for air entrainment/detrainment may be the simplest means available for the analysis of water conduit flow. One-dimensional steady hydraulic analysis of the flow in spillways is usually done using the Bernoulli equation, which is based on the conservation of mass, potential energy and kinetic energy, modified to take into account turbulent (so-called) losses of energy.

The use of all these different computer codes requires a very good understanding of fluid dynamics/hydraulics to appreciate the quality, or the lack of quality, in the computations, and also requires a very good knowledge of the computer code used. The codes generally contain a number of alternatives to be selected by the user, which can change the results significantly. There are also empirical add-ons for difficult situations such as hydraulic jumps, although these have limited theoretical foundation. Validation is therefore of prime importance, but at the same time very difficult. Validation can only be made by comparison with empirical data from measurements of flow in natural-size prototypes or physical hydraulic models.

#### 9.3.1.4 Hydraulic modelling

The major part of the empirical database emanates from laboratory investigations on scale models, where the findings are quite often intentionally reported in a form amenable to practical application for water conduits, often as simple power formulae with some or limited theoretical basis. The coefficients and exponents may then reasonably reflect laboratory results within the range of governing parameters actually investigated, but the formulae generally do not claim to be theoretically or functionally correct. They therefore may have limited predictive or descriptive value outside the investigated parameter range, which is not always reported in full. Moreover, there may be scale effects not known or not correctly accounted for, making it difficult to use the laboratory results directly. Accordingly, considerable caution is called for.

The dissimilitude between prototypes and models, the so-called scale effect, is a subject of considerable discussion and some controversy, in particular with regard to cavitation and aerated flows (Arndt *et al.*, 1968; Baur *et al.*, 1998; Chanson and Gualtieri, 2008; Eickmann, 1993; Ohtsu *et al.*, 2009; Pinto *et al.*, 1992).

Figure 9.9 Conveyance structure at Llyn Brianne Dam: a series of roll waves moving down the spillway, terminating in a stationary hydraulic jump

In hydraulics, as in fluid dynamics, 'steady' flow means that the time averages of the studied variables are independent of time. Turbulence will, however, induce rapid fluctuations in local velocities and pressures, not only in unsteady but also in steady flow. Thus the water pressure against fixed boundaries also varies with time around the mean value usually considered in steady-state analysis. In hydraulic model tests these fluctuations are wide, with frequencies in the range 1–50 Hz. For prototype water conduits frequencies may be near the lower bound of this range.

Hydraulic modelling always contains an element of compromise, where model scales and materials are chosen to reflect the forces dominating the process to be studied, while other forces may not be (very) well represented by the models. Different dimensionless numbers are used to signify the ratio of inertia forces to the forces of gravity, viscosity, pressure and elasticity. These numbers are the Froude number, Reynold's number, Weber number, Euler number and Mach number.

If the relevant dimensional numbers are kept the same in the model as in the full-scale (prototype) structure, the relevant forces are correctly represented and the model will be accurate in that particular respect. Unfortunately it is not possible to retain these number values, as the physical scale is being reduced, unless different materials are used in the model and the prototype, which quite often is not practicable or economical (Figure 9.9).

Physical hydraulic model tests on water conduits with free surface flow are normally done applying full similarity in terms of the Froude number while trying to keep the Reynold's number reasonable but not fully correct and the Weber number large enough to avoid significant influence of surface tension.

Closed-conduit flow should, ideally, be modelled physically by the Reynold's similitude, but this is normally not possible, as a decreased physical scale requires an increased flow velocity to keep the Reynold's number constant. Closed-conduit flow is, therefore, often modelled according to the Froude similitude, while respect for the Reynold's similitude is limited to maintaining the same type of flow in the prototype and the model (i.e. laminar flow or smooth, transitional or fully rough turbulent flow) by not reproducing the boundary roughness according to the chosen physical model scale.

The Weber number also needs to be considered, particularly when modelling two-phase flow, such as aerated flow in surface spillways or closed conduits. However, satisfying all these conditions is not possible. All closed-conduit flows and all aerated flows are accordingly subject to scale effects, which may be both significant and difficult to assess.

### 9.3.2 Types of flow

The flow in water conduits such as free surface spillways may be divided into a number of zones with different types of flow. There may, for instance, be a zone with subcritical flow in the upstream end (the approach canal) followed by an intake with a critical section (at a gated or ungated sill or ogee crest), where the flow changes to supercritical on the chute. Again there is often a stretch of subcritical flow in the downstream end of the water conduit. While the transition from subcritical to supercritical is smooth, the transition from supercritical to subcritical downstream usually leads to the creation of a hydraulic jump with considerable energy dissipation, where intense turbulence transforms kinetic energy into heat.

Computations to locate a transition from supercritical to subcritical flow have to be based on knowledge or assumptions about the flow at both the upstream and the downstream boundary of the water conduit. The result of the analysis needs to be entirely consistent with these boundary conditions. If this turns out to be impossible, the assumed boundary conditions are incompatible, and therefore erroneous, and a new analysis must be performed based on a different set of boundary conditions.

In some water conduits more than one type of flow may be possible at given upstream and downstream boundary conditions. This applies in particular to closed conduits and to spillways with short energy-dissipating basins. Which of the different flow conditions will actually occur may then be determined using the flow conditions immediately preceding the flow situation under investigation (persistence).

Stepped chutes differ from ordinary concrete-lined chutes in that the roughness elements (the steps) tend to be of the same order of magnitude as the flow depth and also to be lined up in parallel, perpendicular to the flow. This leads to flow for which a different set of rules apply and other types of flow need to be distinguished.

Low-level outlets, bottom outlets and water conduits for power stations normally exhibit closed-conduit flow at least in some part. The methods used to analyse steady, developing and developed closed-conduit single-phase water flow are basically the same as for free-surface flow. Tunnels and culverts may, however, also operate partly or fully with free-surface flow, which creates special difficulties of analysis. This is due both to the need to establish the type of flow and due to induced airflows, which may cause macro-instabilities and pulsating flow.

### 9.3.3 Developing turbulent flow

The thickness of the turbulent boundary layer described above increases downstream until it reaches through the entire body of water (Figure 9.10). It develops on spillway chutes downstream of the intake sill, if the water depth upstream of the sill is large and the flow velocities low. On a wide chute this zone of developing flow terminates where the boundary layer has progressed from the invert through the entire depth of flow to reach the free water surface at a point usually called the 'point of inception'.

It is often convenient to express the potential and kinetic energy of flow as an energy level by dividing the specific energy by the acceleration due to gravity.

#### 9.3.3.1 Theoretical basis of developing turbulent flow

Above the boundary layer the flow in spillways can often be characterised as irrotational, potential flow. The kinetic energy head of the flow in the potential flow zone equals the full hydraulic head at the intake $E_{int}$ with no so-called 'head losses' (energy transformed into turbulence and heat) on the chute. In other words, the velocity distribution in the potential flow zone on a wide spillway chute is rectangular and the potential flow velocity $U$ is

$$U = \sqrt{2g(E_{int} - z - Y\cos\varphi)} \tag{9.4}$$

Figure 9.10 Chute flow and the development of the turbulent boundary layer

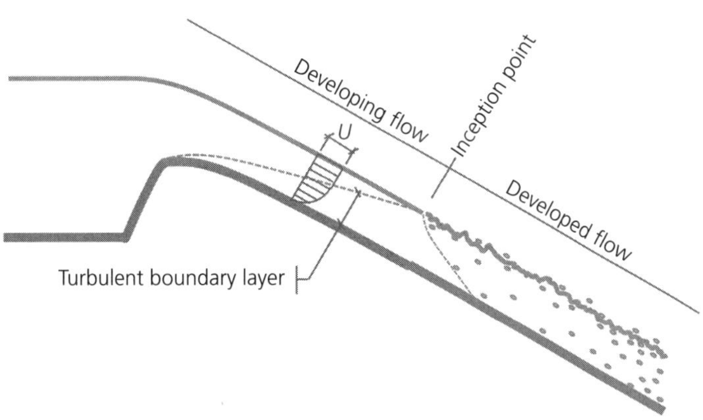

where $z$ is the elevation of the chute invert at the point considered, $y$ is the water depth and $\varphi$ is the local slope of the chute. Inside the boundary layer the time average of the turbulent flow velocity $u$ has been found to increase with the distance $y$ from the fixed boundary as follows:

$$u = U\left(\frac{y}{\delta}\right)^{1/n} \tag{9.5}$$

where $\delta$ is the local thickness of the boundary layer. The constant $n$ normally takes values between 5 and 8 on flat and smooth chutes. According to Chen (1991), the value of $n$ can be approximated by

$$n = K\sqrt{\frac{8}{f}} \tag{9.6}$$

where $K$ is von Karman's universal constant, usually taken as 0.41. The Darcy–Weisbach friction coefficient $f$ demonstrates the dependence on the roughness and the Reynold's number (Castro-Orgaz, 2010).

Several attempts have been made by different investigators to describe the development of the boundary-layer thickness along an ordinary smooth chute by simple power-law equations of the type

$$\frac{\delta}{L} = \gamma\left(\frac{L}{k_s}\right)^{\beta} \tag{9.7}$$

where $L$ is a coordinate *along* the chute, $k_s$ is the equivalent sand roughness, and $\gamma$ and $\beta$ are constants describing the boundary-layer profile.

The different methods yield significantly deviating results. Castro-Orgaz and Hager (2010) proposed that the values $\gamma = 0.0302$ and $\beta = -0.125$ would produce a good approximation both of theoretical results and of all laboratory data available for chutes sloping between 20° and 60° within the range $2 \times 10^3 \leq x/k_s \leq 2 \times 10^5$, and suggested a method with variable parameters that also takes into account the influence of wakes, roughness character and Reynold's number. The simple Castro-Orgaz–Hager equation is very close to the estimate made by Keller and Rastogi (1977), but deviates markedly from some previously used equations.

The effect of the reduced flow velocities in the turbulent boundary layer is, of course, to displace the flow surface upwards compared with its location under assumed potential flow without a turbulent boundary layer. This displacement thickness $\delta^*$ can be estimated as

$$\delta^* = \frac{\delta}{n+1} \tag{9.8}$$

where $n$ is the inverse of the exponent in Equation 9.5 describing the velocity distribution in the turbulent boundary layer, which for $n = 6$ yields $\delta^* = 0.143\delta$. The location of the inception point can then be computed as the point where the boundary layer thickness equals $1/(1 - 0.143) = 1.167$ times the theoretical *potential* flow depth. The reduction in flow velocities in the boundary layer compared with the potential velocities constitutes a measure of the head loss in the developing flow.

The characteristics of turbulent boundary layer development in various other situations have been treated by, for instance, Schlichting and Gersten (1979).

### 9.3.4 Theoretical basis of developed turbulent flow

Head losses in developed turbulent flow are usually calculated using the Darcy–Weisbach or Manning–Strickler formulae. The Darcy–Weisbach equation

$$H_f = C_f L U_m^2/2g \tag{9.9}$$

where

$$C_f = f/4R \tag{9.10}$$

relates the head loss $H_f$ to mean flow velocity $U_m$ and water conduit geometry, and is to be preferred for flow in concrete-lined spillways. $R$ denotes the hydraulic radius, which for wide chutes can often be replaced by the flow depth $y$.

For closed conduits such as unlined tunnels in rock, the Manning equation is often preferred. It may be noted that in a circular conduit $4R = D$, the tunnel diameter.

The roughness coefficients, $f$ in the Darcy–Weissbach equation and $n$ (or $M = 1/n$) in the Manning equation, are both functions of the relative roughness of the water conduit walls and the Reynold's number *Re* of the flow. Experience-based roughness values therefore have to be modified to fit with current conditions throughout the computations. For the Darcy–Weisbach friction coefficient this can be done through a Moody diagram. Experience-based Manning friction coefficients need to be recalculated if applied to water conduits of other dimensions than those that served as the empirical basis.

#### 9.3.4.1 Surface aeration

Downstream of the inception point the spillway chute flow will be spontaneously aerated from the atmosphere providing that the turbulent velocity normal to the flow surface is large enough to break the surface tension and exceeds the rise velocity of entrained small air bubbles – which means a velocity greater than about 0.2 m/s. The flow surface will then be less clearly defined and will exhibit the characteristic white-water appearance.

Diffusion will make the air successively penetrate deeper and deeper into the flow. The partially aerated zone terminates where the air reaches the chute invert and a fully aerated zone follows downstream. Experiments by Chanson (2011) indicate that it may take a distance of the order of 10–15 times the flow depth at inception before any significant aeration has reached the bottom layers in rapid chute flow.

Some further distance downstream on a long chute with uniform slope both air content and flow velocities will reach equilibrium and no longer continue to increase, as the entrainment and detrainment of air balance and the friction losses correspond to the slope of the chute. The flow in this zone can be described as uniform flow (Figure 9.11).

The depth of flow $Y_{90}$ above which the air concentration exceeds 90% is usually used as reference in mixed air–water flows. At higher air concentrations than 90% the air and water do not act like a mixed fluid due to slippage between the two phase components. The characteristic water depth $d_w$ is the depth of flow that the water component *alone* would have had it travelled with the same velocity as the air–water mixture. The flow of the water phase alone at the characteristic depth is often used for hydraulic computations of air–water mixtures, with certain minor adjustments of the material properties of water.

The characteristic water depth is determined using the local air concentration $C(y)$ through

$$d_w = \int_0^{Y_{90}} [1 - C(y)]\,dy \tag{9.11}$$

Figure 9.11 Fully developed turbulent flow at Letsi Dam, Norrbotten, Sweden. (Courtesy of Vattenfall AG)

and the mean air concentration $C$ is defined through

$$d_w = (1 - C)Y_{90} \tag{9.12}$$

An aerated flow with a 30% mean air concentration thus has a characteristic water depth $d_w$ corresponding to 70% of $Y_{90}$, or in other words $Y_{90}$ will be $1/(1 - 0.30) = 1.43$ times the characteristic water depth. In fact there will often be considerable amounts of water splashing above the depth $Y_{90}$, and water conduit wall heights need to be adapted to this fact to avoid overtopping.

Sidewalls also need to take into account instabilities leading to variations in flow depth and cross-waves, due to changes in geometric alignment or cross-section, including the effects of roughness in walls and inverts, as well as the superelevation due to centrifugal forces in bends.

The entrained air has other effects on the flow apart from raising the air–water levels and making the upper surface of the air–water mixture less defined. The aeration also influences the turbulence characteristics of the flow. Prototype and laboratory investigations (Chanson, 1993) indicate that mean air concentrations above some 20% tend to reduce the friction coefficient of the water component significantly for turbulent rough flows.

As noted above, it is common to calculate the flow of aerated flows by first considering the corresponding single-phase flow of water $q_w$ with the characteristic depth $d_w$ using the Darcy–Weisbach equation. A modified friction coefficient $f_e$ may then be used instead of $f$ to approximately compensate for the effect of the mean air concentration $C$ in the flow:

$$f_e = f[1 - 1.6(C - 0.2)]0.2 \leq C \leq 0.7 \tag{9.13}$$

#### 9.3.4.2 Gradually varied aerated flow

A quite significant length of chute downstream of the inception point will exhibit gradually varied, aerated turbulent flow as the uniform flow zone is reached only asymptotically.

Chanson (1993) noted that the development of the mean air concentration $C$ could, under certain assumptions, be estimated for a chute of constant width and constant slope angle $\varphi$ downstream of the inception point on the basis of the water depth at inception, the specific discharge and the bubble rise velocity using the following continuity equation:

$$kX + K_0 = \frac{1}{(1 - C_e)^2} \ln\left(\frac{1 - C}{C_e - C}\right) - \frac{1}{(1 - C_e)(1 - C)} \tag{9.14}$$

$$K_0 = \frac{1}{1 - C_e}\left(\frac{1}{1 - C_e} \ln\left(\frac{1 - C_\star}{C_e - C_\star}\right) - \frac{1}{1 - C_\star}\right) \tag{9.15}$$

$$k = \frac{u_r d_\star \cos\varphi}{q_w} \tag{9.16}$$

where $C_e$ is the mean air concentration, $d_\star$ and $C_\star$ are the water depth and the average air concentration, respectively, at the point of inception, $q_w$ is the specific discharge (of water) and $u_r$ is the bubble rise velocity. The coordinate along the chute from inception $x$ is normalised by division by the depth at inception, so that $X = x/d_\star$.

Chanson (1993) also provides estimates of the bubble rise velocity as a function of the flow velocity at inception.

#### 9.3.4.3 Uniform aerated flow

The average air content at *equilibrium* on a concrete chute of uniform slope angle $\varphi$ degrees may, according to Hager (1991), be estimated as

$$C_e = 0.75(\sin\varphi^{0.75}) \tag{9.17}$$

while the equilibrium bottom air concentration $C_b^e$ may estimated as

$$C_e^b = 1.25\left(\frac{\pi\varphi}{180}\right)^3 \quad \varphi \leq 40 \tag{9.18}$$

$$C_e^b = 0.65\sin\varphi \quad \varphi \geq 40 \tag{9.19}$$

This means that once the equilibrium has been reached in free surface flows on chutes steeper than about 16–23° the (uniform) air content may be sufficient to avoid cavitation damages without artificial aeration.

It is worth noting that the influence on air concentrations of the chute slope reflected in these equations is probably largely caused by the higher equilibrium flow velocities on steep chutes and the correspondingly higher degree of turbulence. The equilibrium bottom air concentration could then also be estimated as $C_e^b = 10^{-7}(U - 7.5)^4C$ based on the mean flow velocity $U$.

### 9.3.5 Developed turbulent flow

A quite significant length of the chute downstream of the inception point will exhibit gradually varied, aerated turbulent flow, as the uniform flow zone is reached only asymptotically.

### 9.3.6 Pressure fluctuations, uplift and cavitation

The pressure fluctuations in stationary undisturbed turbulent flow on spillway chutes are typically positively correlated over an area with a traverse length of the order of four times the local turbulent boundary layer thickness $\delta$ and a streamwise length approximately equal to the boundary layer thickness. In the hydraulic jump zone these pressure fluctuations will be several times more severe in amplitude, and in addition there will be spatial correlation over much larger areas. These fluctuations may cause both cavitation and uplift.

Due to the spatial correlation, a considerable rapidly pulsating force will act on the boundaries, potentially causing vibration or physically lifting slabs of insufficient weight and foundation anchoring. In particular, slabs simultaneously exposed to high pressures underneath by leakage from other parts of the water conduit itself, from the reservoir or from surrounding terrain may be at risk. The pressure fluctuations may also reach the lower face of a chute slab through local cracks and leaky joints, and act on the slabs out of phase with the fluctuation on the topside.

Uplift needs to be considered, especially for low-lying energy dissipators. Tunnel linings exposed to pressure fluctuations may be better off on account of the closed form and the support of the surrounding rock. The possibility of cavitation needs to be taken into account for rapid flow both on chutes and in tunnels, energy dissipators and plunge pools.

#### 9.3.6.1 Cavitation

Cavitation is a phenomenon caused by the implosion of gas (vapour) bubbles in the water, giving rise to shock waves with pressures of up to 1500 MPa according to Kramer (2004). Where cavitation occurs close to fixed boundaries of a water conduit, damages may occur by cavitation erosion or by other processes initiated by cavitation. There is accordingly a need to differentiate between cavitation as a physical phenomenon and cavitation-initiated damages, usually called 'cavitation damage' (Figure 9.12).

Water conduits with high flow velocities may exhibit cavitation where flow conditions create zones with low pressure, for instance due to instabilities, turbulent pressure fluctuations or vortices formed in sheared flows. Vaporous cavitation, which occurs when the local pressure in the flow equals the partial pressure of water vapour, is usually the dominant source of problems in water conduits such as spillways. The potential for vaporous cavitation in undisturbed flow can be described by a cavitation index $\sigma$, sometimes called a Thoma number, which is a local dimensionless number describing the ratio of the mean pressure margin at a specified point to the partial pressure of water vapour, normalised by the kinetic energy of the flow. The cavitation index is often expressed as a ratio of pressures as

$$\sigma = \frac{p - p_v}{\rho U^2/2} \tag{9.20}$$

where $p$ is an absolute reference pressure, (usually, but not always, the mean pressure at the point or section under consideration), $p_v$ is the partial pressure of vapour at the ambient temperature, $\rho$ is the water density and $U$ is a reference velocity (for flow on spillway chutes this is often the potential or mean flow velocity).

Alternatively, the cavitation index can be expressed as a ratio of pressure heads and kinetic heads as

$$\sigma = \frac{H_p - H_v}{U^2/2g} \tag{9.21}$$

As the atmospheric pressure is part of the pressure head margin $H_p - H_v$, it follows that both the elevation of a site above sea level and also the current weather will influence the cavitation potential.

Figure 9.12 Cavitation damage. (Courtesy of BC Hydro)

The cavitation index $\sigma$ of the flow at a certain point is compared with the incipient cavitation index $\sigma_i$, for which vaporous cavitation has been found to start at the given boundary roughness and flow conditions. Cavitation can thus occur if $\sigma \leq \sigma_i$. Even if cavitation occurs it may not necessarily cause significant cavitation damages, as the duration of the moments with low pressures may be extremely short for indices just below the incipient; by definition these are just brief moments during the very largest pressure pulses.

Depending on the construction material, a critical cavitation index $\sigma_c$ is also sometimes used, below which 'damages can be expected' under the circumstances at hand.

As the pressure fluctuations are random in nature, neither of these cavitation indices can be defined precisely; in particular, the critical cavitation is subjective in nature. Moreover, the actual damage process is not sufficiently known and varies from one installation to another. The main difference between incipient cavitation and critical cavitation may just be the time of exposure to cavitating pressures. Critical cavitation may well just be a concept invented by and for engineers in an attempt to bring some clarity to the disturbingly vague picture.

Cavitation due to pressure fluctuations is a danger, especially where there is particularly intense turbulence, for instance in zones of intense energy dissipation such as at hydraulic jumps. Unfortunately, the field of turbulent pressure variations in such zones is spatially variable to the extent that using generalised cavitation indices is not meaningful.

Vortices may lead to very long exposure times to low pressures, and therefore may cause more severe cavitation damages at higher cavitation indices than in undisturbed chute flow.

Cavitation on spillway chutes may be caused in an otherwise undisturbed chute flow by an isolated roughness element, such as a misalignment, ridge, groove (e.g. an open joint) or hole. In general, two-dimensional irregularities (extending cross-flow) are more serious than three-dimensional irregularities. Cavitation may, however, at least theoretically, also occur in flow over perfectly straight chute surfaces with uniform distributed roughness.

#### 9.3.6.2 Cavitation damages

Cavitation damages have, according to Kramer (2004), been reported at a fairly large number of dams in the USA. Other reports come from Tarbela Dam in Pakistan and Karun Dam in Iran.

Cavitation damages occur to structures by a process requiring time. It is therefore important whether cavitating conditions occur frequently or only during brief moments of infrequent major pressure pulses at unusual discharge conditions.

In undisturbed flow down a straight chute with $\sigma \leq \sigma_i$, the duration of cavitating conditions may be quite limited. On the other hand, if $\sigma \ll \sigma_i$ the duration may be considerable.

At present there is no general agreement about how cavitation-initiated damage development should be modelled mathematically. The actual process of cavitation-initiated damage development is highly dependent on the boundary material, its geometry and roughness elements, its resilience (ductility and tensile strength), the flow velocities, the pressure margins, the air content of the flow and the time of exposure to these conditions. The damage process also involves a considerable element of chance due to the random nature of the loading and the presence of micro-cracks and minute weaknesses in the boundary material, the presence of which cannot be determined as they keep developing. Another unknown is how the uniform and isolated roughness of a chute will develop after construction in the future operating life due to wear, abrasion, climatic influence and cavitation. The maintenance (policy) is obviously of importance.

The intensity of cavitation damage (i.e. the rate of damage development) has, for example, been proposed to be proportional to the flow velocity raised to anything from the 6th to the 12th power. Falvey (1990) argued, based on laboratory experiments, that the process has several different phases, starting with an incubation phase during which not much happens, followed by an accumulation phase when the damage rate increases, an attenuation phase and a steady-state phase, although not necessarily in that order. Falvey proposed that the likely rate of damage development $D$ would be proportional to

$$D \propto \left(\frac{1}{\sigma} - \frac{1}{\sigma_c}\right) U^6 \quad \sigma < \sigma_c \tag{9.22}$$

where $U$ is a characteristic flow velocity.

However, the empirical background of this proposal appears to include only damages occurring at cavitation indices in the limited range $0.13 < \sigma < 0.19$, which means that the damage rate could equally well be described by an expression of the type

$$D \propto \left(\frac{1}{\sigma} - \frac{1}{\sigma_i}\right)^2 U^6 \quad \sigma < \sigma_i \tag{9.23}$$

based on the incipient cavitation index $\sigma_i$, without the need to define a critical cavitation index $\sigma_c$.

It has also been suggested that the rate of cavitation damage might be proportional to the logarithm of the operation time; in other words, the rate decreases with time as the distance from the cavitation-initiating feature to the fixed boundary increases successively as more and more material is eroded away. While this is reasonable for pure cavitation damages, it does not mimic other important processes after the initiation of damages by cavitation. These include effects such as water jets impacting directly on rough and damaged surfaces, vibration of exposed reinforcement bars and loss of structural bearing capacity and rigidity.

In fact, cavitation damages may rapidly lead to loss of control of the performance of relatively thin reinforced structures, and therefore also to loss of control of the flow. Other, more substantial, structures may be much more tolerant of cavitation damages.

The damages, therefore, cannot be expressed in a generalised way as a function of operation time, areal extent or depth of erosion, but have to be considered on a case-by-case basis considering both the flow conditions and the structure at hand.

In actual practice the time required for serious cavitation damages to occur may be rather short. The experience from Yellowtail, Glen Canyon and Karun Dams is that serious damages, even in quite substantial structures, can develop in a few weeks with cavitating operation. At Glen Canyon, 11 m of concrete and underlying rock were eroded away in 3–4 weeks.

In many cases it will be sufficient to establish only that the cavitation intensity is high to predict that serious damages will follow shortly if such operation continues. High intensity may be taken to mean $\sigma \ll \sigma_i$.

If the cavitation intensity is assessed as more moderate, Bayesian updating based on site-specific experience could possibly be considered for the assessment of future behaviour (Beck and Au, 2002).

Figure 9.13 Empirical cavitation damages in spillways. (Falvey, 1990)

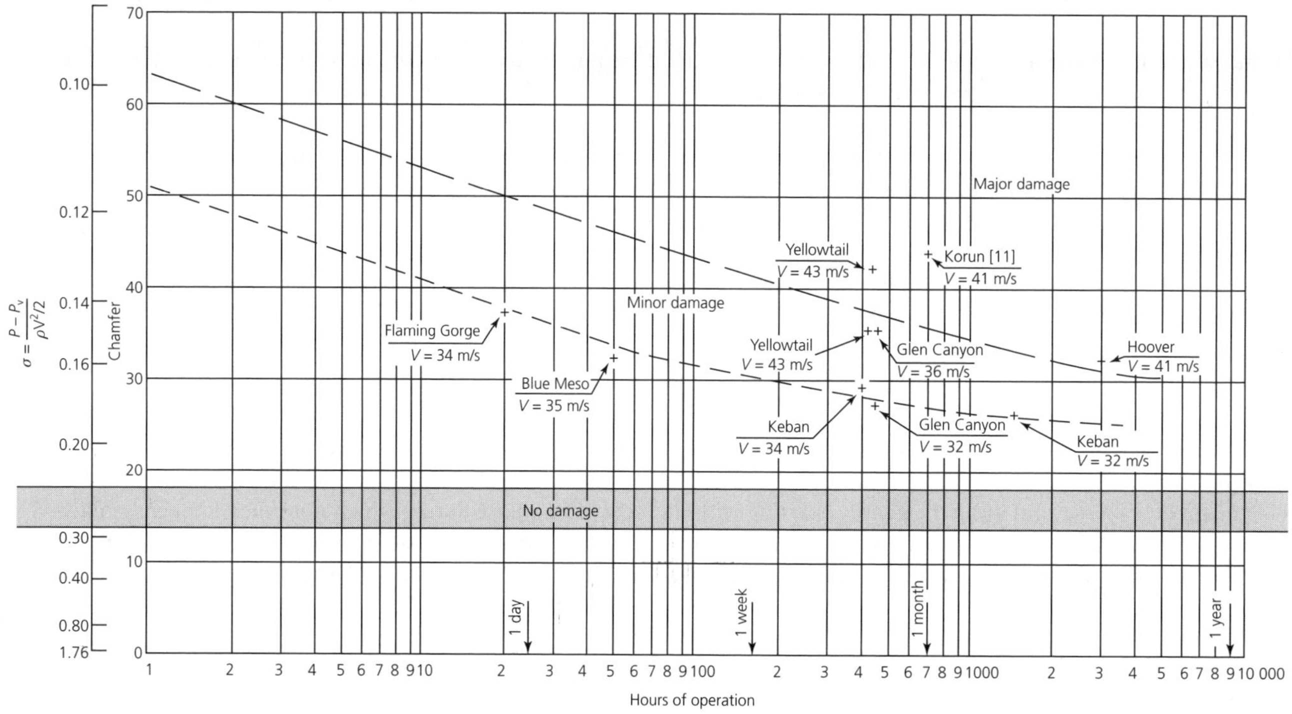


Control or prevention of cavitation damages may be achieved by aerating the flow, by raising the cavitation index (lower flow velocity, better pressure margins) or by better control of boundary shape and irregularities. Falvey (1990) has proposed guidelines for the latter, these primarily being aimed at rectifying or extending the technical life of existing structures with cavitation indices below $\sigma = 0.60$ (Figure 9.13).

It is, however, very difficult to assess and control future surface roughness of a concrete surface due to the effects of wear and the frost–thaw cycle, which may cause spalling. For high dams and high flow velocities, aeration is therefore the preferred method of protection.

#### 9.3.6.3 Spontaneous aeration

Cavitation damages may be prevented by aeration of the flow, as was noted as early as the 1940s by Bradley (Chadwick *et al.*, 2013) and Peterka (1953). An air content of the order of 5–8% in the flow adjacent to a fixed wall will usually serve to do the job according to Peterka's experiments (which only measured average air content in a Venturi-type installation).

Several investigations were subsequently performed to try to verify Peterka's results. Semenkov and Lentiaev (1973) noted that for flow velocities of 22 m/s an air content of 3% by volume would seemingly provide reasonable protection for a concrete surface of 40 MPa strength, as opposed to 10% air content if the strength was only 10 MPa.

In an investigation on cavitation damage prevention on baffle piers by Rozanov and Kaveshnikov (1973), an air content of 1.25–2.5% slowed down the erosion process by a factor of 2–2.5, while a 6–7% air content appeared to largely prevent damages for flow velocities around 25 m/s and $\sigma \ll \sigma_{cc}$.

#### 9.3.6.4 Artificial aeration

Where spontaneous aeration is deemed insufficient to prevent cavitation, air may be introduced artificially, for instance by special aerators shaped as ramps, steps or combinations thereof. The proper function of an aerator is to provide a suitable additional amount of air to the flow in order to prevent cavitation damages. The function depends on the aerator's ability (i) to draw air through a system of air supply channels (pipes) by the suction created by the flow, and (ii) to deliver and entrain the air in the flow against this subpressure. The function accordingly depends on a suitable balance between the supply and delivery systems. An artificial aeration system may, therefore, miss the target by providing too little air at high specific discharges or too much air at low discharges. In the former case, cavitation erosion may occur, while in the latter case the water–air mixture may create misty clouds affecting the proper function of (electric) facilities nearby.

### 9.3.7 Theoretical basis of pressure fluctuations, uplift and cavitation

For artificial aerators the total airflow drawn through the aerator is important, and the aerator efficiency is often expressed as the ratio $\beta$ of the airflow at the aerator to the water flow:

$$\beta = \frac{Q_{air}}{Q_w} \tag{9.24}$$

The function of ramp- and step-type aerators has been investigated extensively (Kramer *et al.*, 2006; Pfister and Hager, 2010a,b; Pinto and Neidert, 1982; Rutschmann and Hager, 1990; Wood, 1991).

Rutschmann and Hager (1990) found that the aeration efficiency of ramp aerators depended on the magnitude of the sub-pressure created under the water jet and the relative jet length, as determined by the Froude number of the flow and the geometry of the aerator:

$$\beta = \beta_{max}\left[\frac{2}{\pi}\arctan(3 \times 10^{-3}\Delta E)\right]^{\zeta} \tag{9.25}$$

$$\beta_{max} = (\tan\theta)^{1.15}\exp[(1.15\tan\alpha)^2](F - 5.4)^{\varepsilon} \tag{9.26}$$

where the (modified) Euler number is

$$E = \frac{U^2}{\Delta p/\rho} \quad \text{and} \quad \Delta E = E - E_{min} \tag{9.27}$$

$E_{min}$ signifies the minimum subpressure below which the aerator is submerged and no air is entrained, that is

$$\beta(E_{min}) = 0 \tag{9.28}$$

$\Delta p$ is the reduction in air pressure under the jet below the ambient pressure. For ramp angles in the range $5° \leq \theta \leq 8°$ and chute slopes $19° \leq \alpha \leq 52°$ Rutschmann and Hager found that the minimum subpressure required to avoid submerging the aerator depended on the step height and the Froude number of the flow; step heights less than 60% of the flow depth and Froude numbers below 5–6 were not effective.

The validity of the conclusions of Rutschmann and Hager (1990) has, however, been questioned based on prototype data from Tarbela Dam in Pakistan and Foz do Areia in Brazil, the contention being that fully rough flows on concrete chutes will entrain more air than flow in scale models with Perspex-walled flumes (see Pinto *et al.*, 1992).

Kramer *et al.* (2006) established empirical data on the decrease in bottom air concentration in downstream aerators for slopes up to 50% and Froude numbers in the range $5 \leq F_0 \leq 12$ at the aerator.

### 9.3.8 Stepped chutes

Hydraulically, the conditions for flow on stepped chutes differs from that for smooth chutes in that the roughness elements (the steps) cannot be considered small in comparison with the depth of flow and are lined up in parallel, which implies different effects of turbulence and flow conditions in general. Horizontal vortices established on the steps imply that there are also three-dimensional effects. The turbulent boundary layer tends to develop more rapidly in flow on stepped chutes than on smooth chutes. Spontaneous aeration is, therefore, a normal occurrence for flow on stepped chutes. Energy dissipation on stepped chutes may be several times larger than on smooth chutes, but the complex flow needs special methods for theoretical analysis and study in hydraulic models.

Cavitation indices for undisturbed chute flow are not applicable to stepped spillways. The full extent of scale effects between models and prototypes has probably not yet been properly established. In general, the use of stepped spillway chutes is dependent on aeration to protect the water conduit from cavitation damages where flow velocities have reached potentially destructive levels.

The use of stepped chutes has been revived in the last few decades, and the hydraulic treatment of such flows is under continuous development. Literature on the subject may, therefore, quickly become obsolete. Field experience is quite limited, especially for specific discharges higher than 15–30 $m^2/s$.

At least three types of flow can be distinguished on stepped chutes, and these are usually called skimming flow, transition flow and nappe flow (Figure 9.14).

Figure 9.14 Flow on stepped chutes: (a) skimming, (b) transition, (c) nappe flow

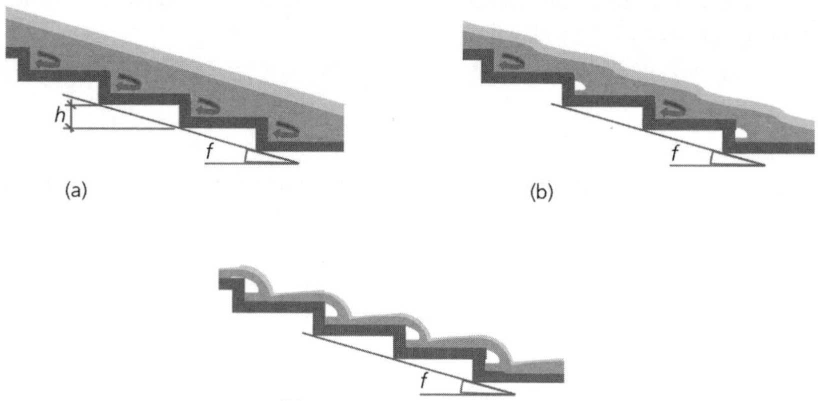


Skimming flow (Figure 9.14(a)) typically occurs for high specific discharges, especially on relatively steep chutes. In skimming flow the outer edges of the steps form a pseudo invert so that the upper surface of the flow exhibits a uniform slope. In this type of flow eddies are formed on each step. Energy dissipation is primarily by form losses rather than friction.

Transition flow (Figure 9.14(b)) is the intermediate region where the upper flow surface is not uniform so that the influence of the steps can be clearly seen while the length of each step is too short for the development of a full hydraulic jump. The exact definition of the transition flow region varies between different researchers; some treat it as a subform of skimming flow, but different from true skimming flow according to Figure 9.14(a).

In nappe flow (Figure 9.14(c)) the specific discharge is so low or the slope of the chute so moderate that a hydraulic jump occurs on each step on a chute. Energy dissipation is by the hydraulic jumps and friction on the steps. Aeration is by the hydraulic jumps. Nappe flow provides the largest energy dissipation, but is severely limited by the low specific discharges possible with this type of flow. For economic reasons, most stepped spillways are therefore adapted to skimming flow.

#### 9.3.8.1 Theoretical basis of stepped chutes

According to Ohtsu *et al.* (2004), skimming flow can on slope angles $\varphi$ between 6° and 55° be distinguished from nappe flow by the equation

$$S_{max} = \frac{7}{6}(\tan\varphi)^{1/6} \tag{9.29}$$

where $S_{max}$ signifies the upper limit of $S$ for skimming flow to occur, where $S$ is the step height $h$ normalised by the critical depth $d_c$, so that $S = h/d_c$. The same investigation separates true skimming flow with a uniform upper surface from and transitional type flow with a wavy upper flow surface, which occurs only at slopes $\varphi$ less than about 19°.

Yasuda *et al.* (2001) proposed that the lower limit of nappe flow could be set to

$$S_{min} = 0.57(\tan\varphi)^3 + 1.3 \tag{9.30}$$

These equations together indicate the difficulty in separating the flow types in a region with step heights between roughly 80% and 200% of the critical flow depth $d_c$.

Proper nappe flow can be analysed based on the fact that there will be a critical section at each vertical step in the canal. Upstream of this a drowned or normal hydraulic jump will occur, depending on the specific flow and the overall inclination of the canal and the step length (or height).

As long as the jumps are not drowned, the aeration may be expected to be higher than that achieved by a single hydraulic jump. The remaining energy will be controlled by the critical sections and the step heights, so that

$$E_e^{res} = d_c(1.5 + S) = \left(\frac{q^2}{g}\right)^{1/3}(1.5 + S) \tag{9.31}$$

provided the horizontal length of each step is so long that a complete hydraulic jump can occur before the next step.

Where the total height of a stepped chute is large enough, developed uniform or quasi-uniform flow will exist in the lower part, implying that the total energy line will be parallel to the overall slope of the chute (skimming flow type A according to Ohtsu *et al.* (2004)).

The Darcy–Weissbach equation, rearranged as for flow on smooth chutes, is normally used to compute energy losses in stepped chutes with developed flow by considering the corresponding single-phase flow of water $q_w$ with characteristic depth $d_w$:

$$f = 8\left(\frac{d_w}{d_c}\right)^3 \sin\varphi \tag{9.32}$$

For type B skimming flow $d_w$ roughly corresponds to the water depth on the horizontal downstream part of the steps. The flow will reach equilibrium (uniform flow) at a height below the crest of the chute of

$$H_E = d_c[(1.30 - 0.0713\varphi + 0.0016\varphi^2)^{-1}][5.7 + 6.7\exp(-6.5S)] \tag{9.33}$$

for slopes in the range $6° \leq \varphi \leq 55°$ and normalised step heights in the range $0.1 \leq S \leq S_{max}$.

This generally implies that for relatively flat slopes a vertical distance of about 10–20 times the critical depth is required for the development of uniform flow, the lower value being valid for flatter slopes and the higher for steeper slopes, providing the step height is as large as possible while maintaining skimming flow.

The residual energy at the toe of the chute at equilibrium (uniform flow) will be

$$E_e^{res} = d_c\left[\left(\frac{f}{8\sin\varphi}\right)^{1/3}\cos\varphi + 0.5\left(\frac{f}{8\sin\varphi}\right)^{-2/3}\right] \tag{9.34}$$

where the true slope angle $\varphi$ is used for true skimming flow and the value $\varphi \approx 0$ is used for skimming flow of the transitional type.

For the gradually varied flow the residual energy may be estimated empirically as

$$E^{res} = d_c\left\{1.5 + \left[\left(\frac{E_e^{res}}{d_c}\right) - 1.5\right]\left[1 - \left(1 - \frac{H_{dam}}{H_E}\right)^m\right]\right\} \tag{9.35}$$

where $m = 4 - (\varphi/25)$.

Ohtsu *et al.* (2004) found that the friction factor $f$ in the Darcy–Weisbach equation for the developed uniform or quasi-uniform flow reaches a maximum at a slope of about 19°–20° for a given value of the normalised step height $S$, and tends to increase with increasing values of $S$ to reach a value of

$$f = f_{max} - A(0.5 - S)^2 \quad 0.1 \leq S \leq 0.5 \tag{9.36}$$

$$f = f_{max} \quad 0.5 \leq S \leq S_{max} \tag{9.37}$$

where different values of $A$ and $f_{max}$ are valid for flat and steep slopes: for slopes in the range $6° \leq \varphi \leq 19°$, $A = -0.15 + 0.064\varphi - 0.0017\varphi^2$ and $f_{max} = 0.032 + 0.016\varphi - 0.00042\varphi^2$; and for slopes in the range $19° \leq \varphi \leq 55°$, $A = -0.452$ and $f_{max} = 0.231 - 0.00275\varphi + 0.0000232\varphi^2$.

Within the investigated range of parameters the investigations point to values of $f$ between 0.07 and 0.18, the latter value being reached where uniform or quasi-uniform skimming flow occurs on slopes between 18° and 23° with a step height in the range 40–50% of the critical depth. Note that the values of $f$ should reasonably include also the effects of spontaneous aeration.

These values differ substantially from earlier published data from prototype measurements, which may be due to difficulties in performing accurate measurements of air–water flow in the field. It may also indicate the presence of some unknown scale effect, although the laboratory studies of Ohtsu *et al.* (2004) were performed at a scale providing for reasonable values of both the Reynolds number ($Re \geq 10^5$) and the Weber number.

Investigations by Chanson (1993) show that the inception point for spontaneous aeration on stepped chutes will be at about

$$L = 9.719k_s(\sin\alpha)^{0.0796}F_h^{0.713} \tag{9.38}$$

where

$$F_h = \frac{q}{(g\sin\alpha k_s^3)^{1/2}} \tag{9.39}$$

and $k_s = h\cos\alpha$ and $h$ is the step height. The air content in the developed uniform flow may be expected to be similar to that in uniform flow on a smooth chute.

## 9.4. Flow in closed conduits

Flow in tunnels of low-level outlets and power lines may occur as closed-conduit flow, as free surface flow or as a mixture of these. The latter type of flow is more complicated to analyse than free surface flow in open canals (Figure 9.15).

Closed-conduit flow differs from open-channel flow primarily due to the effects of elastic waves. While sudden changes in velocity in open channels will create surface waves typically travelling at speeds in the range 5–10 m/s, elastic waves in closed conduits will create pressure waves, which can travel at speeds up to 1440 m/s in absolutely pure water.

Compressibility reduces the celerity. The celerity, therefore, depends on the deformability of the conduit walls and on the amounts of gases, such as air, present in the water, as the water itself may be considered incompressible at normal pressure levels. All water, however, contains at least minute quantities of air in gas form, and even low concentrations of air will influence the celerity, so that the wave speed in closed-conduit flow with 'unaerated' water will often be in the range 1000–1200 m/s. When air concentrations are low and the air is well distributed throughout the flow, it may be treated as a single-phase fluid (water) flow with adjustment of the elastic wave velocity; the higher the air concentration and the larger the deformability is, the lower the celerity will be. To determine the wave velocity precisely, it is necessary to know the exact air content and the deformability of the conduit walls and their connection to the surrounding medium. This means that, for instance, the success of grouting the annular space between a lining and the surrounding material plays a role. In practice, field measurements may be the only reliable method for establishing the true celerity.

Figure 9.15 Penstocks at the Revelstoke Dam, British Columbia, Canada. The BC Hydro hydroelectric generating station at this location was commissioned in 1984. (Courtesy of BC Hydro)

The elastic waves will be reflected at closed and open boundaries of a water conduit and cause significant and rapid pressure changes, which are usually called 'water hammer'. Unsteady flow and transients require different methods of analysis than those applied for free surface flows.

### 9.4.1 Two-phase flow in conduits

Free surface water flow in tunnels and culverts with more than moderate velocities tends to entrain air, and accordingly also creates flows of air above the water surface. Therefore, free surface flow in tunnels need often be considered two-phase flow. The two phases, water and air, may appear as a mixture or as separate phases, or as either of these in different parts of the water conduit. Two-phase flows may be subdivided into a number of subclasses. However, classification varies between different researchers. Sharma (1973) differentiates between, among others, true free surface flow, spray flow, bubble flow and plug or slug flow (Figure 9.16).

One of the challenges with closed-conduit flow is to determine that it is in fact closed-conduit flow. Standard textbooks on hydraulics may give some assistance in the determination of, for instance, whether a culvert runs full or with a free surface.

At intakes from water reservoirs there is often a transition from free surface flow to closed-conduit flow. However, due to vorticity in the flow approaching the intakes, air may also be drawn into a conduit with insufficient inundation and cause instability of flow and operation of hydraulic equipment, such as turbines.

Airflow in a closed conduit may also be brought about by special aeration shafts or by pockets of air trapped at high points in the conduit, when it is filled or when the discharge conditions change. In a conduit with free water surface air flow is caused by the same mechanisms as in open canals, primarily by entrainment of air at turbulent water surfaces and by hydraulic jumps.

The common momentum equation for hydraulic jumps (such as the classic Unwin formula) may be adapted to hydraulic jump situations in horizontal or vertical (shaft) conduits to determine the position of a possible jump in a conduit.

If the pressure level downstream of a hydraulic jump in a conduit exceeds the level of the conduit soffit, the jump creates a transition from free surface flow to 'closed conduit flow'. The jump will however also entrain potentially dangerous air bubbles into the conduit flow, which will then in reality be a two-phase flow, even if the conduit appears to run full. The character of the flow may change downstream to plug or slug flow as bubbles coalesce and gather under the soffit. Analysis of such two-phase flows is difficult and water conduit designs have therefore often been based on avoidance of the flow type.

### 9.4.2 Single-phase closed-conduit flow

*Steady-state* single-phase closed-conduit flow is easier to analyse than the corresponding free surface flow as the size of the cross-section of the flow is generally known and will remain so unless there is a structural change (Figure 9.17).

One-dimensional analysis may be analysed using similar methods as for free surface flows. For developed turbulent flow the Bernoulli equation and the Manning–Strickler or the Darcy–Weissbach equations may be used. Empirical friction coefficients for the Manning equations are available in standard texts but are subject to dimensional restrictions or modifications for use. Important state variables, such as velocity and pressure, can readily be computed in closed form for steady one-dimensional analysis.

**Figure 9.16** Flow in conduits: (a) water flow with air bubbles, (b) spray flow, (c) slug or plug flow, (d) free surface stratified flow followed by hydraulic jump

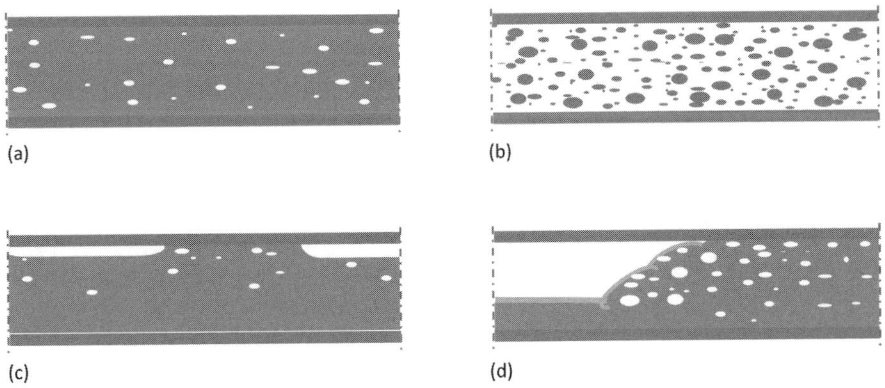

Figure 9.17 Flow in a closed conduit. (Courtesy of OPG)

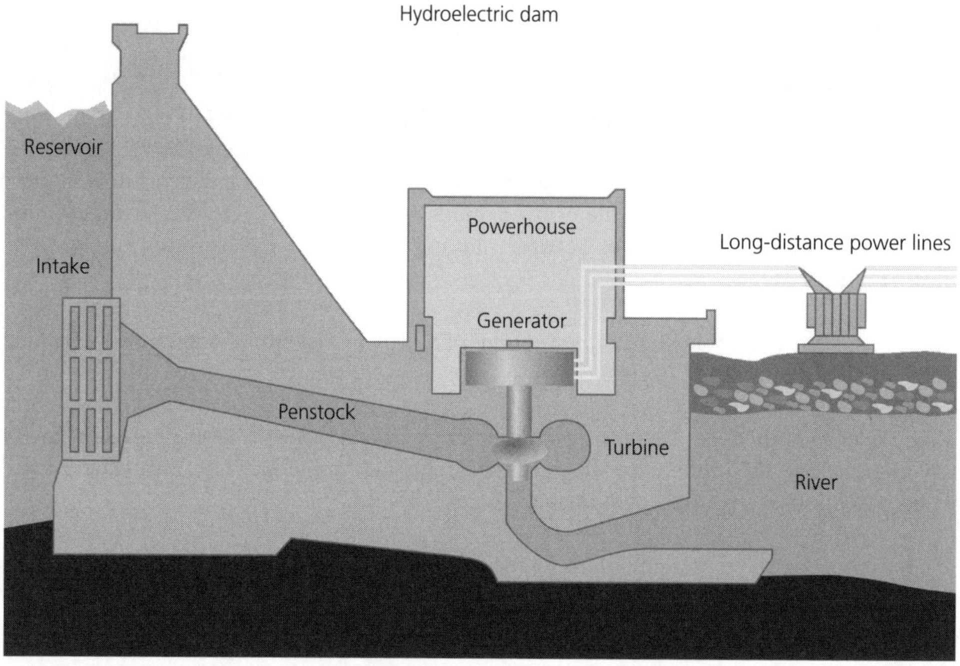


One-dimensional analysis alone does not, however, provide sufficient detail to judge, for instance, cavitation, which may occur due to local curvilinear flow. Three-dimensional simulations may, therefore, be required to establish detailed curvilinear flow patterns at bends and hydraulic equipment, such as turbines and gates, and to assess the consequences of such flow.

In horizontal closed-conduit flow in a water conduit with a vertically symmetrical flow and cross-section the local *pressure level* will be the same at the invert and the soffit. Due to the higher elevation of the soffit the pressure and the pressure margin there will then be less than at the invert and the risk of cavitation will be higher.

*Unsteady flow* occurs whenever the discharge is changed in a conduit, for instance by regulating (opening or closing) a gate, a valve or a turbine. Such changes will always generate elastic waves and pressure surges in the water. *Low-frequency oscillations* (with periods measured in minutes) can normally be simulated with sufficient accuracy as simple one-dimensional mass oscillations governed by Newton's second law, where the pressure difference between the two ends of a conduit is allowed to accelerate or decelerate the moving water volume in the conduit. The velocity of the elastic wave is then, in principle, assumed to be infinite. Typical time steps in the simulations are chosen in the range 3–15 s to suit the conduit under consideration.

This kind of analysis is usually sufficient for the computation of, for instance, the action of surge chambers at hydropower stations. The aim of the calculations is to demonstrate the function and the sufficiency of the surge chamber cross-section (free water surface area) and height by calculating the highest and the lowest pressure (or water) levels accompanying certain possible power station operating patterns, including disturbances and repeated start/stop (load on/load off) sequences and emergency closing. The highest permissible surge level may be governed by the physical layout of the surge arrangement or by adjacent caverns to avoid a danger of flooding, while the lowest surge levels will have to be limited so that, for instance, the turbine draft tube is not emptied, which would cause destructive shock pressures against the turbine–generator when the pressure in the draft tube is rising again after the down surge.

The minimum water surface area (i.e. horizontal area) required for surge stability is known as the Thoma area, which is a theoretical minimum surge area required for small surges to be successively reduced. The need for a detailed analysis can be assessed by comparison of the available surge chamber area with this minimum.

The Thoma area $F_{\mathrm{TJ}}$, as modified by Jaeger (1958), can be calculated as

$$F_{\mathrm{TJ}} = \left[1 + 0.48\frac{z_{\max}}{h_{\mathrm{n}}}\right]\frac{LA}{2gh_{\mathrm{n}}h_{\mathrm{f}}} \tag{9.40}$$

where $z_{\max}$ is the theoretical maximum surge height at a total load rejection in a conduit of length $L$ and cross-sectional area $A$, $h_{\mathrm{n}}$ is the net head and $h_{\mathrm{f}}$ is the head loss in the conduit at a mean flow velocity of 1 m/s (in the conduit). As the equation indicates, long conduits, which are lined to reduce head losses, will require large surge areas for stability.

Surge chamber surface areas of at least around twice the Thoma area in open surge chambers and around 1.5 times the Thoma area in ported surge chambers may be required to stabilise disturbances within reasonable time spans. At a port restricting the flow between the conduit and the surge chamber, the flow is typically allowed to accelerate to 10–15 m/s at large surges to create a head loss, which quickly builds up an extra 5–10 m of pressure head, more rapidly accelerating or decelerating the conduit flow. Open surge chambers may include sections of water conduit with the soffit raised above the water level or be made up of combinations of tunnels and shafts.

In high-head power stations, rock surge chambers are sometime also designed as pressurised air chambers with the required air-tightness achieved by continuous injection of a curtain of water above the chamber. Pressurised chambers may, for stability, need somewhat larger water–air interface areas than open surge arrangements, but avoid the need for the construction of tall shafts.

The maximum surge after a single load rejection from full load can for a conduit with an open surge chamber be estimated as

$$z_{\max} = U\sqrt{\frac{LA}{gF}} \tag{9.41}$$

where $U$ is the initial mean flow velocity before load rejection in a conduit with cross-section $A$, and $F$ is the surface area of the surge arrangement. A port would generally tend to reduce the surge height in the surge chamber.

In the case of *high-frequency oscillations* a more thorough treatment is required to assess the effect of elastic waves. The dynamic equations can be solved by full three-dimensional simulation, but normally one-dimensional simulations are sufficient. These may be done by one-dimensional wave equations formulated in accordance with the method of characteristics for the entire closed-conduit system between free water surfaces of reasonable size. The method of characteristics allows the determination of the state at discrete points in space and time. Typically, time steps in the range 0.01–0.05 s are used for step changes in load (in the time domain). These time steps allow analysis of pressures and velocities at points about 10–50 m apart. The analysis is often performed and presented in the frequency domain. Some computer codes made specifically for the analysis of high-frequency pressure waves are proprietary, and source codes are not readily available. The codes tend to contain empirically based elements.

The high-frequency oscillations in unsteady or transient flow are of paramount importance for the hydraulic, mechanical and electrical performance of a power line in terms of frequency regulation and rapid load following. The analysis needs to include not only water conduits but also the mechanical properties of turbo-generating unit(s) and turbine governors, and for closed-loop analysis the frequency response of the transmission net as influenced, for instance, by voltage regulation must also be included.

Turbine governors are used to regulate the power output by opening or closing the wicket gates of reaction turbines and the needle-valve nozzles or reflection plates of action turbines. Rapid turbine regulation is usually performed with the help of computerised algorithms of PID type. PID stands for 'proportional integral derivative' action of the governor based on the current deviation from the desired value of a certain variable (rotational speed or power output).

The basic problem with unsteady flow in closed conduits, such as water conduits for power stations, is that a disturbance in the form of, say, a small pressure increase in a penstock needs to be balanced by the turbine governor by reducing the opening of the wicket gate in order to keep the generated output constant. This reduction in gate opening will, however, instantaneously result in a positive pressure surge, which will further increase the pressure and the output. The surge will travel upstream at a velocity $c$ of about 1000 m/s, be reflected at the nearest water surface and return as a pressure reduction after a reflection time $T_1$:

$$T_1 = 2L/c \tag{9.42}$$

where $L$ is the length of the water conduit to the first water surface of reasonable size. This negative surge will, in turn, be reflected at the turbine. A full cycle of water hammer will require waves to make the return journey twice before the initial disturbance can begin to be balanced out and the output stabilised. The pressure surges will continue indefinitely unless damped out by friction.

The rapidly fluctuating pressure induced by water hammer between points with an adequate air–water interface (usually free water surfaces) will tax the conduit walls in the water conduits, and is one of the reasons penstocks tend to be equipped with watertight linings of steel, concrete or plastic. Very high and very low internal pressures may cause the linings, or in unlined rock tunnels the surrounding rock, to collapse due to tensile or compressive overstressing, buckling, cavitation, cracking or general deterioration. Gates, valves and turbines placed in closed conduits also need to withstand the pressures that occur during water hammer.

A very significant disturbance to a power line is the sudden loss of the transmission network (net). This will cause the rotational speed of the turbine to rise, and (emergency) closing is therefore started automatically. If the closing is too rapid the pressure effects on the water conduit may become prohibitive. If the closing is slow or does not occur at all, the turbine may increase its speed and its water discharge until a balance condition called 'runaway speed' is reached. Turbines may or may not be designed for operation at runaway speed for a significant length of time; most turbines are not.

For a low-level or bottom outlet there is generally no demand for regulation stability. The analysis will then only have to determine that the pressure surges and the water-hammer effects are acceptable. For power lines regulation stability is usually, but not always, a criterion. Smaller units connected to a stable net may use very slow turbine regulation.

Electricity is a commodity best produced at the moment of consumption. If the demand on a net exceeds the supply at a particular moment, the frequency will start to fall even if countermeasures are instigated immediately. Possible stabilising measures are increased production and voltage and phase angle adjustments.

Different demands on regulation stability will apply depending on the role of the particular power station (e.g. whether or not the station is supposed to participate in keeping the frequency on the net stable). These demands may be expressed as a maximum permissible frequency deviation from the normal net frequency at a certain specified step-load change, when a unit operates on an isolated net. Alternatively, the requirement may be expressed as an open-loop phase lag and amplification factor for disturbances of different frequencies, or as a closed-loop overall negative amplification (damping) factor or stability margin. The damping or stability margin (usually measured in decibels) gives the ratio (on logarithmic scale) between the relative disturbance (load change) and the relative response (frequency change).

It is important to keep the frequency sufficiently stable that different kinds of electric equipment designed for this particular frequency function properly. Among the most sensitive equipment are large thermogenerating units, which may have to be tripped when the frequency on the net falls more than a 1–2 Hz. The tripping would, however, reduce the net frequency and tax the generation–transmission system further, and in order to avoid an uncontrolled collapse of the entire transmission net a sequence of orderly tripping of units and sectioning of the production–transmission system into smaller subsystems are usually initiated at even smaller frequency deviations.

Most or all of the analysis of high-frequency oscillations in closed water conduits and the consequences thereof will have to be performed as a pre-analysis outside the main loop of the system simulation. The performance of a particular power line and the depth of analysis required may be assessed based on approximate, empirically based rules.

Decisive among the regulation properties that can be achieved by a particular power line is the magnitude of the ratio of the unit starting time $T_u$ to the water conduit starting time $T_w$. A minimum ratio of 2–2.5 may be needed for the unit to be able to participate in supporting the system frequency effectively. For stable operation on an isolated net a ratio of around 3 may be required, while really good regulation properties may be expected with ratios of the order of 5–6.

The unit starting time is a measure of the time required for the unit output to accelerate the turbo-generating unit to a synchronous speed of rotation. In metric units it may be computed (in seconds) as

$$T_{\mathrm{u}} = \frac{J\varpi^2}{P} \tag{9.43}$$

where $J$ is the turbine–generator inertia ($J = gD^2/4$), $\varpi$ is the angular rotational speed and $P$ is the turbine output (in watts).

The water conduit starting time is a measure of the time required to accelerate the flow to a particular steady state by gravity alone. It may be computed (in seconds) as

$$T_{\mathrm{w}} = \frac{Q}{gh_{\mathrm{n}}}\sum_i \frac{L_i}{A_i} \tag{9.44}$$

where the discharge $Q$, the acceleration of gravity $g$, the net head $h_n$ and the length $L_i$ and the cross-sectional area $A_i$ should be summed up over the different parts $i$ of the power line between the free water surfaces.

The ratio of unit starting time to water conduit starting time can, if required, be improved by several methods, which include increasing the generator inertia above the natural value, which follows from structural requirements (providing a flywheel, reducing the water conduit length between free water surfaces, etc.). None of these changes is easy to implement in an existing plant.

For reaction turbines like those of the Kaplan or Francis type, the water conduit to be considered in the analysis encompasses parts on both the upstream and the downstream side, while for action turbines like Peltons only the upstream side is relevant.

### 9.4.3 Theoretical basis of single-phase closed-conduit flow

The pressure rise $\Delta h$ in a penstock at a full load rejection (or valve closure) depends on the ratio $\kappa$ of the water conduit starting time $T_w$ to the full stroke unit closing time $T_s$:

$$\kappa = \frac{T_w}{T_s} \tag{9.45}$$

If the flow reduction is linear in time during the closing, the pressure rise can be computed, for example by using Allievi's equation, which is valid for closing times much greater than the elastic wave reflection time,

$$\Delta h = h_n \frac{\kappa}{2}\left(\kappa + \sqrt{\kappa^2 + 4}\right) \quad T_s \gg T_l \tag{9.46}$$

or by using the Sparre equation

$$\Delta h = h_n \frac{\kappa}{1 - \lambda} \tag{9.47}$$

where the value of $\lambda$ varies from $\lambda = \kappa/2$ for linear closing to $\lambda = \kappa$ for 'normal' nonlinear closing.

Alternatively, if the unit closing time is of the same order of magnitude as the reflection time, the pressure rise can be computed as

$$\Delta h = \frac{c\Delta U}{gT_s} \sim T_l \tag{9.48}$$

where $c$ is the elastic wave velocity and $\Delta U$ is the change in conduit flow velocity.

The effect of a surge chamber or a shaft with a free water surface is to improve performance in unsteady flow with respect to both high- and low-frequency oscillations (i.e. frequency regulation and surging). Power stations with short water conduits and no or moderate demands on frequency regulation may not need any surge arrangements at all.

As noted previously, the analysis of unstable two-phase flows is difficult (see Keller, 2006). Moreover, physical hydraulic model testing is not very conclusive, as the size and rise velocity of the coalescing bubbles will tend to model incorrectly. These flow patterns accordingly imply considerable uncertainty.

The net amount of free air $q_a$ spontaneously drawn into the flow by hydraulic jumps in large low-level conduits depends on the discharge of water $q_w$, the flow velocity and the depth of flow, and may, according to Kalinske and Robertson (1943), generally be computed as

$$q_a = 0.0066 q_w (F - 1)^{1.4} \tag{9.49}$$

where

$$F = \frac{u}{(gy)^{1/2}} \tag{9.50}$$

is the Froude number at the vena contracta.

However, if a gate provides only a very small opening in relation to the upstream conduit cross-section, the flow immediately downstream may take on a spray-like appearance. Sharma (1973) investigated the air demand for different types of two-phase flow and for spray flow, and suggested that

$$q_a = 0.2 q_w F \tag{9.51}$$

USACE (1964) proposes the value $q_a = 0.03q_a(F - 1)^{1.06}$ for design purposes, and this is intended to cover both spray and free-flow conditions downstream.

The amount of air may, however, also be restricted by the available conduit size for air transport, so that the actual airflow becomes less than indicated by the above equations (see Sharma, 1973).

### 9.4.4 Two-phase free surface and mixed conduit flow

A free water flow surface in a ventilated conduit will induce airflow by two mechanisms. The water flow will accelerate the air above the surface by friction until the air and water velocity are identical at the interface. The induced airflow then depends on the cross-section of conduit available above the water surface and on the surface water velocity. The second mechanism is entrainment of air into the water flow. This is contingent on (i) the flow being turbulent and (ii) the surface water velocity being high enough to break the surface tension. The required velocity is, however, quite low. Together these mechanisms will create an airflow in the direction of the water flow in ventilated conduits, unless there is a significant chimney effect caused by air shafts above the conduit creating airflow in the other direction.

The limited space available above the water surface or in vents may cap aeration by these mechanisms, which means that they need to be analysed together. In fact, high-velocity water flow close to the soffit may, in some cases, entrain all the available air, and thus lift the mixture and change the flow to closed-conduit mixed flow.

A hydraulic jump in a conduit will also entrain air into the flow. If the water (or, strictly, the air–water surface) does not reach the water conduit soffit, downstream of the jump the entrained air will rise to the water surface and be detrained in the same fashion as in free surface flow in open-channel flow. The air demand is normally higher at the hydraulic jump than elsewhere, and part of the detrained air may, therefore, flow upstream (or reduce downstream air velocities) to balance the air demand in different sections of the water conduit.

If the air–water surface downstream of the jump reaches the conduit soffit, it is no longer possible for the air to detrain in this manner. The air may instead be gathered temporarily just below the soffit.

If the soffit is inclined towards upstream, the air bubbles will be transported downstream continuously with the water flow and be released to the atmosphere at the conduit outlet.

If the conduit soffit is inclined towards downstream and the water velocities are fairly low, the air will accumulate and create a continuously growing air bubble (slug) at the top of the conduit. The pressure in this slug will match the pressure level of the water flow, which may be higher than the local atmospheric pressure. An air slug at the soffit may continue to grow until it becomes unstable, because of increased flow velocities underneath or because it has grown until its downstream end reaches the outlet portal. When the slug reaches the portal, the entire bubble, which may be tens of metres long, will disappear from the conduit instantaneously and rise as one or more huge umbrella-shaped bubbles towards the free water surface outside of the tunnel portal. The process when water rushes in to fill the void left by the air bubble, may be accompanied by potentially destructive shock pressures against the conduit soffit (see Keller, 2006).

If the combination of conduit inclination and water velocity is somewhat different, large slugs of air may instead travel towards the upstream and each slug may potentially lift many tons of water on its way up through an upstream airshaft or intake portal in an event reminiscent of an explosion. Such sudden bubble eruptions have been known to cause casualties among operating personnel.

If sufficient amounts of air are not supplied to a conduit the air pressure upstream of the hydraulic jump may be successively reduced until the supercritical water flow upstream of the jump is lifted to the soffit; large-scale cavitation may result.

## 9.5. Energy dissipation

Energy dissipation may be achieved by the formation of a hydraulic jump creating the required turbulence in a dedicated section of the water conduit denominated energy dissipator or as a drowned jump in a plunge pool (Figure 9.18).

Hydraulic jumps occur where the flow changes from supercritical to subcritical. The position of an unforced jump will vary with the discharge, the flow velocity and other boundary conditions, primarily the tailwater elevation. If the tailwater is too low a jump may not form in the energy dissipator, and if it is too high the jump may move upstream out of the energy dissipator.

Not all river channels will remain stable for all the discharges that an upstream spillway can pass. High discharges may be expected to lead to erosion and/or deposition in the river, changing the downstream rating curve either permanently or temporarily. The simple extrapolation of rating curves based on measurements may, therefore, not always be sufficient.

Figure 9.18 Fully developed turbulent flow in the plunge pool and stilling basin at Letsi Dam, Norrbotten, Sweden. (Courtesy of Vattenfall AB)

A multitude of standard designs have been developed for energy dissipators, including dissipators with sloping inverts, diverging sidewalls or abruptly increased width, and dissipators with different kinds of turbulence-increasing features such as chute blocks as well as shorter basins, where part of the dissipation is achieved by allowing the incoming water jet to hit a surface of concrete or steel. These specific designs behave according to specific rules, which are different from those governing normal dissipators in U-shaped channels for free, unforced hydraulic jumps.

Due to the significant difference in water depth between the two ends of a hydraulic jump, steady-state leakage under the invert slab or through joints and cracks in the foundation rock from the downstream end to the upstream end may destabilise the invert slab in the upstream part.

After the failures of the energy dissipators of the Karnafuli Dam in Bangladesh and the Netzahuacoyotl (Malpaso) Dam in Mexico in the 1960s, intensive investigations were begun into the dynamic pressures acting on the inverts of lined stilling basins (Abdul Khader and Elango, 1974; Bowers and Toso, 1988; Bowers and Tsai, 1969; Farhoudi and Narayanan, 1991; Fiorotto and Rinaldo, 1992; Fiorotti and Salandin, 2000; Güven *et al.*, 2006; Hager, 1991; Hajdin, 1982; Lopordo and Henning, 1985; Sanchez and Capello, 1973).

At Malpaso Dam (Mexico) several invert concrete slabs, each weighing 700 tons or more, were dislodged during the passage of a flood about one-third of the design flood. The ensuing investigations showed that turbulent pressure fluctuations had played a major role in causing the failure. Hydraulic model tests on the Karnafuli spillway (Bangladesh) led to similar conclusions, with pressure fluctuations of up to 11 m of head spread through under-drains being reported.

These pressure fluctuations on the inverts of dissipators are similar to those in undisturbed chute flow, but worse. In simple energy dissipators the fluctuations are roughly normally distributed with frequencies that cover the range from about 1 Hz to 20 Hz or higher. Instantaneous point values of these pressure deviations will actually reach positive and negative values of the order of the incoming velocity head if the stilling basin is in operation for a sufficiently long time. The fluctuations are accordingly 3–5 times higher than in undisturbed turbulent chute flow. The details of the energy dissipator design influence the fluctuations.

In flows that are not sufficiently aerated the instantaneous low pressures may reach the partial pressure of water vapour and initiate cavitation. There is, however, an even more significant spatial correlation in the fluctuations in stilling basins compared to in undisturbed chute flow. Areas around 4–5 times the incoming water depth $y_1$ in the streamwise direction and around 8–12 times $y_1$ in the transverse direction may simultaneously exhibit higher or lower pressures than average. The pressure deviations, therefore, also produce large load fluctuations.

Where there are open joints or cracks in the invert slabs the pressure fluctuations may be propagated also to the lower face of the slabs. The propagation takes place almost instantly, and without appreciable damping unless there is a substantial free drainage capacity. A high pressure occurring at a leaky joint and under the slab may then coincide with a low pressure on a large part of the slab's upper surface, causing even more significant uplift forces.

With low air content in the flow under the invert slabs pressure propagation there would be rapid and the full spectral range of the fluctuating pressures could affect the lower side of the invert slab. The currently preferred counter-measure is the use of heavy invert. Untensioned rock bolts and flexing of invert slabs for load distribution may be looked upon with some doubt due to the possibility of fatigue.

Hydraulic jumps will, however, also entrain air into the water flow on top of the invert slabs, which may alleviate the cavitation risk at a distance downstream of the toe of the jump. Investigations by Rajaratnam (1960) seem to indicate that hydraulic jumps with fairly high Froude numbers will remain aerated close to the bottom over the main part of the dissipator length.

### 9.5.1 Plunge pools

In a plunge pool the flow is jetted from a structure that is usually called a trajectory bucket, flip bucket or ski jump (Mason, 1993; Steiner *et al.*, 2008). After travelling some distance through the air, the jet is allowed to strike a free water surface of a basin or river – the plunge pool.

Energy dissipation takes place both in flight and in the pool of water. A plunge pool has the advantage over a conventional energy dissipator that the turbulence can be created in the water mass at a safe distance from any fixed boundaries, providing the pool is large enough.

There are, in principle, three ways of protecting the bottom of the plunge pool against erosion: by reducing the energy intensity of the free jet, by cushioning the bottom of the plunge pool with an increased depth of water, or by the provision of a tight and resistant surface layer on the invert, preventing the impact pressures from travelling further into the foundation.

### 9.5.2 Energy dissipation in flight

Field measurements collated by Kawakami (1973) and Rae and Castro (1994) have shown that the air resistance shortens the theoretical throw distance of water jets compared with the distances computed using projectile theory by 10–30% for initial jet velocities in the range 20–40 m/s. The effect increases with the velocity because of the speedier break-up and aeration of the jet (Figure 9.19).

**Figure 9.19** Rooster-tail energy dissipation at W.A.C. Bennett Dam, Peace River, British Columbia, Canada. (Courtesy of BC Hydro)

Laboratory experiments by Heller *et al.* (2005) have demonstrated that the jet take-off angle will be smaller than the lip angle of the ski-jump structure, and they have also provided an alternative method of assessing the flight profile and spread of the jet.

Significant amounts of air may be entrained in the free trajectory water jet both before and when it hits the plunge pool water surface. Decisive factors are the state of the turbulent boundary layer development where the jet leaves the ski jump structure, the thickness or diameter of the jet and the length of travel, which is highly dependent on the velocity.

### 9.5.3 Flow conditions in plunge pools

The energy dissipation in the water mass of a plunge pool is best assessed using a hydraulic model, but some attempts have been made to combine physical and mathematical models (Xu *et al.*, 2002). The highly turbulent flow is itself a major obstacle in the calculations, but the main difficulty lies in the fact that the flow may cause significant erosion of the pool bottom and sides, and this in turn will change the flow conditions.

Erosion in rock cannot be accurately pre-calculated due to the variability of rock materials and the lack of detailed data about conditions underground. The tempting method of letting the water jet excavate its own plunge pool may, therefore, be fraught with danger unless perfect control of the discharges is achieved so that the flow can be discontinued at any chosen time.

The eroded rock material may be deposited in locations such that it actually changes the downstream rating curve and increases the water depth. Again, a larger discharge may cause the deposited material to be eroded away, and the water levels would become lower again. The block sizes of the deposited eroded rock will significantly influence this process.

A much better control of the flow conditions and of the function of a plunge pool is possible in pre-excavated pools of such dimensions that the flow conditions do not cause erosion, but this can usually be achieved only at a significant construction cost.

Even for assumed constant dimensions of a plunge pool the calculation of the flow conditions is somewhat complicated. If the plunge pool dimensions are insufficient to create a hydraulic jump the entire pool may be swept clean by the supercritical jet flow, with considerable ensuing erosion. If, on the other hand, the pool dimensions are just sufficient then a jump may occur, but the pool bottom would still be exposed locally to the full attack of the impinging jet and similar erosion might follow.

The sufficiency of the pool dimensions at hand can be checked by initially assuming that they are insufficient, so that the water jet will penetrate to the bottom of the pool and spread out on the invert without significant losses of energy. The water depth and dimensions required to create a hydraulic jump can then be assessed. This hydraulic jump may be one sided, two sided or even circular depending, among other things, on the shape of the water jet and its strike angle. Relevant variations of the original Unwin formula may be developed based on continuity of momentum for the case at hand. For good protection of the pool bottom the water depth needs to be considerably larger than the depth required for a jump so that a postulated hydraulic jump would be completely drowned. Attempts have also been made to back-calculate the energy-dissipation capacity of existing (stable) plunge pools.

### 9.5.4 Energy dissipation in plunge pools

As discussed above, it is important to recognise that the length of travel in water must take into account the possible action of the jet to depress the water surface where it strikes if the plunge pool is insufficient in size. It is equally important to assess how the tailwater depth will be influenced by erosion–deposition.

When a jet passes through the water mass it widens due to diffusion, but the full dynamic pressure, which may be equal to the full water head, is initially maintained as a mean pressure in the centre of the jet. Investigations by Hartung and Häusler (1973) showed that the full head remained until the jet had travelled through the water at least a distance of five times the jet thickness (i.e. $5b$). At longer travel distances the dynamic pressure decreased to half at a travel distance of $10b$ and to a quarter of the full head at $20b$. Other investigations (e.g. Pinheiro and Melo, 2008) looking into the diffusion of a free jet in water with and without aeration have indicated quicker attenuation of momentum and mean dynamic pressure in aerated jets.

Apart from these mean pressures there will also be fluctuations in pressure, as discussed further below.

### 9.5.5 Erosion in plunge pools

A multitude of laboratory experiments and field investigations have been undertaken to try to forecast the extent of erosion of the bed of plunge pools (e.g. Canepa and Hager, 2003; Mason, 1984). All the studies consider erosion in rock containing sets of intersecting joints free from impermeable material (clay) with cohesion. The bed materials may then be treated as friction materials with certain typical grain sizes. None of the developed correlations is entirely convincing, especially with regard to the rate of erosion.

In the first instance, whether the rock foundation is free from open or potentially open cracks is decisive with regard to whether rock erosion will occur. In all other cases (most cases) the dynamic impact pressure will enter open joints or cracks at the jet impact area, spread sideways through intersecting joint sets, and cause uplift and erosion. The mechanism is similar to that which may lift stilling basin invert slabs. Rock that is free from cracks is definitely in short supply, and even if the foundation rock were initially free from cracks the question is whether this will be the case after some time of spillway operation.

The extent of rock erosion is determined by the head (the flow velocity), final block size and density of the rock, the jet and the plunge pool dimensions. The strength of the rock probably has only an indirect effect in that weaker rock will be eroded quicker and may be more susceptible to being ground down into smaller pieces, which are easier to transport away.

Occasionally, the water jets from trajectory buckets are allowed to directly hit a rock surface that is not protected by a pool of water. The ensuing erosion cannot be predicted with any degree of accuracy due to the inhomogeneous nature of rock materials. The forces on a plunge-pool slab protecting the underlying rock have been studied by Melo *et al.* (2006).

### 9.5.6 Theoretical basis of energy dissipation

In a simple U-shaped channel or basin with horizontal invert and parallel, vertical sidewalls a jump will, according to Unwin, occur if the downstream water depth equals or exceeds the minimum required depth $y_2$ over a sufficient length $L$ of the water conduit:

$$y_2 \geq \tfrac{1}{2} y_1 \left( \sqrt{1 + 8F_1^2} - 1 \right) \quad L \geq L_{\text{req}} \tag{9.52}$$

where $y_1$ is the depth of flow upstream the jump, $y_2$ is the required minimum downstream depth and $F_1 = u_1/(gy_1)^{1/2}$ is the Froude number of the incoming flow. The full length of the jump zone may be taken as $L_{\text{jump}} = (6 \pm 0.1)y_2$ for a free jump with $F_1 \geq 4.5$. The Unwin depth condition needs to be met for at least 60–65% of the full length of the jump $L_{\text{jump}}$, and in addition the flow may not be accelerated to supercritical within the full length of the jump.

Hydraulic jumps differ significantly in flow characteristics and energy dissipation depending on the Froude number of the incoming supercritical flow. The ratio of the remaining energy $R_E$ to the total energy $R_T$ of the incoming flow for a free jump may be expressed as

$$\frac{R_E}{R_T} = [(8F_1^2 + 1)^{1.5} - 4F_1^2 + 1][8F_1^2(2 + F_1^2)]^{-1} \tag{9.53}$$

which means that a jump with $F_1 = 3$ will only dissipate one-fourth of the energy, while a jump with $F_1 = 8$ will dissipate nearly two-thirds. In addition, flow conditions in the jump with $F_1 = 3$ will be unstable and produce significant, and sometimes destructive, waves travelling far downstream. Hydraulic jumps in the range $4.5 \leq F_1 \leq 9$ are, on the other hand, both hydraulically stable and efficient.

The amount of air entrained depends on the Froude number just upstream of the jump and may be assessed as follows:

$$q_a = 0.0066(F_1 - 1)^{1.4} q_w \tag{9.54}$$

If the flow is already aerated upstream of the hydraulic jump, the total air content downstream of the jump will be higher than indicated by the above equation.

An investigation by Ervine *et al.* (1997) indicated that the volumetric, relative content of free air $\beta$ for undeveloped or partially developed rectangular jets can be taken as

$$\beta = K \left( \frac{s}{b} \right)^{0.5} \left( 1 - \frac{1.1}{U} \right) \tag{9.55}$$

where $s$ is the length of travel in air to the water surface of the plunge pool, $b$ is the thickness of the water jet (excluding the effect of aeration) and $U$ is the velocity of the water jet when it strikes the water surface. The value of the constant $K$ is estimated to vary from 0.1 for a smooth (undeveloped) jet with $s/b \leq 100$ to 0.2 for a rough (partly developed) turbulent jet with $s/b \leq 50$.

Investigations by Bohrer *et al.* (1998) give some guidance for assessing the effect of the increased turbulence in fully developed jets already aerated throughout at the ski jump. Investigations by Schmocker *et al.* (2008) have revealed the significant effect of pre-aeration of the jet on both the air content and the trajectory of flight.

Mason (1989) proposed that the final erosion depth $Z$ (below the tailwater) in unprotected rock might be estimated as

$$Z = 3.27q^{0.6}H^{0.05}h^{0.15}g^{-0.3}d^{-0.1} \tag{9.56}$$

where $q$ is the specific (unit) flow in the jet, $H$ is the total head to the (local) plunge pool water surface, $h$ is the water depth over the erodible bed, $g$ is acceleration due to gravity and $d$ is the block size of the erodible bed material. Mason's formula is reported to predict real erosion with a coefficient of variation of around 25%.

Canepa and Hager (2003) have proposed that the final erosion depth $Z$ and also the final height of material deposited downstream of the plunge pool can be calculated as

$$Z = 0.37\frac{U(1+\beta)^{0.25}}{\sqrt{\left(\frac{\rho_s-\rho}{\rho}\right)gd_{90}}} \tag{9.57}$$

where $U$ is the flow velocity when the jet strikes the plunge pool surface, $\beta$ is the volumetric free air content of the jet, $\rho_s$ is the density of the bed material, $\rho$ is the density of water and $d_{90}$ is the grain size (90% passing) of the bed material.

It may be noted that both Mason and Canepa and Hager undertook their laboratory studies using a flume where the eroded material could settle immediately downstream, which led to a considerable successive increase in water level where the jet struck. In reality, the blocks may be broken down during the erosion process and be carried away by the water flow, which might reasonably increase the final erosion depth by a factor of 1.5–2 as measured from the initial bottom elevation.

Both formulae tie the erosion depth to the total head, the depth of water and the block size, although partly indirectly. Notably, the most important parameter of the Mason formula, the specific rate of flow, is missing in the Canepa–Hager formula.

Neither investigation, however, describes the effect of significant water depths in an entirely convincing manner. At depths exceeding 20–30 times the jet thickness $b$ the Hartung–Häusler formula may provide more alternative estimates.

The Canepa–Hager investigations indicate that the erosion depth increases with increasing air content in the jet in spite of the quicker reduction in momentum. This has been attributed to the lower density of the jet, which provides more buoyancy and therefore greater ability to transport eroded material out of the erosion pit. The usefulness of aeration to limit erosion has, in other words, not been sufficiently investigated.

Canepa–Hager also tried to assess the rate of erosion. The results are unconfirmed and could be considered speculative. Generally, the different investigations show that erosion is quickest initially; the rate of erosion declines more or less logarithmically and the final erosion depth is approached asymptotically. In practice, this implies that the major part of the final erosion may take place over a short period of time.

## 9.6. Hydraulic system simulation

This final section looks at the inclusion of hydraulics and outflow conveyance within systems simulation models. The interplay between heads, gate openings, outflow discharges, flow velocities and pressures makes these simulations complex. As has been emphasised elsewhere in this book, the spillway demand that these many factors represent cannot be described *a priori* in a single probability distribution, but rather depends on the complicated interaction of many dynamic processes. In addition, many of these dynamic processes depend in turn on the real-time performance of the spillway and conveyance systems.

The incorporation of hydraulics in system simulations is illustrated through two examples of British Columbia dams taken from the engineering records of BC Hydro. The first of these is the W.A.C. Bennett Dam on the Peace River in northern British Columbia, which is one of the larger dams in BC Hydro's portfolio. The second is Strathcona Dam on the Campbell River on Victoria Island. This is a more modest structure at the head of a three-dam cascade.

### 9.6.1 W.A.C. Bennett Dam, British Columbia, Canada

W.A.C. Bennett Dam is a 183 m high earth-fill dam on the Peace River in north-east British Columbia, near the town of Hudson's Hope (Figure 9.20). The dam impounds the 360 km long Williston Reservoir at a maximum normal level of elevation (El.) of 672.08 m. The power station at Bennett Dam is the largest in the province, providing about 25% of BC Hydro's generation. Construction of the dam was completed in 1967. Downstream is Peace Canyon Dam, and a second downstream dam, Site C, is in construction as of 2016.

Figure 9.20 W.A.C. Bennett Dam viewed from the north, showing the power station at the top right and the spillway and conveyance at the bottom along the right abutment. (Courtesy of BC Hydro)

The following is an extract from the 2013 performance evaluation report (BC Hydro, 2013):

> W.A.C. Bennett Dam is rated as an extreme consequence facility and the design inflow is the Probable Maximum Flood (PMF). The PMF was recently updated ... with no inter-basin diversions. [The original project design had assumed future inter-basin diversions from the McGregor River, which have not occurred.] Routing of the summer PMF resulted in the peak hourly spill discharge of 7732 $m^3$/s with a corresponding maximum reservoir level of El. 673.26 m. This updated PMF peak spill discharge estimate is about 11% lower than the peak spill noted ... for the previous PMF peak discharge estimate.
>
> The concrete-lined spillway was constructed in a trench excavated through bedrock on the right abutment of the dam [Figure 9.21] The spillway headworks consist of an ogee section (crest El. 653.79 m) and three 15.24 m wide by 18.9 m high radial gates. The headworks also include nine undersluices which discharge into the spillway chute but are considered out of service. Downstream of the ogee section, there is a 107 m long transition section where the chute width decreases from 54.9 m to 30.5 m with a 4.21% bottom slope. The chute continues for another 445 m at a width of 30.5 m and 0.3% slope. The left and right chute side walls, which are at 4V : 1H, are 18.9 m and 17.4 m high, respectively. The lower section of the chute is sloped at 84% (40°) down to a spoon-shaped flip bucket.
>
> ... The discharge capacity of the three radial gates at the Maximum Flood Level (MFL) of El. 675.13 m is 9218 $m^3$/s. The design discharge of the spillway chute is 9656 $m^3$/s.

At the bottom, the conveyance ends in an upward turn and the creation of a rooster tail, as shown in Figure 9.19.

#### 9.6.1.1 Performance of the conveyance

The contraction through the transition section causes shock waves to form in the spillway chute. The piers, the change in the bottom slope, the geometry of the chute walls and the relative openings of the radial gates all influence the formation and magnitude of the shock waves. The shock waves lead to locally higher water levels, which could cause overtopping of the chute walls at higher discharges.

In 1972, BC Hydro undertook spillway tests in order to understand better the performance of the conveyance. The images in Figure 9.22 show flow patterns for various configurations of gate openings at a discharge of approximately 4950 $m^3$/s. There are three gates. The gate opening ratio is that between the inter-gate opening (numerator) divided by the outer gate openings (denominator). As can be seen, the patterns of flow change dramatically with changes in the opening ratio.

More recently, the outlet structure and conveyance have been modelled using CFD. The geometry files for these CFD analyses are shown in Figure 9.23.

Figure 9.21 Bennett Dam spillway, looking upstream toward the rotating gates. (BC Hydro, 2013)

Representative results are given in Figures 9.24 and 9.25 for a discharge level of 2865 $m^3/s$. Figure 9.24 shows the three-dimensional water surface level along the length of the conveyance, while Figure 9.25 shows the cross-sectional velocities at one particular location along the flow.

Figure 9.26 shows the velocity profile along the conveyance. The dotted line is the top of the retaining walls. The full line is the invert. The light line is the water level predicted by the CFD analysis. The squares are *in situ* measurements of the actual water level during the test, which conform reasonably well with the numerical predictions.

#### 9.6.1.2 Assessment of cavitation potential

Cavitation can occur in high-velocity flow where surface irregularities such as cracks, offsets and surface roughness reduce local water pressure leading to the formation and collapse of vapour cavities. The cavitation index $\sigma$ is used to evaluate the potential for cavitation damage in a spillway, as discussed in Section 9.3.6.

Figure 9.22 Bennett Dam: spillway tests for different gate opening configurations (gate opening ratio = inner gate opening/outer gate opening). (BC Hydro, 2013)

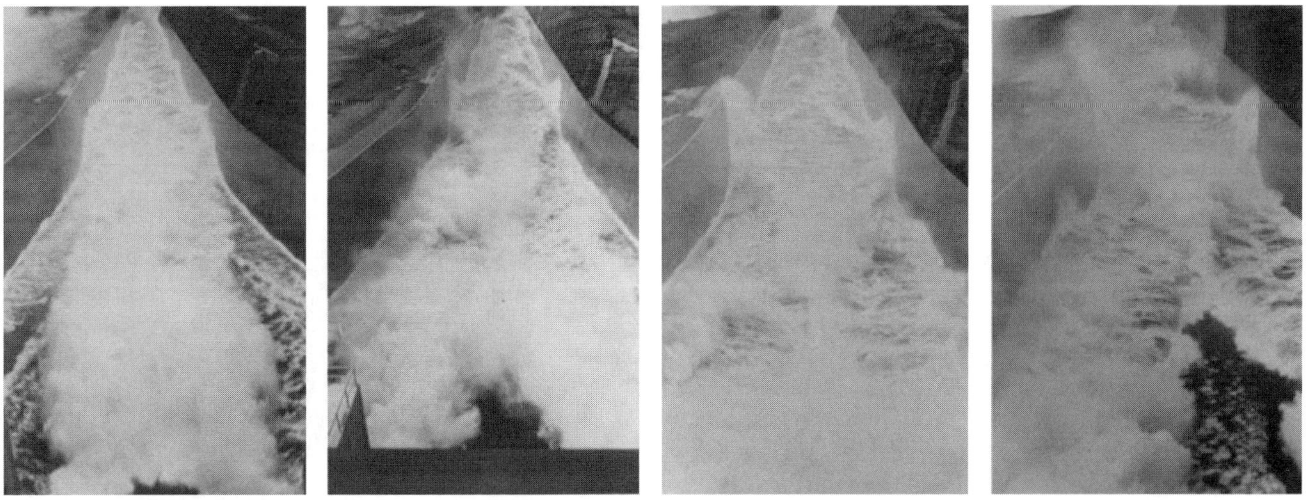

Outer gates open more than inner gate, ratio = 0.66 | Equal gate opening, ratio = 1.0 | Inner gate open more than outer gates, ratio = 1.21 | Inner gate open more than outer gates, ratio = 1.57

Figure 9.23 Bennett Dam: geometry files for spillway tests in FLOW-3D. (BC Hydro, 2013)

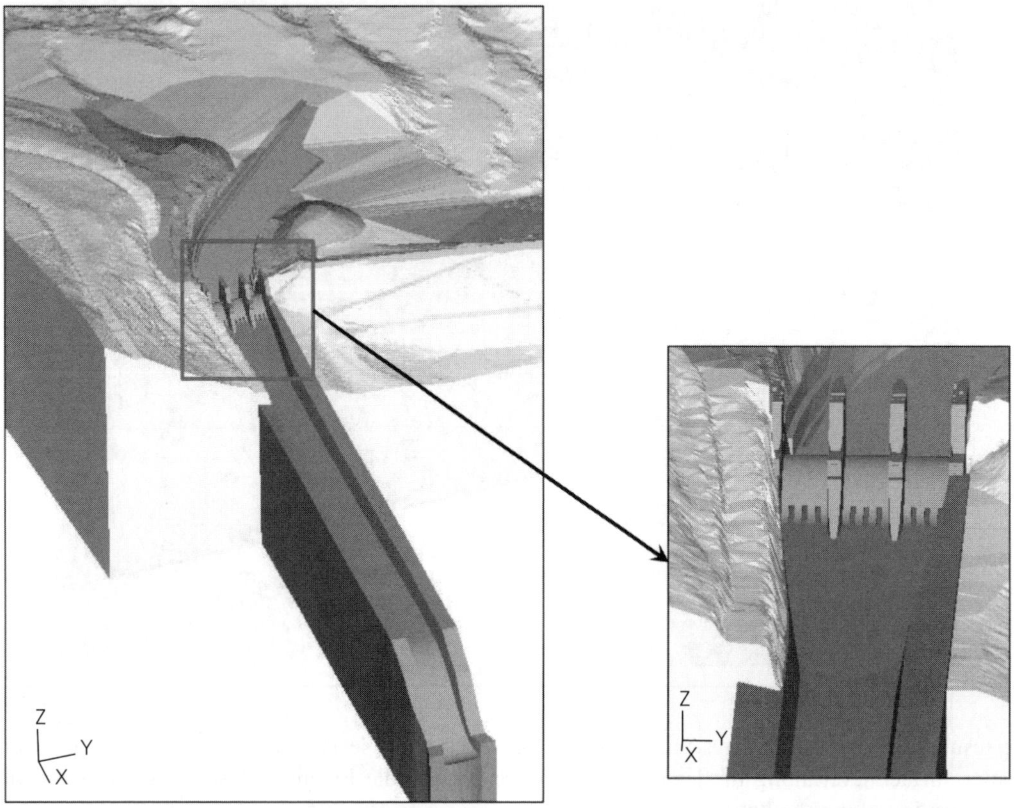


Figure 9.24 Bennett Dam: FLOW-3D water surface in the upper chute for 2865 $m^3/s$. (BC Hydro, 2013)

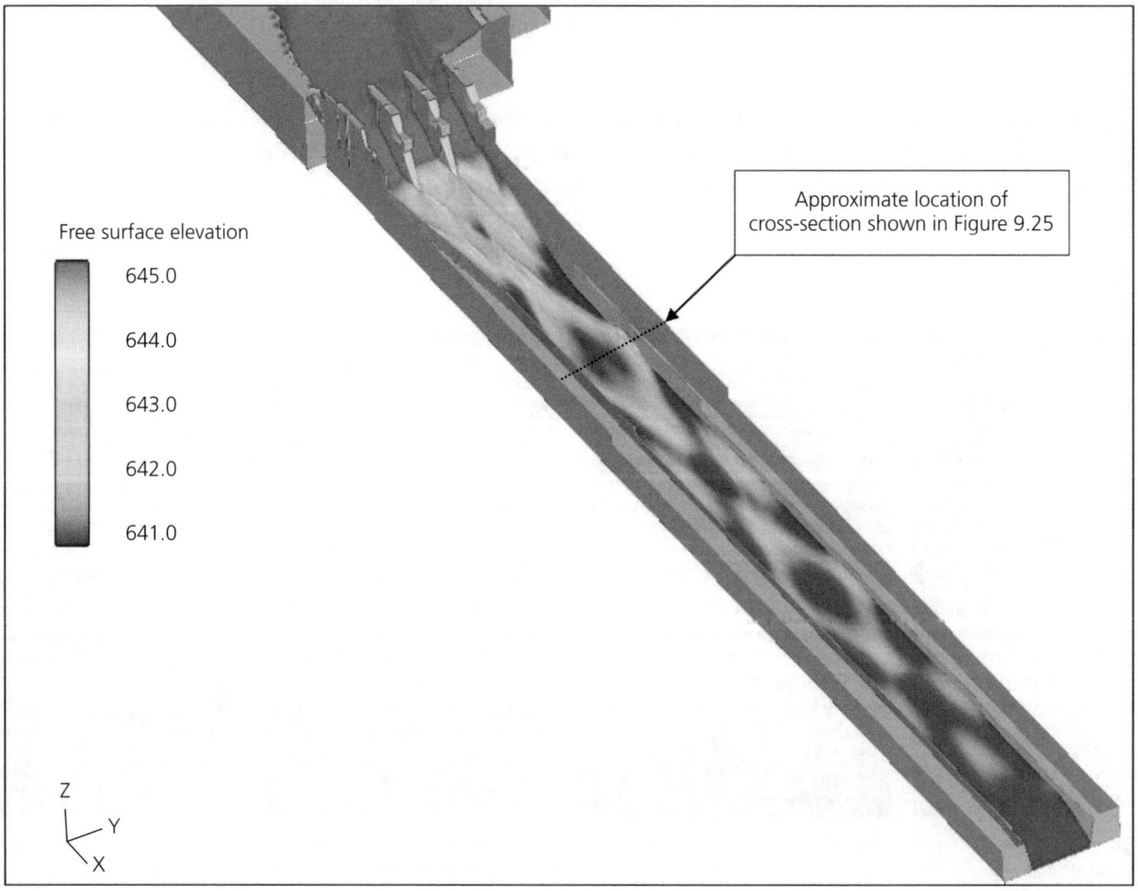

Figure 9.25 Bennett Dam: FLOW-3D velocities at a cross-section 175 m downstream from the spillway crest (near the first shock wave) for 2865 m$^3$/s. (BC Hydro, 2013)

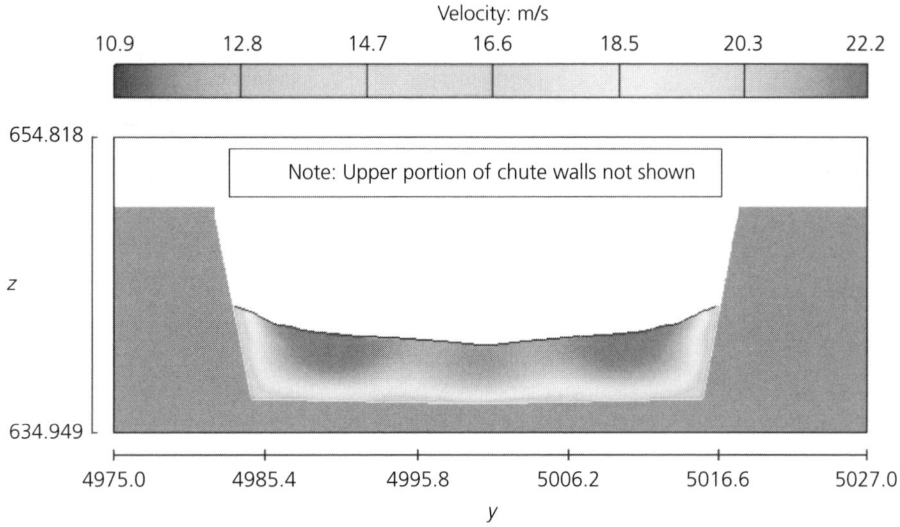


Based on empirical data, Falvey (1990) observed that significant damage has occurred on spillways when the cavitation index is less than 0.2 (see Figure 9.13). Minor damage not requiring repair has been observed for indices greater than 0.2. The extent of damage also depends on the duration of the high-velocity flows, the strength of the surface concrete and adherence to as-built surface tolerances.

To calculate the cavitation indices for the W.A.C. Bennett spillway, the reference pressure $P_o$ was taken as the CFD modelled pressure near the bottom of the chute plus atmospheric pressure.

Figure 9.26 Bennett Dam: FLOW-3D validation – water surface profile along the lower chute for 1477 m$^3$/s. (BC Hydro, 2013)

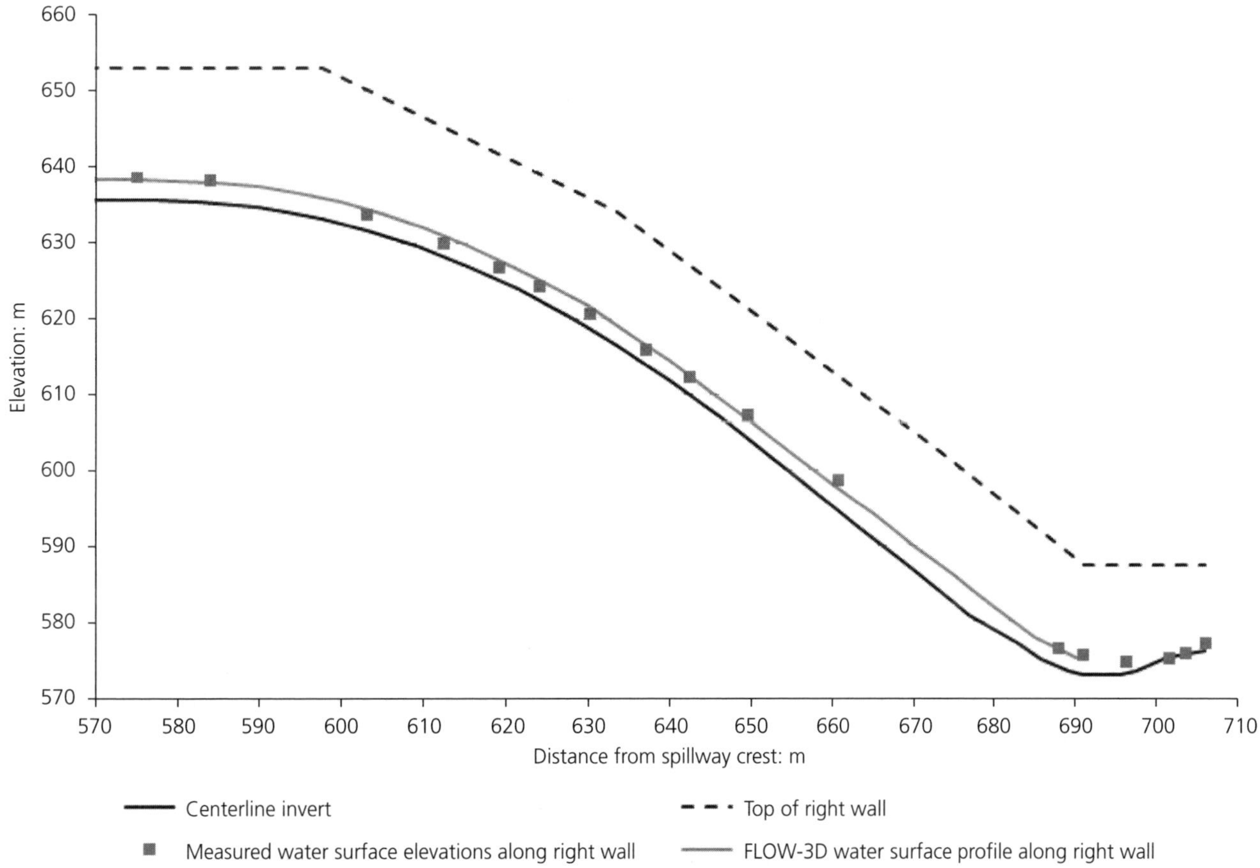

Figure 9.27 Bennett Dam: cavitation index profile along the lower chute for various discharges. (BC Hydro, 2013)

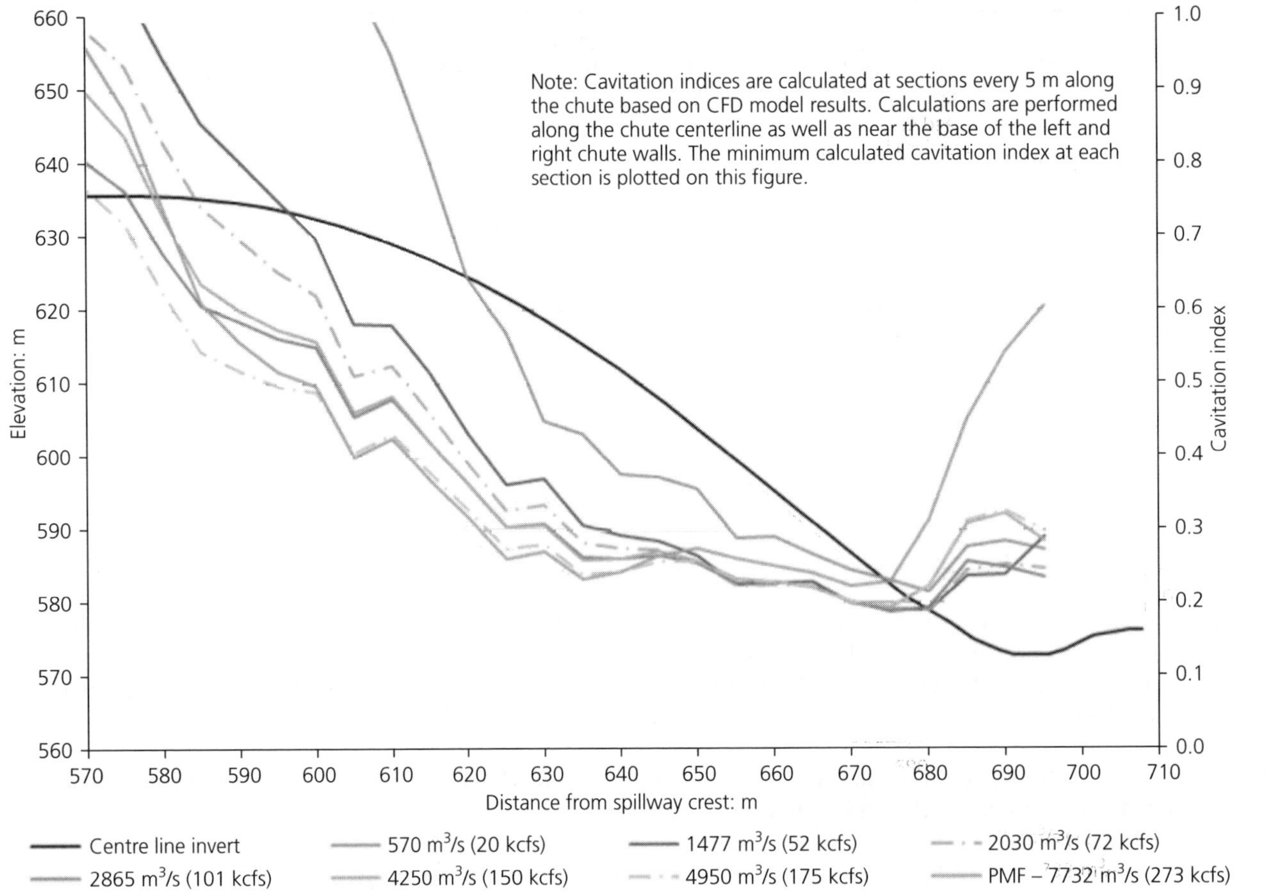


The velocity $V_0$ was taken as the local depth-averaged velocities from the CFD model results. Modelled pressures and depth-averaged velocities were extracted from CFD simulations along the chute bottom centreline and also near the base of the left and right chute walls. Cavitation indices were calculated for spill discharges ranging from 570 m$^3$/s (20 ft$^3$/s) to the PMF peak spill. Cavitation index profiles for various discharges are shown in Figure 9.27. These suggest cavitation damages beyond about 660 m from the gates for discharges larger than about 2000 m$^3$/s.

#### 9.6.1.3 Using CFD models in simulations

The use of CFD calculations in stochastic simulations raises issues in that the numerical calculations are time intensive. The sorts of numerical calculations involved in generating the cavitation index results shown in Figure 9.27 might take many minutes or even hours of computer time. Clearly, these cannot be calculated dynamically in the many tens or hundreds of thousands of simulated runs. Nonetheless, this problem arises in any dynamic simulations of engineering assets and is not unique to simulating dam safety risks.

Typically, the issue of too intensive calculations is overcome by pre-computing a table of results (e.g. Figure 9.27) and then using the table to approximate response surfaces that can be used as look-up tables for the simulations. A response surface is little more than a regression equation that provides, in this case, the cavitation index and its uncertainty as a function of the input parameters (e.g. the discharge at the gates and the distance down the conveyance). Using the look-up table the simulations can proceed without the need to recalculate the numerical results in each cycle. This approach has proven reliable and cost-effective (Vining, 2008).

### 9.6.2 Strathcona Dam, Campbell River, British Columbia, Canada

Strathcona Dam on the Campbell River in Victoria, British Columbia, Canada, is a more modest asset then the Bennett Dam, but is important nonetheless both from the view of energy production and from the view of public safety. The dam sits at the upstream end of a cascade of three dams, and is the largest impoundment on the river. Thus, any safety incident associated with the dam would cause simultaneous problems downstream.

The concrete sluiceway has a 7.2 m long horizontal crest section, which resembles a broad-crested weir, followed by a transition curve and a 30° slope. At the downstream end of this slope there is a concrete apron that extends another 25 m

Figure 9.28 Strathcona Dam: plan view of spillway. (BC Hydro, 2013)

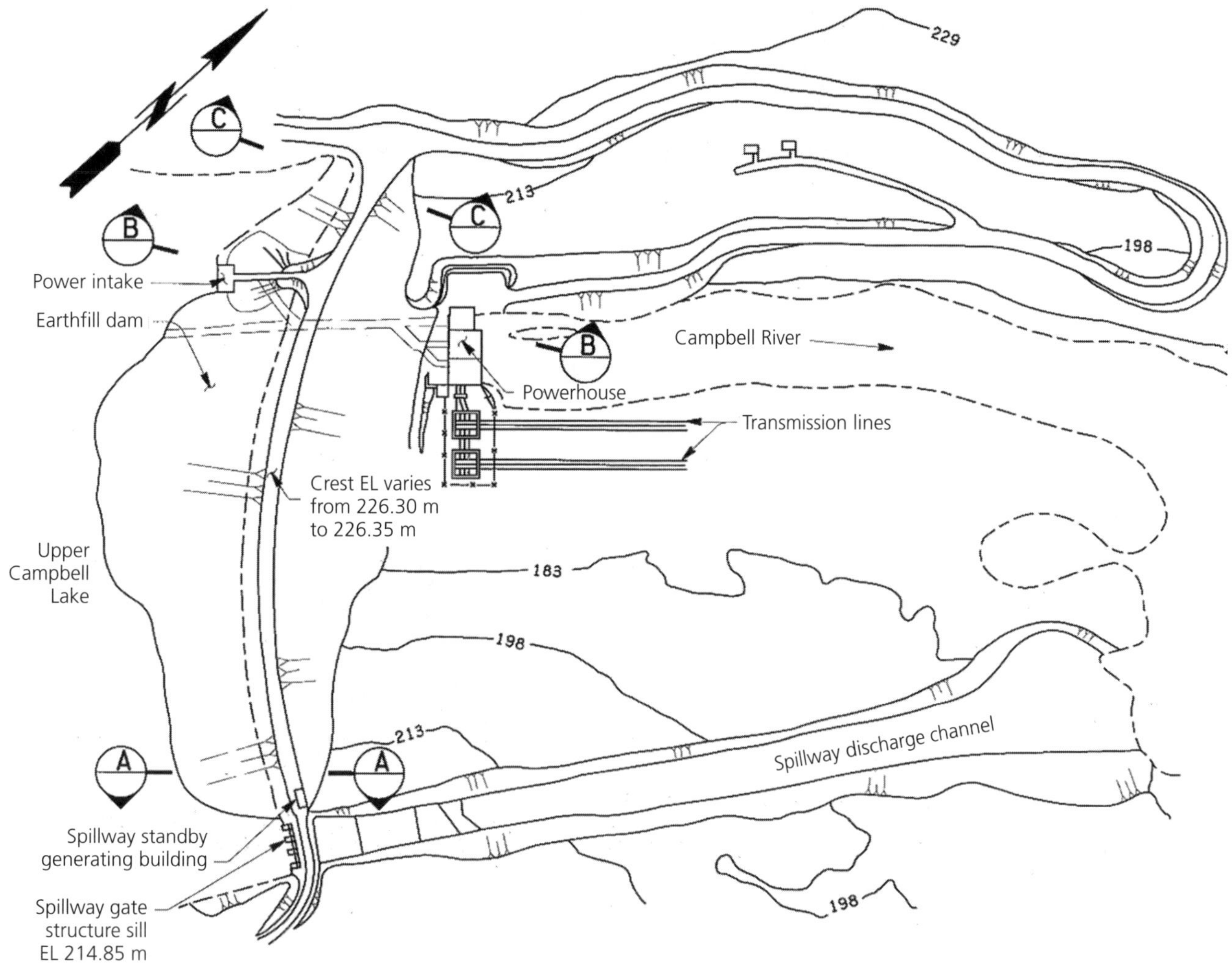


downstream, with a bottom elevation ranging from El. 208.8 m at the upper end to El. 207.3 m at the lower end. There are concrete walls along either side of the concrete sluiceway and apron.

A plan view and section of the Strathcona Dam spillway are shown in Figures 9.28 and 9.29, respectively. The spillway consists of three vertical lift steel gates with a 490 m long chute excavated in rock on the right abutment of the dam. Several issues with the existing spillway arrangement had been identified through historical spill events and previous studies, including the accuracy of the existing spillway rating curves, poor approach flow conditions and the hydraulic capacity of the chute. In addition, BC Hydro's Dam Safety Office had been considering the feasibility of raising the water-retention elevation to provide an increased flood buffer, and the performance of the spillway under such conditions required investigation (BC Hydro, 2015). The geometry files for these CFD analyses are shown in Figure 9.30.

Figure 9.29 Strathcona Dam: spillway section. (BC Hydro, 2013)

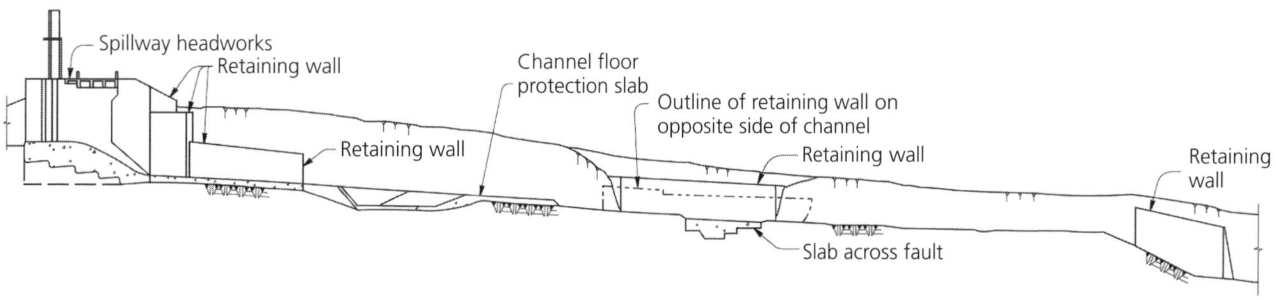

Figure 9.30 Strathcona Dam: 3D CAD geometry for the spillway. (BC Hydro, 2013)

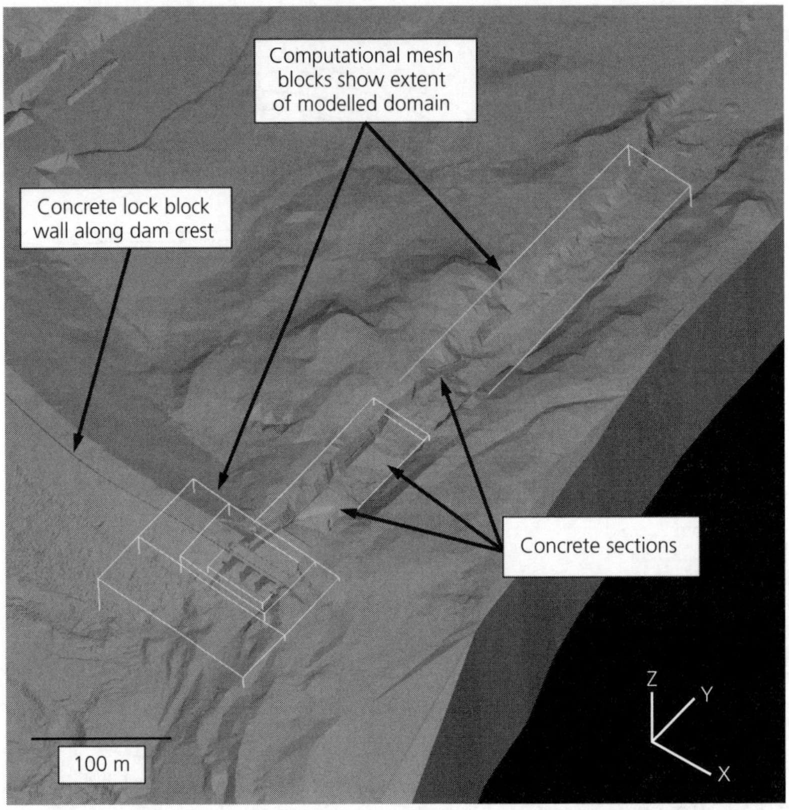


Figure 9.31 Upper Campbell River, Victoria BC, Canada; daily levels 1984–2014. (BC Hydro, 2015)

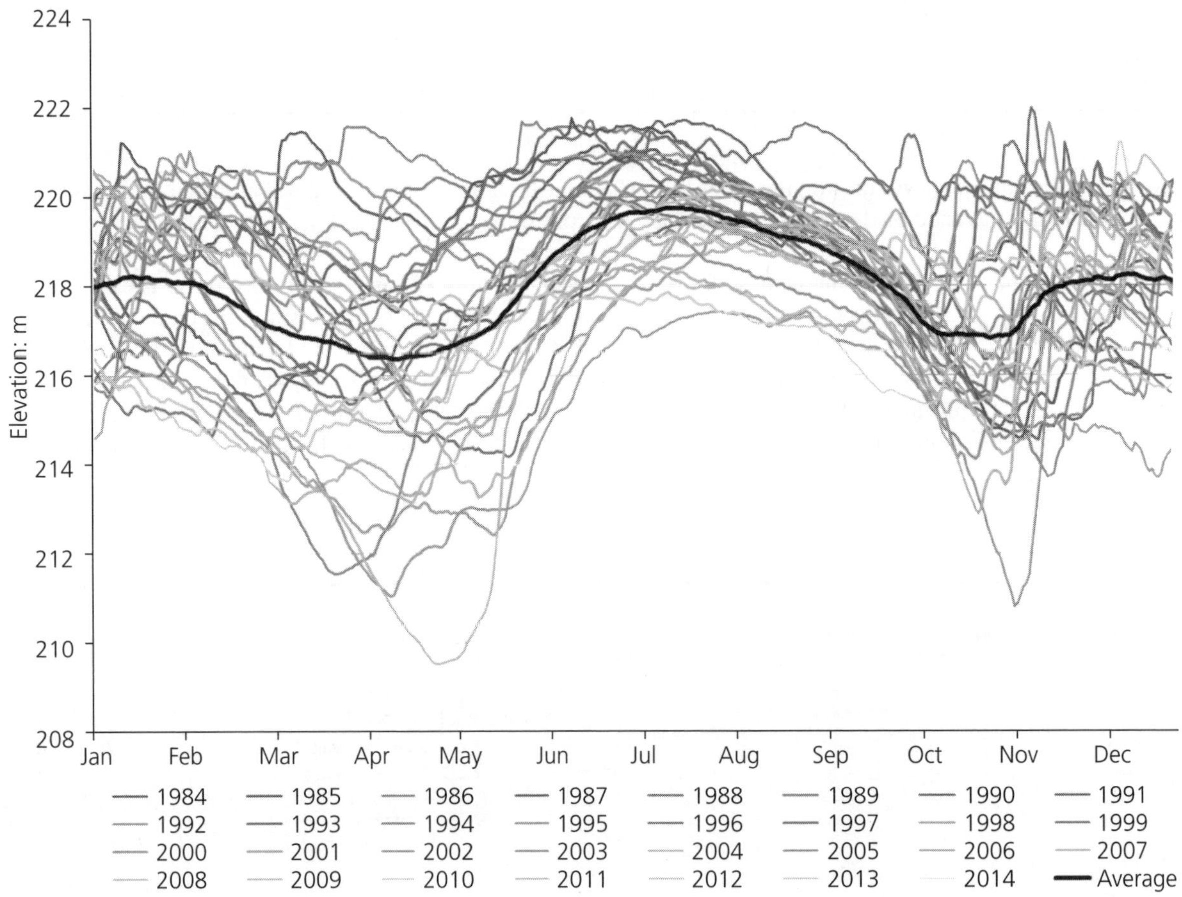

Figure 9.32 Upper Campbell River, Victoria, British Columbia, Canada: spillway daily discharges 1984–2014. (BC Hydro, 2015)

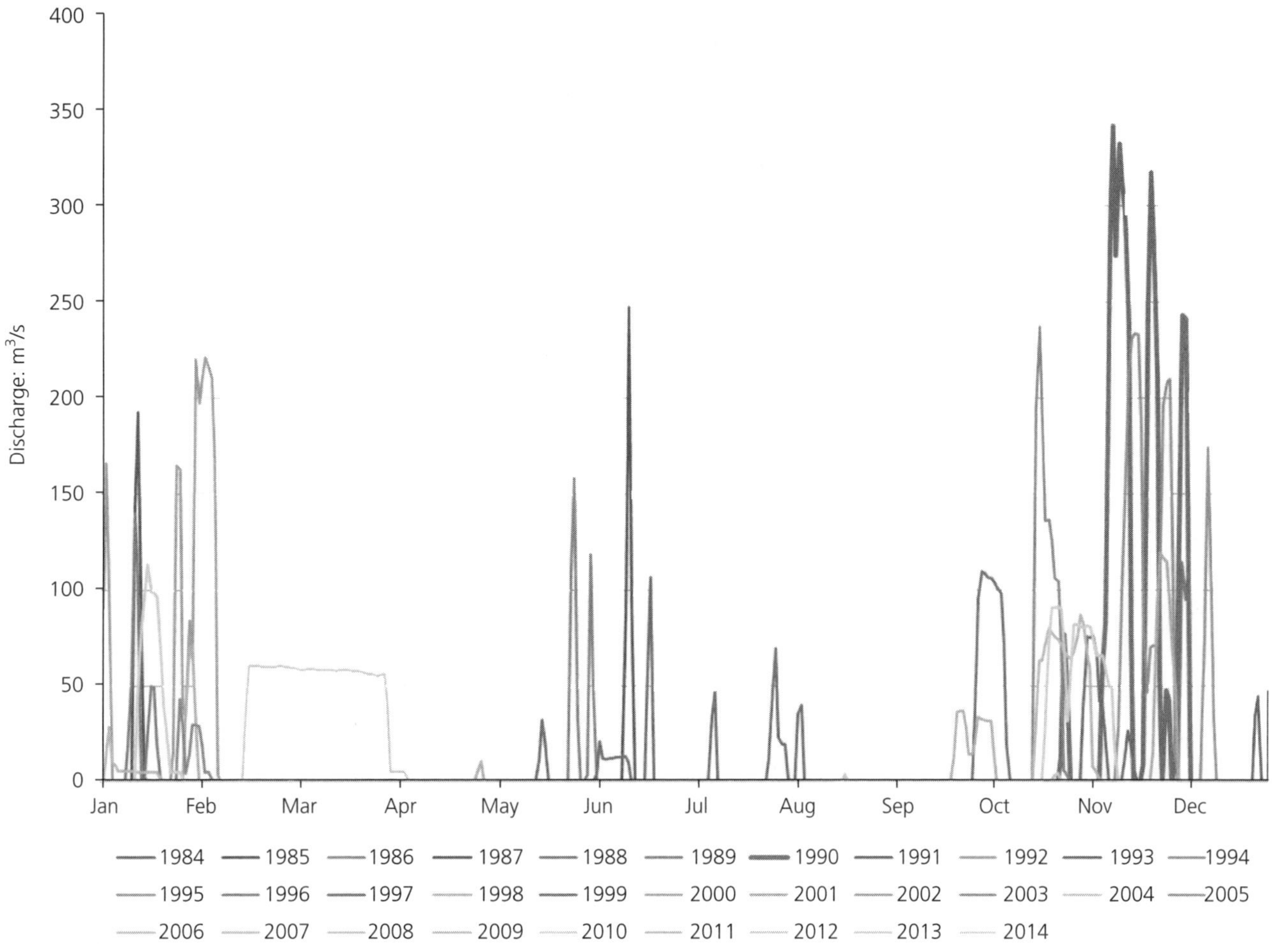


Figure 9.33 Strathcona Dam: modelled results for reservoir El. 228 m with all gates fully open (gates not shown, modelled discharge = 2100 $m^3/s$) – overall modelled area. (BC Hydro, 2015)

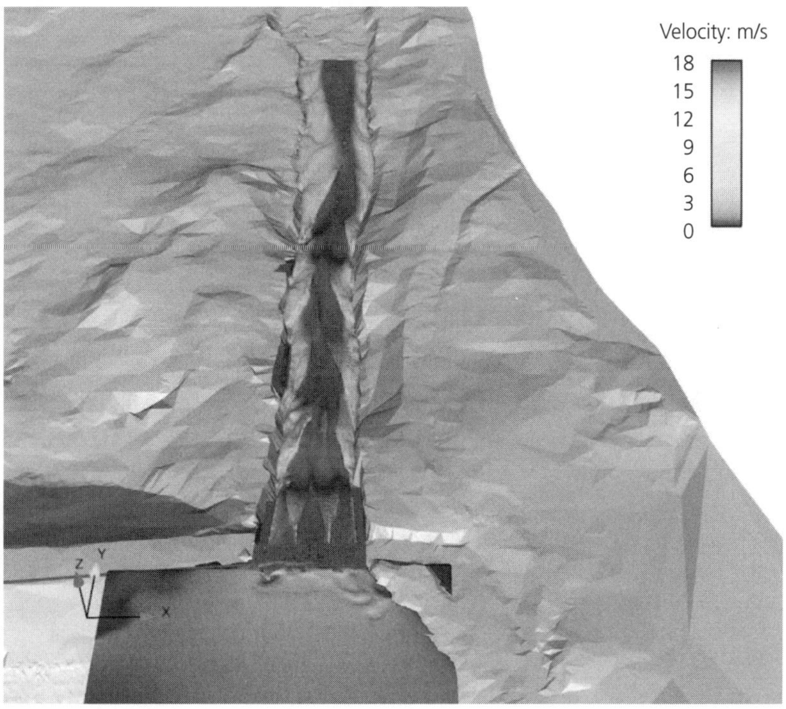

Figure 9.34 Strathcona Dam: modelled results for reservoir El. 228 m with all gates fully open (gates not shown, modelled discharge = 2100 $m^3/s$) – streamlines. (BC Hydro, 2015)

Inflows into the Strathcona reservoir for the 30-year period from 1984 to 2014 (Figure 9.31) were used as input for the simulation studies. The inflows show a mean substantial variations year by year about the mean (indicated by the heavy black line in Figure 9.31). This stochastic time series of reservoir inflows is reflected in the daily spillway discharges shown in Figure 9.32.

Modelling results for the spillway at a pool elevation of 228 m with all gates open are shown in Figure 9.33. Corresponding stream lines through the gates are shown in Figure 9.34, and corresponding fluid pressures in the flow at the centreline are shown in Figure 9.35, from which cavitation index can be calculated.

Figure 9.35 Strathcona Dam: pressure along bay 2 centreline with all gates fully open and reservoir El. 230 m (modelled discharge = 2351 $m^3/s$). (BC Hydro, 2015)

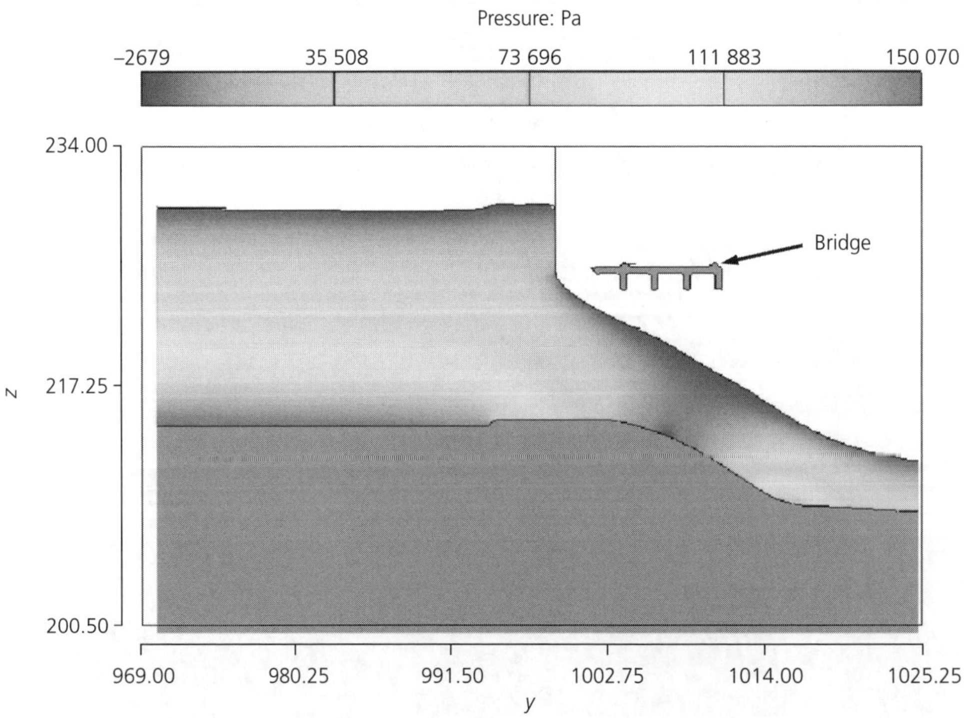

As for the Bennett Dam, the pressure profile and cavitation index can be expressed as a response surface of the flow discharge and the distance down the conveyance, and thus are easily incorporated in the simulation analysis.

## REFERENCES


Abdul Khadar MH and Elango K (1974) Turbulent pressure field beneath a hydraulic jump. *Journal of Hydraulic Research* **12(4)**: 469–489.

Arndt REA and Ippen AT (1968) Rough surface effects on cavitation inception. *Journal of Basic Engineering* **90(2)**: 249–261.

Baur T, Köngeter J and Leucker R (1998) Effects of dissolved gas on cavitation inception in free surface flows. *Third International Symposium on Cavitation*, Grenoble, France.

BC Hydro (2013) *Dam Safety – W.A.C. Bennett Dam Performance Investigation Spillway Chute Capacity*. BC Hydro, Burnaby, BC, Canada.

BC Hydro (2015) *Dam Safety – Strathcona Dam Performance Investigation Spillway Hydraulics Computational Fluid Dynamics Modelling Report*. BC Hydro, Burnaby BC, Canada.

Beck JL and Au S-K (2002) Bayesian Updating of Structural Models and Reliability using Markov Chain Monte Carlo Simulation. *Journal of Engineering Mechanics*, **128(4)**: 380–391.

Bellin A and Fiorotto V (1995) Direct dynamic force measurement on slabs in spillway stilling basins. *Journal of Hydraulic Engineering* **121(10)**: 686–693.

Bohrer JG, Abt SR and Wittler RJ (1998) Predicting plunge pool velocity decay of free falling, rectangular jet. *Journal of Hydraulic Engineering* **124(10)**: 1043–1048.

Bowers CE and Toso J (1988) *Karnafuli Project – Model Studies of Spillway Damage*. American Society of Civil Engineers, Reston, VA, USA.

Bowers CE and Tsai FY (1969) Fluctuating pressures in spillway stilling basins. *Journal of the Hydraulics Division – ASCE* **95(6)**: 2071–2080.

Canepa S and Hager WH (2003) Effect of jet air content on plunge pool scour. *Journal of Hydraulic Engineering* **129(5)**: 358–365.

Castro-Orgaz O (2010) Velocity profile and flow resistance models for developing chute flow. *Journal of Hydraulic Engineering* **136(7)**: 447–452.

Castro-Orgaz O and Hager WH (2010) Drawdown curve and turbulent boundary layer development for chute flow. *Journal of Hydraulic Research* **48**: 591–602.

Chadwick AJ, Morfett JC and Borthwick M (2013) *Hydraulics in Civil and Environmental Engineering*, 5th edn. Taylor & Francis, Boca Raton, FL, USA.

Chanson H (1993) Self-aerated flows on chutes and spillways. *Journal of Hydraulic Engineering* **119(2)**: 220–243.

Chanson H (2011) Bubbly two-phase flow in hydraulic jumps at large Froude numbers. *Journal of Hydraulic Engineering* **137(4)**: 451–460.

Chanson H and Gualtieri C (2008) Similitude and scale effects of air entrainment in hydraulic jumps. *Journal of Hydraulic Research* **46(1)**: 35–44.

Chen (1991) Unified Theory on Power Laws for Flow Resistance. *Journal of Hydraulic Engineering* **117(3)**: 371–89.

Chow VT (2009) *Open-channel Hydraulics*. McGraw-Hill, New York, NY, USA.

Eickmann G (1993) Scale effects for incipient cavitation under consideration of water quality. *Journal of Hydraulic Research* **313(3)**: 347–364.

Ervine DA, Falvey HT and Withers W (1997) Pressure fluctuations on plunge pool floors. *Journal of Hydraulic Research* **35(2)**: 257–279.

Falvey HT (1990) *Cavitation in Chutes and Spillways*. Engineering Monograph 42. National Technical Information Service, Operations Division, Springfield, VA, USA.

Farhoudi J and Narayanan R (1991) Force on slab beneath hydraulic jump. *Journal of Hydraulic Engineering* **117**: 64–82.

Fiorotto V and Rinaldo A (1992) Turbulent pressure-fluctuations under hydraulic jumps. *Journal of Hydraulic Research* **30(4)**: 499–520.

Fiorotto V and Salandin P (2000) Design of anchored slabs in spillway stilling basins. *Journal of Hydraulic Engineering* **126(7)**: 502–512.

Granger RA (1995) *Fluid Mechanics*. Dover New York, NY, USA.

Güven A, Günal M and Çevik A (2006) Prediction of pressure fluctuations on sloping stilling basins. *Canadian Journal of Civil Engineering* **33(11)**: 1379–1388.

Hager WH (1991) Uniform aerated chute flow. *Journal of Hydraulic Engineering* **117(4)**: 528–533.

Hajdin G (1982) Contribution to the evaluation of fluctuating pressure on fluid currents limit areas – based on pressures recorded at several points of the area. In *VIII Conference of Yugoslav Hydraulics Association*, Portoroz, Slovenia.

Hartung F and Häusler E (1973) R3: Scours, stilling basins and downstream protection under free overfall jets at dams. Question 41: *Flow control and energy control during construction and after completion. 11th ICOLD Congress*, Madrid, Spain.

Heller V, Hager W and Minor H-E (2005) Ski jump hydraulics. *Journal of Hydraulic Engineering* **131(5)**: 347–355.

ICOLD (International Commission on Large Dams) (2011) *Dam Safety Management: Operational Phase of the Dam Life Cycle*. Bulletin 154. ICOLD, Paris, France.

Jaeger C (1958) Contribution to the stability theory of systems of surge tanks. *Transactions of the ASME* **80**: 1574–1584.

Kalinske AA and Robertson JM (1943) Closed conduit flow. *Transactions of the ASCE* **108**: 1435–1447.

Kawakami K (1973) A study on the computation of horizontal distance of jet issued from a ski-jump spillway. *Transactions of the Japanese Society of Civil Engineers* **5**: 37–54.

Keller U (2006) *Intermittent Flow in Hydraulic Conduits*. Die Versuchsanstalt für Wasserbau, Hydrologie und Glaziologie (VAW), Zurich, Switzerland.

Keller R and Rastogi A (1977) Design chart for predicting critical point on spillways. *Journal of the Hydraulics Division – ASCE* **103(12)**: 1417–1429.

Kramer K (2004) *Development of Aerated Chute Flow*. Die Versuchsanstalt für Wasserbau, Technischen Hochschule, Zurich, Switzerland.

Kramer K, Hager WH and Minor H-E (2006) Development of air concentration on chute spillways. *Journal of Hydraulic Engineering* **132(9)**: 908–915.

Launder BE and Spalding, DB (1974) The numerical computation of turbulent flows. *Computer Methods in Applied Mechanics and Engineering* **3(2)**: 269–289.

Lopordo, RA and Henning RE (1985) Experimental advances on pressure fluctuations beneath hydraulic jump. *Proceedings of the 21st IAHR Congress*, Melbourne, Australia.

Mahzari M and Schleiss J (2010) Dynamic analysis of anchored concrete linings of plunge pools loaded by high velocity jet impacts issuing from dam spillways. *Dam Engineering* **4**: 307–327.

Mason PJ (1982) The choice of hydraulic energy dissipator for dam outlet works based on a survey of prototype usage. *Proceedings of the Institution of Civil Engineers, Part 1* **72(2)**: 209–219.

Mason P (1984) Erosion of plunge pools downstream of dams due to the action of free-trajectory jets. *Proceedings of the Institution of Civil Engineers, Part 1* **76(2)**: 523–537.

Mason P (1989) Effects of air entrainment on plunge pool scour. *Journal of Hydraulic Engineering* **115(3)**: 385–399.

Mason P (1993) Practical guidelines for the design of flip buckets and plunge pools. *Water Power and Dam Construction* **45(9/10)**: 40–45.

Melo JF, Pinheiro AN and Ramos CM (2006) Forces on plunge pool slabs: influence of joints location and width. *Journal of Hydraulic Engineering* **132(1)**: 49–60.

Ohtsu I, Yasuda Y and Takahashi M (2004) Flow characteristics of skimming flows in stepped channels. *Journal of Hydraulic Engineering* **130(9)**: 860–869.

Ohtsu I, Yasuda Y and Takahashi M (2009) Similitude and scale effects of air entrainment in hydraulic jumps. *Journal of Hydraulic Research* **47(2)**: 285–287.

Pei-Qing L and Ai-Hua L (2007) Model discussion of pressure fluctuations propagation within lining slab joints in stilling basins. *Journal of Hydraulic Engineering* **133(6)**: 618–624.

Peterka AJ (1953) The effect of entrained air on cavitation pitting. *Proceedings of the Joint Meeting of the IAHR and the ASCE*, Minneapolis, MN, USA.

Peterka AJ (1978) *Hydraulic Design of Stilling Basins and Energy Dissipators*. US Bureau of Reclamation, Denver, CO, USA.

Pfister M and Hager W (2010a) Chute aerators. I: Air transport characteristics. *Journal of Hydraulic Engineering* **136(6)**: 352–359.

Pfister M and Hager W (2010b) Chute aerators. II: Hydraulic design. *Journal of Hydraulic Engineering* **136(6)**: 360–367.

Pinheiro AN and Melo JF (2008) Effect of jet aeration on hydrodynamic forces on plunge pool floors. *Canadian Journal of Civil Engineering* **35(5)**: 521–530.

Pinto NL (1994) General report on Question 71: Deterioration of spillways and outlet works. *18th ICOLD Congress*, Durban, South Africa.

Pinto NL and Neidert SH (1982) Model prototype conformity in aerated spillway flow. *BHRA International Conference on Hydrological Modelling of Civil Engineering Structures*, Coventry, UK.

Pinto, NL, Neidert, SH and Ota J (1992) Discussion of 'Air entrainment by spillway aerators' by Peter Rutschmann and Willi H. Hager (June, 1990, Vol. 116, No. 6). *Journal of Hydraulic Engineering* **118(1)**: 114–116.

Rae PJ and Castro DH (1994) Estimation of the free jet trajectory length. In *ASCE Conference on Hydraulic Engineering*, Buffalo, NY, USA, pp. 421–425.

Rajaratnam N (1960) *An Experimental Study of the Air Entrainment Characteristics of Hydraulic Jump*. Indian Institute of Science, Karnataka, India.

Rozanov NP and Kaveshnikov AT (1973) Hidrotechn. Stro. *Hidrotechn. Stro. USSR*, **1**: 29–32.

Rutschmann P and Hager W (1990) Air entrainment by spillway aerators. *Journal of Hydraulic Engineering* **116(6)**: 765–782.

Sanchez B and Capello VA (1973) Turbulence effects on the lining of stilling basins. In *11th ICOLD Congress*, Madrid, Spain.

Schlichting H and Gersten K (1979) *Boundary Layer Theory*, 8th edn. Springer-Verlag, Berlin, Germany.

Schmocker L, Pfister M, Hager WH and Minor H-E (2008) Aeration characteristics of ski jump jets. *Journal of Hydraulic Engineering* **134(1)**: 90–97.

Semenkov VM and Lentiaev LD (1973) *Spillway with Nappe Aeration*. Hidrotechn, USSR.

Sharma HR (1973) *Air Demand for High Head Gated Conduits*. SINTEF, Trondheim, Norway.

Steiner R, Heller V, Hager WH and Minor H-E (2008) Deflector ski jump hydraulics. *Journal of Hydraulic Engineering* **134(5)**: 562–571.

Thun H (2006) *Assessment of Fatigue Resistance and Strength in Existing Concrete Structures*. Doctoral thesis, University of Luleå, Luleå, Sweden.

Urban AL, Wilhelms SC and Gulliver JS (2005) Decay of turbulence downstream of a stilling basin. *Journal of Hydraulic Engineering* **131(9)**: 825–829.

USACE (US Army Corps of Engineers) (1964) *Air Demand. Regulated Outlet Works*. Hydraulic Design Criteria, Sheet 050-1. USACE, Washington, DC, USA.

Vining GG (2008) Adapting response surface methodology for computer and simulation experiments. In *The Grammar of Technology Development* (Tsubaki H, Yamada S and Nishina K (eds)). Springer, Tokyo, Japan, pp. 127–134.

Wood IR (1991) *Air Entrainment in Free-surface Flows. IAHR Hydraulic Structures Design*. Balkema, Rotterdam, the Netherlands.

Xu WN, Liao H, Yang YN and Wu C (2002) Turbulent flow and energy dissipation in plunge pool of high arch dam. *Journal of Hydraulic Research* **40(4)**: 471–476.

Yasuda Y, Takahashi M and Ohtsu I (2001) Energy dissipation of skimming flows on stepped-channel chutes. *Proceedings of the 29th IAHR Congress*, Beijing, China, pp. 531–536.

# Part IV

# Modelling dam and reservoir systems

**Operational Safety of Dams and Reservoirs**
ISBN 978-0-7277-6121-7


http://dx.doi.org/10.1680/osdr.61217.305

# Chapter 10
# Simulating flow control operations

The systems approach to understanding operational safety and reliability has been laid out in Chapter 3, and how simulation modelling can be used to track operations and flow control, and to identify malevolent chains of events has been discussed in Chapter 4. This chapter delves deeper into the concepts and mathematics of systems modelling. Two large-scale examples of the use of systems simulation for dam safety are presented: the Lower Mattagami River Project in northern Ontario, Canada, and the Göta River system in western Sweden.

The problem faced when analysing flow control through a reservoir (in a broad sense, including spillways and generating units) or a cascade of reservoirs is that the whole behaves as a complex *engineered system* the behaviour of which is more than the sum of its parts. Unravelling the behaviour of complex systems is a challenge faced in many modern disciplines.

Complex systems comprise a large number of relatively simple systems interacting together. Each of these relatively simple subsystems can, in principle, be modelled mathematically, at least to a first-order, but their interactions within the larger system typically cannot. It is commonly the case that the quantitative behaviour of the complex system cannot be forecast with precision, although its qualitative behaviour may be structured and understood using contemporary simulation techniques. Stochastic simulation techniques have been used to unravel the mysteries of many engineered and natural systems that had previously proven opaque to analysis. Among these are the electric power grid, communications systems, aircraft performance, nuclear power plants, climate systems and financial markets.

Simulation resolves to a great deal of computer modelling. The relatively simple subsystems are modelled mathematically, and their individual behaviours are then interrelated by means of simulation. Simulation is relied upon because closed-form analytical solution of the overall complex system is usually beyond currently available modelling capabilities. The simulation modelling involves mathematical characterisation of the subsystems, a backbone of network theory, probabilistic cauterisation and differential equations, and a good deal of numerical computation. The result is that we can approximately simulate how the complex system behaves, and we can change things in the input and formulation to see what might happen when the complex system is disturbed. The disturbance might either be by routine variations (e.g. due to human factors or stochastic variations) or by systemic loadings (e.g. as from earthquakes or landslides).

The purpose of the simulation is to gain a holistic understanding of how the complex system works by numerically mimicking its behaviours. Things will go wrong, often by a combination of not-improbable conditions the juxtaposition of which threatens the stability of the system, and in ways that no one thought of beforehand. The modelling leads to an understanding of how management might anticipate or respond to things going wrong in an operation.

For many applications in the natural sciences, social sciences and industry, simulation modelling has become a required tool of analysis and understanding. The use of large-scale stochastic simulation to study complex systems is now common in many scientific disciplines, especially when exploring questions that are poorly suited to traditionally deductive, deterministic analysis. This approach has proven particularly powerful when dealing with systems involving multiple and seemingly incommensurate aspects; for example, in combining factors such as physical science, natural processes, information networks, economics, and human factors. Today, simulation approaches are used in fields as disparate as oil and gas exploration, medicine, materials science, urban planning and aerospace (NSF, 2006). Over the past decade the US National Research Council has published more than a dozen reports on the use of simulation in engineering science, public policy, STEM (science, technology, engineering and mathematics) training and other applications (NRC 2002, 2008, 2010).

## 10.1. System simulation of hydropower operations

In the present book, the system of interest is flow-control operations, the fundamental functions of which are to store, bypass or divert water. These functions are treated as subfunctions of a control system. The operating objective of different kinds of dam systems might be hydropower generation, water supply, navigation control, recreation, etc., or a combination of these. Nevertheless, flow control is the critical enabling system for any of these primary operating objectives.

Here, we are mostly concerned with hydropower generation and discharge control. The natural and engineered systems typical of hydropower can be categorised using the taxonomy proposed by Simonovic (2008):

> *Natural systems* are those regularly occurring in the world, such as hydrological, geological, climatological, or biological processes. These are the settings within which engineered systems are built. Natural systems, by virtue of their evolutionary development, exhibit a high degree of interconnectedness. They typically exhibit cyclic behaviour in that energy, material, and other flows have established themselves over long periods of time and are in balance, until disturbed by externally imposed change.
>
> *Engineered systems* are those designed and constructed by people. They may or may not exhibit the high degrees of interconnectedness characteristic of natural systems. A river basin with a cascade of dams is a system composed of both natural and engineered aspects, and is made more complex by the juxtaposition of these naturally occurring and man-made factors.
>
> *Conceptual systems* exist as concepts and mathematical relations but not in an instantiation in physical reality, that is, in contrast to natural and engineered systems, which do exist in the physical world. Conceptual systems are important in mathematics, computer science, logic, and may fields of study; but are important in the present discussion in the extent to which they may be used in an abstract, analytical capacity to mimic physical systems.

Yet another distinction between systems involves those that are *static* and those that are *dynamic*. A static system, such as a structural frame, responds to forces and other disturbances but does not change or evolve in response to external perturbations. A static system, such as the embankment of a dam, will respond to loads and perturbations, and will generate interactions between its subcomponents, but its structure or interactions will not evolve in response. A dynamic system is one the properties of which do change or evolve in response to the forces or other disturbances applied to them. The riverbed downstream of a spillway and stilling basin is a dynamic system. Its properties may be changed by the loads and perturbations to which it is subjected. The boundary between static and dynamic systems may not be finely drawn.

Systems may also be *open* or *closed*. An open system interacts with its larger environment, and may be adaptive or self-regulating; that is, its functions and interactions may evolve in response to perturbations. A closed system, in contrast, interacts little with its environment, and may be impervious to forces and perturbations imposed from outside. It maintains its functions in spite of the environmental loads and influences to which it is subjected.

### 10.1.1 System states and processes

At its core, simulation is the imitation of salient properties of one process by another. Typically, the first process is something occurring in the natural world, in our case flows of water, information, human actions and possibly other things through a reservoir or series of reservoirs. Typically, the second process is mathematical. We use the term 'salient properties' to mean those that are important, noticeable or conspicuous to the current purpose. The natural system may have many properties that a simulation will not and need not capture, because they are irrelevant to the purpose at hand.

System states are the variables necessary to describe a system at a particular time and for the particular purposes of the study. In principle, there is an infinite number of potential systems states. Thus, system states need to be defined by the modeller for the particular purpose at hand and in a multitude that balances comprehensiveness with efficiency. Defining system states is simply one of the many modelling decisions that need to be made. For a hydropower dam, system states might include the reservoir inflow, the power being generated by each turbine, the open or closed condition of each gate, the discharge of water downstream, and so on.

As a simulation proceeds, the realised values of the system states are tracked at each cycle of time. The values of some of the system states become inputs to other subsystems. The values of others of these (selected) system states may be related to, or may themselves be, output performances of the overall system. Some of these output performances may constitute adverse performance or failure of the system functions. The rate at which these adverse systems states appear in the simulated record may be taken as a statistical estimate of the probability of system failure in particular ways. For example, the rate in simulated time at which the system state 'spillway gate fails to open' can be used as a statistical estimate of gate availability. The rate in simulated time at which the system state 'minimum hydraulic flow pressure on the spillway conveyance' exceeds cavitation limits can be used to estimate statistically the probability of cavitation.

The term 'process' means the sequence of system states as they evolve over time. Because these transitions between states may reflect many intellectual disciplines – hydrology and hydraulics, geotechnology, human behaviour, instrumentation and

**Table 10.1** Categories of mathematical systems models

| System property | Characteristics |
|---|---|
| Linear versus non-linear | In a *linear* model all the relationships between input and output processes, the external constraints on the system and the objectives of the operation (if applicable) are represented by simple linear relationships. This is infrequently the case in flow-control models, where these relationships, constraints and objective functions tend to be non-linear. The behaviour of non-linear models tends to be much less intuitive and predictable than for linear models due to the complexity of interactions between variables. |
| Deterministic versus probabilistic | For the case that each input parameter to the system model can be assigned a specific and definite value, and each computer relationship or submodel within the model translates specific and definite input parameter values to specific and definite output values, a model is *deterministic*. The input values may themselves be uncertain, and varying those parameter values might allow sensitivity computations; but if the values and relationships are specific and definite in any one running of the model, then the deterministic nature of the model still holds. The opposite of deterministic is *probabilistic*, where input and output values are represented by probability distributions rather than specific and definite values. |
| Static versus dynamic | A *static* model is not a function of time. The forces or perturbations on the system achieve an equilibrium condition and remain at equilibrium until conditions are changed or another condition is imposed. *Dynamic* systems respond, usually continuously, to temporal changes in environmental conditions or internal processes. If those conditions and the responses are probabilistic, a dynamic system is said to be *stochastic*. |
| Lumped versus distributed parameter | Models in which parameter values are assumed constant throughout space are usually referred to as *lumped parameter* models. That is, the parameters are lumped at some average or at least defined value and are assumed to be homogeneous in space. They may change over time. *Distributed parameter* models are those where the model parameter values vary over space. The variation may be deterministic or probabilistic. In the latter case, the models might be called *stochastic* or, in special cases, *geostatistical* (depending on the modelling approach). |

control, thermodynamics – the simulation approach lends itself to multidisciplinary analyses. The transitions within individual states or similar sets of states is modelled mathematically. From a mathematical viewpoint, the sorts of systems that we attempt to simulate have characteristic mathematical properties that are useful as a way of categorising the types of simulation that need to be built (Table 10.1).

The simulation models used to replicate and forecast the behaviour of spillway-system reliability are usually non-linear, stochastic and spatially distributed. For example, the hydraulic pressures and cavitation damages induced in outflow conduits or stilling basins depend in highly non-linear ways on spillway discharges and thus reservoir-pool elevations. These pool elevations fluctuate as random time series depending on the hydrological conditions creating reservoir inflows. The reservoir inflows and disturbances from events such as landslides depend in turn on spatially, and possibly temporally, variable basin parameters affecting runoff, debris generation and other processes.

### 10.1.2 Why use simulation for hydropower?

Stochastic simulation approaches to modelling and understanding complex water resource system performance has become increasingly common. For example, for large-scale water and power infrastructure Tilmant *et al.* (2014) have applied simulation to optimising power production in a cascade of reservoirs under variable hydrology, and von Stackelberg and Neilson (2014) have applied simulation to riverine water quality. In addition, Gersonius *et al.* (2012) have applied simulation to incorporate uncertainty and flexibility in the economic analysis of flood risk and coastal management strategies, Regan (2010) has suggested a simulation approach to dam safety, and Billinton and Li (1994) have applied simulation to the assessment of the reliability of power-distribution systems. In applications to other systems safety problems, Blum *et al.* (2010) have applied simulation to the safety analysis of airspace craft design, Leveson (2012) has applied simulation to power-generation systems, Hosse and Schnieder (2012) have applied simulation to highway safety, and Zio (2013) and Blom *et al.* (2006) have surveyed many systems safety applications. Simulation methods are also now widely used in modern financial engineering analyses of risk (Glasserman, 2010).

There are a variety of good reasons for the increasing popularity of simulation approaches to safety and risk analysis:

- simulation allows complex systems interactions to be modelled easily compared with the use of closed-form analytical models
- computer speeds are rapidly increasing, and engineers have increasing access to high-performance computing
- discrete events are readily included in simulations, while they often pose combinatorial problems in fault-tree and event-tree models
- simulation readily allows for the inclusion of and interaction between many and differing types of system response (e.g. physics of failure, communication and control, and human reliability)
- the numerical precision of simulation results is independent of the complexity of the system being modelled, and depends only on the number of simulation trials being performed.

Complex systems, of the type related to operational reliability, share several characteristics that make them suitable subjects for stochastic simulation approaches:

- they involve a large number of parts (components)
- the parts have combinatorially many interactions, both one-to-one and many-to-many
- the input parameter space is large
- the timing of events is important
- uncertainties in the parameters are of many types and are often correlated
- the system behaviours or outputs are also stochastic and are usually unpredictable based on simple abstractions.

Finally, and perhaps most relevant, the output behaviours of these systems are emergent; that is, patterns of performance arise out of the diversity and number of relatively simple component interactions, and those patterns are usually not obvious ahead of time.

## 10.2. Stochastic simulation

A simulation typically consists of three parts:

- Subsystem components, such as reservoir inflow channels, spillway gates, hydraulic conveyances, and data acquisition and control systems. These can be interfaced to model of a particular reservoir system.
- Data acquisition components, which track the progress of a simulation. These aggregate the data from the simulation runs and create summaries.
- User interface and dashboard (i.e. visualisation) components, which allow the data from the simulations to be portrayed to the user.

The modelled system comprises a large number of interrelated subsystem components. Modules can be developed to be free-standing and to 'snap' together as appropriate to form a particular reservoir and flow-control system. The subsystems are designed to be free-standing, and may be redundant in the sense that duplicative types of subcomponents might be developed for different purpose. For example, two or more subcomponents might be developed for reservoir inflow. One might be a simple flood-frequency statistical model using a stochastic time series of river discharges. Another might be a spatially distributed rainfall-runoff model. These two could be swapped for one another depending on the needs of a particular analysis. This is discussed in more depth in Chapter 4.

Each subsystem component is abstracted in the form of mathematical expressions that describe the behaviour of the component for a given input and a given set of disturbances, and specifies the output interactions of the component with other components (Figure 10.1). This set of system components forms a large network of interactions, as illustrated in Figure 10.2.

**Figure 10.1** Schematic system component. SCADA, supervisory control and data acquisition system

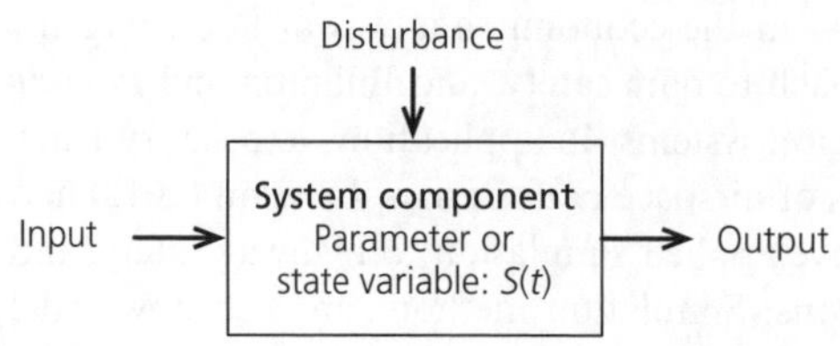

Figure 10.2 System boundary and interactions

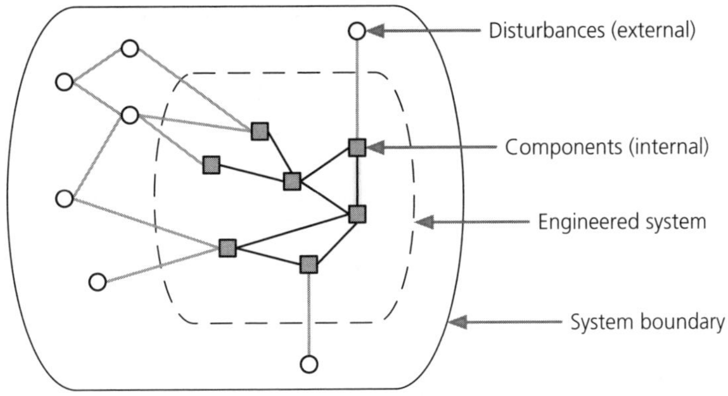


The mathematical expressions that describe the system component contain parameters, the realised values of which are the system states at any one moment in time. Some of these parameters are fixed and unchanging; others change under the influence of the history of inputs and disturbances up to that moment. At the end of any one cycle in the simulation the totality of the parameter values at all the systems components define the system state at that moment.

The total number of system-state parameter values at any cycle can be large – too large to track efficiently. Thus, as described in Section 10.1.1, a subset of a limited number of these is identified as selected system states (e.g. water levels, flow rates, pressure distribution, vibrations and temperature) and the values of these are tracked and compared against performance criteria to identify important functions, hazardous states or failures. These might be tracked as discrete qualities: green for 'acceptable' state, yellow for 'marginal' state, and red for 'unacceptable' or failure state.

As a general rule when modelling systems, the systems states at the respective components may change upon each simulated cycle, but components may also be added, eliminated or their fundamental structure modified (De Weck *et al.*, 2012). In the present case of simulating spillway systems, components may be modified but they are not typically added or eliminated once a simulation in in progress.

From an operational point of view, the critical issue is to develop an interface protocol between the subsystem components such that the output of one subsystem is readily accepted as the input to the next in the chain. This allows any of the subsystem modules to be replaced by a similar module using a different calculation approach (e.g. stochastic reservoir inflows versus rainfall-runoff generated inflows).

Within the simulation, three process flows are tracked:

- physical flows (i.e. water, sediments, debris, etc.)
- communication (i.e. information flows)
- control (i.e. human-action flows).

The principal models along the part of the water flow start with the hydrology model, which generates inflows. Next comes the reservoir module, which translates inflows into pool elevations, and possibly related performance variables. The reservoir model provides inputs to the hydraulic gate and valve modules, which control outflows to the respective waterways (i.e. the surface spillway, powerhouse, bottom drain if any, and auxiliary spillway if any). The gate and valve modules provide input to the downstream conveyances, energy dissipation structure and, ultimately, the downstream channel.

In general, the water-flow modules and path are reasonably straightforward physics-based models of the type common in hydropower engineering. They take naturally occurring weather processes as input, and generate hydraulic process as output.

### 10.2.1 Tracking physical flows

The simulation of water flows was introduced in Chapter 4. Water accumulates in the upstream basin or upstream cascade in a set of hydrology modules. These might be simple time series statistical models, more complex geospatial rainfall-runoff models, or even more complicated synthetic weather models. The hydrology model is coupled with a reservoir routing module that takes reservoir inflows as input and returns pool elevation, and possibly other things, as output (Chapter 5).

These reservoir outputs provide demand inputs to the various waterways (Figure 10.3). The waterway modules, in turn, provide outputs to the downstream channel. The physical characteristics of the water flows are instantiated in states at a

Figure 10.3 Simplified simulation logic. The subsystem modules are designed such that the interface protocols between the modules allow the modules to be replaced by other technologies

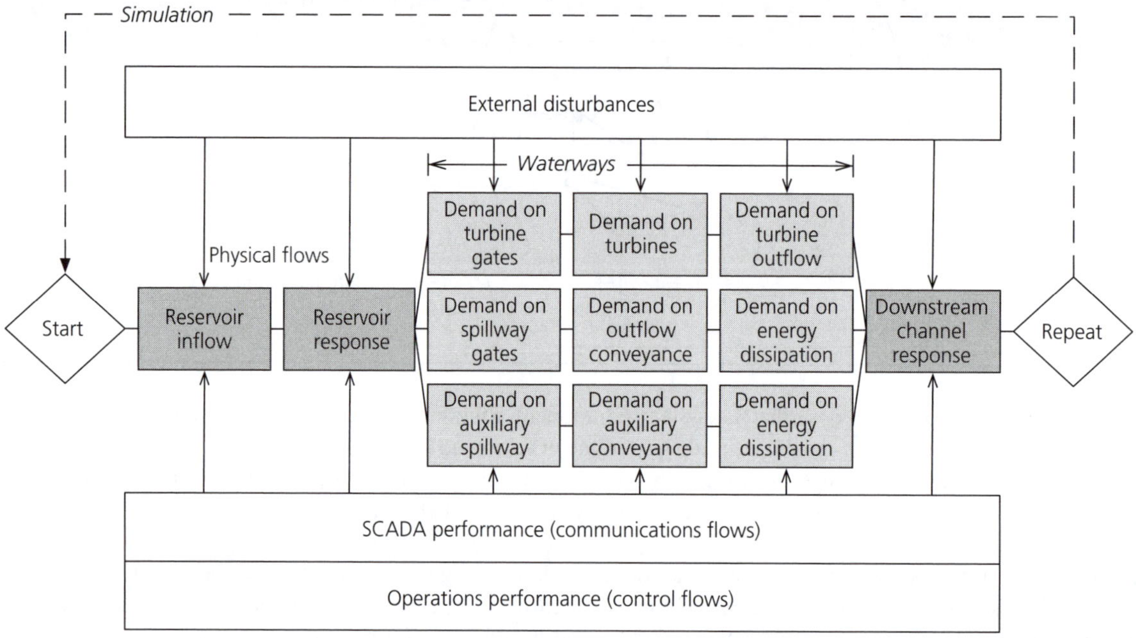


set of predefined system states, and these system states are compared against a set of limiting-state criteria to judge whether adverse performance or failures occur at any time in the simulation.

### 10.2.2 Tracking communication flows

As the physical system operates – as water flows into the reservoir, as waterway gates respond to changing reservoir conditions and demands, as power is generated, and as other parts of the system function – observations are continuously being made to inform automatic controllers and to inform decisions made by human operators (Figure 10.4). These observations form the basis of a communications or data flow from sensors to supervisory control and data acquisitions (SCADA) systems and eventually to operators. From operators at the plant, a series of upward communications move data, reports and other submissions to office staff, management and, possibly, to regulatory authorities. In the current discussion, these outward paths of information are referred to generically as 'communications'. The loop back to the operating plant is closed by a corresponding succession of control actions returning from the hierarchy of operators, staff and management.

From a systems theory view, control necessitates communication to provide the information upon which action (control) decisions can be made (von Bertalanffy, 1969). von Bertalanffy is perhaps best known for his concept of open (versus closed) systems. He argues that open systems (those that are influenced or disturbed from their environment) require feedback loops of communication and control to maintain a dynamic equilibrium, whereas closed systems (those that are not disturbed from outside) settle into a stable equilibrium and absent such feedback. The communication–control feedback loop comprises the lower left-hand side of the logic diagram shown in Figure 10.4.

Figure 10.4 Communication–control loop segment of system simulation. (Adapted from Leveson, 2012)

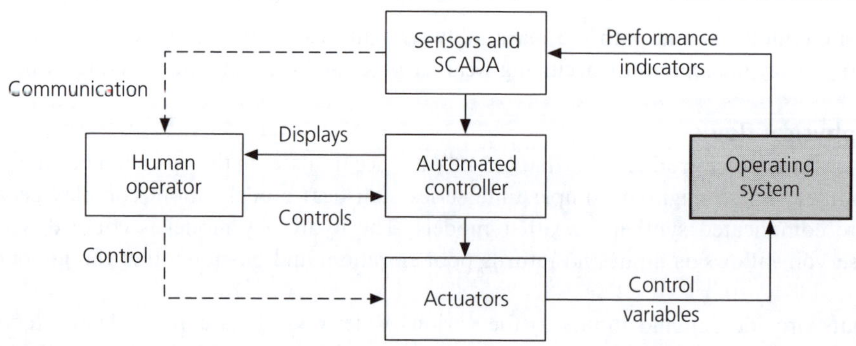

The operating system (the reservoir, dam, waterways, powerhouse, etc.) generates an essentially infinite number of varying system states as it operates. Some subset of these is identified as being important performance indicators. Sensors of one form or another (automatic, manually read, simply observed) are installed to measure these performance indicators. The data then collected by these sensors and tracked in the SCADA system are used to inform automatic controllers, if any, and operator actions. These control actions affect the state of the operating system to achieve some form of equilibrium behaviour.

Sensors are vital components of operating systems; however, they are often the weakest link in system safety because they are less reliable than the operating system they are sensing. The principal causes of sensors being the weak link in system safety is their tendency to interference from the environment, and their susceptibility to bias, drift, scaling and dropout (loss of signal).

Sometimes system accidents or failures occur because the communication information (sensor measurements or their interpretation) is wrong or wrongly perceived. This was a significant cause of the accident at the Taum Sauk hydroelectric power station in Missouri, USA, in 2005 (see Chapter 1) (Hendron *et al.*, 2005), and the fire at the Buncefield oil depot near London's Heathrow Airport in the same year (BBC, 2006). In each case, fluid-level sensors measuring the height of fluid in the reservoir were erroneously interpreted, leading to overtopping and breaching of the Taum Sauk water reservoir, and overtopping and subsequent explosion of the Buncefield fuel storage depot (Box 10.1).

An instrument intended to measure the system state of some performance indicator upon demand is assumed either to perform according to specification with a probability $p$ or to perform abnormally (not according to specification) with the

**Box 10.1** Buncefield fuel storage depot fire, Hertfordshire, UK, 2005

The Buncefield fuel storage depot fire occurred at Britain's fifth largest storage depot. The depot had a capacity of 270 Ml, comprising 30% of Heathrow Airport's fuel supply. The initial explosion occurred at 6 a.m. on 11 December, eventually consuming 20 large tanks. The British Geological Survey measured the blast from the explosion as a 2.4 event on the Richter scale. The cause of the explosion was a fuel–air explosion of high strength. The flames were finally extinguished by the afternoon of 13 December.

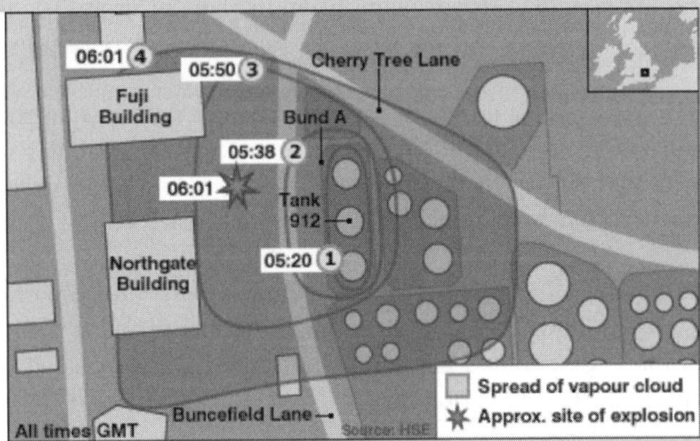


An inquiry was held by the Health and Safety Executive (HSE) and the Environment Agency. According to a BBC (2006) report:

> Starting at 19:00 on the evening of 10 December 2005, Tank 912, towards the north west of the main depot, was filled with unleaded petrol. At midnight the terminal closed, and a check was made of the contents of tanks, which found everything normal. Normally the gauges monitor the level of the fuel in the tank as it fills from the particular pipeline. From about 03:00 the level gauge for Tank 912 began to indicate an unchanging level reading, despite it being filled at 550 cubic meters per hour.
>
> Calculations show that the tank would have begun to overflow at about 05:20. There is evidence suggesting that a high-level switch, which should have detected that the tank was full and shut off the supply, failed to operate. The switch failure should have triggered an alarm, but it too appears to have failed. Forty-one minutes later, an estimated 300 tonnes of petrol would have spilled down the side of the tank through the roof vents onto the ground inside a bund wall.
>
> An overflow such as this results in the rapid formation of a rich fuel and air vapour. . . . By 06:01, when the first explosion occurred, the cloud, which was initially about 1 metre deep, had thickened to 2 metres and had spread beyond the boundaries of the site.

complementary probability (1 – $p$). Within a simulation the error performance of sensors is assumed to be independent from one trial to the next until an abnormal operation is generated. From that time onward, until repaired or replaced, the sensor is assumed to continue to be abnormal.

### 10.2.3 Tracking control actions

The safe operation of any industrial facility is highly dependent on the humans who operate and manage them, and who may be called upon to make decisions in the face of unexpected disruptions or other events. The reliability of human operators in the face of complex technological systems is discussed in Chapter 8. Over the past decades a systematic body of theory and practice has emerged on the types, rates and importance of human operator errors, and this insight needs to be incorporated in simulation models.

Simulation modelling of human reliability is usually approached from one of two directions: (1) treating operators (and managers) as 'humans in the loop', as if they were any other subcomponent of the complex technological systems with associated failure or error rates, and modelling these rates using, for example, fault- or event-tree calculations; or (2) treating operators (and managers) as intelligent agents whose behaviour in a particular special situation is modelled by a set of dynamic probabilities. The second approach uses agent-based modelling as a way of capturing operator performance in a larger systems simulation. The first approach uses simulation methods that have already been discussed, and which are discussed further in Chapter 8.

Agent-based modelling treats individual operators or other humans in the communication and control chain as autonomous actors. Each is assigned a set of potential responses when faced with a given situation or the requirement for a response. The probability distribution over the responses can be varied depending on the maturity of the agent, his or her level of training, the ambient stress of the contemporary environment, and the like.

### 10.2.4 Hydraulic modelling and CFD models as input to system models

Simulations of the water flows in a river system and through dam facilities, spillways and other waterways are usually done using numerical hydraulic models, which can be static or dynamic and are available in one-, two- and three-dimensional (1D, 2D and 3D) set-ups.

For flow simulations through the spillways, computational fluid dynamics (CFD) models have proven to be an effective, and these are now used to a growing extent in many projects. The usual way is to employ CFD models in the pre-design and design stages of a project, and to verify the final design proposal using physical scale modelling.

CFD is a scientific area of its own, and can be described as the art of solving numerical approximations of partial differential equations that model fluid flow. To this extent, it is similar to the use of finite-element models in structural mechanics and geomechanics. As a computational tool, CFD is both powerful and complicated. There are pitfalls related to all parts of the work, and both education and experience are needed to avoid them.

The general CFD principle is to divide the physical geometry into many small cells, and to discretise some set of equations describing fluid flow; that is, to approximate partial differential equations by algebraic equations in order to make them solvable. An example of the application of CFD to a whole spillway section is shown in Figure 10.5. The figure shows spillway parts, simulation of two-phase aerated flow for boundary layer computations.

Figure 10.5 Longitudinal profile of the set-up of a CFD model. The pressure distribution and turbulence on the chute invert is shown along the bottom of the stilling basin indicated by lighter colours

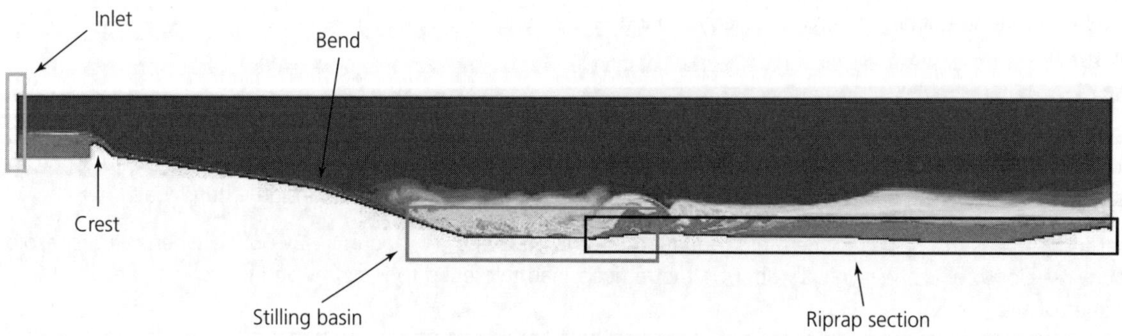

The results of computationally extensive simulations can be integrated into system models. Post-processed data describe important relationships as an input to the system models. Similar approaches are used for other types of physical process modelling that involve computationally intense numerical simulations (e.g. finite-element models).

### 10.2.5 Simulating system response to disturbances

While a simulation is continuously processing the hydraulic flow of water through the system, other events outside the physics of the hydraulic system may occur that have the potential to disrupt water flows through the system. In this book, these are referred to as 'disturbances'. Disturbances are discussed in detail in Chapter 7.

Disturbances include any potential disrupting influence on the physical system, and will vary from project to project depending on regional and local conditions. Typical examples of disturbances are seismic events, floating debris, floating ice or ice loadings on structures, reservoir landslides, fire and power grid disruptions (Figure 10.6). The list of potential disturbances is essentially unlimited, depending on local hazards and conditions.

### 10.2.6 Projecting future scenarios

In the system simulations discussed so far present basin and climatological conditions have been assumed: that is, how does the flow-control system perform given the current basin configuration and environment and its expected variations and extremes? On the other hand, land-use patterns, forestation or deforestation, and other changes in the basin may alter the present conditions in systematic ways so that the 50- or 100-year future of the basin may be strikingly different from what it is today. Climate variations may also occur, fundamentally changing not only the drainage patterns of the basin but also the normal and extreme storms that could impact the basin.

A particular scenario and a particular set of operating policies dictate how the flow-control system may function, generating a set of impacts. Simulation provides a way of experimenting with the basin and the flow-control system response by performing 'what-if' calculations on how changing basin conditions could affect the interactions of the components of the flow-control system. These what-if experiments are based on scenarios describing salient aspects of possible future conditions in the basin. Examples of typical future scenarios that might be explored in the simulation model are suggested in Table 10.2.

There is a set of potential future scenarios of basin conditions that might obtain. One might condition a set of simulations on one particular scenario, or one might treat the range of scenarios as itself being uncertain and assign probabilities within the set of future scenarios. Typically, future scenarios are hypothetical, and may differ dramatically in their impact from one another. Consequently, it is uncommon to assign probabilities to a particular set of assumptions on how the basin or project may evolve over time, but rather individual future scenarios are treated as speculative possibilities describing the future.

A systems dynamics approach is widely used in other fields to project future scenarios, and it is also applicable to dams and dam safety needs. The present chapter elaborates on systems dynamics theoretical considerations, as well as some practical examples.

Figure 10.6 Historical 30-minute power demand for a 1-month period at Madawaska Dam, Ontario, Canada. (Courtesy of OPG)

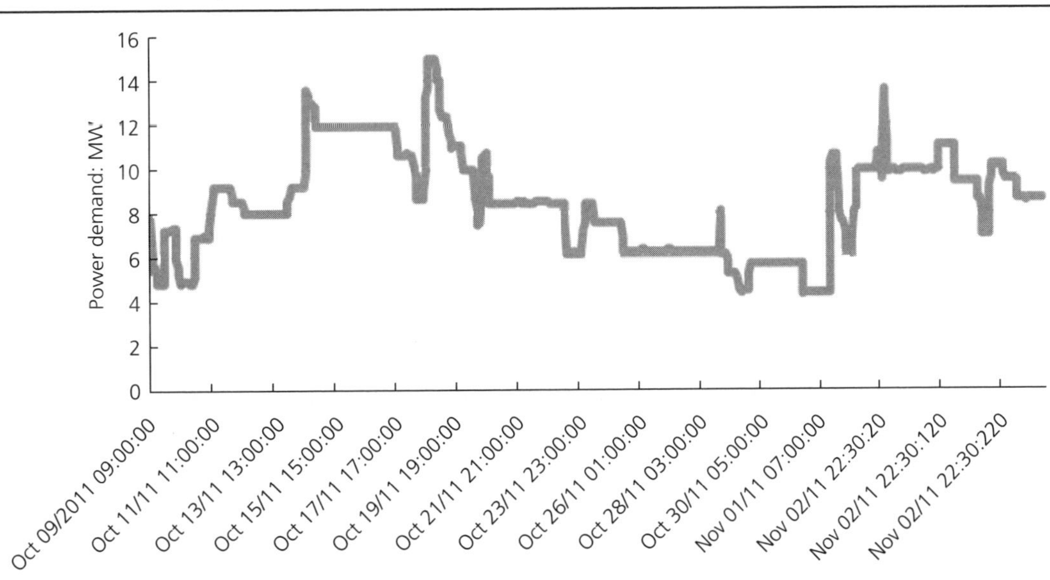

Table 10.2 Examples of future basin and reservoir scenarios

| Scenario of future conditions | Description and comment |
|---|---|
| Aggressive upstream development | Forest cover on certain upstream areas of the basin is cleared for agricultural or residential development. Highways, streets and other paved areas are developed. Each of these changes affects runoff, sedimentation into tributaries, and possibly the electrical grid demands for power production. The expectations, variations and temporal behaviour of reservoir inflows change, possibly dramatically. The frequency of grid impacts on production changes. Changing patterns of water use, both upstream and downstream, force operating-rule changes in terms of power production and downstream releases. |
| Climate change | Changing weather patterns gradually over time lead to changes in annual patterns and quantities of reservoir inflows. The weather patterns may lead to changes in power demand and in water-use patterns, affecting operations. Depending on the direction of change or volatility of weather patterns, demands for flood-control functions of a dam may increase and affect the safety of hydropower operations. |
| Upstream hydropower construction | Dams may be constructed upstream of a particular project, either by the same owner or operator, or by a different entity, introducing a cascade of flows through what becomes a series of dams. The reservoir inflows that previously had been dictated by natural hydrological process now become dependent on the operating rules of upstream producers, flood-control structures or water-supply operations. These may be dramatically different from the natural inflows upon which the original design and operating rules were based. |
| Economic change | Changes in regional economic patterns, demographics or even the financial situation of dam operators may force a reduction in the number or quality of operating staff, the increasing reliance on tele-operation of facilities, or a reduction in the quantity or frequency of environmental or system monitoring. Changing financial conditions may also affect the frequency and detail of maintenance activities. |

## 10.3. Systems model of the Mattagami River

Systems simulation is being applied to an Ontario Power Generation (OPG) project in north-east Ontario, Canada. The project is a cascade of four power stations on the lower Mattagami River. The number of riparians in the river flood plain is few and there is no commercial riverine navigation, so potential loss of life is negligible and operational safety dominates the engineering considerations. The problem facing the engineering analysis was to conceptualise a systems engineering model for the operation of the dams, generating stations, spillways and other components, and then to employ the model through stochastic simulation to investigate protocols for the safe operation of the project.

The system borders two physiographic regions: the Canadian Shield extending from the south and, to the north, the Hudson Bay Lowlands. Several hydroelectric generating stations were built in the basin during the early 20th century (Figure 10.7). In late 1989, Ontario Hydro purchased the plants, and through its successor organisations has operated the facilities ever since.

### 10.3.1 Lower Mattagami Hydroelectric Complex

The Lower Mattagami River Hydroelectric Complex (LMRHC) consists of four hydroelectric generating stations (Figure 10.8). These are located on the Mattagami River, which joins with the Missinaibi River downstream of the project to form the Moose River, which ultimately flows into James Bay 90 km north of the town of Kapuskasing.

*Little Long generating station* at the top of the cascade is a base-load station with four vertical Francis-type units and a capacity of 52 MW. Smoky Falls, Harmon and Kipling generating stations each have two fixed-blade propeller-type units and operate as peaking stations, with station capacities of 136, 140 and 156 MW, respectively. Smoky Falls was the first generating station to come into service in 1931; Little Long, Harmon and Kipling generating stations came into service between 1963 and 1966.

When river flows exceed the maximum power flow of Little Long Dam, a bypass spillway, located 2.5 km east of the station, is used to pass excess water into the Adam Creek channel. The bypass spillway has eight sluices with a total capacity of 4870 cm/s. The bypassed waters flow north and re-enter the river about 17 km downstream of Kipling generating station.

Figure 10.7 Map of the LMRCH generating stations (GS). (Courtesy of OPG)

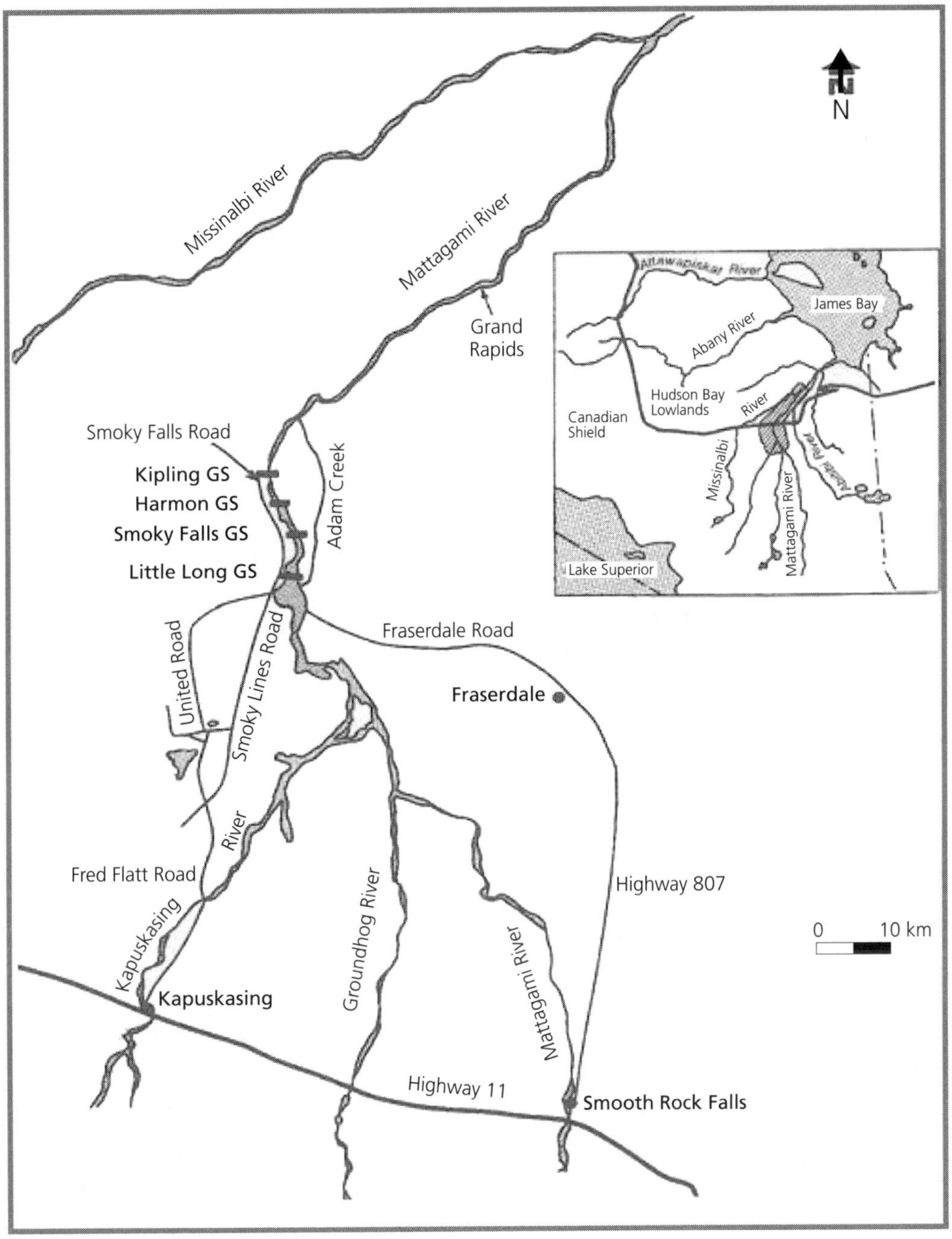


A secondary spillway structure constructed at the dam and releasing into the main Mattagami River channel has a capacity of 1217 cm/s, and provides flow to the downstream stations.

*Smoky Falls generating station* has a concrete dam that incorporates intakes for a powerhouse, a spillway structure to bypass flows in the event of a sudden unit outage, and an earth-fill retaining structure located near the spillway. The head pond extends 7 km upstream, has a surface area of 5.3 km$^2$ and a live storage of 106 million cubic metres (mcm). The existing spillway structure consists of 10 gated sluices, each 8.4 m wide by 9.2 m high, plus a 230 m long overflow crest. The spillway structure was originally designed (prior to the construction of the bypass) to convey what was then the full design flood flow on the river.

*Harmon generating station* has a single concrete dam that incorporates the intakes for the power station and a spillway to bypass flows in the event of a plant outage. The head pond extends 4 km upstream. Its surface area is approximately 3 km$^2$ and live storage is about 6.9 mcm. The operating head is 31 m and the rated flow is 525 cm/s.

Figure 10.8 The four dams and generating stations of the Lower Mattagami River Project of Ontario Power Generation: (a) Little Long; (b) Smoky Falls; (c) Harmon; (d) Kipling

(a) (b)

*Kipling generating station* has a single concrete dam incorporating an intake structure and spillway. The 5.6 km head pond has a surface area of 1.2 km$^2$ and a live storage of 3.2 mcm. The power plant is similar to those upstream but operates at 0.5 m lower generating head.

The Moose River basin encompasses a drainage area of 109 000 km$^2$. The Mattagami River flows in a northerly direction from its headwaters at Mesomikenda Lake and is approximately 418 km long, covering a drainage basin area of 35 612 km$^2$. It is generally a shallow and slow-flowing river with a seasonal flow regime. The long-term average river flow for the Little Long generating station as recorded by OPG is approximately 412 cm/s, based on a period of record from 1926 to 2005. As OPG's hydroelectric stations along the LMRHC are in close succession, intermediate drainage areas are small and the contribution from inflows between the stations is unimportant for planning purposes.

Highway communication for the project is provided by Fred Flatt Road, Smoky Line Road, and Smoky Falls Road. Fred Flatt Road is a 51 km long, two-lane gravel road leased by Tembec Inc. The road is open to the public, and OPG currently

Figure 10.8 Continued

(c)

(d)

contributes financially to its maintenance. Smoky Line Road is a 42 km long, single-lane gravel road owned by OPG. Smoky Falls Road is an 18 km long, two-lane gravel road, also owned by OPG. Highway 643 (formerly Highway 807) links Smooth Rock Falls to Fraserdale via a 73 km two-lane paved road. The 46 km Little Long Road (Fraserdale Road) is a two-lane gravel road that extends from Fraserdale to Little Long generating station, where it crosses the Little Long dam and links up with Smoky Falls Road. Little Long Road is owned and maintained by OPG.

Transmission of electrical energy from the four stations is provided by a 230 kV transmission line from Kipling generating station via Harmon generating station to Little Long generating station, and from there to the Pinard transformer station near Fraserdale (Figure 10.9). Power from the existing Smoky Falls generating station is fed into a 115 kV transmission line that runs directly to the Tembec paper mill at Kapuskasing. Relevant characteristics for each generating station and the nearby Adam Creek watershed are listed in Table 10.3.

### 10.3.2 Objectives of the study and modelling framework

The objective of the simulation was to understand systems interactions in the cascade, and the potential for accidents or failures caused by the interactions. The approach was as follows:

1 A model was formulated and constructed to characterise the hydrodynamics of the cascade, including the dynamics of transport, storage and power generation.
2 The systems modelling framework was applied to understand the systems interactions in the cascade of four dams.

**Figure 10.9** Project layout. GS, generating station; PW, powerhouse flow; SP, spillway flow

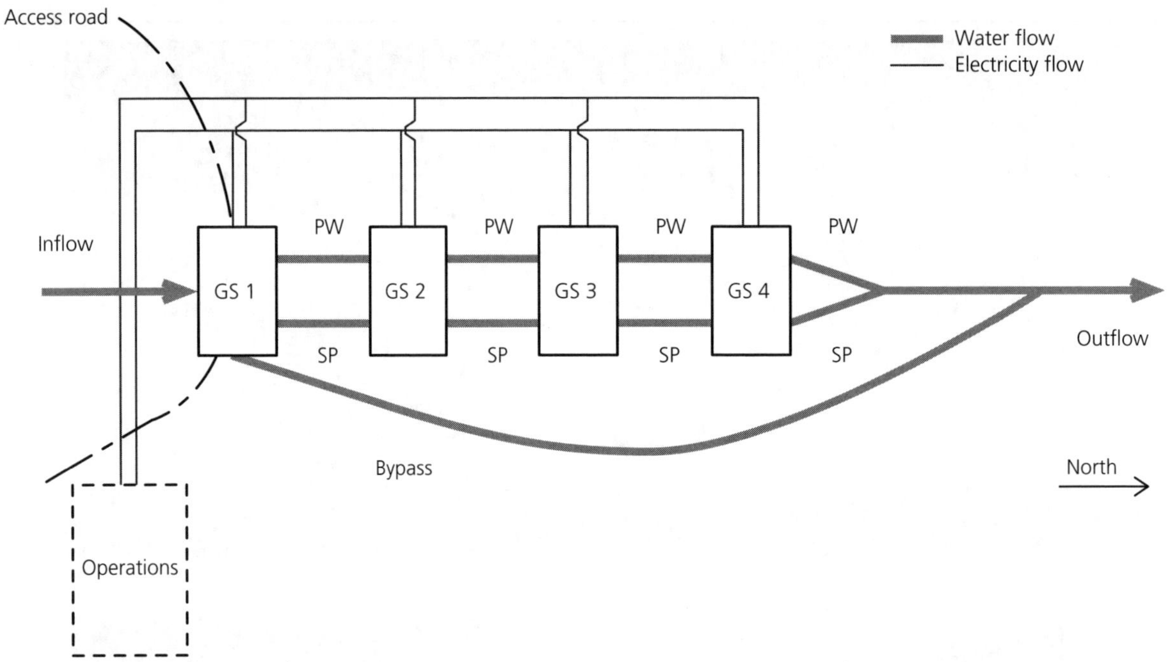


**Table 10.3** Characteristics of the Lower Mattagami Project stations

| Description | Adam Creek | Little Long generating station | Smoky Falls generating station | Harmon generating station | Kipling generating station |
|---|---|---|---|---|---|
| In-service date | 1963 | 1963 | 1931 | 1965 | 1966 |
| Drainage area: km$^2$ | 36 310 | 36 310 | 36 480 | 36 500 | 36 510 |
| Average annual flow: m$^3$/s | – | 412 | 412 | 412 | 412 |
| Maximum monthly flow: m$^3$/s | – | 2865 | 2865 | 2865 | 2865 |
| Minimum monthly flow: m$^3$/s | – | 39 | 39 | 39 | 39 |
| Headwater level: m | 198.1 | 198.1 | 170.3 | 135.1 | 103.0 |
| Tail water level: m | – | 171.11 | 135.8 | 104.1 | 72.5 |
| Gross operating head: m | – | 27.9 | 34.4 | 31.0 | 31.0 |
| Drawdown rage: m | – | 3.02 | 3.05 | 3.40 | 3.02 |
| Surface area: ha | – | 7600 | 530 | 300 | 130 |
| Live storage: 10$^6$ m$^3$ | – | 161.9 | 6.7 | 6.9 | 3.2 |
| Number of units | – | 2 | 4 | 2 | 2 |
| Unit size: MW | – | 68 | 13 | 70 | 78 |
| Station capacity (484 total: MW) | – | 136 | 52 | 140 | 156 |
| Available hours per day | – | 5 | 24 | 5 | 5 |
| Average annual energy: GW h | – | 590 | 376 | 675 | 690 |
| Station discharge capacity: m$^3$/s | – | 583 | 188 | 525 | 585 |
| Nominal capacity factor | – | 0.49 | 0.82 | 0.55 | 0.50 |
| Number of spillway sluices | 8 | 2 | 10 | 2 | 2 |
| Width: m | 12.2 | 12.2 | 8.4 | 12.2 | 12.2 |
| Height: m | 9.2 | 9.2 | 9.2 | 9.2 | 9.2 |
| Total capacity at full supply level: m$^3$/s | 4870 | 1217 | 1182 | 1288 | 1217 |

3 The river basin hydrology, routing of inflows through the reservoir system, operating rules and human factors of operating the spillway, and the dam component fragilities (structural, mechanical and electrical) were holistically integrated.
4 Time series were generated stochastically using the historic inflow time series to forecast inflows for several thousand years and multiple replications to identify unforeseen chains of events that could lead to accidents and failures.
5 Inherent disturbances (lightning, seismic, floating ice, grid disturbances and debris) were modelled via a probabilistic framework.
6 Human reliability analysis was incorporated into the model by using expert judgement and available data to estimate human reliability probabilities on the spillway gate and possibly other operations.
7 The current operating rules were reviewed to determine whether further optimisation techniques could be adopted to improve the power-generation capabilities of the system.

For many engineered systems it is necessary to predict measures such as the system's reliability (the probability that a component or system will perform its required function over a specified time period) and availability (the probability that a component or system is performing its required function at any given time).

Initially, the model had been built on the Matlab™ software platform and its associated Simulink™ package. Nevertheless, with the proliferation of commercial off-the-shelf party simulation engines at least as well optimised for engineering reliability, other alternatives were evaluated. Ultimately, the choice was made to build the model on the GoldSim™ platform. While this platform is less powerful than some of the others reviewed, it has the advantage of simplicity.

The general model interface with the salient aspects of the physical system being modelled was developed. When the simulation is started, the software iterates over many years (perhaps millions) of performance and evaluates the long history of the system, saving selected results for subsequent analysis.

#### 10.3.2.1 The natural system

There is no unique definition of a river system. The definition changes depending on the purpose of the study. What is of interest are the models of river systems. The definition of a river system model can be developed along the following general principles: a system can be defined as a collection of interrelated elements purposely working toward achieving some common objective. Most natural systems can be characterised as dynamic because the collection of elements interact over time. The elements of the system can be subdivided into the three groups: components, attributes and relationships. These, in turn, can be characterised as subsystems that are the operating parts of a system, and consist of input, process and output.

Components take the form of natural or technological artefacts. Each component may assume a variety of values defining the system state. Attributes are the properties or discernible manifestations of the system, its components and the relationships between them. Relationships are the links between components and attributes. At the same time, each component depends on the behaviour of at least one other component in the system. Because of this interdependence, components cannot be divided into independent subsets. Also, if a component is removed or if its characteristics change, the other artefacts in the system alter accordingly, and the relationships between them may also change.

River networks, whether in the natural state or regulated (with human-made structures), can be understood as systems. The behaviour of the river system can be characterised by quantities that vary over time (water levels, flows, velocities at different places within the system, etc.). Such a definition of the model of a river system is not universal and may be inadequate for purposes that differ from the purpose of a particular application.

#### 10.3.2.2 Hydrological modelling and flow routing

The 50-year history of data on reservoir inflows from the lower Mattagami River to Little Long Dam was used to generate stochastic inflows through time-series forecasting.

An autoregressive-moving-average time-series model was used to generate stochastic inflows. A random starting year was chosen within the historical series to serve as the starting point for each realisation. This generated uncertainty in forwarding the historical series.

This procedure shifts the time-series data forward (or backward) by a number of years. For instance, if the actual data set started on 31 July 1989 and ended on 31 December 2011, the simulation start date within the data might be picked at 31 July 1990 but associated to whatever the hypothetical modelling date was chosen to be. To account for uncertainty and randomness while still directly using the historical data, the use of the random starting point option with 'annual periodicity' suited the purpose of the study.

By using a random starting point with annual periodicity and aligning data years with simulation dates, when the end of the data set is reached the data set is effectively replicated and shifted forward by *N* years, where *N* is the number of whole years in the data set. For example, if there were 3.5 years of data in the data set the last data point was on 31 December, and if the simulation extended beyond 3.5 years the data set would be shifted forward by 3 years, so that the first occurrence of 1 January in the data set would follow the last point in the data set.

#### 10.3.2.3 Normal operation and power generation

Under normal operations, water flows for the generating stations are provided from the uppermost reservoir. The water level is normally within the operating headwater level range. The limit of the headwater level is the 'absolute maximum operating level'. The difference between the absolute maximum and maximum operating levels is the flood allowance, which is used to hold water in extreme conditions to reduce downstream flooding. The storage between the absolute minimum and minimum operating levels is used if a system energy emergency occurs. Under normal operating conditions with equivalent discharges at each station, the full operating range would rarely be utilised.

Under normal operating conditions, the outflow from the uppermost reservoir passes through all the generating stations. During any outage of a generating station the spillway at the station experiencing the outage will be operated to pass the flow to the other generating stations. During high river flow conditions (e.g. spring runoff), when the uppermost reservoir is near its maximum limit, the spillway into the diversion is operated in conjunction with the generating stations to pass the full river flow.

#### 10.3.2.4 Gate operations and gate fragility

The simulation includes the effects of hydraulic loads, debris, ice and operator actions. These loads are related to gate availability through fragility-curve relationships. A fragility curve represents the probability of adverse performance or failure as a function of the load on the structure. The inputs to the gate-operations simulation at any time step are the corresponding states of the input and disturbance variables. The outputs are the gate availabilities (probability of use on demand).

Gated spillways are designed to engineering standards. However, with ageing, exposure, preventive maintenance and lack of frequent operations they become vulnerable to non-availability. On-demand failures of gated spillways may be caused by a gate component that can be repaired in minutes to hours, or to a component that may cause complete failure of the gate system and unexpected release of the reservoir containment.

Gated spillways are a complex integration of structural, mechanical and electrical components that must operate on demand. For Little Long reservoir, there are a total of 10 sluices with gates. Two of the 10 sluices (Nos. 1 and 2) are alongside the generating station and open into the river, while eight are 3.2 km upstream and open into the bypass. Sluices 1, 2, 3, 4, 7 and 8 are remotely controlled. Sluices 5, 6, 9 and 10 are locally controlled by agents at the gate. The sluices are numbered from left to right looking downstream.

Gate operations are also affected by instrumentation, SCADA systems and controls. The simulation approach is well suited to uncovering the implications of component performance assumptions on overall systems operations.

Figure 10.10 shows a simulated result over a 12-month period of gate operation, including SCADA performance and gate binding. The inflow rises during the spring freshet in May and June based on the stochastic time series of inflow data. As the pool level rises, the SCADA system generates control instructions to the gates. At two points in late June and early July, for whatever reason in the simulation, instances of gate binding occur. Depending on other factors in the simulation, these two instances of the gate binding may or may not lead to adverse systems behaviour.

#### 10.3.2.5 Mechanical fragility

In risk assessment, limit states or probability of failure serve as yardsticks of system performance. The fragility curve is used to predict the probably of dam failure, given an hydraulic load. A fragility curve is defined by the cumulative distribution function of the system response curve to a load. A typical fragility curve for the probability of mechanical gate failure as a function of gate opening under full pool is shown in Figure 10.11. The fragility curve is the standard way in which structural and geotechnical reliability functions are captured. Fragility curves are pre-calculated, usually using an event-tree or a fault-tree representation. A fault tree is a graphical method of describing the combinations of events leading to a defined system failure.

Structural and mechanical reliabilities of spillway gates are dependent on the level of water on the gates, while the electrical reliabilities are randomly distributed in time with a fixed occurrence rate (here, 0.017 failures/day). In the present model, unavailability due to electrical gate failures is far more common than unavailability due to mechanical failures, and occurs a number of times per year. Structural failures are even less probable in the current model.

Figure 10.10 Simulated effect of disturbances

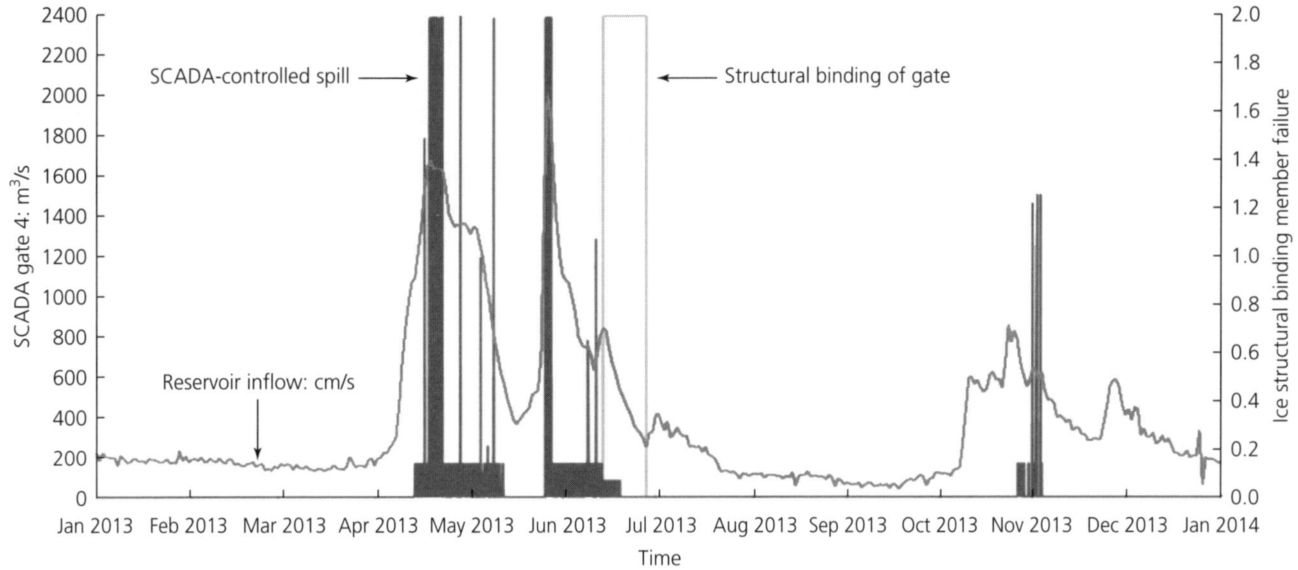


Figure 10.12 shows gate failures for one annual realisation of spillway gate 4. This was a peculiar year. There were several electrical failures and one structural failure. The structural failure was modelled as a gamma distribution with a mean delay time until repair of 6 months and a standard deviation of 2 months. The effect of this random repair-time delay was that the structural failures took a long time to repair whereas electrical gate failures were usual repaired within the same day.

The effect of failure on the performance of the entire system can be observed in Figure 10.13. Spillway gate 4 was inoperable due to its failed state under structural failure and was down for repair sometime before the peak flow for the year (April). Between the end of April and mid-October, spillway gate 4 was down for repair. During this time the SCADA-controlled gates were required to route excess inflows out of the reservoir. Bearing in mind that each gate has a capacity of 608 cm/s, it can be observed in Figure 10.13 that with the failure of spillway gate 4 failure the SCADA gate system was routing about 1800 $m^3$ of water out of the reservoir instead of the total capacity of about 2400 $m^3$ during peak flows.

Figures 10.12 and 10.13 also show another comparison of mechanical failure coinciding with the demand for spillway flow. From the plots, it can be observed that around the start of freshet, there was a mechanical failure at gate 4. This failure lasted for about 7 days in May. It coincided with the start of freshet, meaning the spillway gates were unavailable on demand. The effect of this mechanical failure was the loss of capacity of the SCADA-controlled spillway system. As can be observed, the rest of the spillway gates started operating just as the freshet started to peak with a combined discharge of about 1800 $m^3$ instead of the total capacity of about 2400 $m^3$. This occurred until gate 4 was fixed and added an additional 600 $m^3$ of capacity to the SCADA-controlled gate system.

Figure 10.13 also shows the probabilities of mechanical failure as a function of time for one realisation looking at spillway gate 1. As expected, the probability of mechanical failure is usually higher around the peak of freshet when the elevation

Figure 10.11 Fragility curve for mechanical gate failure

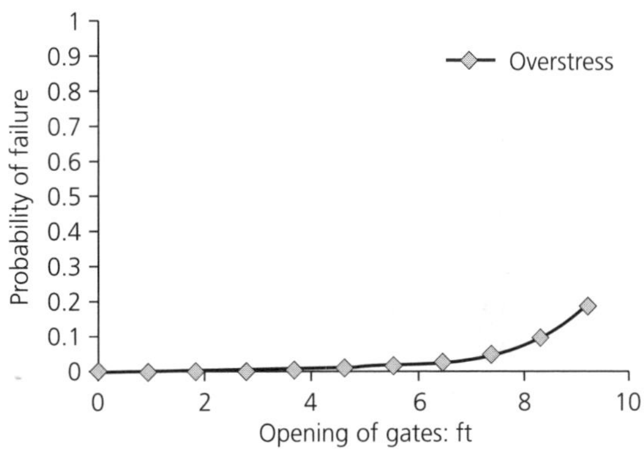

Figure 10.12 Plot of component reliabilities and upstream daily flow versus time

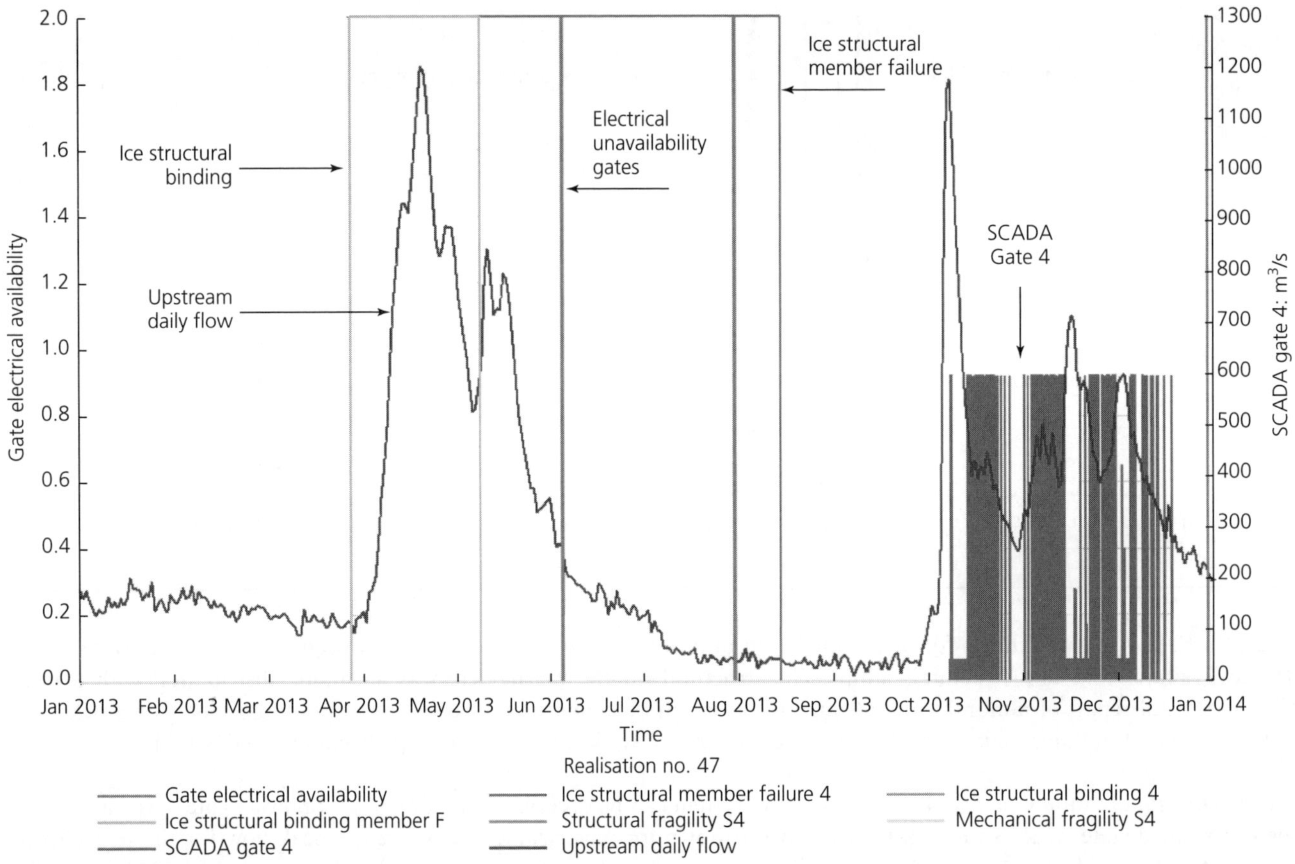


Figure 10.13 Plot of SCADA-controlled spill versus upstream daily inflow to reservoir

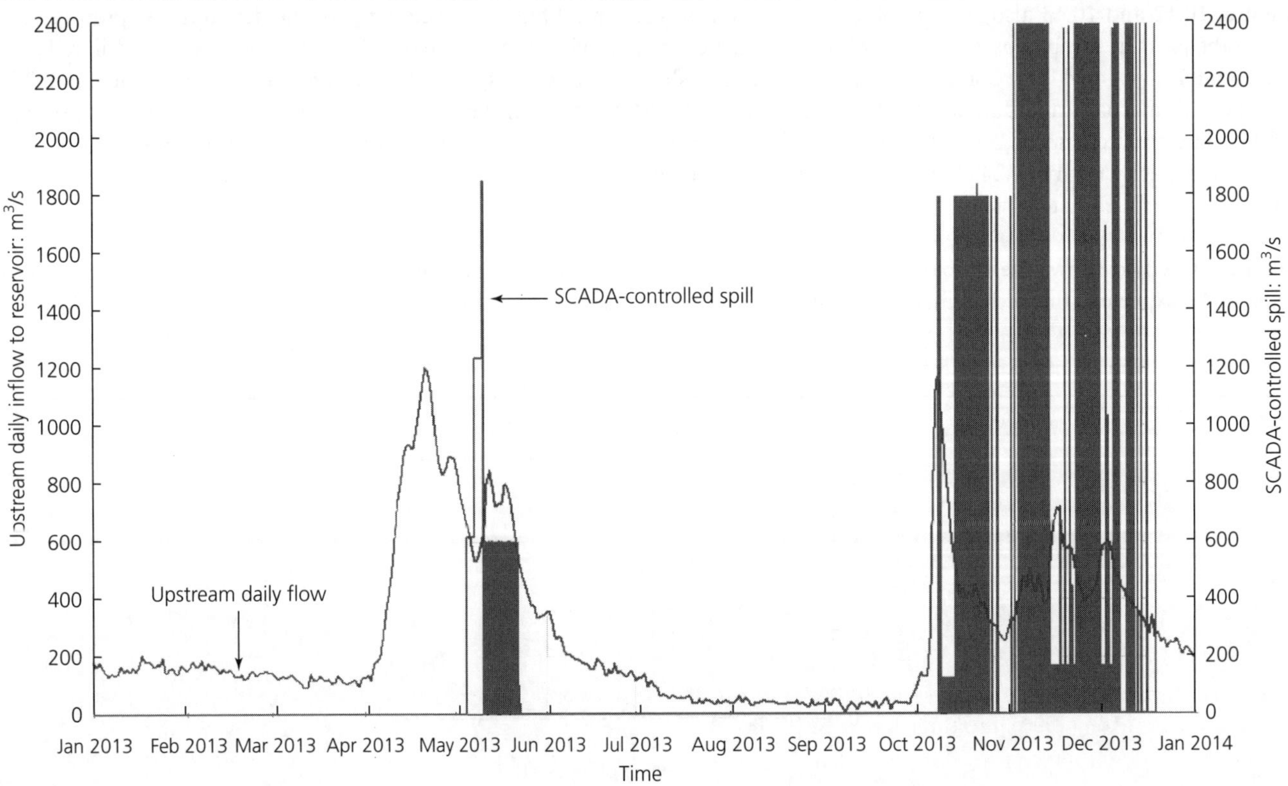

of water on the gates is the highest. Figure 10.14 shows the mean of mechanical failure, spillway discharge and upstream daily flow for a calendar year over many replications. As can be observed, there is a strong correlation between all three parameters.

#### 10.3.2.6 Disturbances

Sometimes during the course of the operation of a dam system an extraordinary event (e.g. an earthquake, an earth or rock slide into the reservoir or conveyance works, or a major fire in the drainage area) occurs at a time that cannot be anticipated, but the event may affect dam operations or even dam safety. Within the simulation such events, and their more modest siblings, are treated as exogenous disturbances. Disturbances can, in principle, include a broad variety of phenomena, both of natural origin, such as lightning strikes affecting power supplies or instrumentation, or of anthropogenic origin, such as the grid being unable to accept power and thus the powerhouse waterway having diminished discharge capacity, or operational incidents or accidents, such as powerhouse fires. The exact definition is a matter of modelling convenience; severe floods or droughts may, for instance, be considered as disturbances or, alternatively, just as aspects of the stochastically modelled catchment hydrology.

A disturbance of major concern in the LMRHC is ice build-up. The time series in Figure 10.15 shows the simulated ice build-up on the gates of any of the dams during one randomly simulated year. The stochastic input for this simulation is the variability in daily temperature generated from a statistical time series identified from temperature data over the past century. Daily temperatures are simulated, and Stefan's equation (USACE, 2002) is used to calculate ice thickness on the reservoir. Stefan's equation uses anticipatory degree-days below freezing with an empirical constant to forecast ice development.

The disturbances included in the present model are limited to external loss of grid availability, lightning (affecting grid availability), floating ice, icing on structures, instrument or SCADA misoperation and human error (Table 10.4). Each is treated as a Poisson process in time, possibly with a corresponding probability distribution of magnitude.

Ice storms can damage structures because of the weight of accumulated ice. Ice storms are known to occur in eastern Ontario and Quebec. On average, Ottawa and Montreal receive freezing precipitation on 12–17 days a year, but this type of precipitation generally lasts for only a few hours. Although it did not occur near the LMRHC, in January 1998 a severe ice storm occurred in eastern Ontario and Quebec; over 90 mm of freezing drizzle fell during a 5-day storm. This magnitude has an annual probability of occurrence of about 1 : 100 (OPG, 2009).

The occurrence of ice storms that affect spillway gate operation was modelled as a discrete process because the only information available on ice storm occurrence was that for a severe ice storm in 1988. Figure 10.16 shows fragility curves for structural member failure and structural binding failure of a gate. The time of initiation of the event is simulated as a Poisson process, such that the expected number of occurrences over some time period $T$ is equal to the product of the rate (0.01/year) and time. The duration is modelled as a gamma distribution for structural member failure and as an exponential distribution for structural binding failure.

#### 10.3.2.7 Floating ice

The ice thickness is a fundamental parameter for practically all ice problems. Usually, the engineer is confronted with the problem of predicting the ice thickness, and often its return period as well, with little information. At the simplest level, one can use empirical analyses based on the number of freezing degree days (FDDs). This can be refined in various ways, especially if ice thickness data are available for calibration.

The ice thickness, $h$, produced by static ice formation is most commonly predicted based on the accumulated FDDs, as given in Table 10.5. The Stefan equation $h = \alpha\sqrt{\text{FDD}}$ is derived by solving the differential equation for the thermal growth rate, and by making various simplifying assumptions (USACE, 2002).

#### 10.3.2.8 Human reliability

It comes as no surprise that human actions are a source of vulnerability for industrial systems, and this has given rise to modern methods and models of human reliability analysis (HRA). HRA aims to deepen the examination of the human factor in the workplace, and is important due to the contributions of humans to the resilience of systems and to possible adverse consequences of human errors or oversights – especially when the human is a crucial part of the large system.

HRA involves qualitative and quantitative methods for assessing the human contribution to risk. There are many and varied methods available for HRA, with some safety-critical domains developing industry-focused methods. Human error has been classified by Reason (1990) into three levels: behavioural, contextual and conceptual. Reason concludes that the first two types include the different errors that can result from the same process, but the conceptual level leads to understanding the causal mechanisms, using more theory than observation.

**Figure 10.14** Plot of SCADA-controlled (gate 4) spill versus upstream daily flow versus time

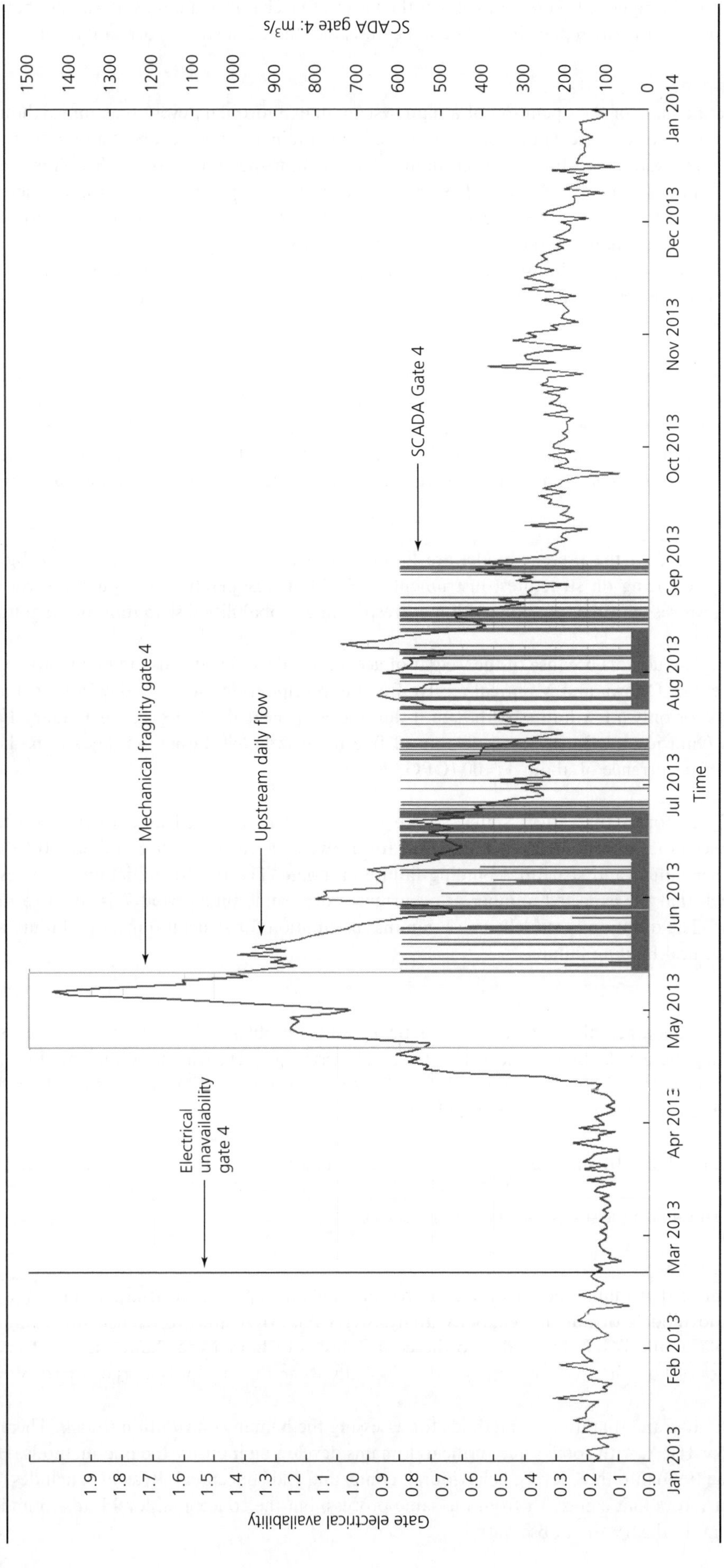

**Figure 10.15** Ice build-up in the reservoir based on Stefan's equation

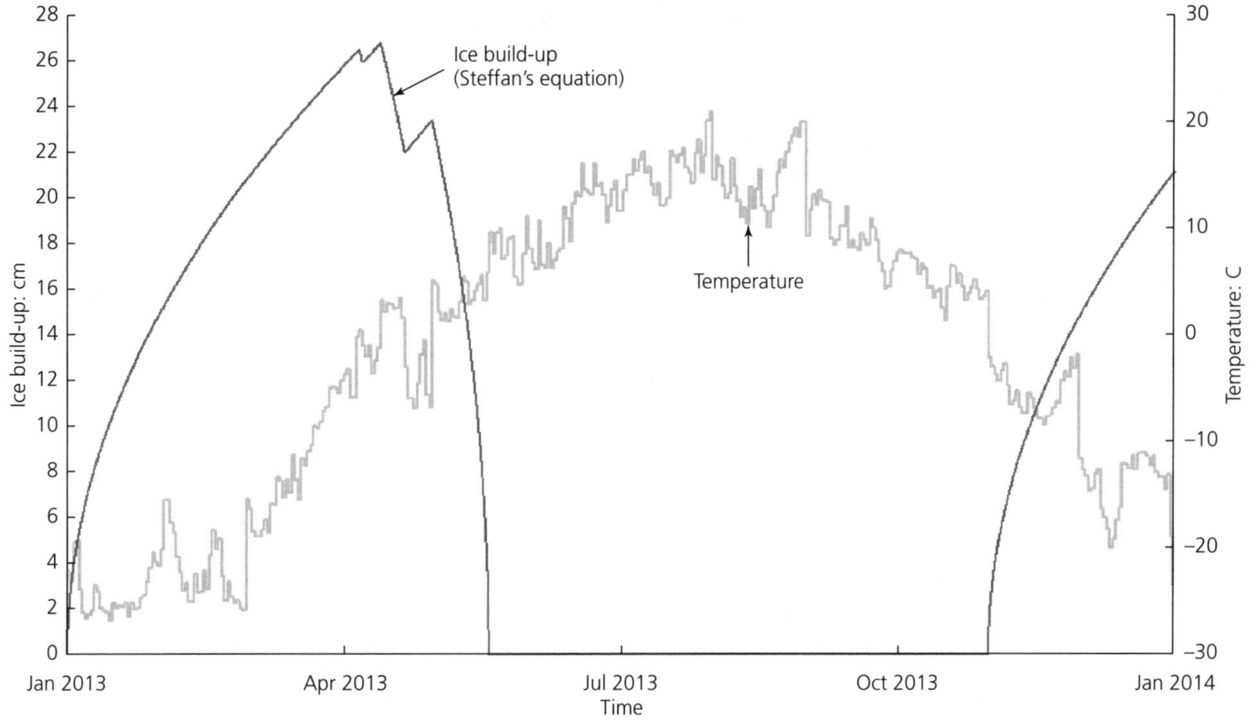


**Table 10.4** Potential external disturbances at LMRHC

| Condition | Principal affected component(s) of the project |
|---|---|
| Flooding | Integrity and function of external structures and systems<br>Integrity and function of dams |
| Ice | Integrity and function of dams and water-intake systems |
| Forest fire | Integrity and function of generating stations and associated facilities |
| Severe weather | Integrity and function of external structures and systems |
| Seismic events | Integrity and function of dams, spillways and powerhouses |
| Climate change | Integrity and function of external structures and systems<br>Integrity and function of operating regime |

**Figure 10.16** Fragility curves for ice loads on the spillway gate

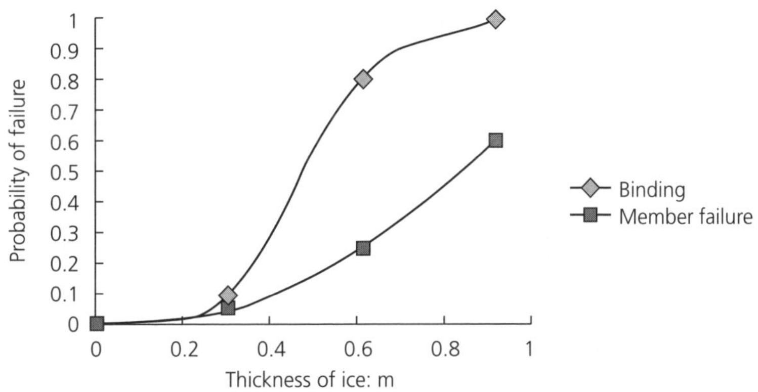

**Table 10.5** Stefan equation values for $\alpha$

| Ice cover condition | Typical values of $\alpha$: °C |
|---|---|
| Windy lake without snow | 2.7 |
| Average lake with snow | 1.7–2.4 |
| Average river with snow | 1.4–1.7 |
| Sheltered small river | 0.7–1.4 |

The simulation effort discussed here uses the SPAR-H approach to HRA (Gertman *et al.*, 2005). A NASA report (Chandler *et al.*, 2006) documents many of the HRA methodologies that have been developed to estimate human error potential. The report compares the methods and provides a selection of four methods that could be used for the NASA risk analysis programme. The report recommends for NASA HRA: THERP (Swain and Guttmann, 1983), CREAM (Hollnagel, 1998), NARA (Kirwan, 1994) and SPAR-H. The report recommends that all four methods be deployed and their results documented to be able to select the most suitable program for HEP estimation for future NASA endeavours. These HRA methods are focused on nuclear power plant and will require significant adaptation before they can be applied to the operation of spillway subsystems.

In the simulations, the drivers to which the frequencies of accidents and failures are most sensitive appear to be: (1) loss of remote control through communications or other breakdowns, (2) incorrect operations decisions that fail to establish the appropriate spill profile for the prevailing conditions, (3) loss of power supplies or other supporting or auxiliary services, (4) loss of access to the dam in an emergency, (5) excessive weather, (6) common cause disruptions such as fire, (7) failure of control or instrumentation, (8) lack of qualified personnel to provide an emergency response, and (9) failure of the gates or supporting structures. Obviously, the observation that these appear to contribute to high frequencies of accidents and failures depends on the input assumptions in the modelling about rates, stochastic processes and fragilities. At this point no generic conclusions are warranted about the causes of operational problems from these results.

### 10.3.3 Model results: operating patterns and flow routing

Flows in the Mattagami River are highly regulated by the presence of hydropower generating facilities and water-control structures that provide electricity generation and flood mitigation.

During the spring freshet, flows in the Mattagami River typically exceed the flow capacities of the generating stations and therefore must be diverted through the Adam Creek diversion. Adam Creek then discharges this overflow into the Mattagami River downstream of the Kipling generating station. Figure 10.17 shows historical daily inflows into Little Long reservoir. When the flow exceeds 583 cm/s the excess water that cannot be used by the Mattagami River generating stations is diverted to Adam Creek. The average peak flow during the spring freshet is above 1500 cm/s and can be variable.

#### 10.3.3.1 Reservoir operations summary

The water flows for the four generating stations are provided from the Little Long generating station reservoir. The water level is normally within the operating headwater level range. The extreme limit of the headwater level is the 'absolute maximum operating level'. The difference between the absolute maximum and maximum operating levels is referred to as the 'flood allowance', which is only used to hold water in extreme conditions to reduce downstream flooding. The storage between the absolute minimum and minimum operating levels is only used if a system energy emergency occurs. Under normal operating conditions with equivalent discharges at each station the full operating range in the Smoky Falls, Harmon and Kipling generating station head ponds would rarely be used and head-pond levels will be significantly more stable during operation.

Under normal operating conditions, the outflow from the Little Long generating station reservoir will pass through all the generating stations. During any outage of a generating station the spillway at the station experiencing the outage will be operated to pass the desired flow to the other generating stations. During high river flow conditions (e.g. spring) when the Little Long generating station reservoir is near its maximum limit the spillway at Adam Creek is operated in conjunction with Little Long generating station to pass the full Mattagami River flow.

#### 10.3.3.2 Power generation

Little Long generating station provides a maximum power flow of 583 cm/s at a head of 28 m. The estimated turbine and generator characteristics for each station are shown in Table 10.6.

With the salient aspects of the model formulated and constructed, the next step is to generate results and analyse whether they accurately replicate the real system. This is part of the model validation process, and hence outputs from the model were compared with data from the LMRHC.

Figure 10.17 Historical daily inflow data for Little Long generating station

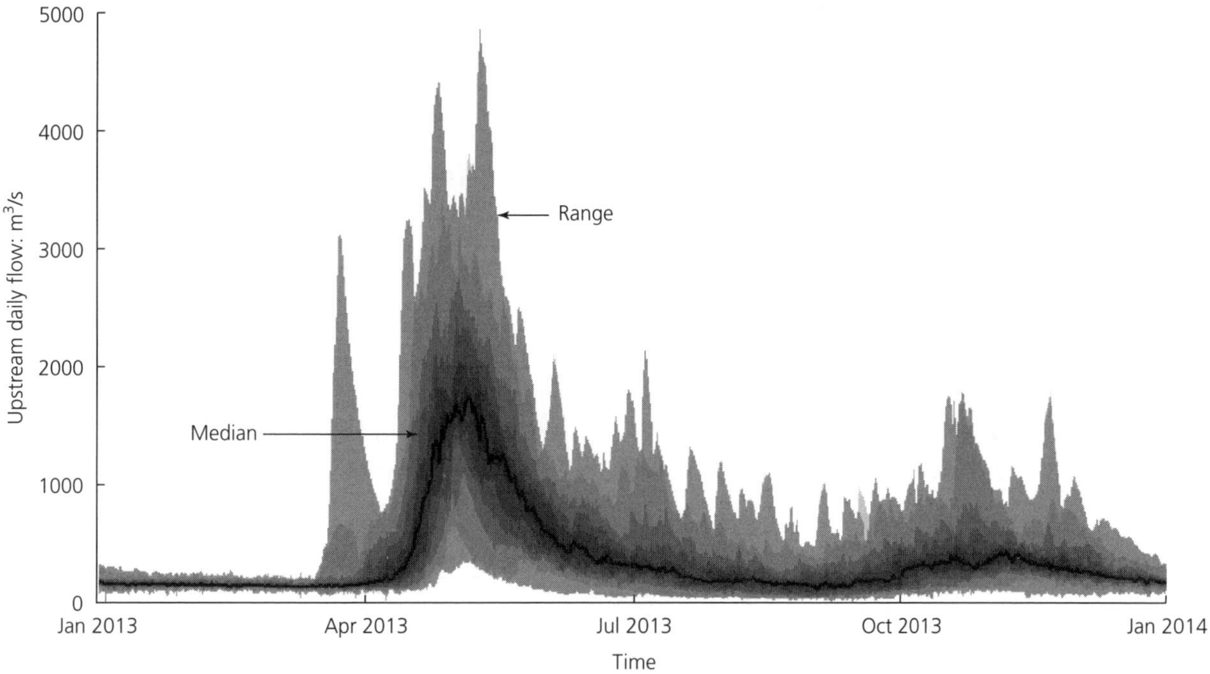


The practice of spilling into the main river is generally discouraged, and the two spillway gates at Little Long only come into use as a contingency when the eight spillway gates that open into Adam Creek are insufficient to route peak flows to keep the reservoir pool elevation within the operating range.

#### 10.3.3.3 Pool and volume capacities

For the 51-year run it is important to analyse how the flow-routing procedures ensure that the reservoirs are mostly operating within the operating range. The operating range for Little Long reservoir is 195.10–198.12 m. The performance of the flow-routing techniques generally keeps the elevation of the dam below the imminent dam failure elevation, but the elevation does occasionally (about once every 2 years) exceed the maximum operating elevation. Of course this does not affect dam operations with regard to power performance or safety as the level still a full metre below the imminent failure elevation.

Figure 10.18 shows the upstream daily flow from the Mattagami River into Little Long reservoir over 50 years. The peaks are the freshet flows each year. The maximum inflow over this 50-year simulation was 4942 cm/s, which is consistent with the maximum inflow from the data set. Figure 10.19 shows the spill into Adam Creek. This follows a similar profile as that of the upstream daily flows: the peak spill coincides with the peak daily inflow as this is the period of peak capacity when the flow is routed through the bypass.

Table 10.6 Turbine and generator characteristics

| Parameter | Generating station | | | |
|---|---|---|---|---|
| | Little Long | Smoky Falls | Harmon | Kipling |
| Number of units | 2 | 4 | 2 | 2 |
| Gross head: m | 27.9 | 34.4 | 31 | 31 |
| Station discharge capacity: cm/s | 583 | 188 | 525 | 585 |
| Station turbine capacity: MW | 136 | 52 | 140 | 156 |
| Unit capacity: MVA/turbine | 68 | 13 | 70 | 78 |
| Best efficiency rate: kW/cm/s | 235.7 | 288.9 | 272.3 | 272.7 |

Figure 10.18 Upstream daily flow simulation results over a 51-year period

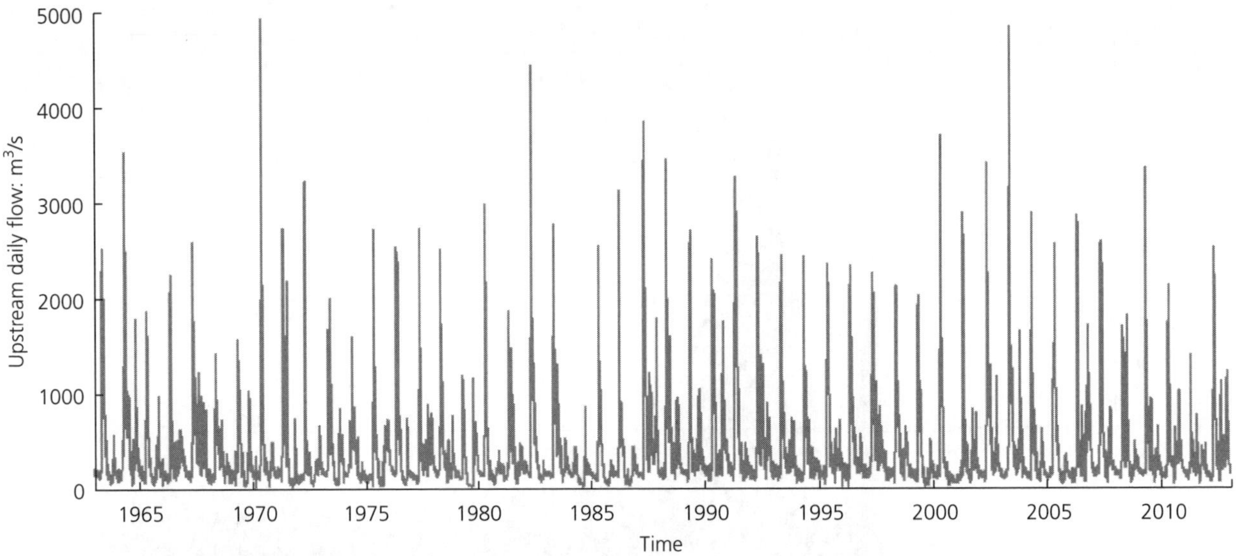


#### 10.3.3.4 Power production

As dictated by the operating rule, the dam is filled to an elevation of 198 m and the calculated head over the turbine intake is 27.28 m. At the maximum reservoir operating capacity of 198.12 m, the maximum head over the turbines is 27.9 m. The difference is to allow for the 4-h time lag required to dispatch operator agents to the station to deal with the contingency. Hence the model is programmed such that Little Long reservoir fills to 198 m. If the inflow into the reservoir exceeds what is required for the most efficient power generation, as provided in the unit-rating table, the excess inflow is routed out of the reservoir through the spillway gates.

Consequently, if the inflow into the reservoir is below that required for the most efficient flow and the reservoir elevation is at the minimum operating elevation, the same head is maintained while the inflow water is routed through the turbine intake sluices. However, if the inflow into the reservoir is less than the minimum for the highest efficiency flow but the reservoir elevation is above the minimum operating elevation, the reservoir elevation is gradually lowered while generating the minimum best-efficiency power until the minimum operating threshold is crossed.

Figure 10.19 50-year simulation run of the mean spill into the Mattagami River and the Adam Creek bypass

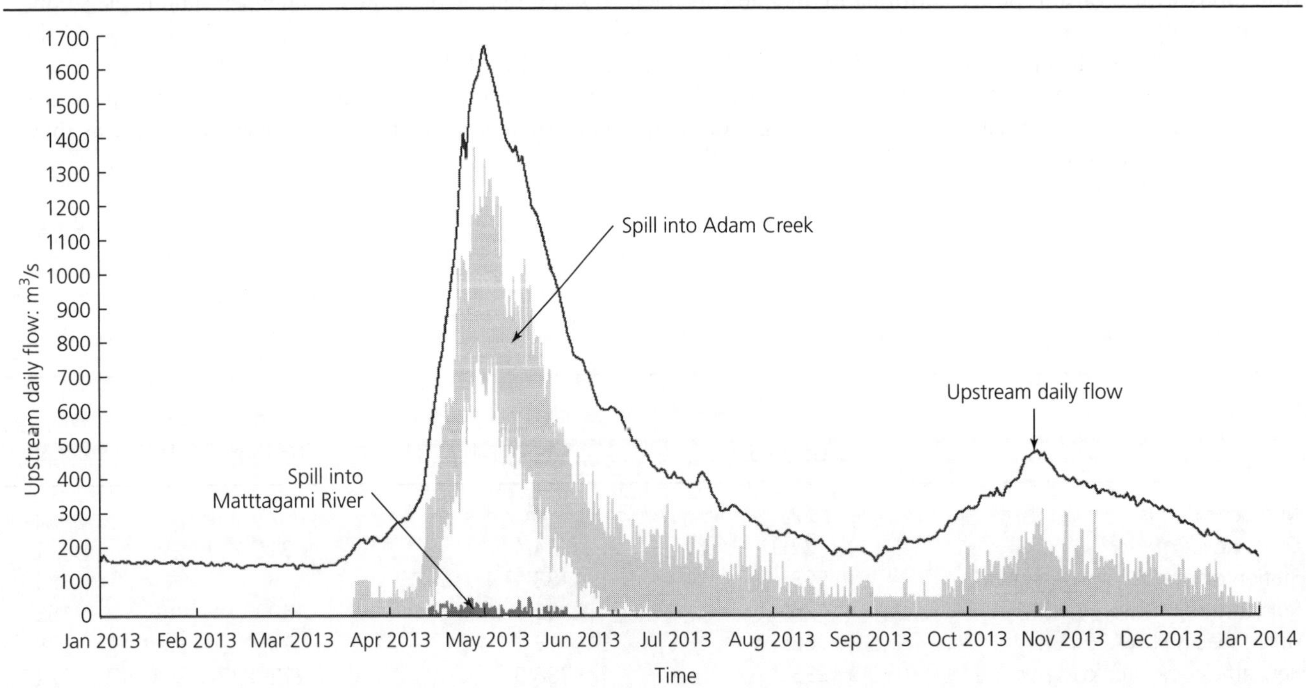

Figure 10.20 The probabilistic seasonal variation of the total power production of all four generating stations

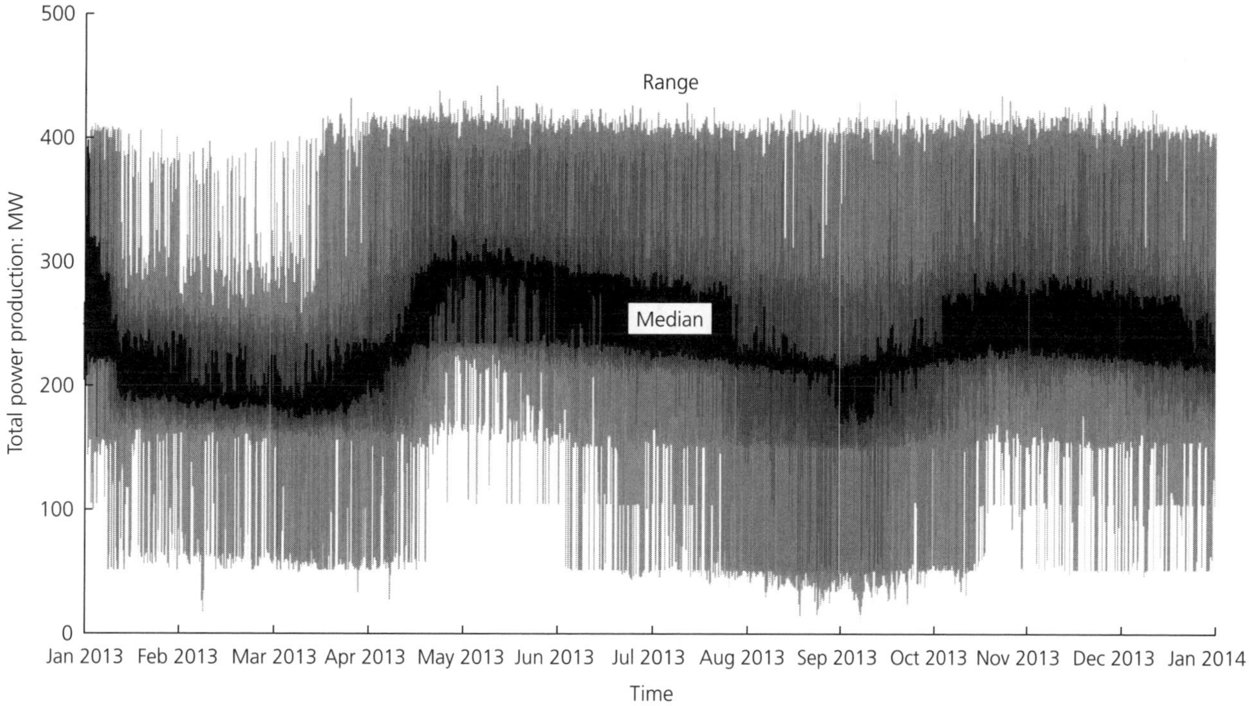


In the simulation, the Little Long generating station is operated for best-efficiency power generation, which is stated in the unit rating table to be when the upstream flow is equal to or greater than the specified discharges for best-efficiency power generation. Where the inflow is more than the best-efficiency discharges, the excess inflow is routed through the spillways, which are automatically controlled through the SCADA system, into Adam Creek (Figure 10.20).

If the four SCADA controlled gates that open into the bypass are insufficient (i.e. when reservoir elevation is greater than 198 m and the upstream flow exceeds the total of the turbine flow and the spilled flow), human operators are dispatched to the site to operate the additional gates that open into the bypass with a mean lag time of 4 h. If all the gates opening into the Adam Creek bypass are still insufficient, the additional two gates that open into the Mattagami River are used to supplement the spilled flow and keep the dam from overtopping.

Figure 10.21 shows a plot of daily power production as a function of time from start of January to end of December for many replications. As can be expected, the profile of the power generation correlates with that of the upstream daily flow.

### 10.3.4 Model results: reliability and dam safety analysis

The main function of spillway gates is to safely pass water from one point to another, usually from a reservoir, through a dam to a downstream river or reservoir. This is achieved by keeping the main components – the rate of flow, the physical conveyance of flow and the kinetic and potential energy of the flow – under control.

The components of water conduits are all natural or man-made structures with civil, mechanical and electrical functions, and have certain capabilities to resist the dynamic and static loads imposed on them. For a spillway to perform its task safely, the flow must be kept to within a range that does not exceed the design capacities of its subcomponents (electrical, structural, mechanical, etc.). Thus, the reliable performance of a spillway system is both a function of time and the loads placed upon it.

Vertical lift gates are a common type of dam gate. Machinery is typically located on a structural feature above the gate. For most dam applications, vertical lift gates must be operated under differential head conditions. The differential hydrostatic pressure can create large transverse forces, which in turn create large friction forces as a gate is being operated. Rollers or other features are needed on the downstream side of the gate to reduce friction between the gate and guides to allow hoisting of the gate. This discussion of vertical lift gates is provided for because all the spillway gates in operation at the LMRHC are vertical lift gates.

Figure 10.21 The mean total power upstream daily flow for a 1-year run and 50 realisations

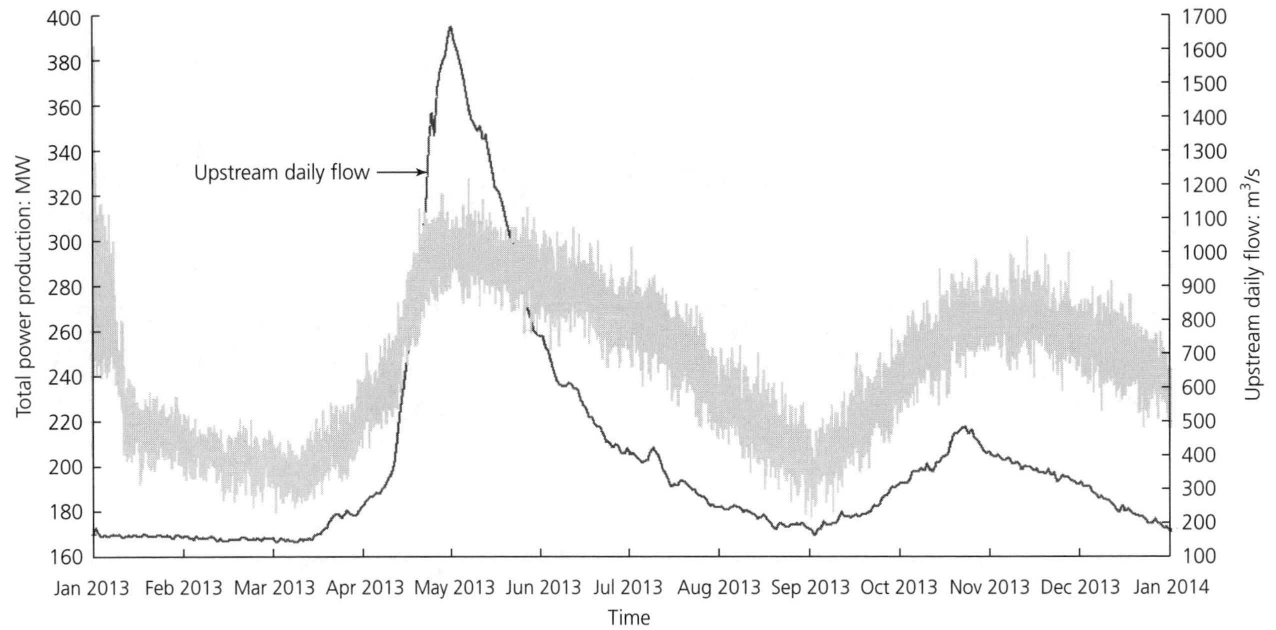


The following load types are applicable:

- *Hydrostatic load.* The hydrostatic load is determined based on site-specific conditions that account for the differential between the headwater and the sill bearing at the spillway crest. The headwater is determined from reservoir elevation computations at each of the dams. For single-section gates, flow is under the gate. No consideration is given to water passing over the top of the gate. The hydrostatic head differential between the headwater and the sill bearing at the spillway crest acts as an uplift on the bottom of the gate when passing flows through the spillway. The net uplift is determined by the combined effects of the downpull forces.
- *Hydrodynamic loads* are based on site-specific conditions. The distribution of wave forces is dependent on the wave height and the depth of water at the structure. The effects of hydrodynamic loads are computed for a range of possible water levels and periods. The effect of these loads was not considered in the simulation.
- *Gravity loads* result from deadweight, ice and sediment. The effect of ice and debris on the performance of the spillway gates is treated as a disturbance. Sediment loads were ignored in the present model.
- *Earthquake loads* are determined based on an operational basis earthquake. Earthquakes are not an issue on the Lower Mattagami Project.

#### 10.3.4.1 Rate of flow and conveyance

The rate of flow in a spillway is controlled at a specific point or cross-section at the intake where the flow changes from subcritical upstream to supercritical downstream. The flow rate is controlled by the inflow into the reservoir, the water level downstream, and the length and height of the spillway opening.

The greater flexibility of operation provided by gated spillways makes it possible to regulate the upstream water level or the water conduit discharge within a narrow range. Thus the pool elevation in a reservoir can be maintained within optimal levels for power generation. The price to pay for the introduction of a movable water barrier is a significant reduction in spillway function reliability, as several components and subcomponents have to come together and function on demand for the gated spillway system to work.

#### 10.3.4.2 Spillway gate operations: flow routing at the LMRHC

During high river flow conditions (e.g. spring) when the Little Long reservoir is near its maximum limit, the spillway at Adam Creek will be operated in conjunction with Little Long generating station to pass the full Mattagami River flow. The duration and magnitude of the spill down Adam Creek is dependent on the magnitude and duration of the incoming flow into the Little Long reservoir. As the Little Long generating stations have a combined capacity of 583 cm/s, an upstream flow in excess of this will be spilled first via Adam Creek and, in rare peak scenarios, through the Mattagami River. The model examples discussed here are for the Little Long dam, as the modelling of the spillway gates at the other three sites follows a similar modelling framework, with modifications of operating rules.

#### 10.3.4.3 Failure mechanisms

Leveson (1995) defined failure as the non-performance or inability of the system or component to perform its intended function for a specified time under specified environmental conditions. Failure conditions are usually brought about by any one of a number of natural hazards or other external or internal disturbances, but eventually failure always comes down to structural insufficiency – either insufficient dimensions to contain the flow or insufficient integrity to withstand the occurring forces, or a combination of these. At a high level of abstraction any occasion when a system does not fulfil its function constitutes a failure of the system.

Failure at a systems level differs from failure at the component level in that failure at the component level may not necessarily cause failure of the entire system. In some cases, a system may still function even though a component has failed, and in fact this is a property of a well-designed system. The high-level definition is not usually sufficient for the purpose of a detailed analysis of the system, but the failure needs to be tied to a particular component or set of components, whether physical objects, resources or procedures. This concept will be explored later on in this chapter with the aim of modelling how component and subcomponent failure affects the entire system at a higher level. For example, a spillway gate might not open on demand due to an electrical issue related to its generator but the other spillway gates may function and the entire system may be unaffected by the failure of one of the subcomponents of that particular spillway gate. It is up to the user to define which particular definition of failure and which failure categories best serve the purpose of the analysis.

#### 10.3.4.4 Failure mode of gates

Insufficient hydraulic capacity of the spillway gates is the classic failure mode. It may manifest as a raised reservoir water level and the final failure may then take the form of a dam overtopping. A lack of availability of potential capacity is also an issue. The hydraulic capacity of spillway gates may be reduced by the influence of floating debris, ice or sediments restricting the free opening size or by gates not opening. Rivers also carry floating debris, especially during floods. Depending on the natural conditions of the catchment, debris may be in the form of uprooted trees, cut timber, branches, bushes, grass, floating mires, dead animals and various man-made objects such as boats and houses. Floating debris can influence the discharge capacity of spillways by sticking to structures such as gates, spillway piers and bridges spanning spillways, to obstruct spillway openings and prevent gate operation.

Ice can reduce spillway discharge capacities in different ways. On the break-up of lake ice or river ice the ensuing ice floes can build up and jam spillway openings, particularly those of limited free dimensions, in the same way as trees or other types of floating debris. Ice or frazil developed in supercooled water can also adhere to gates and spillway piers and prevent gate opening or reduce the size of the free opening, and thereby limit the discharge capacity. Leaky seals are a not uncommon reason for ice build-up between gates and piers.

Ice jams, many metres in height, can also develop where frazil adheres to the boundaries of rivers and channels, and can completely change the hydraulic capacities and rating curves. Fresh water only requires supercooling of the water to fractions of a degree under 0°C to make frazil production possible throughout a water body. Weather conditions promoting supercooling and frazil growth in shallow waters involve those with very high heat losses from the water due to, among other things, radiation, low air temperature, wind and heavy snowfall. Where the effects of ice need to be considered specifically, modelling of the ice conditions in the reservoir and the river upstream and downstream may be required. Ideally, this should be done in coupled models, taking into account the effects of the flow and weather on the ice processes, as well as the effects of the ice on the flow.

Many rivers, especially those running through more easily erodible materials, carry significant amounts of solids. Conversely, sediments may often not be present in appreciable amounts in rivers running through old and stable igneous rocks. Coarse sediments carried as a bed load by a river will initially settle in the upstream end of reservoirs, where flow velocities start to reduce and the carrying capacity of the flow diminishes. Finer sediments carried as a suspended load will continue further into a reservoir and usually tend to create bars sloping at around 1%. Unless regular (annual) flushing is performed of the reservoir these sediment bars may, in time, reach and start to clog intake areas for water conduits, such as bottom outlets, power intakes and sometimes even spillway intakes. Sediments have a similar potential to disturb water conduit operation as do floating debris and ice, but, in addition, the discharge of coarse sediments may also lead to heavy erosion in the fast-flowing water of low-level outlets.

#### 10.3.4.5 Gate operations and results analysis

A high-level model is useful for understanding the behaviour patterns responsible for the unavailability of a spill way gate on demand. Unavailability may induce a high risk of system failure. A simulation from the start to the end of a calendar year was run by sampling a calendar year's worth of inflow data from the historical time series and routing it through the system. Each spillway gate has a maximum capacity of 608.8 $m^3$, and thus the combined capacity of the four spillway gates is about 2400 $m^3$. During peak inflows the combined effect of the four gates is around this capacity. As would be expected, the SCADA-controlled spillway discharges (withdrawals) are strongly correlated with the peak inflows.

**Figure 10.22** Plot of upstream daily flow and operator controlled spill versus time

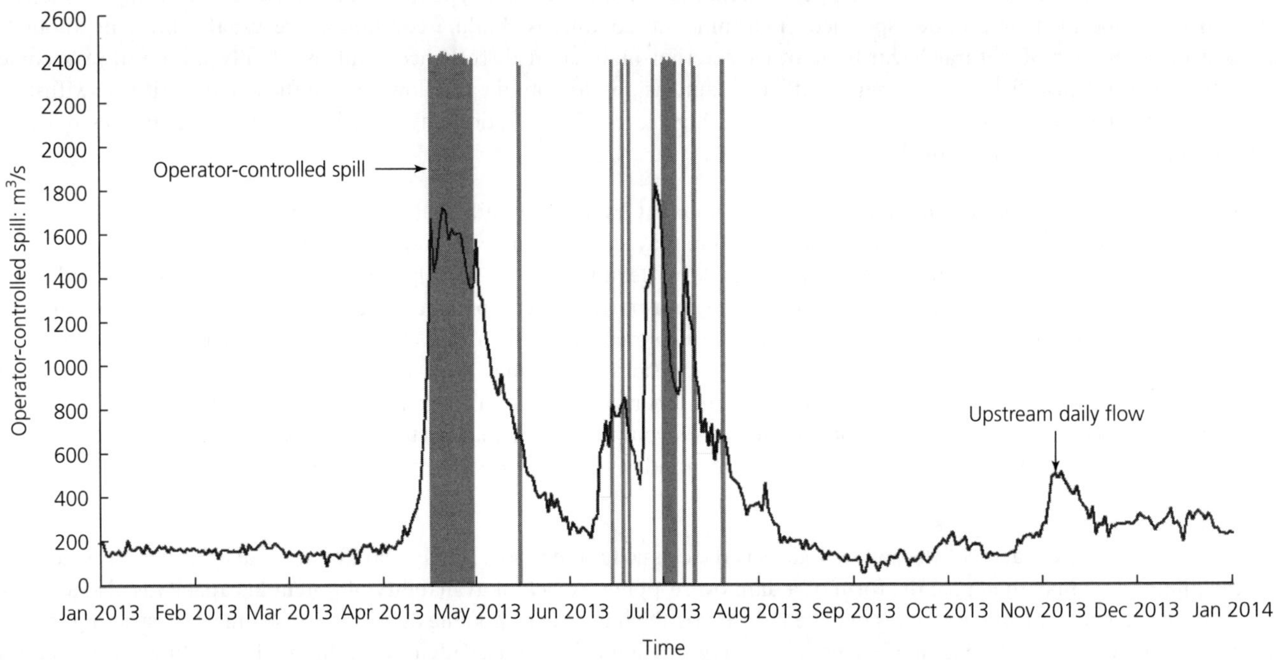


Where the combined effect of the four SCADA-controlled spillway gates is rendered insufficient due to high inflows, operators are dispatched to the site to operate the additional four manually operated gates to add additional spill capacity to the gates and prevent the dam from being overtopped. Figure 10.22 shows a plot of the operator-controlled spill as a function of time, with a graph of the upstream daily flows superimposed on the plot. These gates double the spill capacity of the Little Long generating station.

When the four human-operated gates are also rendered insufficient in routing the excess inflows, there are two additional SCADA-controlled gates that open into the Mattagami River that are instructed to be used in the rare scenarios where the eight that open into the creek are insufficient. Figure 10.23 shows a plot of the total spillway flow as a function of time. It can be seen that there was no need to spill through the Mattagami River as the spillway gates that open into the bypass

**Figure 10.23** Plot of total spillway flow and upstream daily flow versus time

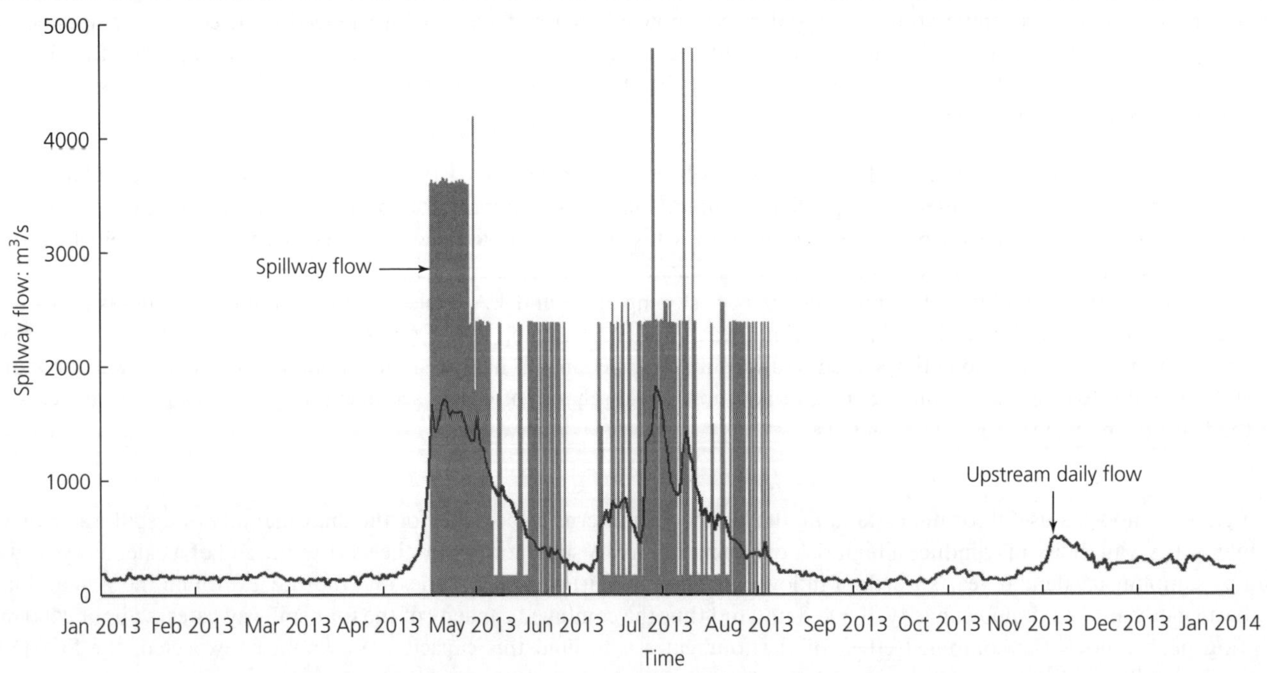

**Figure 10.24** Plot of upstream daily flow and spill into Adam Creek versus time

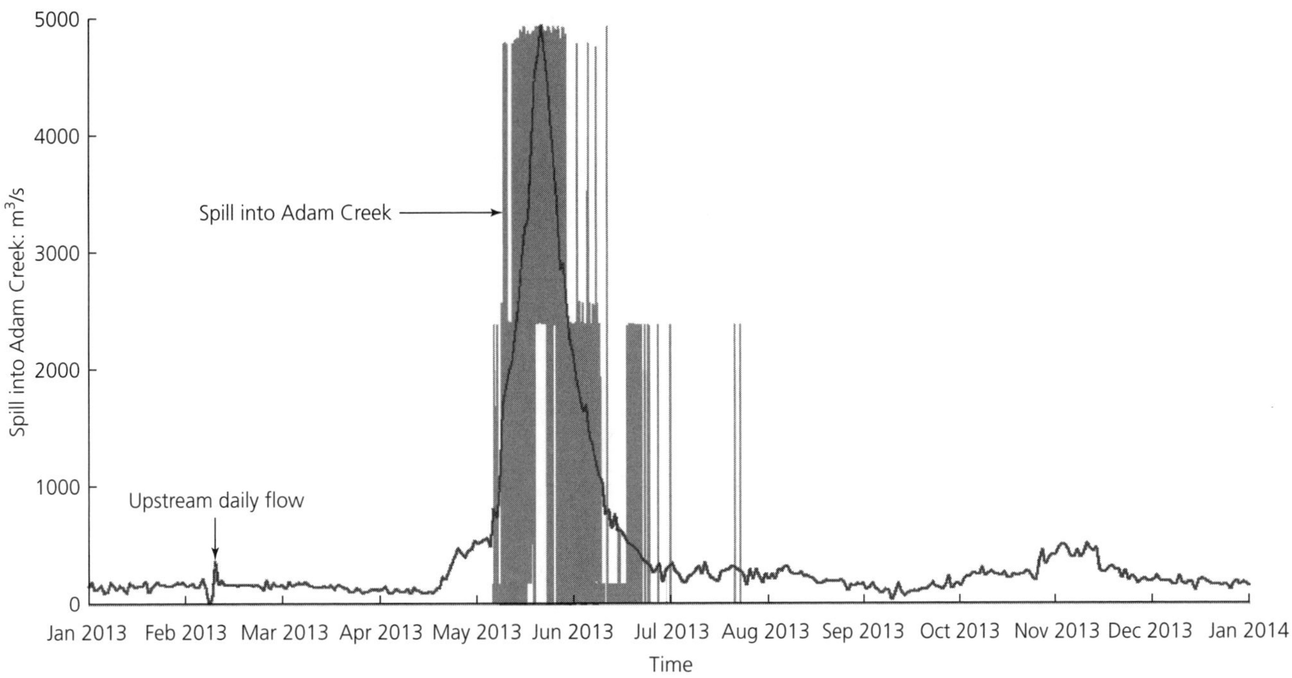


were sufficient to route out the peaking flows. Figure 10.24 shows a plot of the total spill into Adam creek bypass as a function of time.

To demonstrate the effect of gate opening and closing on discharge, a plot of the spillway discharge rate and the gate opening height versus the upstream daily flow was plotted (Figure 10.25). The height of the gate opening is generally at a maximum of 9.2 m during the peak flow periods to enable maximum routing of water from the reservoir.

**Figure 10.25** Plot of SCADA gate 3 discharge and gate opening height versus upstream daily flow

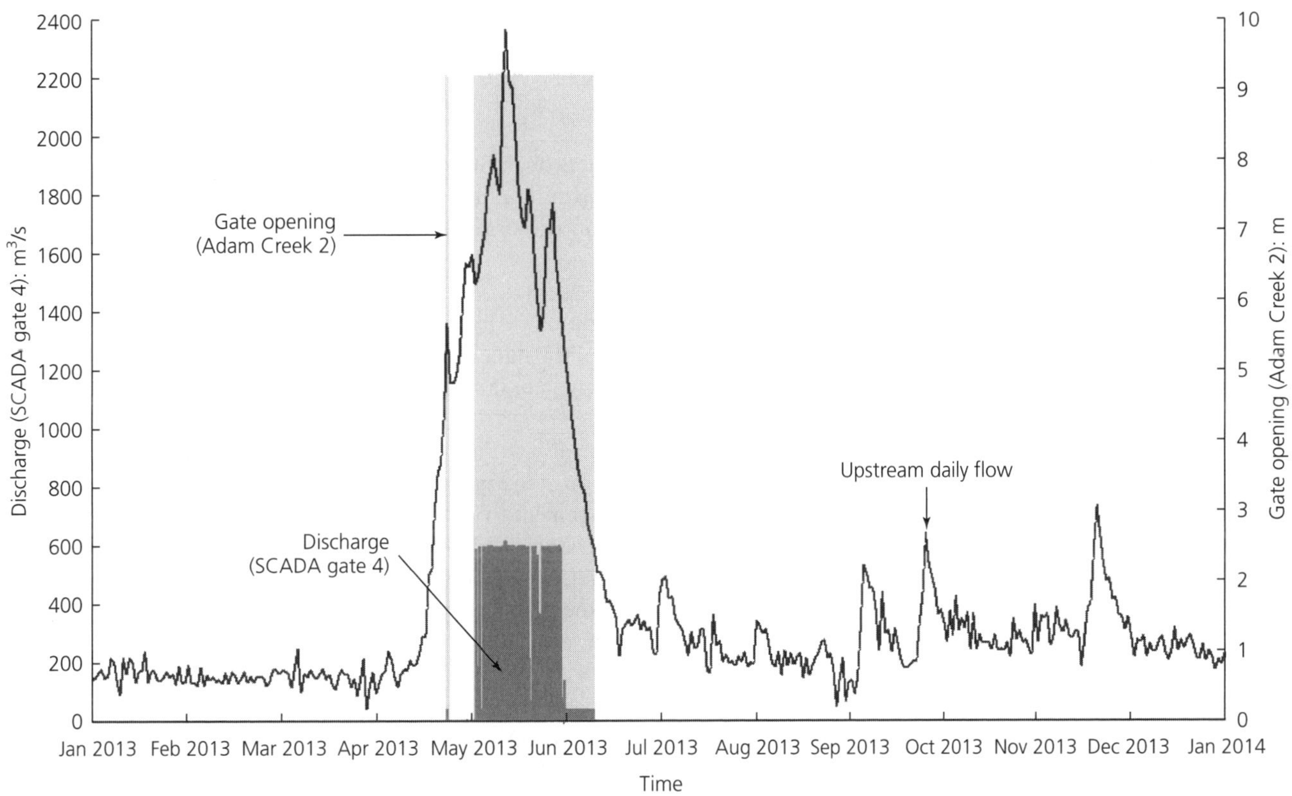

Figure 10.26 Plot of the water elevation on the spillway gates and upstream daily flow versus time

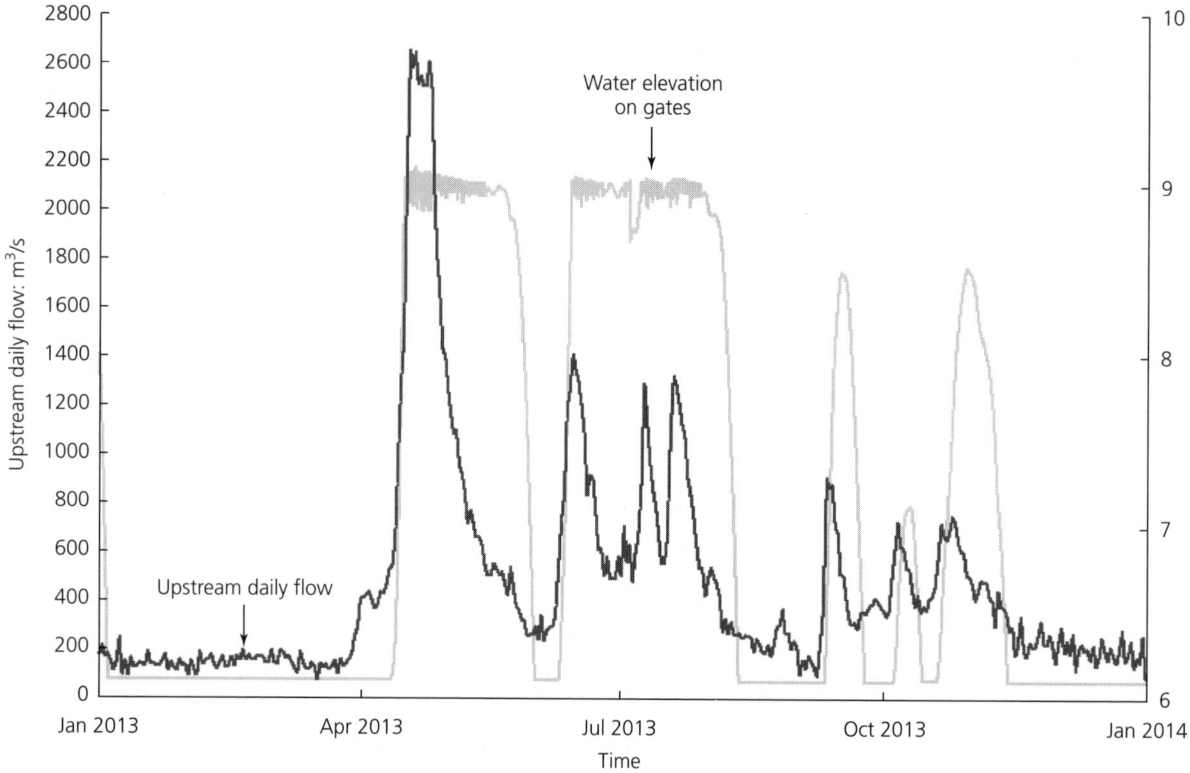


The gate opens at an average of 0.68 m/min and closes at the same rate. Its opening is triggered by the operating-rule requirements; these enable the SCADA systems to open the gate when the requirements are met, and to close them otherwise.

Figure 10.26 shows the correlation between high inflows and the upstream daily flow. The sill of each spillway gate is an elevation of 188.98 m, which means that within the operating range there is always some level of water on the gates. This affects the reliability of the gates because the water on the gates induces both hydrostatic and hydrodynamic forces on the gates. The level of water on the gates is the failure mode control variable for both mechanical and structural failures, meaning that the higher the water elevation on the gates the higher the effect of the aforementioned forces, and consequently the higher the probability of failure.

The remainder of the plots in this section show the means and annual probabilities for the 51 years of the historical time series. Figure 10.27 shows a plot of the mean upstream daily flow, gate opening range and discharge from spillway gate 3 (Adam Creek) for the 51 years of historic data. As expected, the peak inflow correlates with the maximum gate opening ranges and also with the highest routing capacities.

Figure 10.28 illustrates the annual probabilities of gate opening operations. This is important because if a number of gates fail during the period April–July and are not fixed quickly, the chances of the dam being overtopped increase greatly. It is imperative that during this period contingencies are in place to back up any gate failures.

Figure 10.29 shows a plot of the mean operator spill over the 51 years of historical data from the start of January to the end of December. It can be inferred from this plot that operators must be on standby during the start of freshet to travel to the site and operate the additional gates if required. It is important for management to ensure that from the start of April to the end of August all roads to the control centre are cleared of snow or other blockage, and are accessible to human operators on demand so that they can make a swift response. Figure 10.30 also shows that, based on historical data, the probability of operators being dispatched to the site throughout the year is below 0.5, but the probability that annual peak flows would require the action of human operators approaches 0.45. Thus the probability at other times is modest.

Figure 10.31 shows the average of operator-controlled spill and total spill from the start to the end of the year. The difference between these plots – although they both follow the same profile – is the operator-controlled spill. The plot demonstrates the importance of human operators in safely routing out excess inflows from the Lower Mattagami River reservoir. Hence hindrance to their operations or errors by them could be catastrophic in the event of high inflows.

Figure 10.27 Mean statistics for upstream flow and gate opening versus discharge

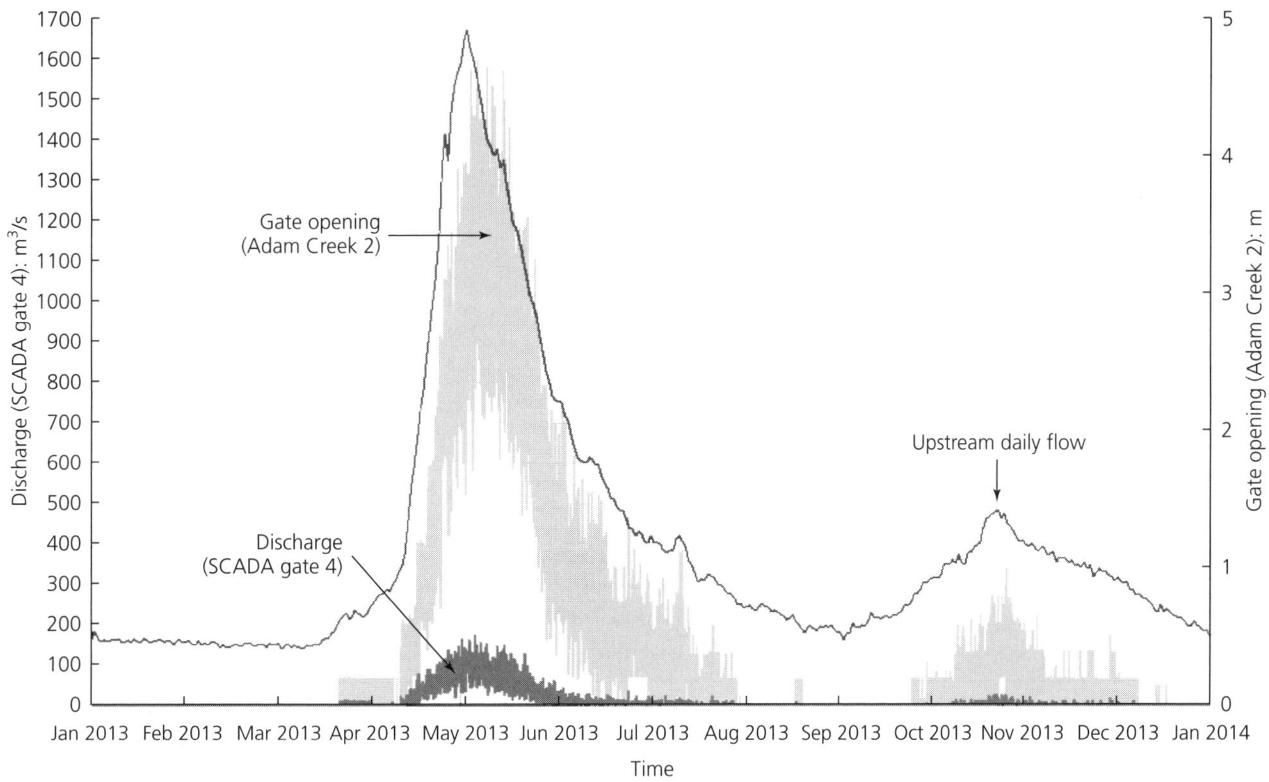


### 10.3.4.6 Reliability and performance of gated spillways

Gated spillway systems are generally designed to a set of defined engineering standards, but the effects of ageing, exposure, preventive maintenance and lack of frequent operations, in combination with human error, seem to make these systems more vulnerable than one would think. The on-demand failures of gated spillways are complex and may be caused by a gate component that can be repaired in minutes to hours or a component that may cause complete failure of the gate system and unexpected release of the reservoir containment.

Figure 10.28 Probabilities for gate opening as a function of time

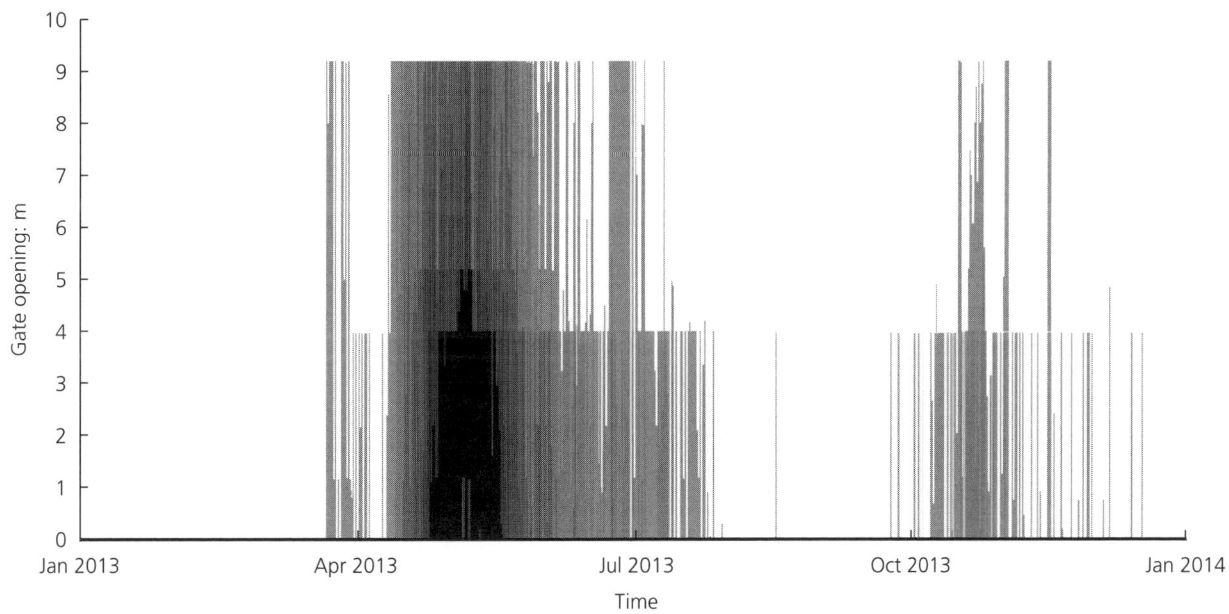

Figure 10.29 Plot of mean operator-controlled spill as a function of time

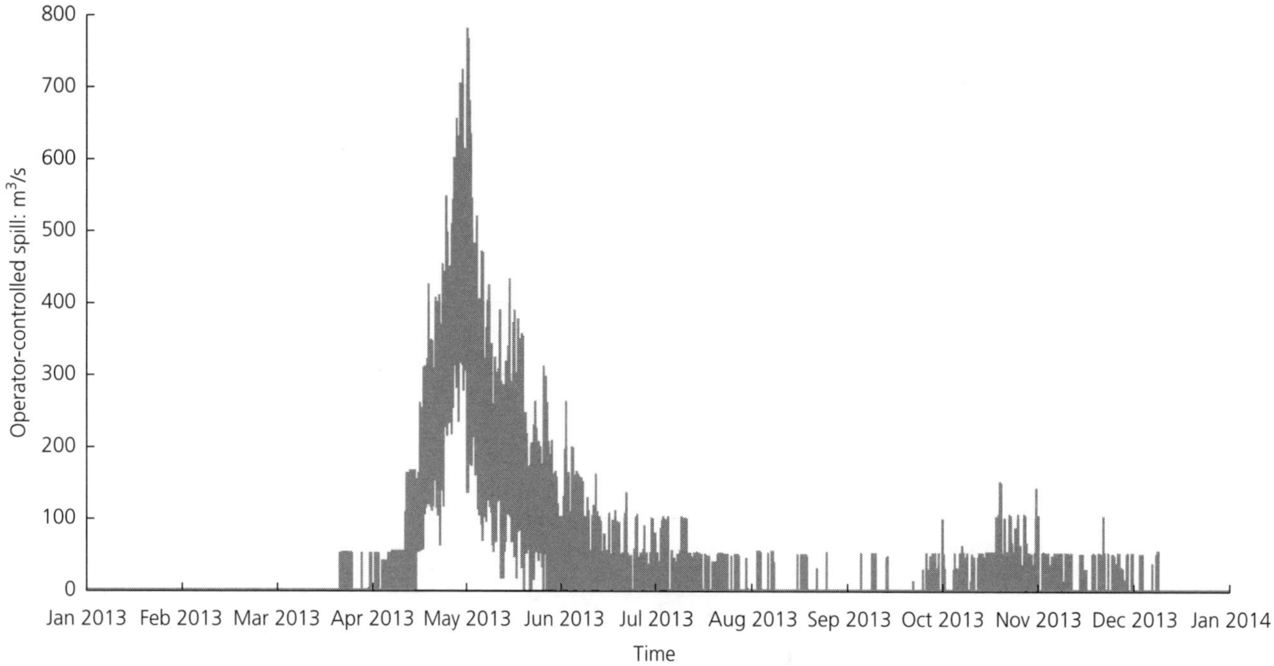


### 10.3.4.7 Flow-routing capacities

Although the four generating stations are in close proximity, Smoky Falls generating station was constructed over 30 years earlier than and to serve a different purpose from Little Long, Harmon and Kipling generating stations. Smoky Falls generating station is a four-unit, baseload station effectively operating 24 h/day with a rated flow capacity of 188 cm/s. The other three stations each have two units and are peaking stations that operate depending on available inflows (OPG, 2009). The different operating patterns under the current situation require that the headpond water levels at Smoky Falls, Harmon and Kipling generating stations fluctuate daily.

Figure 10.30 Probability statistics for operator-controlled spill

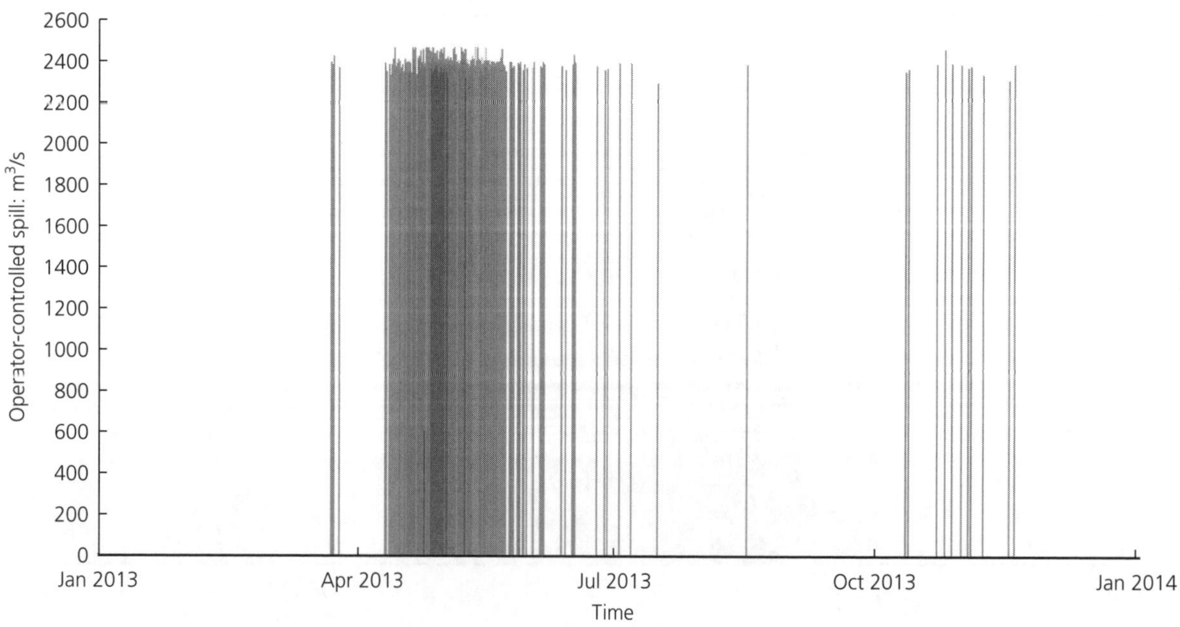

Figure 10.31 Operator-controlled spill and total spill versus time

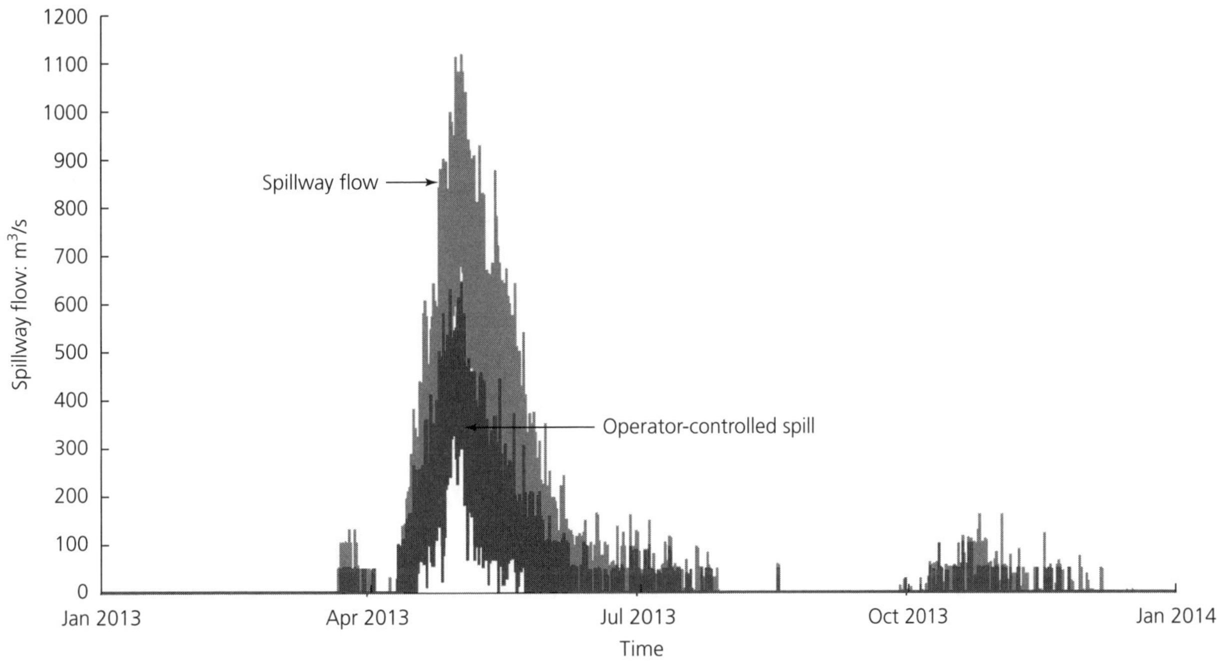


#### 10.3.4.8 Modelling function and failure

During the service life of a dam, operating conditions and natural environmental factors may lead to some deterioration in its structural integrity, mechanical equipment and foundation. A fragility model of a dam provides a tool for rational safety assessment and decision-making by using a probabilistic framework to manage the various sources of uncertainty that affect dam performance. Basic fragility concepts are presented here, and databases required to support the fragility assessment are identified. The method is illustrated using a concrete monolith from the Bluestone Dam in West Virginia, designed in the late 1930s.

Traditional design practices are sufficiently conservative that the probability of dam failure under a probable maximum flood 11 m (36 ft) higher than the original design-basis flood remains small. The reservoir water level on the gates is a key parameter because it affects the loading on the gates (the stresses in gate members, which determines the frictional resistance) and thus the consequence of gate failure (due to the effect on the breach outflow). It is most likely that the reservoir water surface elevation will be at the top of the gates when this potential failure mode is triggered, unless the gates are being operated as part of a routine exercise operation and the reservoir level is down for some reason.

#### 10.3.4.9 Modelling of the Little Long generating station spillway gates

As discussed in Chapter 5, Little Long generating station is operated within an operating range of 195.10–198.12 m, the main aim of the operating rules being to optimise power generation and route flow safely downstream. Each of the gates was modelled independently, as each has independent reliability components at both the component and subcomponent level.

The forebay of the generating station is filled to an elevation not exceeding 198.00 m. After achieving that elevation, any inflow greater than the amount of water required for two-unit operation (583 cm/s) is spilled down Adam Creek. A difference in storage of 12 cm will allow for the 4 h time lag required to dispatch operator agents to the station to deal with the contingency. The maximum forebay level of 198.00 m occurs only during the freshet period. There is no requirement to spill through the main dam, a practice that should be avoided in order to improve the operating efficiency at Smoky Falls generating station during freshet. Sluice gates 5, 6, 9 and 10 are controlled locally by the operator agents at the gates. It takes the operators 2–4 h to reach the site.

#### 10.3.4.10 Communication and human operator modelling

The safe operation of any industrial facility is highly dependent on the humans who operate and manage it, and who may be called upon to make decisions in the face of unexpected disruptions or other events. The reliability of human operators in the face of complex technological systems is discussed in further detail in Chapter 8. From a systems perspective; as water flows into the reservoir, spillway gates respond to changing reservoir conditions and demands, power is generated and other parts of the system function, observations are being continuously made to provide input to the automatic controllers and to inform the decisions made by human operators. These observations form the basis of a communications or data flow from sensors to SCADA systems and eventually to operators. In the present chapter, human operators as pertaining to the LMRHC are treated as 'humans in the loop', that is, they are treated like any other subcomponent of the complex technological systems. This simplistic assumption allows us to model human operator delays as a probabilistic distribution. More detail on human reliability analysis (HRA) is given in Chapter 8.

According to the operating rules for Little Long generating station, operators are to be dispatched to the station when the four spillway gates that open into the Mattagami River are insufficient in routing peak flows out of the dam to keep the reservoir below its operating-range maximum. It typically takes operators 2–4 h to get to the site after they are dispatched, but it may take longer due to inclement weather scenarios, and much longer if roads are blocked. To capture these rare but possible scenarios that lie outside the usual 2–4 h time, a truncated log–normal distribution with an upper limit of 24 h, a minimum limit of 2 h and a mean of 3 h was used to approximate operator arrival time based on historical experience.

#### 10.3.4.11 Reservoir operations

This potential failure mode requires operation of the radial gates to initiate the failure. If the gates remain in the closed position, trunnion pin friction will not be mobilised and the gate members will not be loaded by this mechanism. Reservoir operation levels will only be a factor if the spillway is operated at levels below the top of the gate elevation (or below a level within a foot or two of the top of the gates). If the reservoir level is typically at or near the top of the gates on an annual basis when the gates are likely to be operated or tested, this is a more hazardous situation than if the reservoir frequently does not reach the top of the gates on an annual basis, or the gates are typically tested when the reservoir is low. The likelihood of various reservoir levels at times when the gates will be operated can typically be estimated from the historical reservoir exceedance curves.

#### 10.3.4.12 Electrical failure

The failure rate (also known as the hazard rate) represents the mean, and has dimensions of inverse time. Failures are assumed to be a Poisson point process in time (which implies that, if the rate is constant, the time between events is exponentially distributed). Failures modelled in this way are computed with respect to the time since the simulation was started, and hence time is the failure mode control variable. Electrical reliability, specified as the electrical availability on demand by the system (including operator push-button error and external and internal power failures) is estimated as 0.017 failures/day.

### 10.3.5 Summary

Modelling engineered systems using stochastic simulation is promising. It affords a platform to incorporate feedback about operating procedures and human reliability with regard to systems function, and the ability to fuse models across different technological and human systems. Operational procedures and human decision interventions strongly affect system operations, accidents and failures. These have not usually been accounted for in dam safety risk analysis, and the proposed systems simulation approach described here does a good job of accounting for systems interactions and feedback loops that are generally unaccounted for in contemporary methods.

## 10.4. Systems dynamics

In the domain of *systems*, there are different areas, definitions, applications and tools. The most common expressions used in the systems area are systems thinking, systems engineering and systems dynamics.

The term *systems thinking* is often used in a general manner and can refer to different aspects when dealing with systems. The expression indicates that there is an understanding that nature, organisms and other phenomena encountered in the world, as well as man-made technical objects, should be treated as an integrated whole (i.e. as a system, rather than the sum of its parts). Systems thinkers try to understand the behaviour of a system, analyse its structure, organisation and hierarchy, map complex interrelationships between system parts, understand delay mechanisms, and explain feedback and feed-forward mechanisms. By adopting a holistic approach a better understanding is arrived at of how the individual parts and the system as a whole function, and measures for changing systems behaviour can be targeted more precisely. Moreover, the focus is shifted from design to sustaining functionality.

*Systems engineering* refers to an established technology or model for the design, analysis and synthesis of technical systems and other well-defined processes in a life-cycle perspective of the system. The set of activities within systems engineering is usually described as the process model. There are several standardised process models used by different industries, the Vee process model being among the most familiar ones (INCOSE, 2015). The activities regulated by these systems engineering process models generally involve (in order of application) integrated stakeholder management, system requirements management, functional allocation, design of components, verification and integration. The result of a systems engineering exercise is, ideally, a design of the system that includes optimal capacity and functions that satisfy the requirements and needs of multiple stakeholders.

When we talk about *systems dynamics* we are usually referring to practical applications of systems thinking and systems science for treating real-world systems and analysing of their behaviour within a particular modelling framework. This framework is due to Forrester (1988). There is almost nothing around us that is static and linear. All natural and man-made systems are to some degree dynamic. The degree of this dynamicity varies between systems and between parts within a system. This complexity defines how the system parts and their numerous interrelationships are set up and function, and demonstrates how the states of a system and its parts are influenced, leading to change in the behaviour.

There are several commercial software packages available for system dynamics simulation. The most well-known are STELLA/iThink from iSEE Systems, and Vensim, Goldsim and Simulink Toolbox from Matlab. These software packages differ in their user-friendliness and required computational power.

Dam and hydropower engineering requires the application of all the above system areas. While systems thinking is useful as a general approach, focusing on functionality as a first step makes systems engineering a powerful method for project managing the design or modification of facilities. Sometimes, the engineering effort requires all the above-mentioned system disciplines, and this is a *systems approach*.

### 10.4.1 The dynamic nature of flow-control systems

A hydropower plant as an integrated system is composed of physical and organisational (human) subsystems. The physical subsystem is made of components that can be in different terms of their operational status (i.e. in a normal or hazardous state, or faulty or out of operation for maintenance or replacement). The organisational subsystem consists of humans organised into operational units and social groups performing different tasks for the control and operation of the system. A large amount of physical components being in diverse operational system states controlled by individuals usually organised in hierarchical groups makes the whole enterprise a complex dynamic socio-technical system. The complexity becomes even more challenging due to the fact that hydropower schemes are operated in a natural environment which brings into play random inputs and stresses in the form of external disturbances. More and more electronic components are involved in hydropower operations, control and monitoring schemes. This kind of system involves a time-dependant management of numerous technical and organisational parameters and issues.

It is often the case that new components with new and untested features are added to an old system during upgrading projects. The control mechanisms are usually not completely analysed in connection with that change, and the new components are not always fully tested for all possible load combinations. This applies to especially to control subsystems, particularly electronics. Simulation of these changes might reveal some emergent behaviour in the system that was not possible to foresee during the design or direct upgrading of the system components.

Today, decisions that are being made in this complex environment increasingly need to reflect not only on the physical features of the system but also how the system as a whole meets the requirements and expectations of different stakeholders in the river system. Even decisions made within the company (i.e. decisions that involve only internal stakeholders) are generally reached by trying to balance conflicting interests (e.g. safety standards and environmental obligations balanced against profit demands).

This is where different kinds of system meet, and there should be an integrated way of modelling the life cycle of the facility according to capability requirements synthesised in a top-down approach involving different stakeholders. In this type of work models are essential, and they often serve as a common ground for learning, decision-making and problem solving (i.e. they become the knowledge models of the corporation).

### 10.4.2 A new approach to modelling

Civil engineers working with dams and hydropower facilities are well acquainted with modelling concepts. Models, both physical and numerical, are employed at different stages in the planning and operational phases of the facility. Usually, modelling applications and modelling tools are isolated, serving a particular area of interest. The most common areas where models are applied are hydrology, hydraulics and geotechnical engineering. The nature of these disciplines is complex and dynamic, and in many cases makes a modelling process the only feasible way of achieving complex engineering objectives.

Using modern models in combination with the increasing power of computational resources, today each engineering discipline is capable of generating reliable results as a significant input to further engineering judgement. Nevertheless, there is a new paradigm in engineering science that brings a new way of thinking based on the understanding of the behaviour of complex systems.

Single-purpose models can solve well-defined problems in a satisfactory way. The results of these modelling exercises provide the input and parameters for particular system parts. Examples of these inputs from a modelling exercise are the design flood, the reservoir behaviour, the slope stability, the required capacity of the spillways, the configuration and size of the gates, and the layout of energy-dissipation structures. Nevertheless, there is an essential part missing. While all particular questions seem to be solved in a detailed manner, the challenge lies in integrating these single inputs into the design of the whole facility (i.e. a functioning system). Traditionally, this part of the exercise was, and still is, in the hands of the experienced engineer. As systems become more complex there is a need for better methods and tools for understanding and predicting systems behaviour, and for the ability to control this at both the component and system level.

### 10.4.3 Physical structure and functional analysis of dynamic systems

Every system consists of physical objects. These may be technical components, technical procedures or even humans in the system. Each of these items may have one or several functions attached to it. When undertaking a system breakdown, it is important to keep these two aspects separate. Physical system breakdown serves as a basis for functional analysis, while functional breakdown is more complicated – it uses physical entities as building blocks and attaches relevant functions to them.

In the functional breakdown, one can, for example, find a functional element called 'spill water'. This function may be realised with redundant gates. Thus, when considering functional deficiencies it is important to keep track of both the physical and the functional structure. For example, if there are two gates in the structural (physical) system breakdown which together have the assigned function 'spill water', failure of one gate does not necessarily mean 50% loss of spillway function. It might be that the gates have different capacity or they are purely redundant, meaning that when one gate fails there will still be 100% of spillway function left.

Dynamic models have a certain structure, the building blocks of which are parts of the system connected by functional interrelationships. The hierarchical structure is decided on the functional side. At a higher level, there is always a controller, which through its actuators controls the system and obtains feedback through sensors. This process is a control–feedback loop, a fundamental entity in systems dynamic models (Figure 10.32).

System dynamic models are driven by physical flows, information flows and human actions (which, in turn, are triggered by information flows). The interactions between all the components and the feedback mechanics responding to loads and outputs generate a certain behaviour of the system. Once the system behaviour is depicted in the simulation model, other options of the system behaviour can be analysed.

Figure 10.32 Control–feedback loop, a fundamental entity in systems dynamics

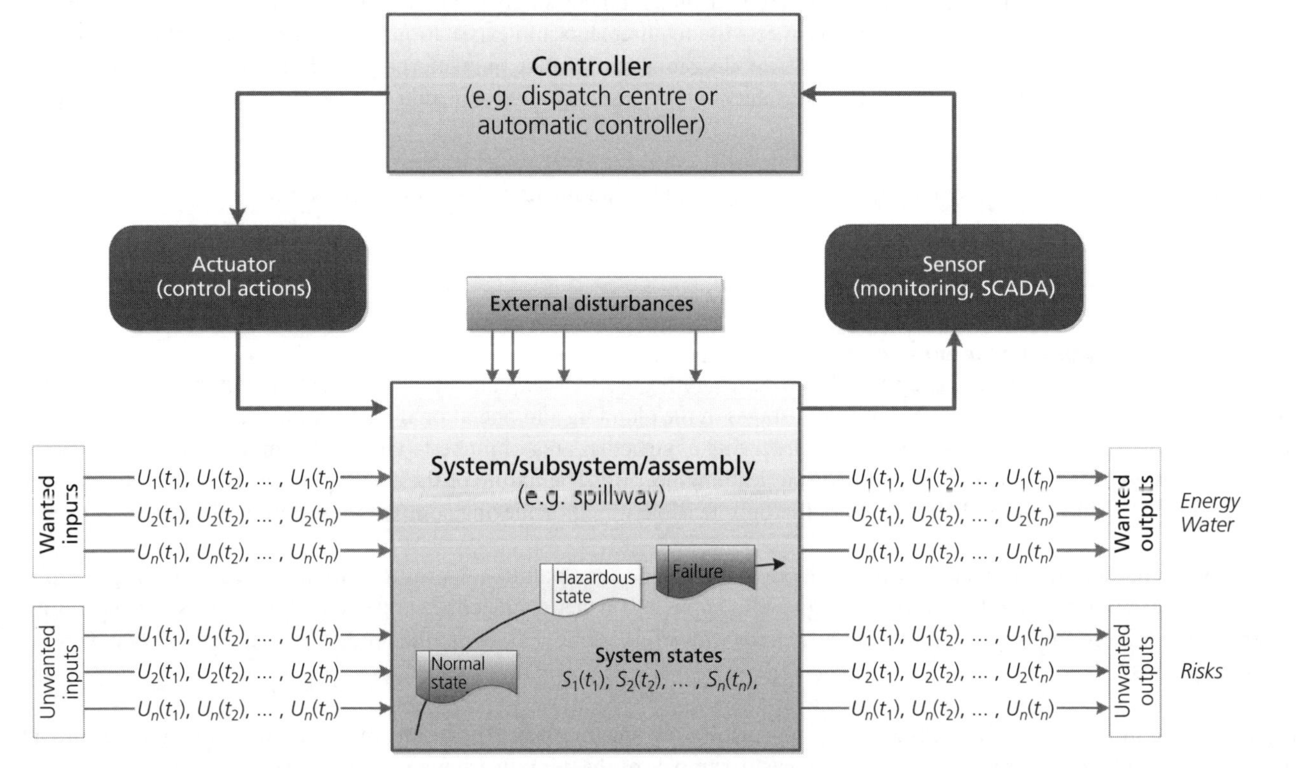

### 10.4.4 Systems dynamics modelling of human performance

Hydropower plant operations nowadays are, to great extent, conducted remotely. This is not completely in line with the original idea about how to operate these facilities when they were designed and constructed back in the 1940s to 1970s. The provision of remote operations required gradual modification of existing spillway systems, and the development and implementation of control subsystems. Recently, these control systems have become fully computerised, and electronic components are now widely used. The combination of remote operations and complex computerised systems creates challenges for the operations personnel in terms of human reliability.

Modelling of human performance is a challenging task, as the interaction between humans and technical systems can be complicated. While technical systems, as long as they are in a fully operational state, in most cases perform as they were designed to, humans represent a different type of degree of freedom – depending on the situation, the freedom to decide can be positive or negative. Human–machine interaction can be addressed using a systems dynamics approach, as systems dynamics models are particularly suitable for simulating interactions between technical and social systems.

## 10.5. Systems model of the Göta River

The Göta River is the longest river in Sweden, its watershed covering approximately 10% of the country's territory. In the middle of the watershed is Lake Vänern, Sweden's largest lake, and at 6500 billion m$^3$ it is the third largest lake in the European Union. Water from Lake Vänern is released into the main river and regulated at the Vargön hydropower plant (Figure 10.33). The Göta River system has three hydropower stations and a flap-gate facility, which are owned and operated by Vattenfall AB, the state-owned power utility. The Göta River ultimately flows to the North Sea west of Gothenburg, where the last branch of the river is tidal. Besides hydropower, there are navigation facilities and ship locks, and many other riparian stakeholders with interests in the river system.

The Göta River and its valley face major geotechnical issues related to quick clays and landslides, extreme water levels in Lake Vänern, flooding, sea-level rise, urban and rural development, water pollution, fish passage and other environmental issues. All these external factors affect the ageing water resource infrastructure on the river.

Dam safety is a major concern. Dams and spillways are in need of refurbishing and upgrading, but many of the riparian stakeholders hold conflicting interests. Balancing these conflicting interests while facing changing future conditions has become a major short- and long-term concern for both the dam owner and those with regional political interests.

Facing these enormous and complex challenges, Vattenfall, as a dam owner, decided to employ a systems engineering process to structure the problems faced in the basin, analyse the reliability of the hydropower facilities, and develop a draft roadmap for the future. This roadmap involved scenarios and plans for hydropower development, dam safety and strategies for solving environmental issues.

Figure 10.33 The Göta River system with hydropower facilities

Figure 10.34 Top-level layer of the Simulink model for simulating the Göta River system

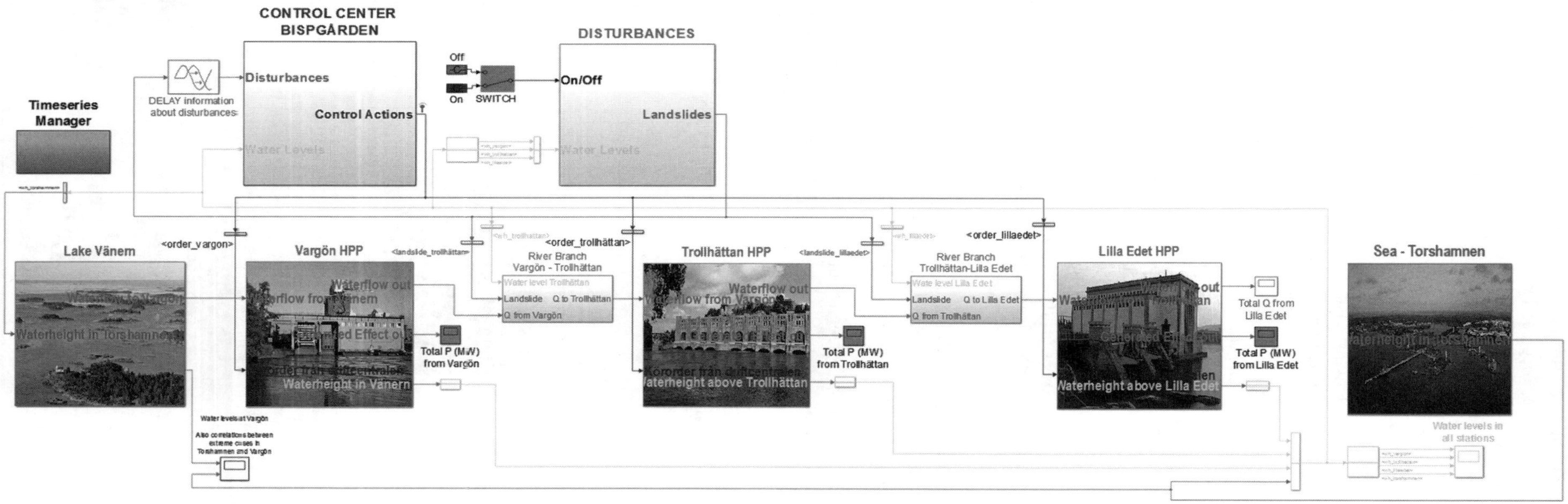

### 10.5.1 Systems dynamics simulation

Vattenfall chose to use a systems dynamics approach to model the river operations (Sterman, 2000). One of the strengths of systems dynamics modelling is that it facilitates participative analysis (USACE, 2008). The visualisation capabilities of systems dynamics models enable improved management of complex word problem statements, and allow numerous stakeholders to be dealt with in a structured way. This approach enables congregation of different stakeholders and experts with the necessary skills.

#### 10.5.1.1 Simulink model

To address the multiple challenges in the Göta River system a simulation model has been developed. The system of interest is a natural human-altered system, with a large lake upstream and tidal water downstream, and can be regarded as a system containing many random components associated with the system inputs and parameters. Inputs to the system are stochastic. Some system behaviours are deterministic and can be simulated using traditional means but others are stochastic. The water-control function is complicated due to the potential release restriction that depends on sea level.

River-system models usually become complicated in terms of the number of components and their interrelationships. To manage this complexity it is advised that the model be arranged in a layer structure using a top-down approach. This means that the model is divided into layers representing different levels of complexity and different subsystems. The overall view of the top-level layer of the Göta River simulation model in Simulink is shown in Figures 10.34 and 10.35.

At the top level a representation is set up of the river system with its main features and facilities. This contains reservoirs, power plants, control centres and disturbances management. Subsequent levels represent single plants with their main subsystems, such as spillways and generating units. At the end there are layers with the smaller subsystems representing assemblies.

For this hierarchical structural decomposition see the system structure in Chapter 4. Some submodels will be reusable. It is an important feature of Simulink that it allows reuse of similar components, and these can be transferred to the modelling of other facilities as well. Simulink models are reusable, and the modelling effort can be distributed to the expertise needed within different subsystems. The Göta River model has six layers. Examples from the second and third layer are shown in Figures 10.36 and 10.37.

Figure 10.35 A closer look at the Göta River Simulink™ model

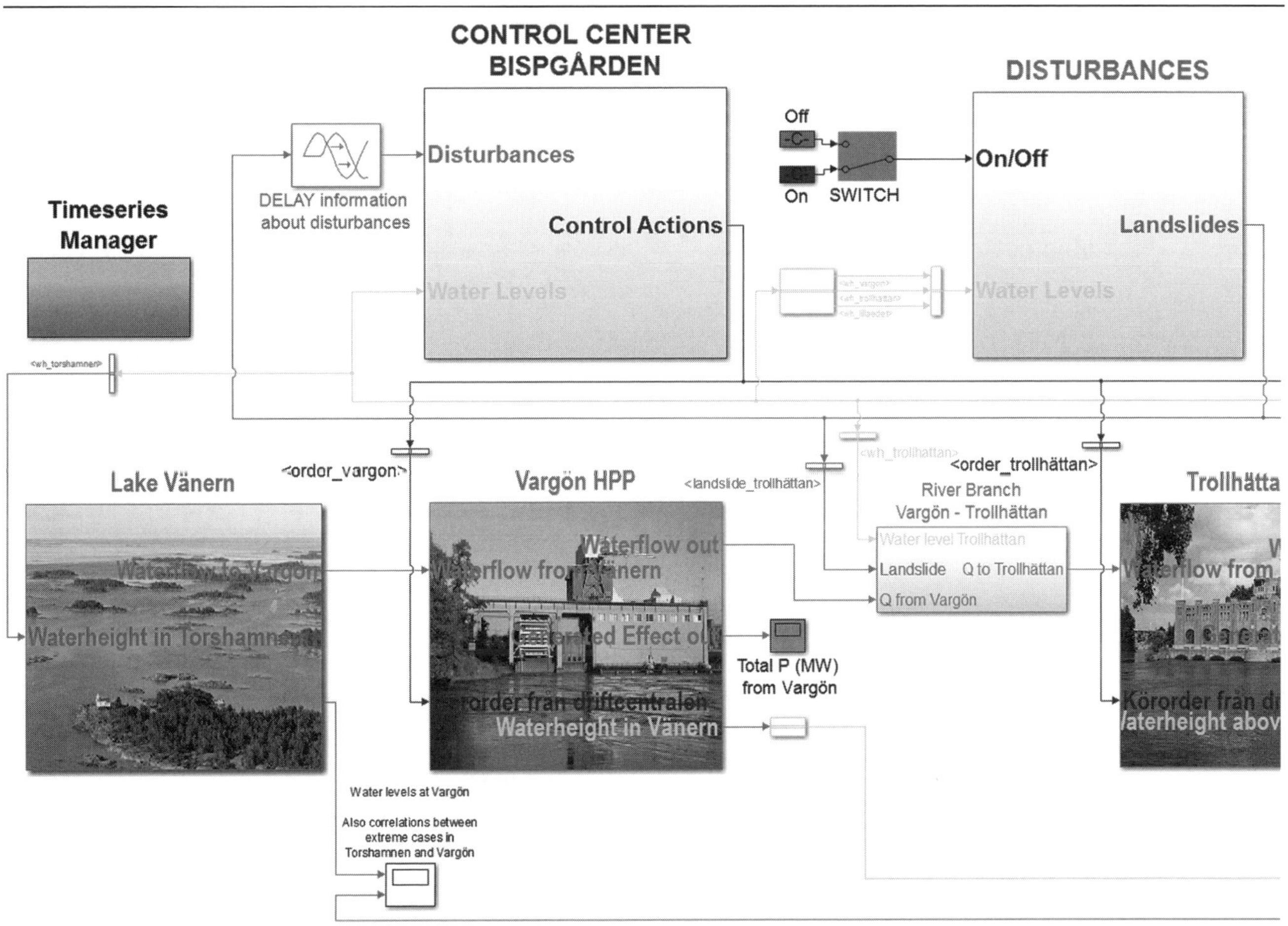

Figure 10.36 Second-level layer of the Simulink model for simulating the Vargön hydropower plant

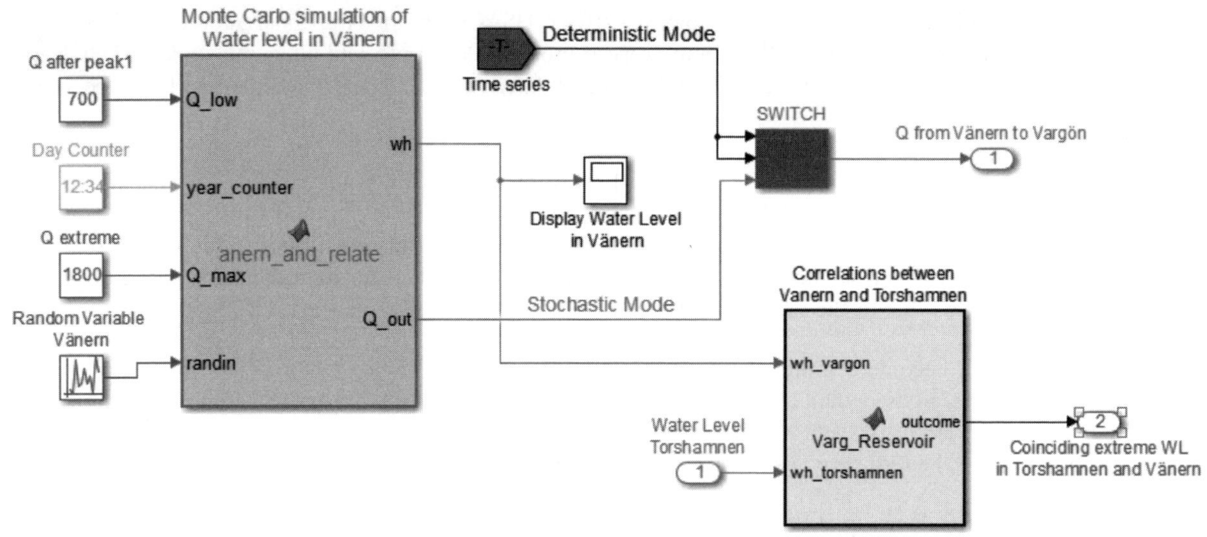


Figure 10.37 Third-level layer of the Simulink diagram for simulating the spillway gates and generators at the Vargön hydropower plant

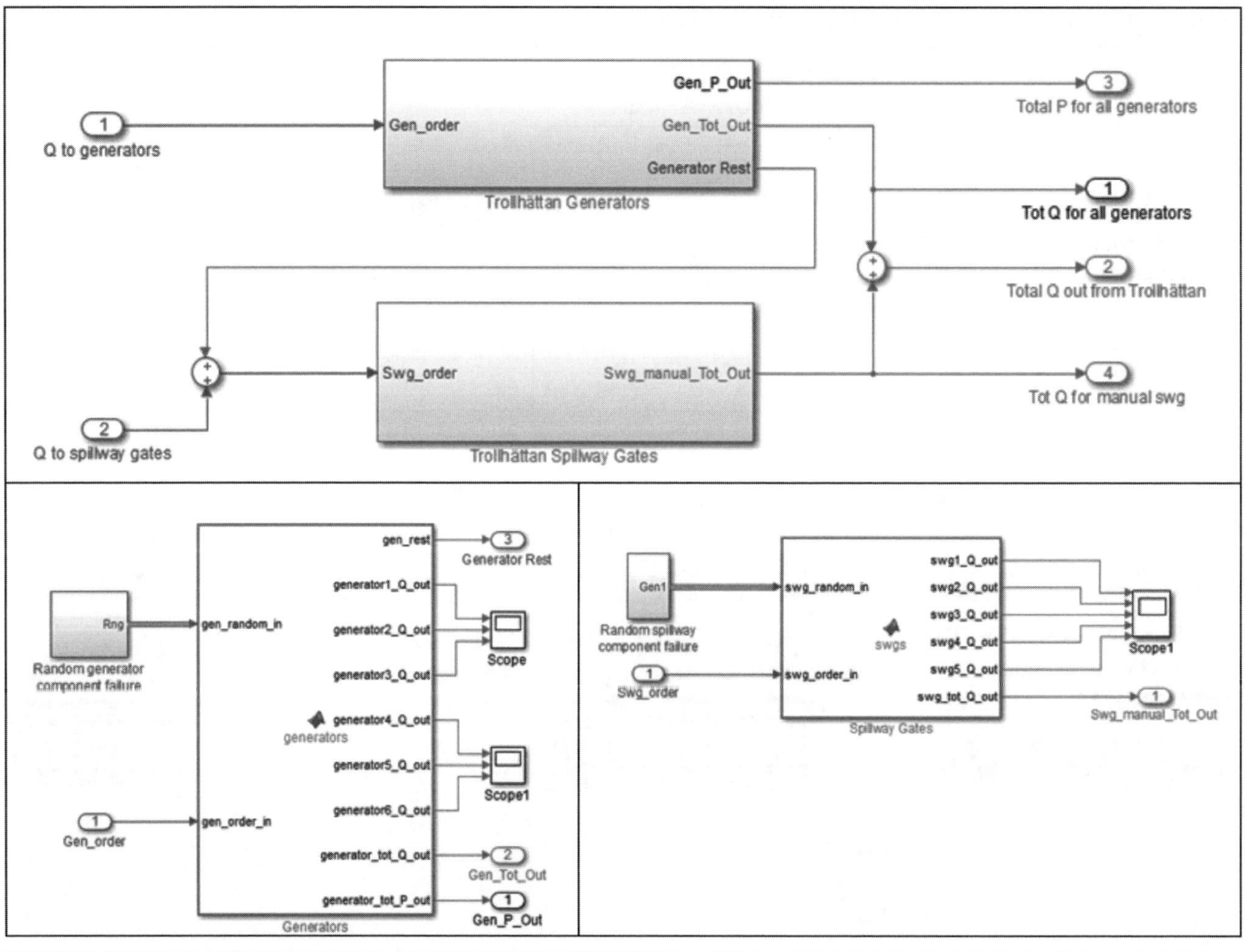

#### 10.5.1.2 Simulation modes

In addition to the layered structure, the Göta River model is equipped for flexible use of its input parameters. The user may choose to run the model in deterministic mode, stochastic mode, hybrid mode or a combined mode. In deterministic mode the inputs are preprocessed time series, for instance from the hydrological model. By switching to stochastic mode, the model uses stochastic inputs that are generated by probabilistic routines such as Monte Carlo models. The deterministic mode is useful, and in many cases is required for model verification and validation tasks.

The stochastic mode is employed for simulation with certain objectives, for example, to learn about the effects of disturbances. The model is equipped with switches, as shown in Figure 10.36, which allow the user to change the mode in which the simulations are run. It is also possible to run the model in a hybrid mode. This means that, where accurate data describing the inputs are available, they can be used instead of randomly generated distributions (e.g. if a fixed power-demand curve is available for the coming year). Similarly, if an environmental restriction on water release is imposed, this would be input to the hybrid simulation model instead of randomly generated inputs.

#### 10.5.1.3 Water-flow and water-level inputs to simulation and flow routing

Water flow in the model is simulated by routing a time series, either preprocessed or generated stochastically. In the Göta River simulation, inflow hydrographs are generated as an input to the system: the total inflows for Lake Vänern, and local inflows between other facilities in the cascade. In the same manner, extreme water levels in the lake and the sea are generated as an input to the simulation.

Flow routing through the system is carried out in the model, and the development of the water flows and water levels at different point of interest can be found and plotted (Figure 10.38). For most of the simulations, inputs are generated in hybrid mode – a combination of time series and stochastic approaches.

### 10.5.2 Simulating the remote control operations

Besides physical flows, the systems are controlled by information or communicative flows. These flows involve the human side of operations and thus are prone to a certain degree of error or corrective actions.

In the Göta River model, the control and monitoring actions are carried out through the control-centre module, a submodel that mimics the operations as they would occur in reality, with the decision delay mechanics, etc. (Figure 10.39). All control actions are based on the information flows with the simulated, assumed or known delays in decisions, information flows and error distribution. System states are reported to the control centre and orders are issued for the plant operation. Some of the control actions are automatic and some are manual. This is also defined in the model.

### 10.5.3 Simulation of disturbances

There are two kinds of disturbance implemented in the Göta River model: extreme sea water levels and landslides. These disturbances affect the whole river system through discharge restrictions. High sea water occurs several times each year, whereas landslides of high magnitude are relatively rare (the last big landslide in the Göta Valley occurred in the 1960s).

Each type of disturbance imposes different kinds of discharge restrictions on the river system. Events can occur separately or at the same time. While the response to the disturbance of extreme water levels occurring individually is historically tested and thought to be manageable due to the short duration of these events, the required responses to disturbances due to landslides is unknown. A coincidence of these two disturbances might lead to a combination of problems that are difficult to depict analytically. A Monte Carlo simulation was therefore used to simulate the possible outcomes in such cases.

#### 10.5.3.1 Extreme sea levels in combination with extreme levels in Lake Vänern

One important external disturbance to operations in the river system are extreme water levels in the downstream area of the river system (i.e. sea water level). Extremes of the sea water level, which can be up to 1.5 m above the normal level, occur due to combinations of tides and wind-induced standing waves, and affect the possibility of discharging water from the last power plant in the system, Lilla Edet. Moreover, a high sea water level creates flooding in the city of Gothenburg, which lies downstream, and therefore any extra discharge from the power plants is undesirable in these circumstances. In situations like this, which occur several times a year, the operator is required to minimise discharge. The combination of extreme water level both of the sea and Lake Vänern can be challenging due to the small safety margins at the lake. Historically, this extreme situation has never been observed. However, it is anticipated that there is a probability of the event occurring, and therefore it is necessary to analyse it. A Monte Carlo simulation was carried out to simulate how this disturbance would affect the whole river system.

#### 10.5.3.2 Landslides

The Göta River valley is prone to landslides due to large areas of marine clays and quick clays. This type of landslide can occur quickly and cause significant damage, including loss of life. For the hydropower plant operator, landslides could block the river, reducing convenience and imposing discharge restrictions.

Figure 10.38 Series of simulated water levels for 100 years at three dam facilities (three upper diagrams) and at sea (lower diagram)

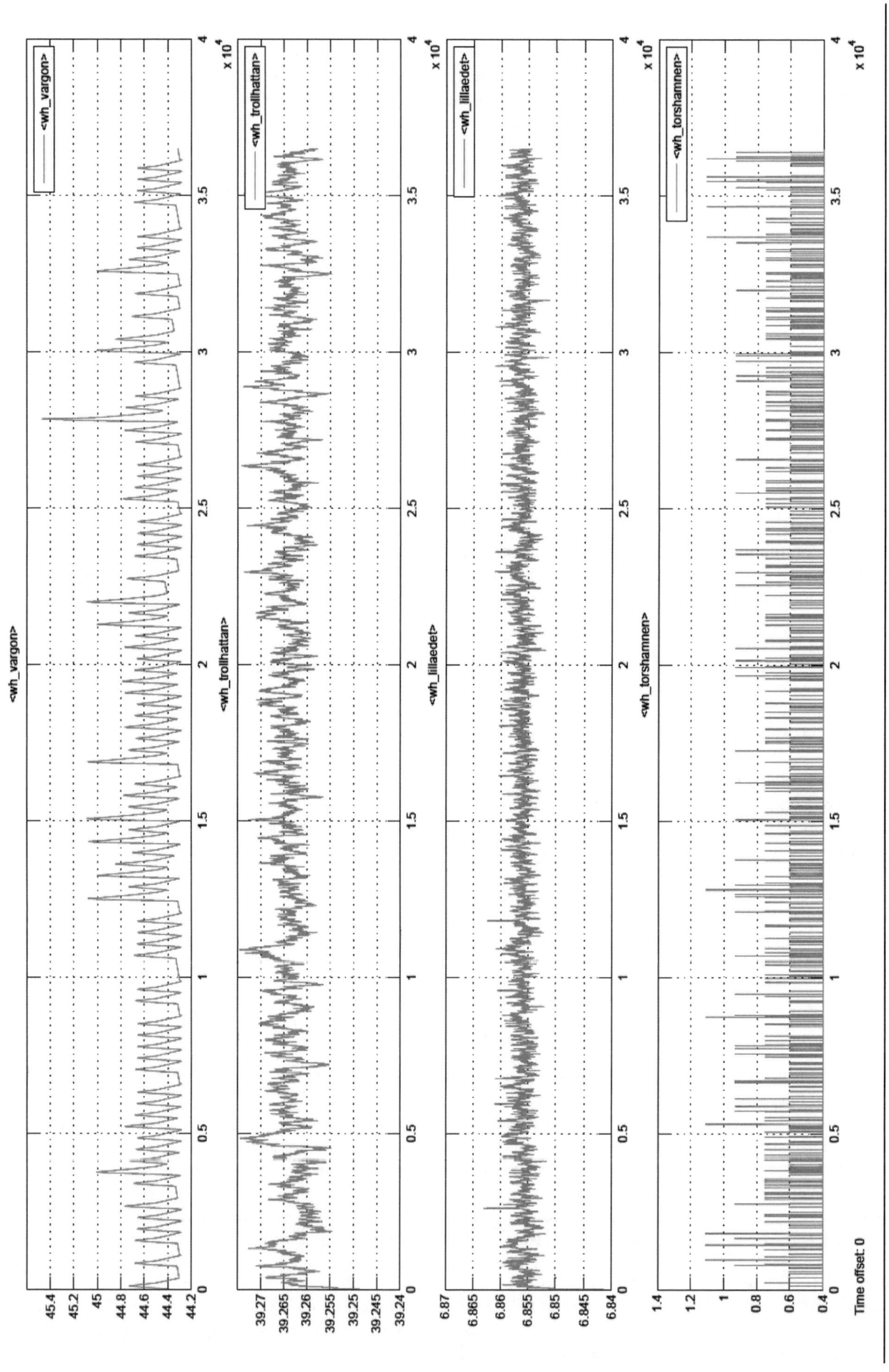

Figure 10.39 Simulation block representing the control centre, which remotely operates the power plants, including the spillway gates

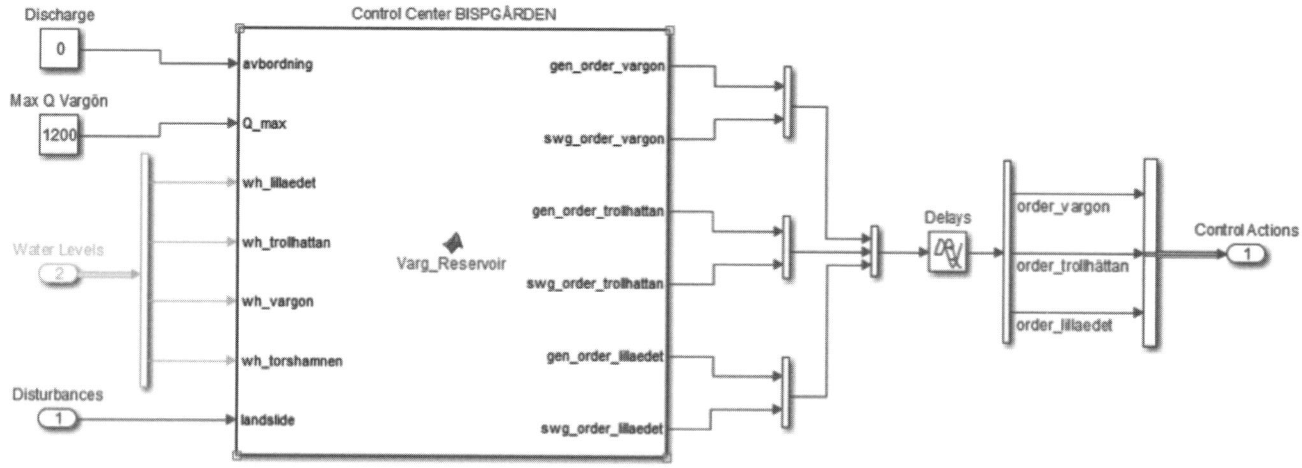

Timing is crucial, as simultaneous landslides may stop water flow and create extreme water levels in the river. This would lead to flooding around Lake Vänern, and in the worst case could lead to overtopping at Vargön. Extreme water levels and landslides are probabilistically (causally) correlated, because rapid water-level fluctuations in the river may induce landslides.

In the model of the Göta River the landslide simulation submodel is placed in the disturbances section. It contains a landslide generation module, landslide effect calculator and connections to other modules (Figure 10.40). The landslide generator creates exponentially distributed random landslides with a given frequency. The exact locations of the slides were determined by mapping risk areas along the river valley, and the associated probabilities were as described in the detailed investigation carried out on the Göta River by the Swedish Geotechnical Institute (SGI, 2012). The severity of landslides was calculated and grouped into five classes according to magnitude. The landslide effect was then simulated in the effect module. The effect is Pareto distributed. Distribution parameters were obtained by fitting data from the SGI landslide database using the distribution-fitting tool in MATLAB.

Figure 10.40 Landslide simulation module in Simulink

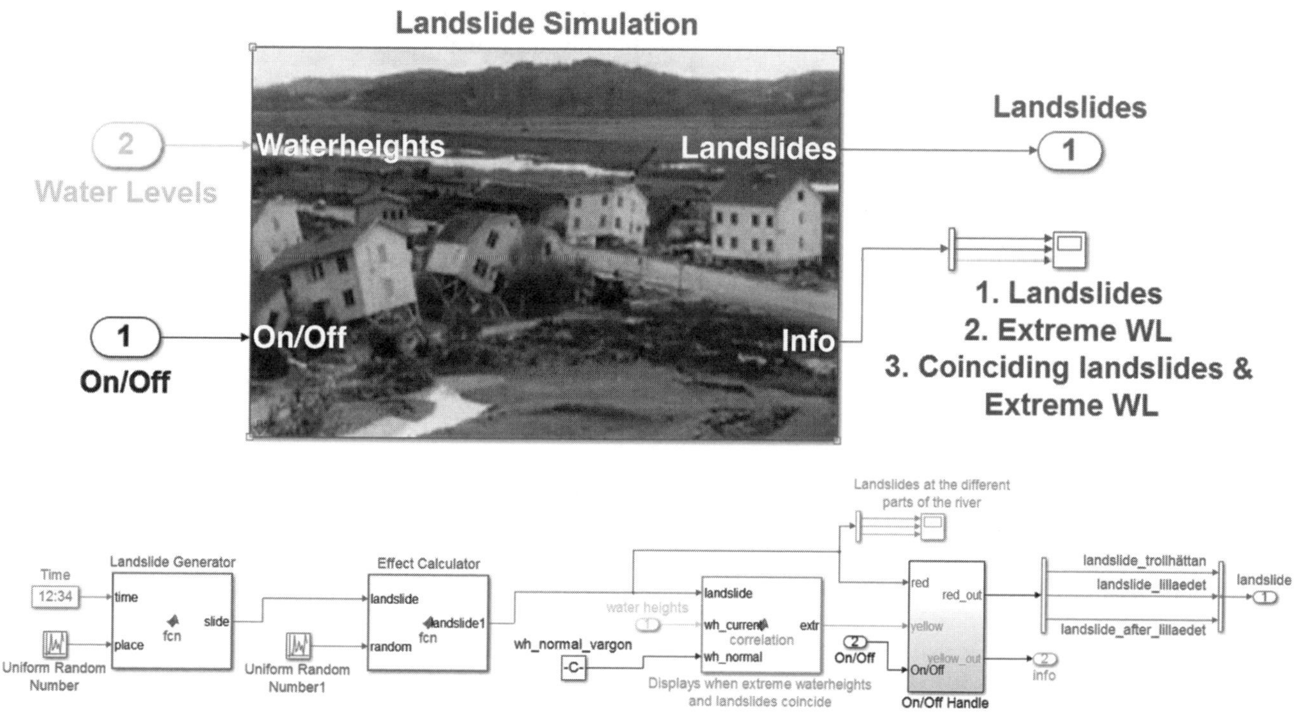

The simulation of disturbances to the system by landslides generates random landslides at exact locations, and their magnitude, occurrence and duration. In real life, once a slide occurs information must somehow reach the control centre for action to be taken. This can occur in two ways. For smaller landslides, human observation is necessary and contact will probably be made with the authorities and, in turn, with the operator of the power plants (in practice, Vattenfall's control centre). For bigger landslides, the control centre will notice an unusual behaviour of the water levels along the river; the water level will rise upstream and decrease downstream on the branch where the landslide has occurred.

In both cases it will take a certain amount of time for information to be received and decisions to take action to be made. This delay is simulated in the model. The disturbance signals are transferred to the control centre, where the decision will probably be to stop the discharge into the river as a remedial action. This ceasing of discharge may last for a long time, depending on the landslide size and duration, which is also simulated in the model. For large slides occurring in the river system the river will be blocked completely, resulting in landslide-created dams that eventually will breach. A flood wave would then overtop the downstream facilities.

All this random landslide disturbance is integrated in the simulation, providing realistic assumptions and simulating the effects on the river system operations. None of this effort can be achieved effectively by any risk modelling techniques other than simulation.

### 10.5.4 Discovering new hazardous states and failure modes through simulation

By running different scenarios with and without disturbances, insights can be generated about the behaviour of the river system. Simulations are usually set up and run for a 10 000-year period. The results presented here are for 100-year runs.

#### 10.5.4.1 Scenario A: Current conditions compared with an increased release of water from Lake Vänern

Climate change is creating the need to release ever more water from Lake Vänern for relief of the flooding around the lake and to increase the safety margin of the dam. Extreme water levels downstream in the sea are working against this need. How does the water level in the lake relate to high water levels in the sea? What would happen if the design discharge was increased? This simulation was run as a basic scenario to obtain answers to these questions, and to learn about system behaviour and eventual emergent properties. The results are presented in Figure 10.41. The diagrams show the development of the water level at three locations along the river length (Lilla Edet, Trollhättan and Vargön) over a 100-year period for two scenarios.

In the top diagram, the higher lighter coloured spikes show the current situation, indicating that the water levels at the Vargön hydropower plant are dangerously near the critical level with a small margin to the dam crest. The different colours represent different system states: the darker colour shows normal operations, and the lighter colour shows a hazardous system state. A similar situation like the one represented here was in fact experienced in 2000–2001.

The other two traces in the top diagram show the results for the alternative of increased discharges from Lake Vänern. The lighter trace represents the normal system state (i.e. normal water levels in the lake) and the darker trace represents hazardous system states. The situation might seem satisfactory in this case. Initially, the magnitude of the hazardous states is significantly lower, providing a better safety margin.

The middle and lower diagrams show the water levels at the downstream facilities Trollhättan and Lilla Edet. It can be seen that water levels there are normal (i.e. there is a normal system state in both alternatives). These results show that extreme water levels in the sea and following discharge restrictions do not affect the system in a critical way if other disturbance factors are ignored.

Extreme water levels in the sea are always present in the simulation, as this natural force cannot be controlled in the river system. Perhaps in the future flood gates will be installed at Gothenburg, but they do not exist today. Such gates might change the situation, potentially in either way.

#### 10.5.4.2 Scenario B: Higher increase of water release from Lake Vänern

This simulation was run to test the situation of even greater discharge from Lake Vänern in order to answer the question of whether it would be possible to lower the hazardous levels even more.

The release of more water from Lake Vänern would lead to more water in the river system. As can be seen in Figure 10.42, at the beginning the water levels (i.e. system states) are almost always in the normal range. After 30 years, there is an increase in hazardous states in both scenarios, but they are manageable. Nevertheless, when randomness is introduced in the simulation combinations of the values occur that are high and hazardous. At the same time an extreme sea level, also generated randomly, coincides with high water levels in Lake Vänern. This unlucky combination of individually manageable events leads to a major accident, as can be seen in the simulation results in the bottom diagram – the Lilla Edet facility is overtopped.

**Figure 10.41** Results for simulation Scenario A

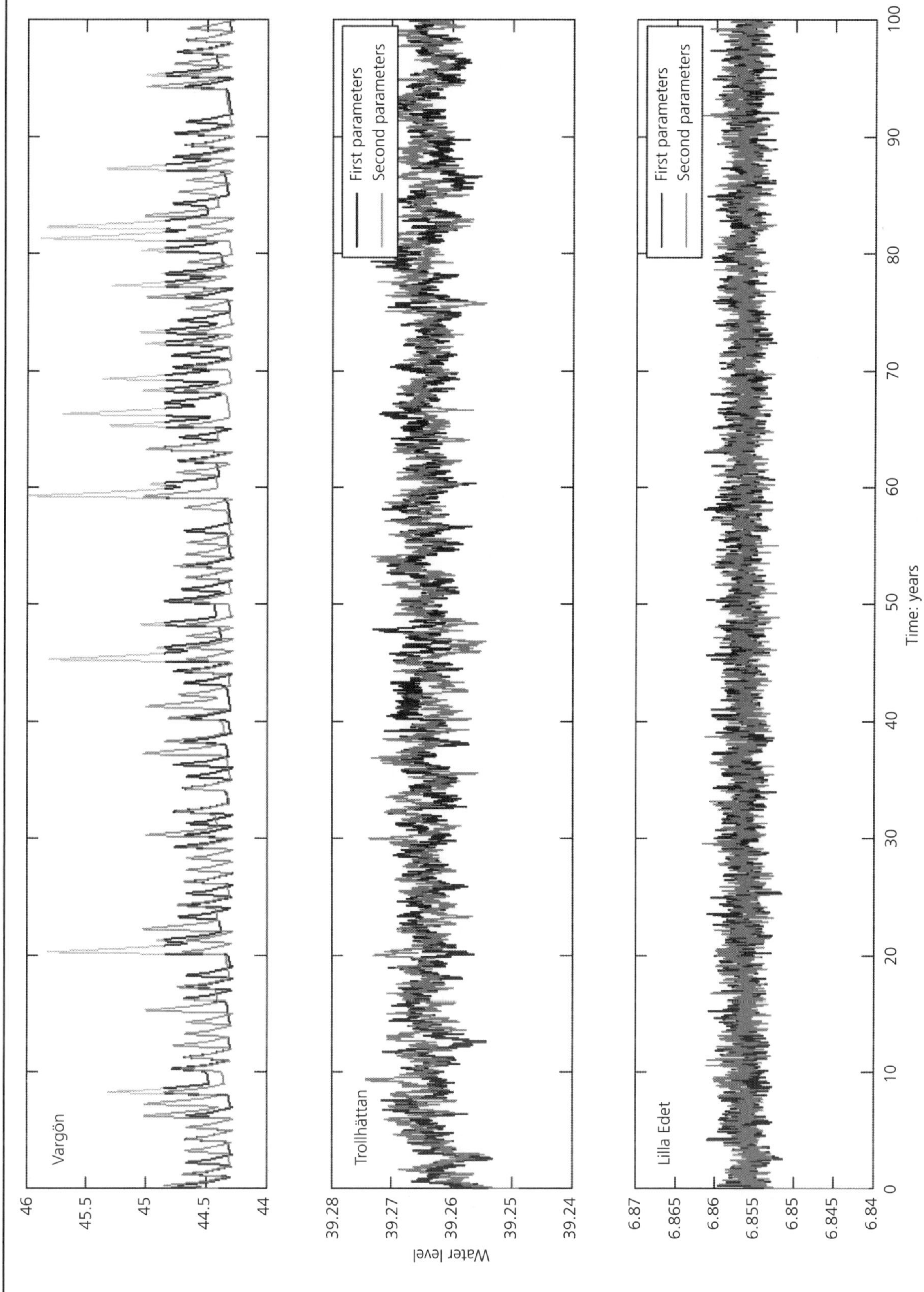

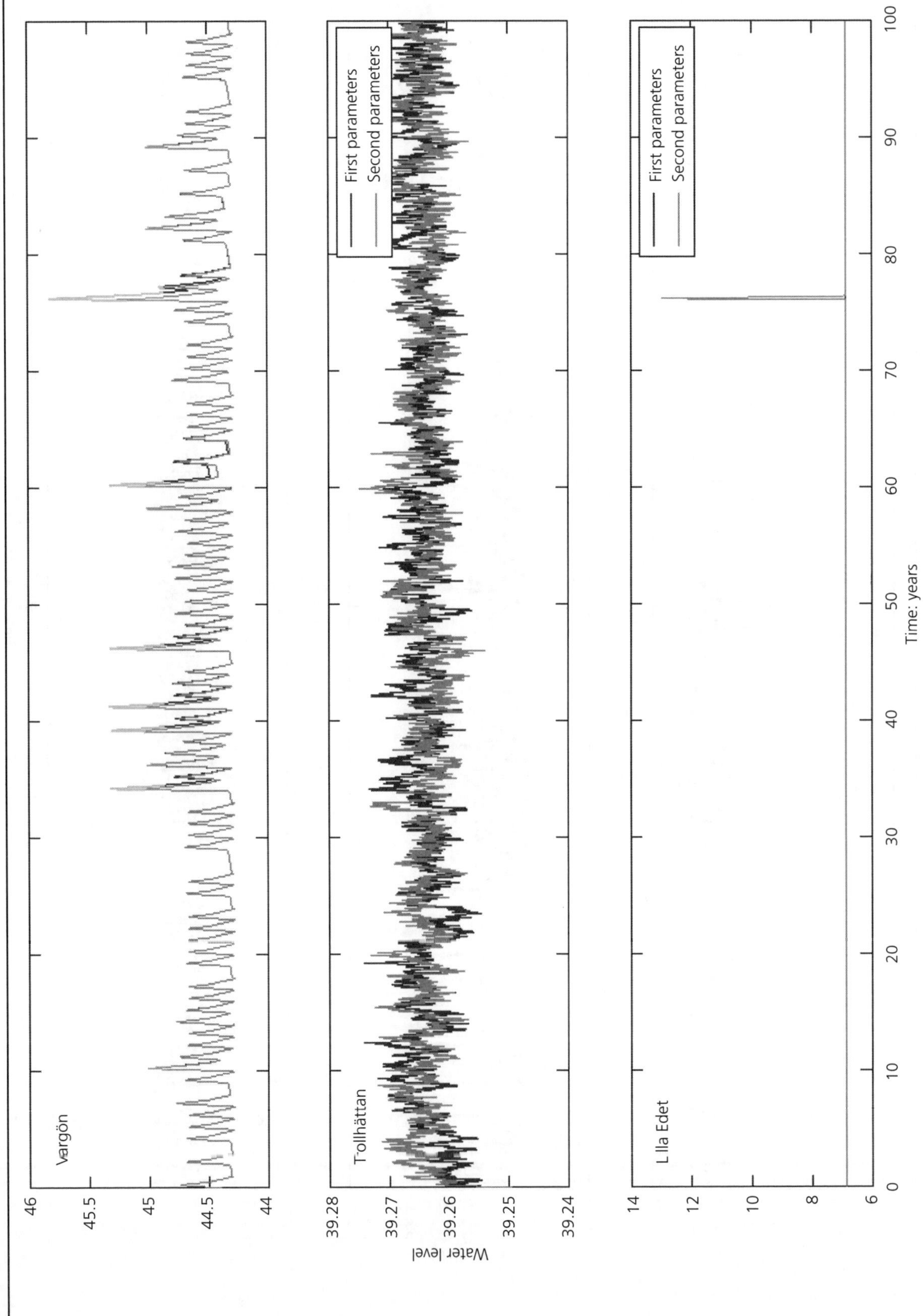


**Figure 10.42** Results for simulation Scenario B

This simulation tells us to move back to a more moderate increase in the discharge from the lake. Still, this is not the whole picture. The Göta River valley is one of the most landslide-prone areas in Sweden, and this risk cannot be ignored. So another simulation run must be carried out with the integrated simulation of landslides as a disturbance to the flow control in the river. The results of this simulation run are described below.

##### 10.5.4.3 Scenario C: Moderate increase of water release from Lake Vänern and the landslide disturbance model activated

Landslide risk cannot be ignored, and this scenario analyses the system with this disturbance activated. The results are shown in Figure 10.43. As expected, the effect of landslides, in combination to other extremes that could be managed if they occurred individually, is the overtopping of both downstream dams (see the overtopping peaks in the middle and bottom diagrams). Again, unlucky combinations of individually manageable loadings lead to adverse systems performance.

Moreover, it is interesting to note that in this simulation the probability of this system failure is $p = 0.01$ per year for Trollhättan and even more frequent for Lilla Edet. The frequency of this anticipated failure can, of course, not to be ignored by the dam owner, and more investigations must be made. This is a concern the owner might not be completely aware of, and even if they were the magnitudes are unclear. The simulation brings this information forward so that the necessary actions can be taken.

##### 10.5.4.4 More simulations!

Other hazards can be addressed in the similar manner using the dynamic system model. The previous examples show that river systems and their dam facilities must be addressed from a holistic perspective. Isolated analysis of single events gives the impression that it is possible to manage situations, which might not be the case in reality. Many hazardous states arise, some of which lead only to local incidents, while others interact in some unlucky combination and lead to failure of the whole system (i.e. mission and life loss). Event-tree methods, with one initiating event, lack the possibility to mimic systems behaviour in a realistic dynamic way with their constantly changing system states and functional demand. The Göta River model will evolve, and simulations are planned to cover shipping accidents as disturbances, power system blackouts, malfunctioning of Ormo flap gates and human error in operations.

## 10.6. Verification, validation and accreditation

Verification and validation are independent procedures that are used to check whether a modelling endeavour fulfils its objective. Each procedure is fundamental to the success of a simulation. Verification answers the question: 'Does the simulation correctly perform the calculations intended and correctly implement the desired algorithms?' Validation answers the question: 'Do the predictions made by the simulation confidently reflect what happens in reality, at least for the purpose at hand?' (Figure 10.44). A mathematical concept is implemented or programmed into discrete computational code. Whether this code faithfully generates a solution in consort with the mathematical concept relates to its veracity. The computational code is used to generate a forecast of how the physical system works in reality. The accuracy of this forecast relates to its validity. Verification does not imply validation and nor does validation imply verification.

Verification and validation in stochastic simulations present unique challenges (Oberkampf, 2010). To a large extent the aim of the verification and validation of simulations is to increase the confidence one has in a model to indicate qualitative modes of performance of a system rather than to demonstrate absolute quantitative accuracy in a model's outputs. A model, of course, is never 'correct', it is only a simplified approximation, and thus there is no such thing as absolute validity (Robinson, 1997). In a Karl Popper worldview, the aim of verification and validation is not so much to prove that a model is correct as to demonstrate that it cannot be proved incorrect.

Accreditation or certification of a model is an official determination that the model is fit for purpose for some domain of application. That is, accreditation is a decision or finding. For example, the US Army Corps of Engineers (USACE) has a formal process for certifying planning and engineering models (USACE, 2009), as do many other technical organisations (NASA, 2008). Nevertheless, a concern sometimes arises when organisations insist that only accredited models be used for certain studies but those models may be inappropriate compared with alternatives that have not yet been fully accredited. Accreditation processes are also sometimes criticised for placing too much emphasis on verisimilitude and too little on transparency and efficiency (Davis, 1992).

### 10.6.1 Verification

The aim of verification is to ensure that a model has been implemented correctly, that algorithms are correctly formulated, and that the model contains no errors or omissions. The goal of verification is not to ensure that the model solves a real problem in practice, meets a given set of requirements or is well calibrated against reality.

Verification involves reasonably common sense procedures of practice. Among these are (1) having the computational routines checked by someone other than the programmer, (2) creating a detailed logic diagram of the code, (3) examining intermediate

**Figure 10.43** Results for simulation Scenario C

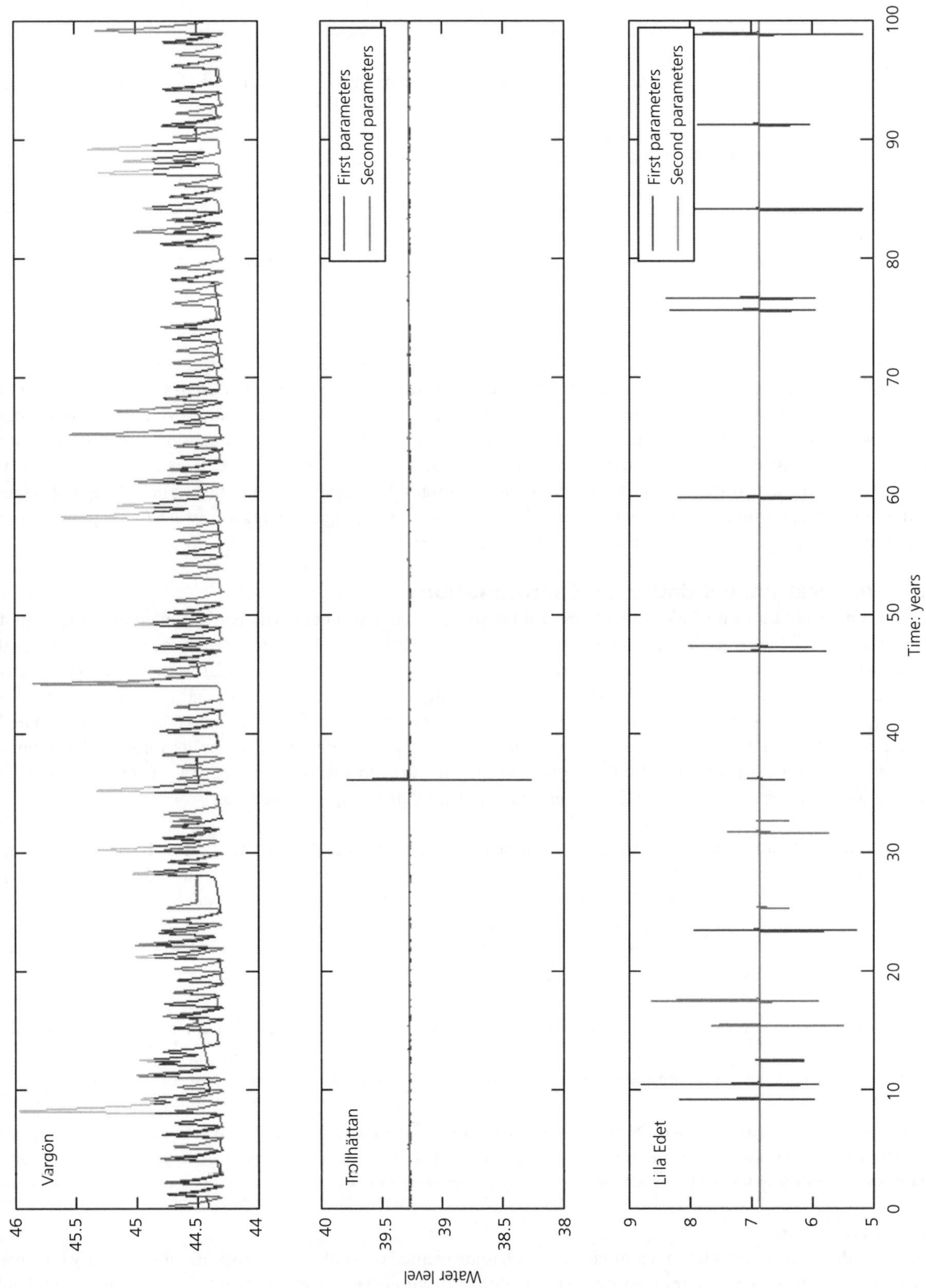

Figure 10.44 Concepts of verification and validation

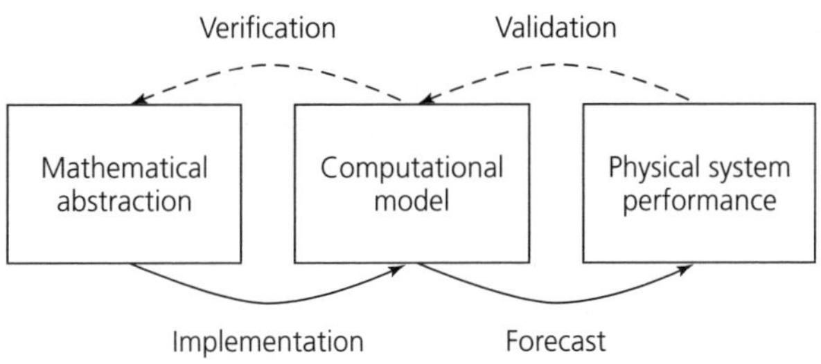


and final output for reasonableness under a sensitivity analysis, (4) creating detailed printouts or visualisations of inputs and outputs, and (5) fully documenting the model. None of these things is particularly surprising.

Typically, verification is divided into two parts: code verification and calculation verification (Thacker *et al.*, 2004). The aim of code verification is to confirm that the software implementation is working correctly. This means identifying programming or implementation errors within the software, and establishing the correctness of the numerical algorithms used in the code. This is sometimes accomplished by comparisons against well-defined benchmark cases. The aim of calculation verification is to establish the magnitude of the numerical error in code calculations. This means addressing inaccuracies caused by spatial or temporal discretisation, convergence tolerances and similar numerical aspects of the code. This is sometimes accomplished through code-to-code comparisons.

### 10.6.2 Validation

The aim of validation is to ensure that a model is well calibrated against reality. That is, that predictions accurately correspond to the behaviours of systems in the world being modelled. The approaches to validation are varied, and there is no formal protocol for establishing validity (Figure 10.45).

There are, furthermore, subtle distinctions between concepts of validation (Davis, 1992). Validity, in a descriptive sense, means that the model is capable of explaining how a system behaves and the reasons behind certain behaviours, even if it cannot accurately forecast quantitative behaviour. Validity, in a verisimilitude sense, means that the model's components, parameters

Figure 10.45 Taxonomy of verification, validation and accreditation of models. (Davis, 1992)

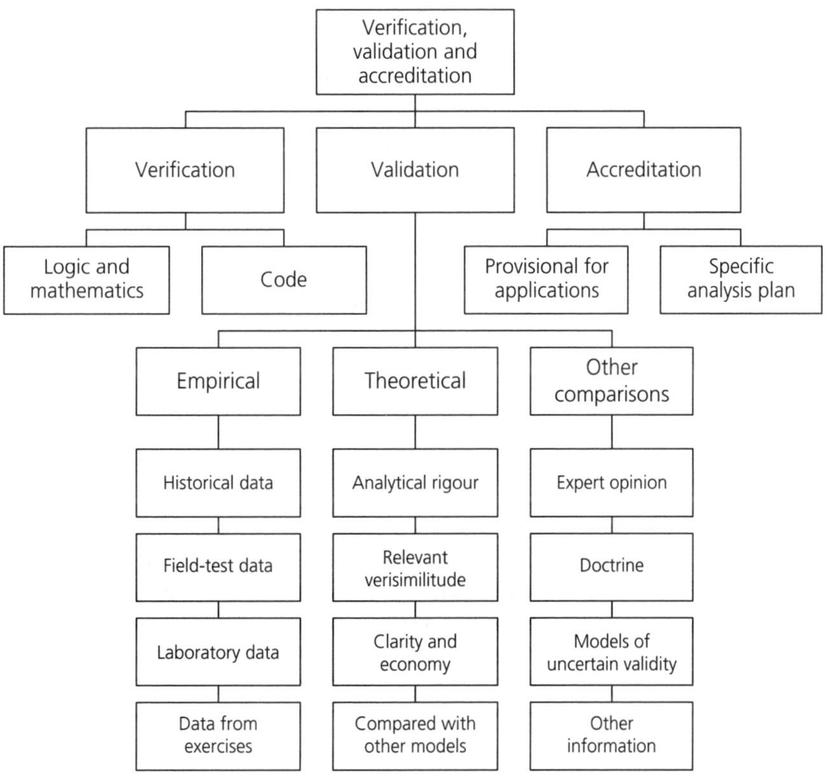

and interactions are parallel to those in the real world. Finally, validity, in a predictive sense, means that the model accurately and precisely forecasts observed behaviours.

The most straightforward approach to validating a simulation is to compare its forecasts against systems performance observed in the field. If the model accurately (and, to the extent possible, precisely) mimics real-system behaviour, then is it validated. The purpose of validation is to ensure that the model addresses the right problem, provides accurate forecasts and provides relevant information about the system.

On the other hand, several challenges affect attempts to validate models of complex systems. First, there may not be enough time or sufficient resources to validate a model fully. Second, real-world data may not be readily obtainable, available data may only be a sample of actual behaviour, and thus themselves subject to statistical variation, or the data are themselves biased for some unknown reason.

In many cases there may simply be no real-world system to compare predictions against. This is the case when a model is being used to forecast the behaviour of a system that is yet to be built. It is also may be the case, as with spillway systems, that rare behaviours are being forecast that have a low probability of occurring and for which, therefore, there may be little historical precedent. Often our ability to test predictions against real-system behaviour is severely limited. The real system may involve a hazard such as an extreme flood or large earthquake that has yet to occur, or may involve long periods of time that have yet to be observed.

Some ways of grappling with the unavailability of good test cases is to use:

- the review and judgement of subject-matter experts
- Turing tests, in which experts are asked to compare real-world and simulated performance data and choose which is which
- comparisons with other mathematical models
- comparisons with scaled physical models
- comparison with only partially parallel field cases.

In the limit, if validation at the whole-system level is not possible it may still be possible to validate individual components of the simulation against individual parts of the real-world system, although this may ignore important component interactions in the whole system.

### 10.6.3 Accreditation or certification

Accreditation or certification are procedures for ensuring that a model meets a set of explicit standards. The USACE (2009) criteria for certification provides an example.

> *Criteria for certification.* Technical soundness is the primary criterion on which certification/approval decisions will be based. Technical soundness reflects the ability of the model to represent or simulate the processes and/or functions it is intended to represent. The performance metrics for this criterion are related to theory and computational correctness. In terms of the theory, the certified should: 1) be based on validated and accepted 'state of the art' theory; 2) be consistent with … policies and requirements; 3) properly incorporate the conceptual theory into the software code; and, 4) clearly define the assumptions, sensitivities, uncertainties, and known limitations inherent in the model. In terms of computational correctness, the certified/approved should: 1) be free of computational errors and employ proper numerical/mathematical methods to estimate functions and processes represented; and, 2) adequately characterise, estimate and forecast the actual parameters it is intended to represent. A certified/approved will stand the test of technical soundness and theory, computational correctness and usability.

Examples of high-level criteria for accreditation are given in Box 10.2.

A verified, validated and accredited simulation model will ideally serve as an integrated part of an organisations' learning and decision chain, providing useful information and insights. A model with a clever design and flexible structure will be able to employ a great deal of diverse operational data (big data) and convert it into useful information, leading to improved knowledge and shifting the organisation towards data-driven and wiser risk management.

Box 10.2 Examples of accreditation criteria (US Army, 1999)

Criteria guidelines

1. The levels of [components] and interaction[s] have sufficient fidelity and resolution.
2. The [model] is suitable for the overall intended use (e.g. training, explanatory, predictive).
3. The [model] output/results may be used clearly, adequately and appropriately to address the problem.
4. The configuration management policy is in effect and responsive to the anticipated needs of the [model] users.
5. All required data values are well defined and data sources for obtaining accredited data have been identified.
6. The [model] runs may be accomplished and results analysed within the project timelines.
7. There is availability of baseline scenarios, ... and performance data for the [model].
8. The algorithms ... and environment representations are functionally adequate.
9. The clarity, fidelity, complexity and level of detail of the simulated entities are acceptable for its intended usage.
10. The documentation, user training, and user help are adequate.
11. The [model] stability has been investigated and found acceptable for the hardware and software platforms which will be utilised.
12. [The model] demonstrates appropriate sensitivity to data perturbations and response at boundary (limiting value) cases.

## REFERENCES


BBC (2006) Buncefield tank 'was overflowing'. *BBC News*, 9 May.

Billinton R and Li W (1994) *Reliability Assessment of Electrical Power Systems Using Monte Carlo Methods*. Springer, Berlin, Germany.

Blom HAP, Stroeve SH and de Jong HH (2006) *Safety Risk Assessment by Monte Carlo Simulation of Complex Safety Critical Operations*. Nationaal Lucht- en Ruimtevaartlaboratorium National Aerospace Laboratory NLR, Amsterdam, the Netherlands.

Blum DM, Thipphavong D, Rentas TL, He Y, Wang X and Pate-Cornell ME (2010) Safety analysis of the advanced airspace concept using Monte Carlo simulation. *American Institute of Aeronautics and Astronautics Guidance, Navigation, and Control Conference*, Toronto, Canada, pp. 2–5.

Chandler F, Chang J and Mosleh A (2006) *Human Reliability Analysis Methods Selection Guidance for NASA*. OSMA Technical Report. NASA, Washington, DC, USA.

Davis PK (1992) *Generalizing Concepts and Methods of Verification, Validation, and Accreditation (VV&A) for Military Simulations*. Rand, Santa Monica, CA, USA.

De Weck OL, Roos D and Magee CL (2012) *Engineering Systems: Meeting Human Needs in a Complex Technological World*. MIT Press, Cambridge, MA, USA.

Ebeling RM, Fong MT, Wibowo JL and Chase Sr A (2012) *Fragility Analysis of a Concrete Gravity Dam Embedded in Rock and its System Response Curve Computed by the Analytical Program GDLAD_Foundation*. Engineer Research and Development Center, US Army Corps of Engineers, Vicksburg, MS, USA.

Forrester JW (1988) *Principles of Systems*. Productivity Press, New York, NY, USA.

Gersonius B, Morselt T, van Nieuwenhuijzen L, Ashley R and Zevenbergen C (2012) How the failure to account for flexibility in the economic analysis of flood risk and coastal management strategies can result in maladaptive decisions. *Journal of Waterway, Port, Coastal, and Ocean Engineering* **138(5)**: 386–393.

Gertman D, Blackman HS, Marble J, Byers J, Haney LN and Smith C (2005) *The SPAR-H Human Reliability Analysis Method*. US Nuclear Regulatory Commission, Washington, DC, USA.

Glasserman P (2010) *Monte Carlo Methods in Financial Engineering*. Springer, Berlin, Germany.

Hendron AJ, Ehasz JL and Paul K (2005) *Taum Sauk Upper Dam Breach*. Federal Energy Regulatory Commission, Washington, DC, USA.

Hollnagel E (1998) *Cognitive Reliability and Error Analysis Method: CREAM*. Elsevier, Oxford, UK.

Hosse RS and Schnieder E (2012) *Analyzing German Vehicle Traffic using STAMP/STPA*. MIT Press, Cambridge, MA, USA.

INCOSE (International Council on Systems Engineering) (2015) *INCOSE Systems Engineering Handbook: A Guide for System Life Cycle Processes and Activities*. Wiley, Hoboken, NJ, USA.

Kirwan B (1994) *A Guide to Practical Human Reliability Assessment*. CRC Press, Boca Raton, FL, USA.

Komey A (2015) *A Systems Reliability Approach to Flow Control in Dam Safety Risk Analysis*. MS thesis, University of Maryland, College Park, MD, USA.

Leveson N (1995) *SafeWare: System Safety and Computers*. Addison-Wesley, Reading, MA, USA.

Leveson NG (2012) *Engineering a Safer World: Systems Thinking Applied to Safety*. MIT Press, Cambridge, MA, USA.

NASA (National Aeronautics and Space Administration) (2008) *Standards for Models and Simulations*. NASA, Washington, DC, USA.

NRC (National Research Council) (2002) *Modeling and Simulation in Manufacturing and Defense Acquisition: Pathways to Success*. National Academies Press, Washington, DC, USA.

NRC (2008) *Behavioral Modeling and Simulation: From Individuals to Societies*. National Academies Press, Washington, DC, USA.

NRC (2010) *Simulation Methods for Reliability and Availability of Complex Systems*. National Academies Press, Washington, DC, USA.

NSF (National Science Foundation) (2006) *Simulation-Based Engineering Science*. NSF, Arlington, VA, USA.

Oberkampf WL (2010) *Verification and Validation in Scientific Computing*. Cambridge University Press, New York, NY, USA.

OPG (Ontario Power Generation) (2009) *Lower Mattagami River Hydroelectric*. OPG, Toronto, Canada.

Reason JT (1990) *Human Error*. Cambridge University Press, Cambridge, UK.

Regan P (2010) *Dams as Systems – A Holistic Approach to Dam Safety*. US Society on Dams, Sacramento, CA, USA.

Robinson S (1997) Simulation model verification and validation: increasing the users' confidence. *Proceedings of the 1997 Winter Simulation Conference* (Andradóttir S, Healy KJ, Withers DH and Nelson BL (eds)), Atlanta, GA, USA, pp. 53–59.

SGI (Swedish Geotechnical Institute) (2012) *Landslide Risks in the Göta River Valley in a Changing Climate*. SGI, Lincöping, Sweden.

Simonovic SP (2008) *Managing Water Resources: Methods and Tools for a Systems Approach*. Routledge, New York, NY, USA.

Sterman J (2000) *Business Dynamics: Systems Thinking and Modeling for a Complex World*. McGraw-Hill Education, New York, NY, USA.

Swain AD and Guttmann HE (1983) *Handbook of Human Reliability with Emphasis on Nuclear Power Plant Applications*. US Nuclear Regulatory Commission, Washington, DC, USA.

Thacker BH, Doebling SW, Hemez FM, Anderson MC, Pepin JE and Rodriguez EA (2004) *Concepts of Model Verification and Validation*. Los Alamos National Laboratory, Albuquerque, NM, USA.

Tilmant A, Arjoon D and Marques G (2014) Economic value of storage in multireservoir systems. *Journal of Water Resources Planning and Management* **140(3)**: 375–383.

USACE (US Army Corps of Engineers) (2002) *Ice Engineering Manual*. USACE, Hanover NH, USA.

USACE (2008) *The Shared Vision Planning Primer: How to Incorporate Computer Aided Dispute Resolution in Water Resources Planning*. USACE, Alexandria, VA, USA.

USACE (2009) *Assuring Quality of Planning Models*. USACE, Washington, DC, USA.

US Army (1999) *Verification, Validation, and Accreditation of Army Models and Simulations*. HQ Department of the Army, Washington, DC, USA.

von Bertalanffy L (1969) *General System Theory: Foundations, Development, Applications*. George Braziller, New York, NY, USA.

von Stackelberg N and Neilson B (2014) Collaborative approach to calibration of a riverine water quality model. *Journal of Water Resources Planning and Management* **140(3)**: 393–405.

Zio E (2013) *The Monte Carlo Simulation Method for System Reliability and Risk Analysis*. Springer, Berlin, Germany.

**Operational Safety of Dams and Reservoirs**
ISBN 978-0-7277-6121-7


http://dx.doi.org/10.1680/osdr.61217.357

# Chapter 11
# Postscript

This book was prompted in the mid-2000s by the recognition that the focus on the operational reliability of spillway gates that emerged in the early 2000s, which focused on the demand for gate availability, did not address the broader question of the reliability of the spillway discharge function. Furthermore, it was recognised that this focus on gate availability omitted consideration of the vitally important matter of characterising the full spectrum of demands on spillway gate function, as general practice considered only the demand for spillway function in terms of the need to pass floods. At the same time, BC Hydro and Vattenfall had initiated work on a systems engineering approach to spillway-gate reliability, and it appeared that systems engineering thinking would be beneficial in addressing the matter of discharge function reliability. Precisely how this would be achieved was unclear initially, but clarity developed as the work progressed. Throughout, the questions 'What is the probability of demand for spillway function?' and 'What is the probability of successful functioning?' remained the focal points for the work.

## 11.1. Development of risk analysis in dam safety

Prior to the early 1990s, quantitative probability and risk analysis were unrealised concepts in dam safety practice. However, the calamitous failure of Teton Dam, Idaho, in 1976, followed by a series of less spectacular but more deadly dam failures in the later 1970s, led to national interest in the USA for improved dam safety policies, and the potential to use risk analysis as an instrument for improvement became of interest. The US Bureau of Reclamation (USBR), having been party to the Teton Dam failure, took up the charge. By the mid-1980s, the USBR had pioneered new thinking about quantitative risk analysis for dam safety. This was reflected in a series of published works (USBR, 1986; von Thun, 1985). The 1986 USBR report *Guidelines to Decision Analysis* was seminal among these.

*Guidelines to Decision Analysis* laid out a comprehensive, quantitative risk approach to dam safety based on contemporary benefit–cost thinking (USBR, 1986). The report built on and fundamentally extended the thinking reflected in the National Research Council reports of 1983 and 1985 (NRC, 1983, 1985). The NRC reports had focused on extreme flood and earthquake loading criteria; that is, on probable maximum floods (PMF) and maximum credible earthquakes (MCE). These safety criteria for natural hazards were essentially deterministic. The USBR's approach replaced these deterministic loading criteria with probabilistically specified hazards; that is, with annual exceedance probabilities of levels of inflow or water levels, and of magnitudes of seismic events. This was a fundamental break with traditional practice, but it would be a decade before this changed perspective was to enter practice. While the new probabilistic thinking appeared in the 1986 USBR report, there was little progress until the early 1990s, when BC Hydro, with the support of the USBR, took the lead in implementing the proposed USBR process in a way that specified 'a consistent quantifiable method for evaluating the safety of all dams on the basis of tolerable risk' (BC Hydro, 1993). In the USA, the powerful influences of both the US Army Corps of Engineers (USACE) and the Federal Energy Regulatory Commission (FERC) had not embraced probabilistic thinking, and dam safety decision-making involving tolerable risk. However, in parallel, interest in the implementation of a risk approach was evolving in Australia, culminating in the publication of *Guidelines on Risk Assessment* by the Australian National Committee on Large Dams (ANCOLD, 1994).

The USBR's *Guidelines to Decision Analysis* (USBR, 1986) was progressive for the time as it adopted probabilistic criteria for the extreme loads of hydrology and earthquakes. Nonetheless, adverse conditions during normal operations, for example, due to slope instability or internal erosion, were not tackled with the same vigour. The Guidelines note only that the 'loading condition for internal forces is a function of the reservoir level and rate of change … and of gravity loadings'. They go no further to explain how these failures under normal conditions might be appraised.

ANCOLD defined risk assessment as 'The process of deciding whether existing risks are tolerable and present risk control measures are adequate and if not, whether alternative risk control measures are justified or will be implemented'. Australian regulatory practice was also strongly influenced by that country's nuclear safety initiative, and specifically by the work of D.J. Higson (Higson, 1978, 1985, 1990). These early risk-based guidelines and the following guidance of 2003 were

Figure 11.1 Number of papers listed in the ASCE library having the keyword 'risk'

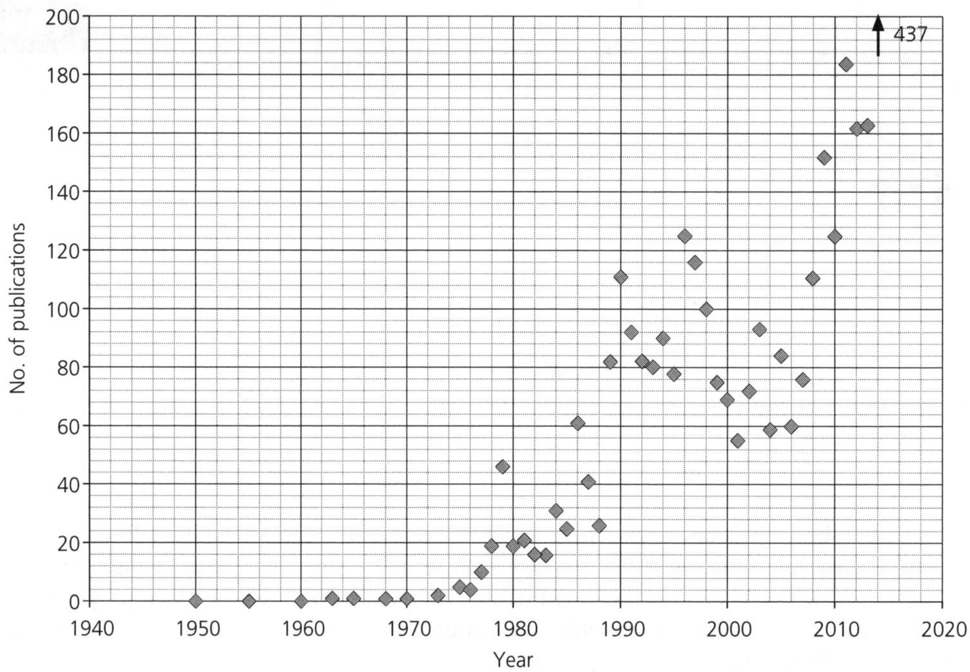


instrumental in promoting the practical use of quantitative risk criteria in dam safety around the world (ANCOLD, 2003).

The Dam Safety Interest Group, affiliated with what was then the Canadian Electricity Association, undertook at this time a comprehensive study of quantitative risk methodology. This resulted in the publication by Thomas Telford of the book, *Risk and Uncertainty in Dam Safety* (Hartford and Baecher 2004), which captured and extended the aforementioned simultaneous developments in the USA, Australia and elsewhere during the 1990s.

Interest in quantitative risk analysis across the civil engineering industry was simultaneously expanding during this period. Figure 11.1 shows the number of papers published in journals of the American Society of Civil Engineers (ASCE) from 1950 to the present that have 'risk' as a keyword. In the 1970s and early 1980s, interest in risk analysis was confined principally to academia. Wider interest in the topic only began to manifest in the late 1980s and into the 1990s.

The tragedy of Hurricane Katrina can be argued to have been the triggering event in bringing USACE to the practice of quantitative risk analysis. Following Hurricane Katrina, USACE undertook a determined and ultimately successful transformation to risk-informed methodologies. By this time, the groundwork that had been laid by the USBR was becoming ever more mature, and USACE built on this groundwork to develop its own approach. This approach is outlined in the publication *Safety of Dams – Policies and Procedures* (USACE, 2014).

## 11.2. Effectiveness of current risk-analysis approaches to dam safety

Current risk-analysis approaches to dam safety have been effective in transitioning the enterprise from its deterministic underpinnings as late as the 1980s when the National Research Council reports were drafted. While, to some extent, these new risk-analysis approaches primarily provided a probabilistic veneer to the earlier deterministic thinking, they went a long way to incorporating probabilistic descriptions of hazard and reliability-based thinking about structural and geotechnical fragility. This was an important progression and has served the industry well. These new approaches led to a different and more subtle appreciation of the effect of uncertainty on dam safety, and it can be argued that they led to a much richer understanding of the potential failure modes of dams. Indeed, the potential failure modes analysis (PFMA) process pioneered by the FERC, USBR and USACE is a major achievement of this body of work (USBR and USACE, 2012).

A parallel development over the same period of time, and strongly aligned with the movement toward quantitative risk analysis, was an increased focus on loss of life as the driving criterion for dam safety. Until that period of history commencing with the failure of Teton Dam in 1976 and ending with the catastrophe of Hurricane Katrina in 2005, 'the potential loss of life in dam failures was considered unacceptable and to be avoided regardless of the cost'. It was reflected by the use of physically limiting events (PMF and MCE) as standards that were used almost everywhere in North America and elsewhere when lives were at stake. This has changed profoundly, and while the change is not necessarily due to risk-based thinking, nonetheless the two have evolved simultaneously.

A further parallel development over the same period of time, and one that is clearly associated with risk-analysis thinking, was the increasing emphasis on quantifying expert engineering judgement. This goes under several names, the most common of which is 'expert elicitation'. In an earlier time it was more often called 'subjective probability assessment', but the concept is the same (Vick, 2002). The decision-analysis community had used subjective probability assessment for at least two or three decades prior to the advent of modern risk analysis for dam safety (Raiffa, 1968), but it was only with the advent of risk analysis for dam safety that these technologies started to be accepted in civil engineering practice. The past 20 years have seen extensive and intensive use of quantified expert opinion, the use of which has now extended widely beyond dam safety applications.

Subjective probabilities and their application to dam safety risk are principally directed toward epistemic uncertainties. Today, we distinguish between *epistemic* uncertainties, which have to do with the state of knowledge, and *aleatory* uncertainties, which have to do with natural variations in time or space. Epistemic uncertainties are properties of the mind. They have to do with how much we know. If we invest time and effort and learn more about a situation, our corresponding epistemic uncertainties can be reduced. Aleatory uncertainties, on the other hand, are properties of nature. They are what they are, and more effort invested in knowing them can make their assessment more precise, but they can never be reduced. The advent of risk analysis in dam safety drove home the distinction between these two types of uncertainty, which had not been well understood previously (Hartford and Baecher, 2004).

## 11.3. The next generation of risk analysis for dam safety

Despite the successes of the risk-analysis techniques introduced in the 1990s and early 2000s, these methodologies continued the earlier focus on a small number of extreme loadings as design events. Today, as evidenced in *Safety of Dams – Policies and Procedures* (USACE, 2014) and in the later Federal Emergency Management Agency (FEMA) report on (US) *Federal Guidelines for Dam Safety Risk Management* (FEMA, 2015), the principal failure modes addressed in these risk analyses are of only three types: hydrological overtopping due to reservoir inflows exceeding available discharge capacity, structural or geotechnical instability due to extreme seismic ground shaking, and normal operating failures due to internal erosion or piping. While, in principle, other failure modes such as spillway-gate unavailability might be included within these critical failure modes, in practice they seldom are.

We know today, by studying past dam accidents and failures, that the chains of events leading to accidents and failures are seldom so simple. There is almost never a simple root cause of an event. If we study accidents and failures of dams we find that they are usually the result of systems interactions. An accident or failure may occur following a modest hydrological event, not because that event was so extreme in itself, but because an unfortunate chain of other events also occurred, and when taken in combination they led to a failure. That is, an uncommon combination of common events when occurring in a certain way may have bad outcomes. Examples of some of these were described in Chapter 1. They are much more common than accidents or failures due to extreme hazards.

It is usually the case that these unfortunate chains of events involve operational and other considerations that would not normally be accommodated in traditional dam safety investigations or contemporary probabilistic risk analysis. This is not a criticism of traditional dam safety or contemporary risk analysis, merely an observation that these systems-engineering type failures involve considerations that we have not heretofore included in our models and analyses. This book attempts to show that we have the capability of treating and analysing the systems-engineering aspects of failures. We can do this by incorporating operational aspects within a dynamic simulation. This is not entirely different from attempts to develop dynamic fault-tree models for application in nuclear safety, or process-based safety analysis in the aerospace industry. The needs and constraints of dam safety risk analysis differ somewhat from these other industries, but the concepts presented here are very similar. These systems-engineering focused approaches are not replacements for our current generation of dam safety risk-analysis tools but are extensions to them.

## 11.4. The goal of this book

The purpose of this book was threefold. First, it was intended to point out that most accidents and failures of dams occur not because of simple extreme loadings but because of the systems interactions of hazards, disturbances, mechanical and electrical systems, and human operators. It is the interactions between these many things, influenced by organisational practice, that cause accidents and failures. This is not unique to dam safety, but applies across the spectrum of technogenic risks.

Second, when one approaches dam safety from a systems-engineering view, many individual technological considerations must be brought to bear. The breadth of these considerations is much greater than in contemporary dam safety risk models. The considerations involve the dynamic time-varying aspects of hydrology, weather and other factors.

They include mechanical and electrical equipment, and instrumentation and SCADA systems, as well as disturbances arising external to the system. In addition, and importantly, they include the performance and behaviour of human operators.

Third, advances in computation, numerical modelling, statistics and other technologies have allowed us to develop sophisticated simulation models with which to combine these many considerations. These models allow us to incorporate dynamic aspects of loads, fragilities and operations.

As will be clear to the reader, we are in the early stages of learning how to develop and apply system simulations of dam operations. The present book suggests how these problems can be approached, but it is too early yet to provide packaged tools. These will be developed over the coming years, as will our understanding of how to apply the results of systems simulations to practical problems. As noted in the Preface, the 'new approach' outlined here may not strike the journeyman engineer as immediately practical, just as computational fluid dynamics or critical-path scheduling did not strike the previous generation of engineers as immediately practical. But both those methods are today valued and routinely applied in practice, and we anticipate that the 'new approach' laid out here, and the extensions by others that grow out of it, will enjoy a similar fate.

## 11.5. Summary

In the USBR's 1986 *Guidelines to Decision Analysis* it was noted:

> During normal operation, loading conditions may generate a failure mechanism by acting on structural defects inherent to the dam. Examples of failure mechanisms that may occur during normal operation of embankment dams include slope instability and internal erosion due to piping. The loading condition for internal forces is a function of the reservoir level and/or rate of change (e.g. high level steady-state, rapid drawdown) and gravity loadings.

Thirty years ago, and subsequently, the focus has been on risk associated with structural defects and effects of extreme natural hazards. In the modern context, dam safety has a broader focus, and knowledge, science and technology have advanced to the stage that a means of addressing risk due to operational factors can now be incorporated in the endeavour of dam safety management.

REFERENCES

ANCOLD (Australian National Committee on Large Dams) (1994) *Guidelines on Risk Assessment.* ANCOLD, Sydney, Australia.

ANCOLD (2003) *Guidelines on Risk Assessment.* ANCOLD, Sydney, Australia.

BC Hydro (1993) *Guidelines for Consequence-Based Dam Safety Evaluations and Improvements.* H2528 Interim. British Columbia Hydro and Power Authority, Vancouver, Canada.

FEMA (Federal Emergency Management Agency) (2015) *Federal Guidelines for Dam Safety Risk Management.* FEMA, Washington, DC, USA.

Hartford DND and Baecher GB (2004) *Risk and Uncertainty in Dam Safety.* Thomas Telford. London, UK.

Higson DJ (1978) The development of safety criteria for use in the nuclear industry. In *6th National Conference on Chemical Engineering – CHEMCA 78*, Surfer's Paradise, Queensland, Australia.

Higson DJ (1985) Nuclear reactor safety goals and assessment principles. *Nuclear Safety* **26(1)**: 1–13.

Higson DJ (1990) Nuclear safety assessment criteria. *Nuclear Safety* **31(2)**: 173–186.

NRC (National Research Council) (1983) *Safety of Existing Dams: Evaluation and Improvement.* National Academies Press, Washington, DC, USA.

NRC (1985) *Safety of Dams: Flood and Earthquake Criteria.* National Academies Press, Washington, DC, USA.

Raiffa H (1968) *Decision Analysis: Introductory Lectures on Choices under Uncertainty.* Addison-Wesley, Reading, MA, USA.

USACE (US Army Corps of Engineers) (2014) *Safety of Dams – Policies and Procedures.* ER 1110-2-1156. USACE, Washington, DC, USA.

USBR (US Bureau of Reclamation) (1986) *Guidelines to Decision Analysis.* ACER Technical Memorandum No 7. Guidelines. USBR, Denver, CO, USA.

USBR and USACE (2012) *Best Practices in Dam and Levee Safety Risk Analysis.* USBR/USACE, Denver, CO/Washington, DC, USA.

Vick SG (2002) *Degrees of Belief: Subjective Probability and Engineering Judgment.* American Society of Civil Engineers, Reston, VA, USA.

von Thun L (1985) *Application of Statistical Data from Dam Failures and Accidents to Risk-based Decision Making on Existing Dams.* Engineering and Research Center, US Bureau of Reclamation, Denver, CO, USA.

**Operational Safety of Dams and Reservoirs**
ISBN 978-0-7277-6121-7


http://dx.doi.org/10.1680/osdr.61217.361

# Appendix
# Weibull data for gate components

Flow-control in hydropower and related dam systems is a complex system involving civil, mechanical, electrical, communications and control components. These systems components interact in complex ways, and failures of flow-control systems involve not only reliability failings of physical components but also the interactions of these components with communications and control, and with human actions and errors. An analysis of the performance of the system needs to incorporate all these aspects.

Commonly, fault tree analysis (FTA) for mechanical and electrical gate components uses Weibull distributions to model the reliability or availability of interim or top events (Patev *et al.*, 2013). FTA commonly uses a two parameter Weibull distribution on time to failure, $t$,

$$f(t \mid \beta, \alpha) = \frac{\beta}{\alpha}\left(\frac{t}{\alpha}\right)^{\beta-1} e^{-(t/\alpha)^{\beta}} \quad \text{for } t \geq 0 \tag{1}$$

and zero elsewhere. The parameter $\beta$ is a shape factor or the slope of the Weibull pdf, which determines which member of the family of Weibull failure distributions best fits or describes the data. The parameter $\alpha$ is a scale factor or the location of the pdf which determines the central tendency of the distribution and the mean time between failures.

The slope or shape parameter, $\beta$, indicates which class of failures is present:

$\beta < 1.0$ implies infant mortality,

$\beta = 1.0$ implies random failures independent of age, and

$\beta > 1.0$ implies wear out or old-age failures.

The characteristic life or alpha, $\alpha$, of a component is related to the Mean-Time-To-Failure (MTTF) and the failure rate, $\lambda$. This relationship is,

$$\text{MTTF} = (1/\lambda) \equiv \alpha\Gamma(1 + 1/\beta) \tag{2}$$

Note that the relationship between the characteristic life $\alpha$ and MTTF depends on $\beta$. For $\beta = 1$, MTTF $= \alpha$. That is, for random failures independent of age, the failure rate, $\lambda = 1/\alpha$; thus for this case, $\alpha$ might be thought of as a 'return period'. The characteristic life, $\alpha$, relationship is dependent on the value of $\beta$ as follows:

$\beta < 1$, MTTF $< \alpha$

$\beta = 1$, MTTF $= \alpha$

$\beta > 1$, MTF $< \alpha$

$\beta < 1$, MTTF $> \alpha$

$\beta = 0.5$, MTTF $= 2\alpha$

Typically, characteristic life $\alpha$ is based on assumptions such as the components having similar maintenance practices, no replacement of smaller internal parts, consistent or protected environmental and operating conditions and that all components are composed of materials that were properly selected and designed.

For $\beta = 1$, the characteristic life is that time for which 63.2% of the units under test have failed, sometimes called the B63.2 life. The cumulative distribution function (CDF) of Eq. (1) is,

$$F(t \mid \beta, \alpha) = 1 - \exp\{-(t/\alpha)^{\beta}\} \tag{3}$$

which equals 0.63.2 at $t = \alpha$.

Patev *et al.* (2013) reports that,

> The current USACE mechanical and electrical (ME) reliability methods use generic component failure rate data from US Department of Defense Military Standard 756B (DoD MIL-STD-756B) documents. These failure rate data are processed for components that function in different operating environment, different failure modes, and different maintenance practices than at USACE navigation projects. Therefore, the reliability of the ME system from this data set yields conservative results and very often overestimates the time-dependent reliability of the entire ME system.

In response, USACE has undertaken work to develop improved reliability models for the ME gate components at inland navigation and flood risk management facilities. Table A shows a list of typical causes for failure events in the ME fault trees for spillway gates (Patev *et al.*, 2005).

As noted in Section 8.2.3, failure data were collected by Schultz (2013) for mechanical and electrical components at 295 USACE flood control projects. These data have been processed for use in FTA for spillway projects. An estimate of the shape parameter, $\beta$, and characteristic life (in years), $\alpha$, using both traditional Weibull plotting methods and Bayesian inference methods are shown in Tables B and C. These are representative estimates of the rates of gate component failures for both mechanical and electrical equipment (Mosleh and Zhu, 2013). For a description of the procedure for estimating the parameters of the Weibull distribution using the Weibull Plotting method see Kumamoto and Henley (1996), and for Bayesian inference see Dodson (2006).

**Table A** Causes of failure events for fault trees of gated spillways (Patev *et al.*, 2005)

| Failure event | Cause |
|---|---|
| Concrete deterioration | ■ Corrosion of R-bar<br>■ Cracks from overload<br>■ Salt (chloride) or<br>■ Sulphate reaction |
| Inadequate spillway capacity | ■ Changes in PMF<br>■ Changes in run-off as a result of increased development in basin<br>■ Political or public opposition to raising dam height or modifying spillway<br>■ Lack funds to make necessary changes 0 Poor maintenance of equipment<br>■ Silt/debris build-up an front of gates and on the bottom of the pod!<br>■ Gate entrance blacked from ice dams or debris<br>■ Failure of upstream barrier |
| Broken linkages | ■ Failure of gear box<br>■ Failure of cables/chains<br>■ Failure of chains and cable attachment to gate<br>■ Failure of driveline couplings (came apart) |
| Foundation instability | ■ Terrorism<br>■ Overload from overtopping |
| Logs | ■ Logs that hang up on top of gates and prevent opening<br>■ Logs that could prevent gate closure<br>■ Logs that could hang up on gates and cause hydraulic unbalance, vibration, and failure of gate |
| Ice | ■ Ice that causes gates to be frozen closed<br>■ Ice from leakage and wave splash that builds up on gate and prevents gate opening as a result of excessive hoist load<br>■ Ice that freezes gates in the open position<br>■ Ice build-up on walkways, gate operating equipment that prevents access to and operation of controls |

**Table A** Continued

| Failure event | Cause |
|---|---|
| Failure of trunnion arm | ■ Debris falling on arm<br>■ Failure due to corrosion<br>■ Icing<br>■ High trunnion friction<br>■ Overtopping |
| Failure of gate leaf/skin | ■ Corrosion<br>■ Physical damage from debris<br>■ Brittle fracture |
| Failure of metal anchor | ■ Corrosion<br>■ Fatigue<br>■ Torsional loading<br>■ Overloading |
| Failure of concrete piers | ■ Concrete deterioration |
| Failure of steel support | ■ Corrosion<br>■ Overloading<br>■ Fatigue<br>■ Fasteners |
| Concrete support | ■ Concrete deterioration |
| Drive shaft coupling | ■ Decoupling<br>■ Keyway failure (key and keyway)<br>■ Torsional failure |
| Gear motor | ■ Failure of gears<br>■ Shaft failure<br>■ Bearing |
| Gear reducers | ■ Bearing failure<br>■ Gear failure (due to lack of lubrication, fatigue)<br>■ Shaft failure<br>■ Housing failure<br>■ Sprocket failure |
| Chain linkage failure | ■ Loss of snap rings<br>■ Corrosion |
| Gate connection | ■ Connection itself<br>■ Corrosion |
| Cable end socket | ■ Corrosion |
| Hoisting brake | ■ Keyway failure<br>■ Shoe failure<br>■ Solenoid failure<br>■ Torsional failure of spring |
| Transmission lines | ■ Ice<br>■ Lightning<br>■ Tornadoes |
| Utility substation equipment/transformers | ■ Short circuit<br>■ Lightening |
| Control circuit failure | ■ Relays<br>■ Control buttons |
| Motor failure | ■ Insulation failure<br>■ Bearing failure<br>■ Windings<br>■ Control buttons |

**Table A** Continued

| Failure event | Cause |
|---|---|
| Inadequate seals | ■ Weathering failure<br>■ Abrasion due to debris |
| Differential monolith settlement | ■ Foundation settlement and rebound earthquake<br>■ Seepage and piping<br>■ Excessive friction |
| Trash and Debris | ■ Trees, logs |
| Galvanic corrosion | |
| Cables/chains | ■ Corrosion, mineral deposits<br>■ Seizing of chains |
| Trunnion anchorage | ■ Corrosion<br>■ Brittle fracture |
| Trunnion bearings/pin | ■ Inadequate lubrication |
| Trunnion arms | ■ Paint failure |
| Inadequate design | ■ All load conditions not assessed |
| Operational procedure | ■ Human error<br>■ Inadequate hydraulic calculation capabilities<br>■ Inadequate surveillance<br>■ Motor on overload and operator fails to recognise it<br>■ Annual gate operations test |
| Excessive friction | ■ Bearing failure<br>■ Inadequate lubrication<br>■ Misalignment<br>■ Differential movement (resulting in binding)<br>■ Alkali-aggregate reaction (vertical lift gates) |
| Vandalism | ■ Due to sabotage of mechanical parts<br>■ Due to sabotage of electrical controls/motors<br>■ Due to sabotage of structural parts<br>■ Maintenance Procedures<br>■ Lubrication of trunnion bearing |
| Maintenance procedures | ■ Lubrication trunnion bearing<br>■ Gear reducers<br>■ Wire ropes/sockets inspections for damages protection for gates<br>■ Heaters in gearboxes and controls<br>■ Gate-side seal heaters<br>■ Gate-air bubbler system<br>■ Trash removal from in front of gate<br>■ Insulation testing (electrical) |
| Icing | ■ Gates frozen in dosed position<br>■ Gates frozen in open position |
| Lifting devices | ■ Access<br>■ Icing/adverse weather conditions |

Table B Weibull parameters for mechanical components at gated spillways

| Components | Total inventoried | Weibull plotting | | Bayesian (uniform prior) | |
|---|---|---|---|---|---|
| | | Characteristic life: $\alpha$ | Shape parameter: $\beta$ | Characteristic life: $\alpha$ | Shape parameter: $\beta$ |
| Air compressor | 51 | 47.22 | 10.37 | 66.94 | 8.94 |
| Bearings (bronze bushing type) | 2014 | 74.96 | 6.751 | 81.93 | 7.29 |
| Bearings (roller type) | 3557 | 132 | 4.102 | 129 | 5.18 |
| Bearings (self-lubricating type) | 87 | NA | NA | NA | NA |
| Brake (springs and pads) | 997 | 93.78 | 3.898 | 102 | 3.26 |
| Bridge crane | 150 | 90.39 | 4.748 | 97.2 | 4.39 |
| Butterfly valves | 126 | 88.16 | 2.833 | 90.49 | 3.91 |
| Chain (link type) | 514 | 63.8 | 5.115 | 63.09 | 8.71 |
| Chain (roller type) | 465 | 73.44 | 6.039 | 75.88 | 6.37 |
| Check valves | 737 | 68.28 | 5.698 | 71.73 | 5.05 |
| Clutch (jaw) | 56 | 104.8 | 2.154 | 99.29 | 3.26 |
| Couplings (flexible) | 1160 | 71.46 | 7.981 | 77.73 | 8.99 |
| Couplings (rigid) | 2147 | 140.3 | 4.228 | 141.98 | 4.67 |
| Cylinders | 1260 | 113.8 | 2.25 | 110.56 | 2.51 |
| Flexible hydraulic hose | 975 | 44.1 | 4.5 | 52 | 3.87 |
| Gear reducer – parallel gears | 1101 | 125 | 4.25 | 132.87 | 4.71 |
| Jet/Howell bunger valve | 24 | 45 | 0.76 | 55.02 | 1.33 |
| Jib crane | 64 | 124 | 1.8 | 128.49 | 2.07 |
| Lifting stems | 790 | 100.5 | 2.857 | 106.64 | 2.67 |
| Manual control valves | 798 | 96 | 2.68 | 88.84 | 3.27 |
| Pipes (carbon steel) | 1887 | 117.1 | 3.021 | 105 | 3.51 |
| Pipe (galvanised) | 52 | NA | NA | NA | NA |
| Pipes (stainless steel) | 211 | 90.62 | 1.517 | 94.42 | 2.11 |
| Pressure relief valves | 294 | 82.97 | 5.028 | 80.21 | 5.94 |
| Pumps (fixed disp.) | 422 | 75.42 | 4.245 | 80.16 | 3.93 |
| Pumps (var. disp.) | 50 | 51 | 10.2 | 54.57 | 10.15 |
| Reservoirs | 228 | 89.98 | 4.384 | 102.47 | 4.03 |
| Right angle gear box | 484 | 230.6 | 2.29 | 244.75 | 2.69 |
| Fixed wheel/roller gates | 90 | NA | NA | NA | NA |
| Roller train for caterpillar gates | 662 | 99.5 | 2.44 | 90.76 | 2.2 |
| Caterpillar gates | 180 | 69.66 | 18.34 | 106.99 | 5.76 |
| Rotating shafts | 1240 | 103.4 | 7.958 | 111.89 | 8.68 |
| Screw actuator (electric) | 359 | 70.8 | 4.049 | 83.98 | 3.35 |
| Screw actuator (manual hand wheel) | 213 | 71.43 | 4.098 | 85.57 | 3.33 |
| Sector-bull gears | 907 | 576.4 | 1.98 | 2119 | 2.19 |
| Sheave gears | 141 | NA | NA | NA | NA |
| Slide gates | 690 | 144.7 | 3.657 | 144.44 | 3.98 |
| Sluice gates | 532 | 134 | 2.584 | 123.42 | 2.9 |
| Solenoid control valve | 457 | 56.97 | 6.076 | 62.72 | 5.11 |
| Stem nut | 912 | 144.6 | 2.222 | 153.24 | 2.36 |
| Spur-pinion gears | 1139 | NA | NA | NA | NA |
| Sprockets | 436 | NA | NA | NA | NA |
| Sump pumps | 211 | 79.73 | 1.357 | 65.66 | 1.75 |
| Trunnion pin and bearing | 954 | 81.2 | 5.71 | 89.1 | 5.32 |
| Wire rope (carbon steel) | 1288 | 82.85 | 2.06 | 79.59 | 2.17 |
| Wire rope (drum) | 515 | NA | NA | NA | NA |
| Wire rope (flat) | 369 | 56.81 | 3.261 | 59.97 | 4.33 |
| Wire rope (multi-part sheaves) | 376 | 105.4 | 4.843 | 112.82 | 4.75 |
| Wire rope (stainless steel) | 2049 | 77.35 | 2.349 | 74.92 | 3.04 |
| Wire rope (sheaves) | 1848 | 66.12 | 10.52 | 5496 | 0.62 |
| Worm gears | 373 | 71.02 | 7.242 | 92.18 | 7.69 |

**Table C** Weibull parameters for electrical components at gated spillways

| Components | Total inventoried | Weibull plotting | | Bayesian (uniform prior) | |
|---|---|---|---|---|---|
| | | Characteristic life: $\alpha$ | Shape parameter: $\beta$ | Characteristic life: $\alpha$ | Shape parameter: $\beta$ |
| Brakes (DC rectifier) | 902 | 74.95 | 5.171 | 80.86 | 5.18 |
| Control cables (fibre optic) | 24 | NA | NA | NA | NA |
| Control cables (multi conductor/twisted pair) | 1342 | 66.88 | 5.94 | 72.63 | 4.36 |
| Control panel | 1190 | 81.96 | 4.37 | 74 | 5.57 |
| Circuit breaker (fused disconnect) | 2341 | 75.1 | 3.388 | 80.77 | 3.23 |
| Electric motors | 1979 | 91.43 | 4.047 | 93.46 | 3.88 |
| Encoders | 190 | 56.62 | 4.063 | 54.35 | 4.32 |
| Generators | 402 | 48.84 | 3.454 | 49.97 | 3.21 |
| MCCs | 346 | 83.42 | 3.249 | 89.53 | 3.64 |
| Motor starter (full voltage) | 1502 | 78.96 | 4.329 | 79.01 | 4.4 |
| Motor starter (reduced voltage) | 156 | 59.6 | 10.35 | 483.02 | 0.57 |
| Panel board | 431 | 82.39 | 4.958 | 83.45 | 4.95 |
| Push button switches | 4410 | 78.73 | 4.525 | 87.5 | 3.6 |
| Power cable (in conduit) | 1203 | 70.03 | 6.345 | 73.07 | 5.08 |
| Power cable (buried) | 129 | 85.25 | 2.914 | 84.89 | 3.12 |
| Power cable (in duct tray) | 90 | 74.8 | 4.432 | 73.07 | 5.08 |
| Power cable (overhead) | 46 | 105 | 1.72 | 112.5 | 1.84 |
| PLCs | 105 | NA | NA | 817.27 | 0.64 |
| Rotating cam switches | 255 | 51 | 19.4 | 91 | 7.76 |
| Rotating limit switches | 1717 | 77.65 | 6.414 | 82.07 | 6.87 |
| SCADA | 62 | NA | NA | NA | NA |
| Selysn indicator motor | 154 | 53.2 | 4.007 | 58.67 | 3.48 |
| Switchboard | 74 | 65.38 | 6.071 | 70.99 | 5.14 |
| Switchgear | 55 | 78.84 | 3.788 | 82.63 | 3.83 |
| Transfer switch (automatic) | 130 | 57.33 | 3.54 | 57.57 | 3.63 |
| Transfer switch (manual) | 229 | 64.25 | 3.758 | 70.79 | 3.28 |
| Transformer | 360 | 70.48 | 3.298 | 71.14 | 3.26 |
| Travelling nut limit switch | 43 | NA | NA | NA | NA |


**REFERENCES**

Dodson B (2006) *The Weibull analysis handbook*. ASQ Quality Press, Milwaukee, Wis.

Kumamoto H and Henley EJ (1996) *Probabilistic risk assessment and management for engineers and scientists*. IEEE Press, New York.

Mosleh A and Zhu D (2013) *Bayesian Weibull Analysis of USACE Equipment Life Data*. US Army Corps of Engineers, Washington, D.C.

Patev RC, Buccini DL, Bartek JW and Foltz S (2013) *Improved Reliability Models for Mechanical and Electrical Components at Navigation Lock and Dam and Flood Risk Management Facilities*. Engineering Research and Development Center, Vicksburg, MS.

Patev RC, Putcha CS and Foltz SD (2005) *Methodology for risk analysis of dam gates and associated operating equipment using fault tree analysis*. US Army Corps of Engineers, Engineering Research and Development Center, Vicksburg, MS.

Schultz R (2013) *Survey of USACE flood control projects* (unpublished). U.S. Army Corps of Engineers, Louisville District, Louisville, KY.

**Operational Safety of Dams and Reservoirs**
ISBN 978-0-7277-6121-7


http://dx.doi.org/10.1680/osdr.61217.367

# Index

Page numbers in *italics* and **bold** indicate figures and tables, respectively, if outside the text.